C000129048

THE
SMALLHOLDER
ENCYCLOPÆDIA

Edited by

WALTER BRETT, F.R.H.S.

CONTENTS

FOREWORD

IN this new and completely revised edition, *The Smallholder Encyclopædia* is able to present the standard practices in small-husbandry, or small-scale food production, as tested by fire—the fire of the greatest world war, as re-shaped in the light of the vast store of fresh knowledge our national fight against starvation has given us, and as adjusted to suit to-day's conditions.

The purpose of the new edition remains the same as that of the original—to serve as a comprehensive ready-reference work on matters concerning:

Poultry-management in all its branches and on all scales;

Commercial and home-use vegetable and fruit production;

The production of flowers and plants for market and home sale;

The rabbitry in its modern form as a provider of meat and pelts;

Goat-keeping in its Continental sense—as an economical means of providing milk, butter and cheese;

The smallholding proper, with its few cows, its piggery, some sheep, its moderate area of "farm" crops;

The apiary—no longer valued only because of its honey-production aspect but also because of its important aid in the matter of fertilizing fruit and vegetable blossom.

In my foreword to the original edition I said that in a life-time spent on *The Smallholder* my work had to a large extent consisted of solving thousands of personal reader's problems and that this had given me an unrivalled insight into what kind of information the small-scale food producer wants and the form in which he wants it.

For every question I answered in any past year I have answered a dozen in these years of war through which we have lived—years when food-production efforts have been not solely for personal profit but for a Nation's salvation, years when maximum yields and complete success have been not a gratification but a stark necessity. So my store of knowledge of what the small-scale food producer wants to know in a general way is enriched by a deep insight into what he wants to know in the new and strange conditions prevailing to-day.

My hope is that, as a result, the new *Smallholder Encyclopædia* will be to the present generation of "almost farmers" what its forerunner was to the past, and may play some part in helping those who live, wholly or in part, by the land to produce from it the utmost of which it is capable.

Of a certainty the *need* for maximum production will be greater in the years to come than ever it has been at any time in our history.

WALTER BRETT.

THE SMALLHOLDER ENCYCLOPÆDIA

POULTRY

ACCREDITATION SCHEME. An official scheme designed to improve poultry in this country, the stock being examined and approved as regards health and quality, and the stock must not only pass the blood test conducted by the State Veterinary Service but must be free from disease. This means control over stock and eggs offered for sale. There are also accredited hatcheries, which obtain the whole of their eggs from accredited farms.

ACCREDITED BREEDER. See Breeder.

ACCIDENTS. Accidents will happen in the best-run poultry plants, but fortunately nearly all forms of breakages and wounds in poultry are curable if they are treated at once.

Birds should be isolated during treatment. The best "hospital" is an ordinary wire show pen fitted up in a warm, protected but light and handy shed. Whatever is used must be scrupulously clean to prevent infection from outside sources, and bedded down with ample straw or hay. Three tins must be provided, one for mash and corn, one for water and one for shell.

Whilst in hospital the patient must be lightly fed; a laying mash mixed with warm water being the chief food. Skimmed milk is the best drink when available.

Broken Legs (Shanks). The injured bird is seen to be hopping on one leg or resting often on the ground. In bad fractures the broken leg is plainly seen to be injured.

If the scaley part is broken through, bathe carefully with warm disinfectant solution. If there is no outside wound, this will not be necessary. Splints must then be made to fit as exactly as possible to hold the two parts of the fractured limb in as near as possible the position they were in before the accident. The splints are easy to cut out of thin box-wood, using a sharp knife. The splints should be of such a shape and so fitted that they do not interfere with the circulation or the nerves. Ideal splint sizes are 1½ins. long,

½in. wide and about ⅛th to ¼in. thick. Two will be required—one each side of the shank.

The splints should be held in place with a cotton bandage 18ins. long, 1½ins. wide. This bandage should afterwards be painted with thin glue so that it is automatically held in place.

If there is a wound, treat as for wounds below. If all goes well the bandage may be taken off in 10 days, but the fowl should be kept in hospital 6 days longer.

In the early stages of treatment it may be necessary to tie the bird up to prevent excessive movement and disruption of the bandages.

Broken Thigh. A bird having a broken thigh is unable to stand on the leg, hops on the other and painfully drags the broken one behind.

Pluck a few feathers away to make examination easier. Follow the procedure as given for broken legs. Splints will have to be longer (4ins.), broader (½-1in.), and more curved. Three may be necessary. Bind firmly, but only just tightly enough to keep the splints in place. When the bandage is on paint it with glue, as advised for a broken leg.

Examine frequently to see that the flesh is not swelling or becoming discoloured (bluish or greenish), in which event it will be necessary to undo the bandage at once and re-bind.

Broken Wings. The wing hangs down or the bird walks awkwardly sideways.

Pluck the feathers round the fracture and examine the nature of the break so as to gain an idea as to the correct shape of splint to use. Take great care over the making of the splints. Frequent fittings to the wing and further scrapings out with the knife as required will ensure a perfect fit. Before setting the bone take a look at the sound wing to see exactly how it is shaped. Thus, you will ensure that the bones are set in exactly their proper position. Be sure the bones remain correctly set whilst fitting the splints.

After fitting the splints bind as advised for a broken leg. After binding, ensure that the wing is held closely to the body

A*

1

by tying a wide bandage round the whole body. Remove the bandages after 10 days and let the bird gradually find use of its wings for a further 5 days in hospital.

Dislocated Hocks and Wings. A dislocation is not a fracture of the bone but a displacement of a joint. After an examination endeavour to replace the bone in its correct position and here an examination of a normal joint will be of assistance. Where there is swelling and a bad stretching of the tendons it will be necessary to bind up the limb, not in splints, but with a bandage.

Dislocation often shows what appear to be similar symptoms to fractures, but a close examination reveals a loss of use of that limb from the joint.

Broken Toes. Indicated by blood specks in litter, nest-boxes and on droppings-board. In severe cases the bird limps, but rarely so.

In show birds it will be necessary to set the toes correctly, and this is best done by binding a piece of thin cardboard round the limb and holding it in place with a glued or adhesive bandage. The position of the toe-nail can be kept as desired by plugging the end of the tube with cotton-wool.

Previous to binding up, ample bathing in warm disinfectant is necessary, for there is generally profuse bleeding if the skin is torn. Friar's Balsam or tincture of iodine should be painted on after bathing. The bird should not be allowed to walk for 10 days.

Simple Cuts. These must not be allowed to go untreated, for not only may the fowl die of blood poisoning, but cannibalism may occur.

If a simple cut contains no foreign matter, then painting with tincture of iodine will effect the necessary safeguard. If there is clotted blood or dust or dirt round the wound it must be washed with a warm disinfectant solution and then painted as above. Wounds which are in awkward places and which may be opened by movement will necessitate the placing of the bird in a hospital cage and probably bandaging the wound for a while.

Deep Wounds. These must be well washed as for simple cuts. If the bleeding persists they should be plugged with cotton-wool held in place with adhesive tape. The following day this can be removed, the wound being then washed carefully and dusted with boracic powder or covered with boracic lint held in place with adhesive tape.

The treating of all wounds in poultry is remarkably rapid. As soon as a scab has formed and is quite dry, the bird can be put back with the flock.

AGE OF HENS, TO DETERMINE.
The following signs are generally sufficient to indicate the age of a bird, provided it is not more than 27 months old. Pullets have soft and silky feathering; their bones are soft and pliable, and the shanks are smooth and supple, with little trace of spurs. The bones of a two-year-old are hard and set; the scales on the legs are often rough, hard and well defined, and the knees, or elbows, of the legs are almost flat.

The lower half of the beak of a pullet is not so firm as in an old bird; it can be moved backwards and forwards quite easily. The pullet's eyes, too, are bolder and brighter than those of the hen.

AILMENTS. It is essential that every poultry-keeper, whether his birds are numbered by tens or hundreds, be something of a poultry doctor. The majority of ailments and diseases from which the birds may suffer will quickly respond to treatment if they are taken in hand at once, but may assume serious forms and spread rapidly, even resulting in an epidemic, with heavy loss of life among the flocks, if allowed to go on.

Signs of Disease in Adult Birds. Some diseases carry very obvious symptoms, but in others they are not so obvious. Generally speaking it may be taken that *something* is wrong when a bird has ruffled, bedraggled plumage, when the feathers round the vent are soiled, when a bird is not active but stands about moping, when she limps, when she has no appetite, when the droppings are abnormal, and when the comb is shrunken or pale.

Signs of Disease in Chicks. The most obvious signs of trouble in chicks are stunted size, ruffled plumage, poor feathering, a humped-up appearance, drooping wings, a lack-lustre eye and inactivity. The fit chick will be full of life, strong on its legs, noisy with its chirpings, clean in its plumage.

The symptoms and causes of, and the remedies for the various poultry complaints are given under their respective headings below.

Examining Newly-bought Birds for Signs of Disease. When any bird is bought, be it fowl, duck, goose or turkey, young or old, male or female, it should be isolated for a week before it is allowed to associate with other birds. In countless instances, new birds have brought disaster to a poultry-keeper. Time and again they have come to him carrying the germs of some deadly disease that has spread like wildfire among his existing birds, killing them off wholesale.

During the isolation week take up each bird individually and give it a thorough examination.

First of all assure yourself as to cleanliness about the head. The comb and wattles should be free of sores, warts, scabs or floury deposits.

Next the beak should be opened wide and the mouth and tongue and throat looked at. As far as you can see these must be free of all deposit. When opening the mouth take note of the hen's breath. It should smell very little. The tongue should be free from horny covering.

Now examine the eyes. These must be dry, with no bubbles in the corners and no swellings round about. Bright, "alive," clean eyes are an indication of health.

The earlobes come next. These must be clean, smooth, even and free from warts and scabs. In the white-lobed breeds, they must be a clear white, not an unhealthy yellow.

Passing down the neck, the feathering should look tight and there should be no bare patches on the head when viewed from the back. Bare places suggest the presence of persistent parasites.

The crop will be empty if a long railway journey has been made; a full crop after such a journey is a bad sign. When handled, the crop should not be watery—unless, of course, you have just given the birds a drink.

The breast should be straight and free from scabs.

The vent feathering must be dry, clean and free from lice eggs. A healthy back region will always be clean and fluffed out.

The legs must be clean and have no upraised scales near the ankle joint.

Abscess. The most common form of abscess in poultry is a swelling on the ball of the foot, giving rise to a condition known as Bumble-Foot—which see. Internal abscesses are comparatively rare, and can be dealt with only by an operation, which should be performed by a skilled person. See also Tumours.

Anæmia. Common complaint of layers and other fowls due to poorness of blood. Paleness of headgear, lack of energy, and, when long-continued, a thin breastbone, are the symptoms.

In intensively-kept birds, anæmia is due to lack of direct sunlight; otherwise, wrong feeding, hereditary weakness, inbreeding, lack of green food, and presence of lice.

Give a 50-50 mixture of Parrish's Chemical Food and cod-liver · oil—one teaspoonful twice a day. As a pick-me-up for a flock, give 1 teaspoonful of Chemical Food to every pint of drinking water.

Aphonia, or Wheezing. Common complaint of fowls, not to be confused with bronchitis; is indicated by wheezing and loss of normal "voice." It is due to some "foreign body" (for instance, a hay stalk) causing an obstruction in the windpipe, roup or catarrhal cold.

Examine windpipe for diphtheric sores; if such are present, remove them and paint the places with iodine. If due to an obstruction, this may be felt at the front of the neck. Give one teaspoonful of glycerine per day. Sometimes the obstruction can be moved by working it up with the fingers.

Apoplexy. Is a fairly common summer complaint of poultry. There is struggling, and often the sufferer is found dead on the nest.

It is due to cerebral hæmorrhage causing partial or complete insensibility, hereditary influences, overfeeding, excessive heat or great excitement.

Remedy the above faults. If the bird is alive when found, the vein under the wing may be cut a little to let out some blood. Also see Vertigo.

Appetite, Loss of. Assuming that birds are being properly fed, loss of appetite may possibly be a symptom of a disease or ailment, such as crop-binding. etc. It may also, however, be due to "natural causes" in which event the appetite can be stimulated by one or other of the following methods: (1) leaving out the grain for one day; (2) giving cheap, separated milk to drink for one day; (3) feeding *unlimited* chopped fresh green food in troughs at noon for two days; (4) giving a noon feed of soaked grain in troughs, having cut out the morning grain feed; (5) changing the litter to one of a different colour; (6) giving a spice; and (7) wetting part of the mash if fed dry.

Aspergillosis. Disease of the breathing organs occasionally affecting adult fowls and frequently occurring among chicks, especially those reared artificially.

Signs are moping, head drooping, eyes half-closed, feathers loose and lustreless.

Really due to a fungus which is found in damp places and on food.

There is no certain remedy, and some affected birds may die. The most effective plan is to expose birds to fumes of heating tar. Birds that recover should be given 40 grains of citrate of iron and quinine crystals to the gallon of drinking water.

Asthenia. Rare ailment of fowls. Affected birds lose weight but keep up a good appetite.

Due to a germ, which only gains entry into a bird weakened by poor feeding and lack of care.

Give 1 tablespoonful of Epsom salts per 12 hens in the mash twice a week. Three days after the first dose give 1 teaspoonful of iron sulphate to each gallon of drinking water.

Asthma. A common ailment affecting the respiratory system.

Good appetite but loss of flesh. Spasmodic and wheezy breathing, especially during dry, frosty autumn weather. The face takes on a sunken appearance.

It is often chronic, or due to a fungus.

No medicine will effect a cure, but relief can be given by making bird inhale fumes from heated tar. Give bird plenty of fresh air.

Bacillary White Diarrhœa (B.W.D.) is one of the most dreaded diseases of the chicken-rearer, for once it gets a hold amongst youngsters, there is very little one can do to stop its rapid spreading and ultimate decimation of the flock. There is no known cure for the disease.

Fortunately, science has recently discovered a method of preventing B.W.D. —by blood-testing all birds used for breeding. See under Blood Testing.

No poultryman who has blood tested all birds he proposes to breed from and excludes those that the test shows to be carriers of the disease, should ever be bothered with B.W.D.

Poultry-keepers who do not hatch their own chicks should always stipulate that the chicks they buy are from blood-tested parents.

If, in spite of all precautions, an outbreak of B.W.D. does occur, every possible endeavour must be made to prevent it spreading. The germ is quickly taken by healthy chicks and is spread by dirty food and appliances, contact with anything that has become infected, through the droppings of other infected hens or chicks, through the incubator (chick tray), in the yolks of infected eggs, and by allowing a " carrier " broody hen to sit on eggs.

The symptoms of B.W.D. are similar to those that occur in practically all chick ailments. That is to say, the chick looks ill, its eye is dull and half closed, its wings are held low, its back appears hunched up and the feathers are loose and ruffled. The abdomen appears too large for the body and there is a sticky whitish diarrhœa.

Points which give more definite hints that the trouble is B.W.D., are that the deaths usually occur most frequently when the chicks are from 3-10 days old, are very numerous and the chicks do not respond to any treatment.

It should be remembered that strong chicks are more able to resist B.W.D. than are chicks whose upbringing is not all it should be. Predisposing factors are mistakes in brooding, causing overheating or chilling ; faulty incubating, causing a weakening of the embryo chick ; wrong feeding and exposure to bad weather and draughts.

Bent Breastbone. See Deformities.

Blackhead. This is a disease that causes a very considerable number of casualties among turkeys. It is so called because of the purplish colour of the head gear, but is really inflammation of the intestines and liver caused by a minute animal organism. Infection is carried from bird to bird by the droppings, which quickly affect food and water near by. The most susceptible age is 10 to 14 weeks.

There is first drowsiness and then wing drooping, feather ruffling and a diarrhœa of a greenish yellow colour. An important indicator whether a bird was suffering from blackhead or not is obtained if the dead bird is cut open and the blind guts examined. If one *only* is inflamed, enlarged and hard to the touch, you may be sure that you have found a case of blackhead, and rigid disinfection of all runs, coops and houses, etc., must be carried out at once.

Give the following mixture dissolved in 1 gallon of drinking water, allowing no other drink: Sodium sulpho-carbolate, 15 grs., bichloride of mercury, 6 grs., citric acid, 3 grs. Supply in earthenware vessels. Feed 2 per cent cod-liver oil in the mash and give skim milk to drink.

To ensure freedom from blackhead trouble, use only eggs from adult stock guaranteed free from blackhead. Fresh ground for rearing each year is a further vital precaution.

Blindness. Poultry do not often suffer from total blindness (and where they do are at once to be detected so that they can be put out of their misery as chicks), but blindness in one eye is of comparatively common occurrence. It may be caused by an injury when fighting, by contact with a thorn in a hedge, or by a bad attack of roup.

Loss of sight in one eye is denoted by the bird's habit of turning her head sideways to look in front of her or as she walks ahead. The eye may or may not be closed.

If an eye has an odd-shaped pupil, or an iris that is partly changed in colour, or is pearly or greyish orange, or if there is a bluish mask to the pupil, that is not an indication of blindness, but *does* denote the likelihood that the bird is suffering, or will suffer from fowl paralysis (see Paralysis).

Partially-blind birds (unless the loss of sight is definitely due to fighting or accident) should never be used for breeding.

Blood in Droppings. It is a common occurrence, especially in early summer, for the droppings of poultry to contain blood. It may be due to feeding an excess of fish-meal or meat-and-bone meal in the mash, to the feeding of too much fibre in the food constituents, or to internal rupture in the cloaca due to the passage of a large egg. Forced and wrong feeding is a further common cause. All these causes indicate their own remedy.

Blood-Stained Eggs. Frequently eggshells, especially those of the first eggs from a pullet, are noticed to be stained or smeared with blood.

The cause is obvious when it is mentioned that the organs of the newly-laying pullet are taxed severely by the strain of passing eggs, which each week become larger and larger till they reach full size. The stretching of the organs inevitably causes occasional bleeding.

The trouble will soon right itself. Should the birds be slightly anæmic, give steel drops—half a teaspoonful, with enough water to fill up the spoon, every morning for a week.

Bronchitis. A disease affecting the respiratory system of fowls, ducks, etc. The breathing is rapid and wheezy (as detected by holding the ear close to the body). Wattles and comb are often bluish and there is a " sneezy " cough.

Bronchitis is due to exposure to weather, damp quarters, draughty or stuffy night housing, or by irritant vapours such as an excessive use of dusty quicklime on the dropping boards.

First isolate the bird by removing it to warm quarters. Let it inhale steam vapour from hot water in a jug into which one teaspoonful of tincture of iodine is poured. Give soft foods. A quick cure is 2 grains of black antimony twice a day mixed with the food. If the attack is severe, give 12 drops of spirits of turpentine in a teaspoonful of olive oil. Relief from severe gasping or wheezing is obtained by giving the bird 10 drops of Ipecacuanha wine.

Bumble-Foot. A very common complaint with fowls. An abscess on the ball of the foot, causing painful walking and limping.

Probably due to a wound into which a germ has gained access. Tubercular birds have a strong tendency to this complaint.

If soft, poultice until " ripe," lance, squeeze out the matter and anoint with boracic ointment. If hard, paint with tincture of iodine. A bandage may be desirable for a few days.

B. W. D. See Bacillary White Diarrhœa.

Cancer, or Carcinoma. Internal growths in the body of a bird are not uncommon, but not all are malignant. As it cannot be determined whether an internal growth is malignant or not without operation, a suffering bird—unless valuable—is best killed, the carcass being burnt. See also Tumours.

Canker. Common complaint in all poultry.

The symptoms are similar to those in roup, which see. In addition there is a clogging of the mouth with false membranes.

The cause and remedy are as for roup.

In addition, remove those membranes that appear loose—with the aid of a blunt match dipped in disinfectant—and burn them. Paint the places and the whole of the inside of the mouth with tincture of iodine every three days. Give two per cent cod-liver oil in the mash.

Cannibalism. Name given to the vice of toe- and vent-pecking among chicks, which often leads to the loss of stock. See Toe-Pecking.

Catarrh. One of the commonest poultry complaints.

The birds appear dull, they sneeze, and breathing through the nostrils is made difficult by the presence of mucus. Sometimes the eyes are watery and the bird has the appearance of having a chill, as in human beings.

It is due to draughts, exposure to cold and wet.

Bathe the whole of the head and face, eyes and inside mouth and nostrils with a fairly strong solution of boracic acid. Keep warm, feed well on a spiced mash, give a 2-grain quinine sulphate pill every night. Care must be taken to differentiate between this and contagious catarrh, i.e. roup, and immediate action therefore must be taken when " running " eyes and nostrils are noticed. For flock treatment, give Visol (a proprietary preparation) in the drinking water in the proportion advised by the makers.

Chicken-Pox. Highly contagious disease, in which there are warts on the unfeathered parts of the head.

Immediately isolate the affected bird. First wash the whole head with a disinfectant soap, such as carbolic or Izal. This loosens the scales, which fall off and reveal the actual seat of disease. Then soak the sore places with Izal fluid (one part to 100 parts water).

Chills. There may be no running at the eyes or nostrils, but the bird looks generally out-of-sorts and has a regular " hang-dog " appearance. Usually affects either the lungs or the liver. The mouth should be examined for spots or growths.

Remove spots in mouth with a pointed match-stalk and touch up with iodine. Make up some pills according to the following recipe : Take cinnamon, 3 oz. ; ginger, 10 oz. ; gentian, 1 oz. ; aniseed, 1 oz. ; and carbonate of iron, 5 oz., and mix thoroughly. A teaspoonful should then be taken and made into eight pills with flour and fat and one given night and morning. Visol (a proprietary preparation) or permanganate of potash in the drinking water is a good preventive.

Cholera. A virulent disease that may attack adult birds, but is more common in chicks. It is due to a germ, *Bacillus aviceptus.* Other chicks become infected through the drinking water, contaminated by droppings of suffering chicks or by food.

The symptoms are almost the same as in Bacillary White Diarrhœa, i.e. mopiness, huddling and some pasting of the vent. See B.W.D.

Severe disinfection and isolation of infected broods are necessary. All diseased bodies must be completely burnt at once, not buried. Give as an internal antiseptic : Sodium sulphocarbolate, 15 grains ; bichloride of mercury, 6 grains ; citric acid, 3 grains. This quantity is dissolved in 1 gallon of drinking water and no other water allowed. Give in earthenware vessels.

Cloacitis. See Vent Gleet.

Coccidiosis. This is a highly infectious disease which is caused by sporeforming parasites, known as coccidia. As well as fowls, turkeys and peafowls may be infected, but the disease is rare in ducks.

The disease occurs in two forms—an acute type in chicks (between 2 to 12 weeks is the most vulnerable period) with a quick and heavy death-rate, and a chronic form in adult stock. While not so quickly fatal this form may be the cause of many deaths in a flock over a period, and a source of infection to younger stock.

The coccidia, in the form of oocysts or fertilised eggs, are picked up from the ground by the birds, and the hatched parasites are liberated in the birds' intestine. The parasites eat their way into the inner wall of the intestine and rapidly increase in numbers. Later, oocysts are voided in the droppings ; thus the disease is passed on.

Treating Chicks. When chicks are attacked they become dull, listless and huddled. At first they eat ravenously, then they lose appetite and condition. Usually there is a greyish-white diarrhœa stained with blood. In the event of an outbreak of Coccidiosis, every possible precaution must be taken to prevent its spread.

First of all, isolate all sick birds, if possible in a separate house, but otherwise by dividing off the existing brooder-house, taking great care that no litter, food or appliances pass from one to another.

Disinfect the appliances and the brooder house floor with caustic soda, soap and hot water, and afterwards syringe with a 5 per cent solution of carbolic acid. An even more certain method is to go over everything carefully with a blow lamp.

There are now several methods of treating coccidiosis. The Harper Adams Agricultural College has had excellent results with the following :

A stock solution is prepared of iodine resublimate, ½ oz. ; potassium iodide, 1 oz. ; and water, 25 oz. One part of this is added to nine parts (by volume) of separated or whole milk and the whole heated to 70 degs. Centigrade until it becomes white. One pint of the resulting solution is added to each gallon of the chicks' drinking water.

Another method of treatment is to give hydrochloric acid, 1 teaspoonful of the acid being added to each quart of drinking water in glass or earthenware vessels, the birds being fed lightly on grain only.

There is every reason to believe that abundance of milk to drink or in the mash aids in the rapid recovery of the chicks. By " abundance," is meant up to 25 per cent dried milk in the mash and milk to drink, but this large amount should be used only when Coccidiosis is feared, never in a normal way—and not during wartime.

As a precautionary measure many poultry-breeders now make a point of rearing their chicks in pens with wire-netting floors (see Sun Parlours) so that there is no possibility of them coming into contact with the soil.

The pens are raised well above the ground so that the chicks cannot even eat the grass that grows beneath. Of course, precautions are taken not to give them grass or greenfood obtained from ground infected with the germs of the disease.

Treating Adult Birds. The chronic form of coccidiosis affects birds of all ages. Sufferers usually lose flesh and become little more than skeletons, though they eat ravenously. Again there is blood-stained diarrhœa.

In advanced cases treatment is not worth while. The bird should be killed and the carcase at once burnt.

When the disease is observed at an earlier stage the sufferers should be isolated. To expel the coccidia add Epsom salts to the mash—1 oz. for six birds. Repeat the dose six days later. To control the diarrhœa and bleeding add vinegar to the drinking water, at a ratio of 1 to 80, for a week.

Clean the droppings board as early as possible each morning and cover with sand or fine ashes to facilitate the removal of the next days droppings. As a further precaution against re-infection fasten under the perches a piece of wire netting large enough to cover the droppings board. The litter on the house floor should be changed frequently—three or four times a day is not too often.

Some birds appear to have a natural resistance ; they harbour the disease but show no outward symptoms. Such carrier birds are a source of danger to other stock as they constantly pass oocysts in their droppings. There is no method by which carriers can be distinguished.

Colds. A common complaint with all poultry.

There is running at the eyes and nostrils, accompanied by slight gurgling in the throat; may lead to roup, or wasting away in a few instances.

Treat any with clogged nostrils as follows : Dip the head in a bowl of strongish permanganated water and wipe out the nostrils and mouth with a piece of rag dipped in the solution. Take a small quill feather, dip it in eucalyptus and insert in each nostril and in the " cleft " of palate; take another quill and smear over the feathers under the wings (where you will probably see soiling from the catarrhal discharge while birds are asleep). Finally give a teaspoonful of olive oil. Three " goes " as above should suffice. See also Catarrh.

When there are colds about make a point once a week of adding a teaspoonful of ammoniated tincture of quinine to each pint of the first lot of drinking water given in the morning. For the remainder of the day, pure water should be supplied. This not only generally stops incipient colds, but prevents birds that are unaffected from getting smitten.

Comb, Frost-Bitten. It may quite likely happen, during a severe winter that a bird's comb and wattles, particularly those of the cock, may be frostbitten. The degree of severity depends upon (1) the humidity of the house ; (2) the physical state and health of the bird, and (3) the action of the heart. Wattles are affected more than comb. Birds in confined but cold houses are most affected. The parts swell up, darken in colour, and, if neglected, become ulcerous, die and fall away.

Prevention by rubbing, when particularly cold nights are suspected, with vaseline. Remove the bird to a warm but not artificially heated room and paint the parts either with tincture of iodine or carbolised glycerine. For show birds prevention is even more important, for the comb is often disfigured for life when once it has been frost-bitten.

Congestion. A condition of the lungs arising in severe colds. There is difficulty in breathing, general dejection and drooping wings.

The only remedy is to cure the cold in its early stages. Once congestion sets in a cure is almost impossible.

Constipation. Common ailment of fowls, caused by lack of exercise, lack of greenfood and incorrect feeding. There is straining, and the droppings are small in quantity and dry.

For flock treatment give Epsom salts, ¼lb. per fifty birds dissolved in the drinking water. For an individual bird add enough Epsom salts to the drinking water just to make it taste, or, in serious cases, dose with a dram of castor-oil applied by means of a fountain-pen filler.

Consumption. See Tubercolosis.

Corns. Occasional complaint from which all poultry may suffer. The corns generally assume the form of a swelling (not large) situated directly beneath the foot on the pad.

Due to rough perches or continued walking upon sharp surfaces such as cinders.

There is a hard core to the corn. Paint with tincture of iodine and then pull the core clean out with the finger and thumb. Paint again with iodine and again 3 and 6 days later.

Cough. Very common complaint with fowls. The trouble starts with a slight running at the nostrils and then, perhaps, there will be a sneeze. After that, you get head shaking, accompanied by a sort of guttural cough. The bird's head will be held half-down, the neck extended to give greater breathing action and the head to one side. What is happening is that the bird is endeavouring to get rid of phlegm.

Due to a neglected cold.

Treat as for a cold. See Colds.

Cramp. This is a complaint of a rheumatic nature. The joints swell, the legs are hot and the pain in the joint causes disinclination for walking by the bird. The toes are drawn up and the muscles of the legs are rigid and quite hard.

Cramp is due to standing about on damp ground and exposure to inclement weather.

Hold the legs in warm water, then massage them with embrocation. Finally place in a coop well littered with hay. Feed liberally on a good stimulating mash. Strong doses of Epsom salts should be given first, followed with cod-liver oil and Parrish's Chemical Food, half and half—1 teaspoonful morning and night.

Crooked Breasts. See Deformities.

Crooked Toes. See Deformities.

Crop Binding. Common complaint with all poultry.

The bird mopes, has a sort of stiff neck and indicates her discomfort by wagglings of the chest.

The trouble is due to faulty feeding or to long grass having become matted in the crop.

Give the bird 3 teaspoonfuls of glycerine and massage the crop to soften the mass. Then give weak solution of bicarbonate of soda, and massage the crop again. The bird's head should hang downwards whilst the crop is worked. The contents of the crop will then be dislodged. The remedy of cutting open the crop, removing the contents, then sewing up the wound is very seldom practised nowadays, but is readily practicable.

Crop Impaction. Occasional complaint of chicks. The affected chick is distressed, and if the crop is full, it will be hard and tight.

Crop impaction is caused by overfeeding, new grain, eating of straw fibre or indigestible material, or sometimes by disease lower down the digestive tract.

Give olive oil and work the crop about until soft, holding the bird as in tympany (see Tympany), to see if contents of the crop can be pressed out. Finally, swill out crop by injecting bicarbonate of soda solution.

Cyst. Occasional complaint in all poultry, taking the form of a swelling containing a collection of natural fluids within the body. The swellings vary in size. Generally found in over-yeared hens.

The remedy is surgical. If the cyst is external, cut it away, bandage and wash out wound with disinfectant until healed.

Debility. Debility is not a disease ; it is a general run-down condition resulting either from illness, from poor-feeding, from a long spell of hard-laying or, just as with human beings, from the need for a tonic or pick-me-up.

One of the very best methods of preventing a run-down condition is to give the flock cod-live oil (see Cod-liver Oil, under Foods).

It is frequently found in spring that the layers are run-down as a result of heavy winter-laying. A good pick-me-up is provided by sulphate of sodium solution. Make up a stock solution of 4 oz. sulphate of sodium in 1 pint warm water, bottle and cork tightly. Give the birds a dessertspoonful of this per pint of drinking water. Repeat the dose a week later.

Depluming Mite. See Insect Pests.

Diarrhœa. Trouble from which both chicks and adult birds are likely to suffer.

There is looseness of the bowels, contents of which stick round the vent. Droppings may be any colour from white to brown. There may or may not be mopiness but invariably there is inactivity.

With chicks, the cause is usually a chill—maybe as a result of the brooder lamp going out or of carelessness in moving the birds from incubator to brooder. Also caused by exposure to cold winds and by feeding on sloppy or wrongly balanced foods. In adults possible causes are bad food, or a sudden heatwave.

First wash clean the fluff around the vent, using warm water tinged red with permanganate of potash. Then bring the patient into a place where it can be warm and dry, and feed it on good food over

which a thin sprinkling of powdered chalk has been dusted. Give two or three feeds of boiled rice if wrong feeding is the cause of the diarrhœa.

Also see Bacillary White Diarrhœa.

Diphtheria. A very contagious disease, likely to spread rapidly unless an affected bird is at once isolated.

The symptoms resemble those of roup (see Roup), and in addition false membranes will be seen in the mouth. Patient has a throaty, choking cough, with noisy breathing.

Diphtheria is due to a germ, as in roup.

Remove membranes that appear loose and burn them. Paint the places and the whole of the inside of the mouth with tincture of iodine every three days, and the head and eyes daily as recommended for catarrh. See Catarrh.

Dizziness. See Vertigo.

Down-behind. An ailment causing part of the internals to protrude from the vent. See Prolapsus.

Dropsy. Fairly common poultry complaint which may well have serious results. It causes debility and anæmia. The bird walks tenderly. The abdominal skin is tense and, when the abdomen itself is pressed, liquid can be heard inside.

It is due to obstruction of the circulation by old age, or liver or kidney disease, with collection of fluid in abdomen.

A veterinary surgeon can make a small incision in the body to drain away the liquid, but this is hardly worth while except with a valuable bird. With ordinary utility stock the best plan is to destroy the sufferer.

Dysentery. A fairly uncommon complaint of all poultry. There is profuse diarrhœa of varying colours, accompanied by frequent discharges of blood.

It is due to putrid food, foul water or other similar cause.

The best treatment is to give a dose of Epsom salts, add a little granulated vegetable charcoal to the food and give 3 grains of subnitrate of bismuth daily in the mash. This treatment, although the best known, is not always successful. It is always touch and go in a case of dysentery.

Eczema. Name given to a roughened condition of the skin, most frequent in ducks that are kept under confined conditions. Epsom salts in the drinking water, and adequate supply of green food, and an antiseptic lotion applied to the affected parts, will cure the trouble.

Egg-Binding. A very common complaint of fowls and of laying ducks.

The affected bird pays frequent visits to the nest, walks slowly with lowered abdomen and tail, and often has an appearance of illness, but is bright of eye. It frequently appears to be straining.

The cause is lack of "lubricant" after the moult, constipation, or the formation of a very large egg.

The bird should be left on a quiet nest for 3 hours. If no egg is passed by then, give the bird 2 teaspoonfuls of *castor* oil and also pour into the vent 2 good teaspoonfuls of *olive* oil. This will move the egg in normal cases.

If it fails, the next plan is to subject the abdomen to heat. The bird may be held in front of a fire for 10 to 15 minutes, watching all the while to see that she is not heated too much, causing faintness. A better method is to obtain a bowl full of hot water (not injuriously hot) and hold the abdomen in for 15 minutes. The heat treatment should be immediate; that is to say, as soon as the bird is noticed to be egg-bound.

If the patient does still remain egg-bound, smear carbolised vaseline on the first finger, push this carefully into the egg channel (in an upward direction) and work it gently round the egg.

If by any chance you break the egg in the passage, don't rest content until every particle of shell has been got out by means of the vaselined finger ; otherwise the bird will die.

Egg Sterility. A not very common complaint with fowls. An affected hen permanently ceases to lay. The trouble is due to degeneration of the ovary through atrophy or physiological impotency. Hereditary influences due to faulty in-breeding and wrong feeding are two of the main causes.

One may try giving vigorous exercise, e.g. scratching, and alter feeding, but it is often useless.

Emphysema. This is air-sac swelling, an uncommon complaint of fowls.

There are puffy swellings in neck or thighs, due to obstruction of air passages, rupture of air-sacs or to caponizing (which see).

The remedy is to prick the swellings by threading a sterilized needle with silk and drawing it through the top of the swellings. Paint with tincture of iodine and give a mineral mixture containing potassium iodide.

Enteritis. A complaint of all poultry, resulting in inflammation of the intestines.

There is watery diarrhœa and, in some instances, signs of blood. Then intense thirst and prostration.

Enteritis is due to bad foods, salt, poisons or internal diseases.

If the disease is caused by infectious internal parasites, death may occur in a few hours. As food and drink, give skim milk with the white of an egg added. Dissolve a teaspoonful of sodium sulphocarbolate in half a pint of milk; give this liberally until the bird seems eased.

Fatty Disposition. This is a common trouble, especially with elderly poultry.

It is suggested by the heavy, idle appearance of the hen. The abdomen feels like a bladder of lard, with no movement when pressed. Probably due to giving starchy foods to poor or non-laying birds.

Make the birds scratch in deep, loose litter, and give a 10-day course of Epsom salts (1 teaspoonful to the pint of water). Cut out maize, barley and potatoes from the ration.

Favus. See White Comb.

Feather Bound. A plainly noticeable condition of the plumage of a moulting fowl, showing a lot of feather stems. On closer examination one finds that the stem-tubes still imprison part of the feather.

Is due to a low condition during the moult. With cockerels it is often due to under-feeding.

Adjust feeding, give iron tonic and pull off the more noticeable and prominent feather quills by hand.

Feather Picking. Often chicks and fowls are to be seen picking, and often eating, feathers from pen-mates. It is probably due to idleness, lack of fresh green food or of animal matter in the ration.

Remedy faulty conditions. Rub bare places plentifully with boracic ointment. If a particular bird is very bare and sore, isolate her until her wounds heal.

Fleas. See Insect Pests.

Foot Swelling. A common complaint with all poultry, causing limping, holding up of one foot, swelling of ball of foot.

Due to tuberculosis tendencies, thorns, injury.

If soft, lance with disinfected knife, squeeze out matter and anoint with boracic ointment and bandage to keep out dirt. Repeat each day till healed. If the swelling is hard, paint with tincture of iodine.

Fowl Paralysis. See Paralysis.

Frost Bite. See Comb, Frost-bitten.

Gapes. Very common complaint in young chicks.

Due to the gape-worm, which lodges in the chick's windpipe and, if neglected, results in suffocation.

The symptoms are gasping, stretching of the neck, wheezing and gaping. Affected birds appear dull and lose their appetite and their feathers become ruffled.

Obtain a box large enough to hold the chick and on the floor of it sprinkle the following mixture : finely powdered camphor, one part ; fine chalk, two parts. Place the chick inside and shake the box gently, the object being to make the bird inhale the "snuff." The worms will then be coughed up, when they should be collected and burnt.

When there is a risk of chicks running on infested land, add sodium salicylate to the drinking water at the rate of three tablespoonfuls to the gallon.

" Going Light." See Tuberculosis.

Gout. An uncommon complaint of poultry. There is evident pain in leg and wing joints, some heat in the affected parts and puffiness at the joints. Rheumatism is more chronic than gout, coming and going with weather changes and there is " gnarling " of joints rather than swelling, with no heat.

Give a good dose of salts first, keep the patient " up," and follow the salts with 10 drops of colchicum wine daily for 10 days.

Growers' Mite. See Insect Pests.

Heat Stroke. Not an uncommon complaint during hot summers. The bird collapses. Usually occurs when bird is on the nest and only on an exceptionally hot day, when the temperature is over 90 degrees in the shade.

As soon as the bird is noticed to be ill, at once place it in a cool, dark place and feed on a light mash for 3 days. The complaint is not infrequently fatal.

Indigestion. Common to all poultry. There is loss of appetite and vitality, and a tendency to sit about the house. Due to lack of sufficient green food or grit ; improper feeding generally, bad drinking water and insufficient exercise.

The diet should be changed, and for the moment reduced ; salts may be useful if the trouble is constipation, or ground chalk if looseness ; and a tonic is desirable. Induce exercise by giving corn in scratching litter.

Inflated Crop. Disorder usually caused by the eating of stale food. See Tympany.

Internal Laying. Internal laying is a complaint resulting in a bird laying her eggs internally instead of passing them through the egg passages, in the normal way. A bird may be suspected of internal laying if she possesses keen, large, bright eyes, large abdomen, good appetite, proper activity—all indicating good egg-production—and is yet a non-layer.

The causes of internal laying are various. An overfat condition will sometimes bring it on ; so will an internal rupture caused by rough handling or high perches. A cyst that obstructs the passage of the egg through the oviduct may also be responsible.

There is no cure for the complaint. The only possible course is to kill off the bird.

Keel Disease. A fairly common complaint with ducks.

The symptoms resemble those of Bacillary White Diarrhœa of chicks. The trouble is probably transmitted through the eggs of the breeding duck, and is rapidly fatal.

The remedy is the same as for B.W.D. in chicks. Kill and burn all seriously ill ducklings and, as a medicinal aid for those not quite so bad, give 15 grains of the following compound in 1 gallon of drinking water—equal parts of finely powdered sulpho-carbolates of zinc, calcium and sodium.

Lameness. May be due to Rheumatism, Cramp or Bumble-foot, which see.

Layer's Cramp. Usually a spring ailment, due to digestive trouble combined with vigorous egg-production and a lack of direct sunlight. It usually attacks the heaviest layers.

The symptoms are that the bird is unable to stand up, and yet has a brilliant comb and wattles and a good appetite.

Treatment consists in giving the following medicine, at the first sign of trouble : Take a perfectly clean mug and half-fill it with water and pour in ordinary Epsom salts until a strong solution is made. Fill a 2-oz. bottle almost to the top with this solution, add 1 teaspoonful of Steel Drops, and thoroughly shake. As soon as you notice a lame bird, put her in a hamper of hay near a fire. Feed her on soaked oats and give 2 teaspoonfuls of the mixture morning and night.

Leg Weakness. A disability more common among young birds, though adults may suffer as a result of wrong conditions.

There is inability to stand up, moping, and a very erratic appetite.

When wrong feeding is the cause, look to the mashes. As treatment, give skim milk to drink and add 5 per cent of bone meal to the mash. Let the birds run on short grass if possible. It also helps if each suffering bird is given a quarter of a teaspoonful of syrup of hypophosphites after a feed.

When overheated and wrongly designed brooders are the cause, resulting in debilitated air, the brooder or hover should be raised on a platform 2 in. high and made of ½-in. mesh wire netting.

When a too dry atmosphere under the brooder is the cause, spray the litter sparingly underneath with a very weak solution of disinfectant every other day.

In growing birds the trouble may be caused by insufficient exposure to direct sunlight.

When overcrowding and foul air are the cause, the remedy is obvious—run fewer chicks in the quarters concerned.

To help the birds to recover, give fish meal impregnated with cod-liver oil and let the chicks run out on clean soil as often as possible.

When dampness, caused by leaky houses, by undrained soil and filth, is the cause, again the remedy is obvious and again cod-liver oil as above should be given.

Legs Swollen. Due either to Ring Sloughing or Rheumatism. See under separate headings.

Lice. Insect pests that freely infest all poultry. See Insect Pests.

Limber Neck. Caused by ptomaine poisoning due to affected birds picking up germs in putrid food. The neck muscles are paralysed and sufferers can lift their heads only with great difficulty, if at all.

Treatment consists of giving the birds a good dose of Epsom' salts. Confine them to a light, well ventilated house, and keep them on a very light diet until they recover.

Liver Troubles. In addition to the serious diseases such as Aspergillosis, Coccidiosis, etc., which affect the liver, a bird may have a disordered liver, caused by excess of animal matter or extreme heat. The symptoms are a blue appearance about the tip of the comb and an inclination to mope. The remedy should at once be applied, to prevent an acute stage.

Dissolve 4 oz. of sodium sulphate in 1 pint of warm water and of this give 1 dessertspoonful to every pint of drinking water. Individually, give 12 drops of Turkey rhubarb per bird per day.

When other symptoms of illness beyond the blueing of the comb and mopiness are noted, the bird should be examined for more serious trouble than simple liver disorder.

Loss of Appetite. See Appetite, Loss of.

Loss of Weight. A symptom of many common poultry diseases, particularly Tuberculosis, in which case it is often referred to as " going light."

Mites. Parasites of poultry. See Insect Pests.

Oviduct, Rupture of. Not very common complaint of fowls, but sometimes occurs during the flush of spring laying. In the rapidity of egg-making, the walls of the egg organs are weakened and a rupture occurs. Forced feeding encourages the trouble.

Unfortunately, cure is almost impossible, for if inflammation sets in, the bird dies. The alternative is internal hæmorrhage and death on the nest. Prevention is the only thing, and the best plan is to feed on common-sense lines.

Paralysis. Fowl Paralysis has assumed very serious proportions in the last few years and is reckoned as one of the most disastrous troubles the poultry-keeper has to contend against. It accounts for a great number of losses in the year—on a single farm such losses have amounted to 1,400 in a year.

There is some doubt as to the actual cause of the paralysis (Neuro-lymphomatosis is the correct name), but it appears that large numbers of minute cells, called lymphocytes invade the tissues, particularly of such active organs as the liver and ovary. The paralysis may be due, however, to other causes. It must not be confused with the ordinary forms of paralysis or partial paralysis. It usually occurs in young birds (cockerels and pullets equally) of from 2 to 18 months of age. Often an outbreak occurs just when pullets are on the point of laying.

Signs of Paralysis. The signs of attack in this deadly complaint are as follows :

1. *"Pearling" of the Eye.* By "pearling " is meant the paling of the coloured part, or iris, of the eyes. Either partial or complete blindness may be caused. Not every bird suffering from the complaint exhibits this symptom.

2. *Lameness.* A slight lameness of one leg is usually the next warning sign, and this increases until the bird lies on its side with the legs stretched out and helpless. Often the legs are both stretched out behind or else one to each side at right angles to the body. The toes are generally drawn up in a clutching position.

3. *Wing-Dropping.* Alternatively, a drooping of one wing may be the second sign. The bird may carry on with this defect for a month or more. Later, both wings will droop and touch the ground.

4. *Twisting of the Head.* Later symptoms show a twisting of the head and neck, which is lowered. The bird generally assumes a sitting position when the neck is twisted. The head may be turned backwards or to one side, but it is always returned to its abnormal position even when righted. In severe cases the birds may endeavour to somersault.

5. *Body Condition.* From the onset the bird begins to lose condition rapidly, so that if it lives a month it is practically fleshless.

There is so far no proved cure for paralysis. The moment an outbreak is suspected one of the suffering birds should be killed and the carcase sent up to an authority for post-mortem examination.

On suspicions being confirmed, all badly affected birds should be killed. With less advanced cases a cure may be attempted with one or other of the proprietary remedies.

Some success in the treatment of paralysis, under experimental conditions, has been obtained by injections into the abdomen—in front of the thigh and behind the last rib—of 5 cubic centimetres of a 10 per cent solution of potassium iodide in distilled water. One injection is given on each of two successive days.

There is less likelihood of an attack of

fowl paralysis if the following rules are observed :

(1) Not to give more than 20 per cent yellow maize meal nor more than 30 per cent bran.

(2) One should endeavour to provide conditions that maintain the highest vitality. Ventilation at night *must* be ample, but without draughts.

(3) All land should be drained to prevent the thriving of bacteria ; foul land should be limed ; and it is best to maintain runs of fresh grass.

(4) The perches provided should give resting posture for the birds' claws.

(5) Great care should be taken with the breeders, not using any birds that have abnormal eyes. Only fully mature breeding stock should be used.

Parasites. See Insect Pests.

Peritonitis. Somewhat rare complaint among poultry. Suffering birds seem to be in pain, walk uneasily, appear ill and may die suddenly.

The cause is internal inflammation of the serous membrane (lining the abdominal walls) caused by foreign matter, internal derangement, fracture by rough handling and disease.

An affected bird should be kept apart, and given half a salt-spoonful of Epsom salts. The mash should be mixed with water to which a little disinfectant has been added.

Pests. See Insect Pests.

Pip. Term applied to an affection of the tongue that commonly occurs among young birds. The end of the tongue can be felt as a hard, piplike substance.

The trouble seems to be due to excessive breathing through the mouth when troubled with difficult breathing, especially when the nostrils are blocked by roup, etc.

The hard end should *not* be removed, but be liberally moistened with glycerine and vaseline, using a small camel-hair brush. If the cause of the keeping of the mouth open is removed the cure will be speedy and complete.

Pneumonia. A serious and often fatal disease. Chicks are more susceptible than adult birds. The symptoms are the same in each : Rapid and painful breathing, loss of appetite, thirst, and often constipation. With ducks the causes are chills, neglected colds, faulty brooder or hover construction or management.

Bad ventilation, overheating and lamp fumes are the causes of what is known as brooder pneumonia.

There is little hope of a cure ; all that can be done with chicks, is to keep the birds in a place with a warm and even temperature, and remove cause of trouble. As a strengthener, surviving birds may be given 40 grains of citrate of iron and quinine crystals per gallon of drinking water.

With adults pneumonia is most likely during and just after the moult on wet, cold days. A cure is unlikely, but it is worth while to try the effect of giving every hour ½ teaspoonful of brandy in which 2 drops of spirit of camphor are dissolved. Tincture of iodine may be painted over the lower sides of the chest, and the bird should be kept in a very warm place—e.g. a box by a kitchen fire.

Prolapsus and "Down Behind" are frequently linked together as one and the same thing but actually the complaints are different. Prolapsus is eversion of the cloaca—the organ at the end of the oviduct or egg tube in which the shell is manufactured, and of the egg tube itself. In severe cases a part of the large intestine protrudes.

Due to too stimulating foods (such as immoderate quantities of spice or excess of fish and meat meal), or unusually large eggs, causing straining, or severe constipation.

First wash the parts in warm water tinged a light red with permanganate of potash. Dry them gently and anoint with boracic acid ointment. Then very gently ease them back into place with ointment-covered fingers.

If the trouble is very severe and there is difficulty in putting the protruding parts back, the bird must be supported in a sling, her wings and legs being fastened securely to prevent flapping, the bathing and ointment treatment, of course, being performed first. Twelve hours in the sling will usually ensure normal organs.

All birds that have had successful treatment should be fed sparingly afterwards and given doses of Epsom salts every other day for 3 days.

In the case of Down Behind the symptoms are much the same. Affected birds have a sprawling gait, resembling that of a duck. Older birds which have laid well are usually the sufferers. There is no cure for Down Behind. Sufferers should be killed for the table.

Red Mite. See Insect Pests.

Rheumatism. Common complaint of all poultry.

Lameness and painful joints, the feet and hock joints being generally affected. Swellings appear at the side and may bleed and fill with matter. The birds lose flesh, the feathers are roughened and death may follow if the trouble is neglected. It must not be confused with Cramp or Leg weakness, in the former of which the toes are drawn up, and in the latter of which the bird still remains healthy and tight about the head.

Rheumatism is due to exposure to cold and wet, and overheated roosting quarters.

Camphorated oil or strongest tincture of iodine will ease the legs. The quickest and best treatment is 5 grains of aspirin 3 times a day.

Rickets. Fairly common trouble among young chicks. The symptoms are soft bones and crooked breasts, due to wrong feeding.

Give 2½ per cent cod-liver oil in mash, access to sun rays and ample fresh green-food.

Roup. A serious and all too common complaint with fowls. It is found in many forms, viz., One-Eye Roup, Contagious Catarrh, Diphtheria and some forms of Canker. The general winter form is either as nasty mucous patches around and inside the mouth and eyes, or as diphtheric growths inside the mouth.

Roup is highly contagious and, if neglected, will sweep through a flock.

Treatment must be immediate. The sick birds should be isolated at once and given easily-digested soft food. The head and mouth and eyes must be bathed twice daily with disinfectant solution, used to the strength advised by makers, or boric acid, 15 grains to the oz. of water.

The mucous type of roup can be reduced with aspirin as for rheumatism.

As a tonic and antiseptic, give the following solution : 2 oz. of copper sulphate and 1 oz. of iron sulphate mixed in 1½ pints of water. Of this give one full tablespoonful to a quart of water. This is also a fine preventive solution to give the rest of the flock.

As an appetiser and tonic when the bird has pulled round, use the following powder : iron sulphate, ½ part ; ginger 2 parts ; gentian ½ part ; aniseed ½ part and liquorice 2 parts, all the components being in a finely-powdered state and thoroughly mixed. As a dose, give 1 tablespoonful to 24 hens or, if only a few are being treated, sprinkle and mix into the wet mash so that it smells aromatic.

Burn or bury deeply in lime all fowls that die of this disease.

Scaly Leg. See Insect Pests.

Slipped Wing. Complaint mostly affecting young birds. The bird seems to find one or both of its wings too heavy to carry well up, the result being that the wing or wings will droop whilst the bird is walking.

Get some strong, broad tape and, folding the wing correctly, tie a length fairly tightly round it just where the quills disappear into the small feathers bordering them. Knot another piece of tape into the middle of this, tie on the upper side, and run it from here over the bird's shoulder, at the " arm-pit," and under the wing again, to be tied finally in the middle of the first tape on the under side. Repeat this on the other wing and com-plete by tying two final tapes across the bird's back, one at the shoulders, the other at the flight ends (where first tied). These two last tapes should be adjusted so that the wings, when finished with, may rest just a shade higher than they do in a normal position. Keep the harness on for three weeks or a month, and, meanwhile prevent the bird from being bullied.

Sour Crop. Common complaint of poultry caused by lack of greenfood, stale and unsuitable food, etc.

Empty the crop of its water content by pressing when the bird is held head downwards. A teaspoonful of the following mixture should then be added to each quart of mash, daily, for a week or 10 days : Gentian, 4 oz. ; saltpetre, 1 oz. ; ginger, 1 oz. ; and iron sulphate, 2 oz. These ingredients should be finely ground and well mixed.

Stonebelly. Uncommon complaint of fowls.

An affected bird is heavy behind as if the abdomen has sagged because of too short a breast-bone. In a prolonged attack, the bird will walk curiously with a side-to-side gait. When acute, she will appear ill and drowsy. On handling, the abdomen will feel as if a hard substance were inside, and this may vary in size.

The trouble is due to escape of egg matter into the abdominal cavity.

The only treatment is surgical, which is doubtfully successful.

Throat, Rattle in the. Probably due to a piece of hay or other foreign matter having lodged in the throat, the remedy being to remove the obstruction by inserting an oiled feather down the windpipe. Rattling is also a symptom of Bronchitis, which see ; of Ulcerative Pharyngitis (generally troublesome at meal times and noticeable in the form of sudden gasps) and Diphtheric Roup (see Roup).

To cure Ulcerative Pharyngitis, swab the ulcers with iodine and remove them with a blunt-pointed wooden pick.

Toe Pecking. A common vice among chicks. The chicks peck at one another's toes to such an extent as to draw blood and cause sores sometimes followed by blood poisoning. May either be caused by a lack of animal food, a lack of green-food, or lack of litter. In the latter event the chicks' toes, shining pink—especially in bright sunshine—prove an irresistible temptation to their fellows, which probably regard the toes as something good to eat.

To remedy, use good chick mashes (see Feeding) containing the correct quantity of animal food—that is, fish meal. Always litter the floor with clean bright yellow straw chaff as this neutralises the colouring of the chicks' toes.

Tuberculosis (or Consumption). Deadly disease affecting all poultry.

The signs are wings protruding beyond the body, loss of weight, weakness, disinclination to scratch, although the bird has a good appetite and bright eye.

Predisposing causes of tuberculosis are : Filth, bad housing and feeding, and inbreeding.

It is well to ensure plenty of sunlight, and to give cod-liver oil and mineral salts, especially calcium, but if the bird is not worth a great deal, it is best to kill it and burn the carcase. If an attempt *is* made to cure a sufferer, it must be rigidly isolated, as the disease is highly infectious. Disinfect the remainder of the birds thoroughly.

Tumours. Abdominal tumours are fairly common in poultry. They may be felt quite easily—as a hard mass—if the hand is placed round the abdomen. When feeling, care should be taken to differentiate between a tumour and an over-fat condition. In the latter, the fatty deposits will be felt in layers, whereas a tumour is a distinct entity. If it is a cyst, a sloppy, watery condition will be heard and sometimes even felt. When tumours occur externally, they can easily be seen.

Remedial measures should not be attempted : any affected bird should be destroyed. See also Abscess.

Tympany, or **Inflated Crop.** Common complaint, symptoms of which are dullness and occasional stretching of the neck. Occasionally bird may bring up a watery mass. The complaint is caused by bird having eaten stale, heated or other unsuitable food.

The crop should be filled with a cherry-red solution of permanganate of potash, and worked about. The bird should then be held upside down to force out the liquid by pressing the crop. Repeat the process and don't feed for 24 hours, but give milk in which bicarbonate of soda is dissolved at the rate of 1 teaspoonful to the pint.

Under-Nutrition. Complaint affecting young geese, possibly due to keeping the birds on poor grass and at the same time not providing them with sufficient food of other kinds.

The symptoms are thinness on breast and mopishness.

Feed properly. As a quick pick-me-up, mix cod-liver oil with mash—1 tablespoonful per bird.

Vent Gleet. Common complaint with fowls. Soreness and redness of the vent, and discharge are symptoms.

Due to digestive disturbances and also a germ. Often communicated by the male.

In ordinary cases, bathe with a warm disinfectant solution and then swab with tincture of iodine every 3 days. If ulcers have developed right inside the vent, inject, by means of a syringe, a 1 per cent solution of iron sulphate 3 times daily.

Vent Pecking. Common complaint among chicks, allied to Toe-pecking, which see. Chicks peck at one another's vents, causing bleeding and sores. The cause is as for Toe-pecking.

Paint the wounds with tincture of iodine after having cleaned them with weak disinfectant. Give birds clean bright litter and feed on correct rations. (Birds given milk to drink will seldom indulge in the vice.)

Vertigo. Occasional complaint of fowls denoted by a staggering walk, often backwards.

Caused by blood pressure due to high feeding, heat or excitement.

Keep the bird in the shade and give it a strong dose of Epsom salts until it is thoroughly cleansed. To calm her, give her a 4-grain dose of sodium bromide every 2 days for a week.

Wheezing. See Aphonia.

White Comb. One of the most infectious of all poultry ailments.

There is white scurf or scales on the comb and face, which, in time, extend over the body.

To remedy, anoint the affected places with a mixture of 1 part red oxide of mercury and 8 parts of lard. Washing with disinfectant water and afterwards dressing with carbolic oil and glycerine—in the proportion of 3 drops of the former to a tablespoonful of the latter—is also excellent. After an outbreak disinfect the house thoroughly.

Worms. Intestinal worms are a very common complaint with poultry and account for much " mystifying " loss of condition and laying ability in a flock. The symptoms are ruffled plumage, extraordinarily big and capricious appetites, and general restlessness when the sufferers are roosting at night.

Treatment to prevent worms should begin as soon as chicks are removed from the intensive house or brooder house. Give a daily dose of tobacco dust. This cannot possibly do any harm, because nicotine is not a cumulative poison : it is passed out of the system.

At about 10 weeks old give the youngsters ¾ lb. of tobacco dust in every cwt. of mash, weighed in its dry state. A month later increase this to 1¼ lb. ; at about 4 months increase it once more and allow 2 lb. per cwt. of mash. Keep on with this allowance ever afterwards.

In the event of an attack of worms in adult birds, dose the sufferers with : Santonin, ¼ grain per bird (¼ grain per bird equals ¼ teaspoonful per 60 birds), given in the evening after starving for

8 hours. Mix the medicine in mash. Give castor oil in the morning. Repeat both doses a week later.

AILMENTS. *Health Tonics.* Occasionally even the healthiest of birds will be all the better for a tonic.

General Pick-me-ups (useful when birds require toning up). The simplest is to add a crystal of sulphate of iron, about the size of a pea, to each pint of drinking water. Or sulphate of iron may be dissolved in cold water—1 oz. of the sulphate to 3 oz. of water—a teaspoonful added to each gallon of drinking water being a safe form of iron, but rather slow in action. Tincture of perchloride of iron is excellent. A teaspoonful to each two pints of water makes a suitable dose, being neither too strong nor too weak. Phosphate of iron, too, may be used in the same way and in the same proportion.

Douglas Mixture. This is an excellent and reliable general tonic. See Douglas Mixture.

A Spring Pick-me-up. (For birds suffering from the strain of heavy winter laying). Make up a stock solution by dissolving 4 oz. of sulphate of sodium in a pint of warm water. Bottle, and cork tightly. Twice a month give a dessertspoonful of the stock solution to each pint of drinking water. The second dose should follow the first within 4 or 5 days.

A Tonic for Moulters. Some birds "hang fire" unduly after the moult. to these give : Cascarilla bark, 2 oz. ; aniseed, ½ oz. ; pimento, 1 oz. ; malt dust, 2 oz. ; carbonate of iron, 1 oz. Mix above in with four times its weight of sugar, molasses or similar substitute, and add just a sprinkling to the mash of about 20 birds. One good handful should suffice for 150–200 birds, given every day for a fortnight.

A Tonic for Turkey Poults (to be given occasionally during the summer). Take powdered cassia bark, 2½ oz. ; ginger, 10 oz. ; carbonate of iron, 4½ oz. ; and gentian, 1 oz.

The ingredients should be mixed thoroughly and stored in a dry, air-tight container. Once a week give a teaspoonful to every dozen growing birds, mixing it with their wet mash.

AIR-SACS. The bladders which, in conjunction with the lungs, constitute a bird's breathing system. Sometimes the air-sacs swell, giving rise to a trouble known as Emphysema—which see.

APPROVED BREEDER. See Breeder.

ASPARAGUS CHICKEN. The name given to surplus cockerels hatched out in February, March and early April, and fattened for sale as table chickens, See Fattening.

BALANCER MEAL. War-time ration allowed to domestic poultry keepers. Such meal serves to supplement or "balance" mashes prepared from house scraps, etc. The composition of "balancer" varies greatly but broadly it is composed of processed house waste, wheat-feed, meat or fish meal, dried potato, oats, damaged flour, etc. For methods of feeding "balancer" see under Feeding Poultry.

BARRING. Term applied to the broad marking across the plumage of a fowl, as in the Barred Plymouth Rock.

BATTERY POULTRY KEEPING. This method of keeping poultry may be defined as permanent confinement singly in small cages arranged in tiers. They are commonly constructed of galvanized sheeting and galvanized wire, and are used for rearing, for fattening, or laying hens.

Battery Brooders. The battery brooder is a "revolutionary" device for the rearing of chicks. In the United States it is in very extensive use but in this country opinion is still divided. Battery brooding was at first a failure, but improvements and modifications have resulted in almost universal success.

How Battery Brooders "Work." The system is as follows. The chicks are reared in total confinement in a series of cages or "trays" which together make up the battery. The cages vary in size according to the age of the chicks and are floored with wire netting or woven wire, the mesh of this also varying according to the age of the chicks. Below each cage is a droppings-tray, and on the outside of the front of each cage are the feeding and water troughs. In these cages the chicks eat, drink, sleep and take their exercise ; in fact, live either permanently or up to a specified age.

The tiers of cages or batteries are housed in a shed that can be kept at the correct temperature for the age of the chicks, and all attention is given in that shed.

Advantages and Disadvantages. The advantages claimed for battery brooding are : (1) much labour is saved, as there is no walking about over fields to feed, water and attend to the chicks ; (2) space is economised, thousands of chicks being reared in a very small space ; (3) feeding is done under cover ; (4) cleanliness is assured because cleaning-out is quickly and easily done ; (5) costs are reduced because litter is not required ; (6) better health is ensured because the droppings fall through the wire floor and are not trampled upon or pecked over ; (7) the air in which the birds sleep is sweet because ventilation is assured ; (8) there is no huddling or night-sweating ; (9)

almost complete immunity is ensured against coccidiosis, B.W.D., etc., because the chicks do not come in contact with tainted ground.

The disadvantages lodged against the system are that it is costly, that so much care is required in feeding and management that there is apt to be trouble, and that it is impossible with battery brooding to keep the chicks in the brooders after they have attained any size.

In regard to the latter point, opinion is fairly unanimous that chicks ought to leave the battery brooders after they are 6 to 8 weeks old, the birds then living a normal outdoor life until maturity. Later, they may be brought back, if desired, to the battery cages for fattening or laying.

Methods of Heating. There are three methods of heating the batteries commonly employed : (1) by heating the whole room ; (2) by heating the battery only with pipes; and (3) by small heated compartments like brooders. In the first method an extra stout shed is used with double glass windows. In the second a stove and pipes are employed ; and in the third a small heater of the Putnam type is used. In this last method the chicks have a heated hover and a run with a wire netting bottom where they feed and drink.

Any existing shed has been found to do for housing the battery if it is properly ventilated, of sufficient height, properly lighted, and made of sound material.

Where the batteries are heated by pipes or are housed in a heated shed, a cooler shed or room is needed into which the chicks are " weaned " at, say, 3–4 weeks. When from 6–8 weeks, as already mentioned, they go outdoors.

Feeding Battery Chicks. Feeding is on dry mash, which must be freshly mixed and of the best samples. Moreover, it must be specially composed. An ordinary chick-rearing mash will rarely do except some of the proprietary mashes. A recommended mixture is as follows : Middlings 9, Sussex ground oats 4, yellow maize meal 2, maize-germ meal 3, bran, 4, alfalfa meal 1, dried milk 1, fish meal 1½, parts by weight, and cod-liver oil 2 per cent.

Grain is not given, but plenty of clean water must be available at all times.

Laying Battery. A group of cages for housing layer birds. The layers live their whole life in their cages, each having a cage to itself. The birds are none the less happy and lay practically as well as birds kept* in runs. The advantage of the cage method is its economy of space. Batteries may either stand in the open or be housed in a light and well-ventilated shed. In the former case, weatherproofing is, of course, all important.

The dimensions of laying battery cages vary to some extent. A cage 14 in. or 15 in. wide (from side to side), 18 in. from front to back, 18 in. high in front and 14½ in. at the back will be satisfactory.

The floor should extend not less than 6 in. from the front of the cage proper and the clearance between the bottom of the front wires and the floor should be 2¼ in. If the space at this point is too large, the birds can reach their eggs and may start egg-eating.

As a further safeguard, the front of the wire floor may be in the form of a shallow trough. The front (bob) wires should be 2¼ in. apart or the fronts may be of slats if desired, attaching them to the top of the cage with hooks and eyes. They should, of course, swing inwards.

The floor may be of 1-in. mesh (16-in. gauge) wire-netting, the sides and back of ½-in. mesh. It should fall 3½–4 in. in the depth of the cage, which is 18 in.

Droppings boards should be fixed 6 in. below the floor at its lowest point.

Food troughs may be clipped or hooked on the framing or supported by shelf brackets. They should be quickly detachable.

It is better to have the drinking troughs about 6 in. above the food trough. Four-inch half-round spouting running the length of each block is recommended.

Almost any shed can be made suitable for use as an indoor battery. Nothing is wanted beyond a cover and protection from draughts. An open front will do if it does not get much cold wind.

Any hens may go in cages, or pullets at four to five months old, but indifferent stock should never be used.

The system is a forcing one, and can be likened to a glasshouse with plants. Therefore, stock of good stamina only should be used, for weaknesses are soon discovered.

Feeding follows on normal lines except that mash ingredients should be too small in size for the birds to pick and choose, or pick up and throw out.

Moulting birds may be left in the cages to recover. They may moult quickly, but vary considerably in the time taken to come back into lay. No change need be made in feeding them.

With wire-netting floors, the feathers and droppings get matted, and must always be removed. In a scientifically designed floor the feathers will fall through, and no trouble be experienced.

Sometimes a bird's excreta will be very watery, and such a bird is unlikely to do well if retained.

Caged birds always have paler combs and wattles than their outdoor sisters, but this is a natural result of their environment and need not cause any worry as long as the ventilation and light are right.

BEAN. Term applied to the peculiar-shaped black tip on the bill of some ducks —the Buff Orpington, for example.

BEARD. Term applied to the tuft of feathers found under the throat of some fowls—Faverolles and Houdans, for instance. Term is also applied to the tuft of rough hair on the breast of an adult turkey cock—it is sometimes called the tassel.

BLOOD TESTING. It has long been known that the dread disease, B.W.D. (Bacillary White Diarrhœa), which kills off thousands of chicks every year and, in the past, has spelt ruination for not a few poultry farmers, was communicated to the doomed chicks by the hens laying the eggs from which they were hatched. It has also long been known that certain hens are "carriers" of the B.W.D. disease whilst others are not.

Until recent years, however, no means had been discovered of ascertaining which members of a breeding flock were the "carriers" and which the safe birds so far as breeding is concerned. It is now possible to have every breeder tested and to know for certain which hens may safely be used and which are likely to produce B.W.D.-infected offspring.

There are two tests commonly used. The latest, which also is the one now approved and recommended by the Ministry of Agriculture, is that known as the "Rapid" or Antigen Test.

The Antigen Test. Also known under such names as the "hotspot," "plate," and "slide" test. A little blood is withdrawn from the bird's comb or a vein underneath the wing and placed upon a glass slide or porcelain plate. To this is added a spot of antigen. If the blood has been taken from a bird affected with B.W.D., the mixture "clumps" or curdles, little granules appearing. If the bird is free from the disease the mixture remains turbid, like normal blood. The tests are conducted on the farm—samples of blood do not have to be sent away for testing.

No doubt in due course it will be possible for any experienced poultry man to do his own blood-testing. At the time of writing all testing is done under the auspices of the Ministry of Agriculture by a qualified veterinary surgeon. Also the privilege of having stock tested by the Antigen method can only be enjoyed by accredited and approved breeders. To some extent restrictions are due to the fact that it is of primary importance that the antigen be obtained from a reliable source, it being entirely misleading to trust to unreliable material.

The Laboratory Test. This involves the taking of a blood sample from each

bird in a tube and sending it to one of the recognised laboratories. (Addresses will be supplied on application to *The Smallholder*, Tower House, Southampton Street, London, W.C.2). The charge is usually 1½d. per bird. The scientists examine the blood and state whether the bird from which it was taken should be bred from or not. The terms used in their reports are "Negative" and "Positive." "Negative" birds can be bred from; "Positive" birds must be discarded. The elimination of even only one "carrier" bird from a breeding flock, of course, will easily save far more than the cost of having a flock of breeders tested.

Write to your selected laboratory and state number of birds required to be tested. You will then be sent a box of glass tubes—one for each bird—a special needle about 4 in. long and a paper form in duplicate. The tubes are about 2½ in. long, fitted with a rubber stopper and bearing a numbered label. Sometimes you will see a little colourless liquid in the tubes. This is a preservative to protect the blood sample in the interval before its examination and therefore must not on any account be emptied out.

The first step is to starve the birds (which should include all those you desire to breed from) for twenty-four hours.

An assistant is required to hold the bird whilst the sample of blood is being secured. He grasps the two wings with one hand and the legs with the other so that the bird lies on the table on its right breast with the wing open.

The operator stands facing the inside of the wing. His first task is to locate a prominent vein under the wing near where it joins the body. This vein crosses the junction of the first two bones of the wing. The vein located, a few small feathers should be plucked away so that it stands out clearly; then it should be swabbed liberally with a pad of cotton-wool soaked in methylated spirit, not enough spirit being used, however, to leave drops to get into the blood sample.

Now one of the tubes is taken in the left hand and the needle in the right. The rubber cork is removed from the tube by gripping it between the little finger and the palm of the right hand and held there. The open tube is held close against the skin below the vein described above and simultaneously the needle is used to pierce the vein. Blood will commence to flow from the vein after the first drop or two has exuded, and the flow should be caught in the tube.

When the tube has collected from ½ to ¾ inch of blood, it should be withdrawn, and the stopper replaced, and on the label on the tube should be written the number (or leg-band colour) of the bird.

Naturally, every bird to be tested must have an identification ring (numbered or coloured) ; otherwise you will have no means of knowing afterwards which sample came from which bird, which birds are certified by the bacteriologist to be safe and which " carriers " of Ailments.

Usually the blood will cease to flow a few moments after piercing. If the flow *does* happen to continue after you have collected as much as required, a little pressure applied to the tiny wound made by the needle will stop it.

When all the blood samples have been secured, fill in the duplicate forms sent you by the bacteriologist and dispatch samples and forms without delay, together with the small fee chargeable.

For treatment of cases of Bacillary White Diarrhœa, see under Ailments.

BLOOM. Term applied to the gloss or sheen on the plumage or legs of poultry.

BRASSY. Term used to denote the yellow tint sometimes seen on birds— particularly Barred Plymouth Rocks, Rhode Island Reds, Red Sussex and many white breeds. It is a fault.

BREED. Term used to denote a distinct race of fowls, ducks, etc.—e.g. White Leghorn, Aylesbury Duck, Black Norfolk Turkey.

BREEDER, ACCREDITED. A breeder who has accepted certain conditions as to breeding, under an official scheme known as the Accreditation Scheme, which see.

BREEDER, APPROVED. A breeder whose stock has been approved, as sound foundation stock, by a local panel, and are allowed supplementary rations under the war rationing arrangements; but there is no official control over stock or eggs offered for sale.

BREEDING. The making up of the breeding pens—the selection of the hens and cocks that are to' produce the eggs required for hatching—is one of the most important items in the poultryman's calendar. Like begets like. As the breeders are, so will the chicks be. Unless every bird in the breeding pen is carefully chosen with full regard to merits there can be no assurance that the resulting progeny will be of good class.

The poultry-keeper with only a few birds may, if he wishes, merely turn in a rooster among his flock of hens in January or February, and in March or April use as many of the eggs as required for hatching—though he would be better advised to buy sittings of eggs from a poultry-farmer who specialises in quality sittings. The

only practical procedure for the commercial poultryman, however, is to set aside selected birds for the production of sittings, penning these apart and treating them purely as breeders.

The number of hens to allot per male in these pens is 6-7 for heavy breeds, 9-10 for light breeds.

The Birds to Choose for Breeders. As already mentioned, the hens must be selected with the greatest care.

With utility birds their record as egg-producers must first be taken into consideration. It must be beyond reproach ; every bird must have been a good and consistent layer of eggs of 2 oz. and over in weight. With show birds, of course, the dominant factor is the possession of good show points.

Apart from a good laying record, the hens must have good breeding points— show that they possess the ability to produce the right type of offspring—and also be thoroughly healthy.

To this end hens should be of medium size for their breed, active, keen after their food, full of energy, tightly plumaged, possessing fine (medium-thin) bone-framework with a back that combines breadth with length. A back that has length without breadth predisposes the future chicks to the laying of small eggs and a weakly constitution. The eye of every bird should be full, round, bright and bold, and set high in a head that also boasts full quality comb and wattles.

Birds that possess too fine a framework, although quite possibly very good layers, probably lay small, thin-shelled eggs and are certainly more disposed to sickness and disease than the heavier-boned ones.

In regard to age, so long as the birds are active, lean in body condition and lay early enough for breeding in the new year, age makes little difference.

Probably the best mating is 2-year-old hens to vigorous and fit 12-months-old cockerels.

All the breeders must be in good body condition. An over-fat hen or a sluggish, fat cockerel may produce many infertile eggs.

In regard to the selection of the cockerel, he must possess equally good breed and health points as the hens. He should also be a bold, dominating fellow, truly masculine in his manner, appearance and treatment of the hens. The cockerel that always takes second place in a batch of males will never make a good sire.

When the Breeders Should be Mated. The following table shows the best mating times for the various purposes :

For pullets for the early shows, mate in early December.

For July-laying pullets, mate in early December.

For early spring table chickens, mate in early November.

For winter-laying heavy-breed pullets and stock males, mate at the end of December.

For winter-laying light-breed pullets and stock males mate in late February.

First of all, the hens should be penned together and given a few days in which to become accustomed to one another. Then the male should be introduced to the pen. Do this at night, placing him quietly on the perch among his wives.

Eggs will be fully fertile and may be used for hatching 8 days after the cock has been introduced into the breeding pen.

If there is need to change a male it will be necessary to allow 18 days before using the eggs; only thus can you be sure that the influence of the first male will not be reflected in the progeny.

Feeding the Breeders. One must put into the hen the foods that she in turn can place into the egg to ensure a big store of rearable chicks. Improper feeding may lead to weak germs and chicks dead-in-shell. If one is used to feeding poultry on one of the excellent proprietary laying or breeding meals, this may be continued, but for those who prefer to mix ingredients at home, here are recommended mashes (normal, subject to wide war-time variation):

For wet mash use 5 parts by weight of middlings, 2 of maize germ meal, 1½ of Sussex ground oats, 1¼ of fish meal, 1 of clover meal, ¾ of broad bran and ½ of linseed meal. Add 2 per cent cod-liver oil and 2 per cent mineral mixture.

For dry mash a guarantee of strongly fertile eggs is—3 parts by weight of middlings, 1½ each of Sussex ground oats and maize germ meal, 1 each of fish meal, broad bran and clover meal, and ¼ of linseed meal, giving cod-liver oil and mineral salts as for wet mash.

The breeder's grain feed is ⅓ each kibbled maize, wheats and oats.

As for menus, normal procedure should be followed. For wet mash, scratch feeds of grain may be given morning and noon; and, half-an-hour before sunset, a large feed of wet, crumbly mash. For dry mash feeding, a small scratch feed is desirable morning and noon and then a big grain feed last thing, the dry-mash hoppers being open always, with ample shell available.

The rooster should be well fed. Sometimes he will not touch food until his wives have satisfied their appetites, by which time the troughs are empty. If so, he should be given a meal apart.

Causes of Infertile Eggs. Every egg from a properly selected and managed breeding pen should be strongly fertile. If there is infertility trouble, the causes may be:

(1) Weak or too small or too large male, or windy weather, may cause serious infertility, the remedy being to change the male or protect from the wind.

(2) Spasmodic infertility may be due to windy weather or careless storage of eggs, the remedy being to protect birds from wind or collect eggs more often and store properly.

(3) Eggs from one particular hen may be infertile because the cockerel dislikes the hen or the hen won't mate. The remedy is to change the hen.

(4) Eggs may have weak germs because the male is underfed. Remedy: Feed male separately.

(5) Fertility of eggs may lessen as season progresses because of too many hens to the male. The remedy is to reduce size of breeding pen to correct proportions.

The Storage of the Hatching Eggs. The eggs should be collected regularly every day—never allowed to lie for any time in the nests. In frosty weather, it is best to collect every two hours. The eggs should be stored by laying them on their sides in shallow boxes of bran kept in a room with an even temperature of 50 deg. F. The eggs must at no time be jarred or shaken.

Mating Ducks. The rules as regards selecting birds of health and stamina are the same with ducks as with fowls. For the rest, the chief mating rules are: (1) To choose only birds of standard weight, (see Ducks); (2) Neither to underwork nor overwork the males, the standard numbers of females per male being, Indian Runner, Coaley Fawn and Abacot Ranger, 7; Khaki-Campbells, Magpies and Crested, 6; Orpingtons, Cayugas, Penguin, 4-5; Aylesbury, Pekin, Rouen and Muscovy, 3. (3) To feed the breeders well on an ordinary laying mash. (4) To provide swimming water only if the birds have been accustomed to it before mating. If ducks in the habit of swimming have no pool at breeding time there is a danger of infertile eggs at the start. (5) To employ breeders anything between 2 and 5 years old but to have neither drake nor ducks under 2 years of age.

Mating Geese. Geese should always be mated in the autumn, for they take a long time to become accustomed to each other.

Special care should be taken to choose a first-class gander. Normally, a good one costs anything from £2 to £4; it is better to pay at least £3, because it is money well spent. A first-class male can be used for years, and as he will throw a large number of goslings each season the expense is fully justified.

The best type of gander to select is one that is true to breed type, is medium-sized, alert, active, with bright eyes and good, crisp plumage, while fierceness is a good characteristic for him to possess.

As regards the age of the geese, when early-hatched goslings are required, it will be found necessary to breed from young geese of the previous year's hatching, but, if possible, they should be at least 12 months old. Geese give high fertility up to 7 years old and often older; the older they are, the later they come into lay in spring. Highest possible fertility is obtained from two-year-old geese.

All the breeders, both geese and ganders, should be of normal weight (see Geese).

The number of geese to allow per gander is 5 with the Toulouse and Embden, 3 with all other breeds.

Mating Turkeys. For the turkey breeders birds of medium weight are the best choice, so long as they are well formed generally and have long, straight breast bones. If early turkey chicks are required, they can be bred from the previous year's hatched hens so long as they are quite sturdy, healthy and well-grown. The correct number of females to allow per male is 8-10. The age of the birds is unimportant so long as it is not less than two years, for a turkey takes two years to mature properly. The birds should be mated as early in the season as possible. Turkeys are slow to start breeding and late mating leads to infertile eggs at the outset.

Cross-Breeding. Term applied to the mating of two different breeds of fowls, ducks, etc. A bird bred from pure-bred parents of two different breeds is spoken of as a first-cross; a bird bred from first-cross birds mated with pure-bred birds is spoken of as a second-cross.

Cross-breeding is freely practised for the production of table fowls, it being found that better table birds are produced by crossing two breeds than by mating birds of the same breed together. Popular table crosses are : Light Sussex male and White Wyandotte females ; Faverolles male and Light Sussex females ; Faverolles male and White Orpington females ; Faverolles male and White Wyandotte females ; White Bresse male and Light Sussex females ; White Bresse male and White or Buff Orpington females ; Indian Game male and Light Sussex females.

Cross-breeding is also freely practised for the production of sex-linked chicks. See Sex-linkage.

Double-Mating. The practice of using two separate breeding pens, one for the production of exhibition cockerels and another for exhibition pullets. Where only one pen is used for the production of both exhibition male and female birds, this is single-mating.

In-Breeding. Breeding from related fowls, such as brother and sister. In-breeding is undesirable, resulting in the weakening of the stock. It must not be confused with the deliberate mating of a parent bird with its own offspring, in order to establish desirable qualities. The latter is known as Line-breeding.

Line-Breeding. A method much practised by exhibition poultry breeders. It consists of mating father to daughter or mother to son, the object being to " fix " particularly good points that have been noted in that family.

BROILER. Term applied in America to a young chicken marketed at 9-12 weeks old, when weighing about 1½ lbs. In England the term usually means a fowl too old for roasting.

BROODERS. Brooders are appliances for the " artificial " rearing of incubator-hatched chicks, ducklings, etc.

Often the terms brooder and hover are used indiscriminately, but though their purpose is the same there is some difference between the two appliances.

A *brooder* is made with a metal or wooden floor and sides, exit for the chicks being allowed at one place only, that place being closed when desired by a sliding door. A *hover* is made with strips of felt forming the walls.

Some poultry-keepers declare that the hover is preferable, (1) because it is " nearer to nature," (2) because it is better ventilated, (3) because it is cosier, and (4) because the chicks can get into it at any spot, instead of having to find the one and only entrance.

Other poultrymen favour the brooder, their reasons being that it scores where rats and other ground vermin are troublesome, enabling the chicks to be shut up at night quite safely from attacks.

Both brooders and hovers, unlike a foster-mother (which see) have to be installed in a building ; a foster-mother can be operated out-of-doors.

A brooder is a domed construction, made of metal or wood, with a lamp installed to maintain an interior temperature as required for the safe " rearing " of chicks. The stated capacity is commonly 100 chicks but the actual capacity is only 80.

Points to remember when buying a brooder are that it must not have a lamp, (1) that cannot be trimmed without frightening the chicks ; (2) that cannot be filled without unscrewing the burner ; (3) with a " patent " chimney-less burner that will not turn low without smoking ; (4) that has not a big enough oil-tank to last well over 24 hours.

Given a sound building (see Brooder House), the best place for the brooders to stand is towards the end furthest from the entrance, so that the rest of the floor space can be used for the chicks to run over when not in the brooder. The brooders should stand 8 inches away from the walls of the house.

How to Avoid Lamp Trouble. It is all-important to manage the brooder lamp correctly. If it smokes, or the flame is too high or too low the chicks may well be " roasted " or frozen.

At the beginning of each season, tip out any old oil, rinse out with new and fill, not *right to the top*, but to within ¾ in. with the best paraffin. The wick must be long enough to reach the bottom of the tank with 2 in. to spare or instead of running for 30 hours, as it should, the lamp will go dim after about 22 hours and the chicks will become chilled.

Trim new wicks by cutting straight across with scissors and then snipping off the corners ; trim old wicks by rubbing only. Don't spill one drop of paraffin.

Light up and turn the wick three-quarters high. Regulate the flame until the thermometer registers a steady temperature of 80 to 85° F. Not until the temperature is rock-steady at the right degree must chicks be put into a brooder.

After re-filling the lamp, always turn the wick slightly lower than normal, for it is bound to creep and so attain its correct height. Always carry out all alterations to the lamp in the morning, so that you can watch throughout the day and leave it at night knowing that everything is as it should be.

Installing Chicks in a Brooder. Before chicks are installed in a brooder the floor should be well littered with clean, dry, dustless straw chaff to a depth of ¾ in. The chaff should be laid on 1 in. of damp, loose earth.

The best time to instal the chicks is in the morning, so that the heat can be watched, and any chicks that get out can be put back into the warm.

Place all the chicks in as gently but as quickly as possible, and inspect the temperature in an hour's time. It will probably have gone up a little, so that the wick must be regulated slightly. The ideal temperatures are 85° to 90° F. for the first week ; 80° to 85° the second, and 75° to 80° the third.

The chicks *must* remain under the brooder for the first 12 hours. If they get out they are apt to contract " brooder pneumonia," for which there is no remedy.

The depth of litter in the brooder should be increased as the chicks grow older— to 1 in. at a week old, 2 in. at three weeks old and 3 in. at six weeks.

BROODER HOUSE. Any building, so long as it is well built and rat-proof, will serve as a brooder house. A wooden floor is best, but other types are good so long as they are damp- and draught-free and well lighted into every corner. Wind and rain must be kept out by storm boards, but the chicks must be able to bask in the direct rays of the sun. This is very important and is ensured by fitting one of the special ray-passing glasses or glass substitutes, or by providing sun parlours—that is, wire-netting-floored runs projecting into the open air so that the chicks may exercise there in sunny weather.

The house should be divided into compartments each of which will contain one hover or brooder. A suitable size for the compartment is 10 ft. by 5 ft. ; this would accommodate 200 chicks.

BROODINESS. Broodiness is a purely natural and normal desire on the part of a fowl to reproduce her species by hatching her eggs. Unfortunately, from profit point of view, a bird that evinces this desire goes out of lay, and, if allowed to sit, stays out of production for from 6 to 10 weeks. This may mean a loss of 20 to 50 eggs.

Heavy breeds, such as Sussex, Wyandottes and Rhode, are the worst offenders for broodiness; light breeds, such as Leghorns and Anconas, rarely go broody.

Fortunately, broodiness can be cured —that is, the bird can be dissuaded from the desire to sit and brought back to lay *within a week*—if it is taken in hand at once. Broodiness can also be *encouraged*.

Causes of broodiness are : (1) the feeding of excessive quantities of maize meal in the mash, i.e. over 25 per cent, (2) the housing of the birds in close confinement, (3) lack of ventilation during the day, (4) the influence of a male bird, (5) warm weather, and (6) lack of regular water supplies.

How to Recognize Broodiness. It has been mentioned that to cure broodiness quickly it must be taken in hand at once. This means that the poultry-keeper must learn to recognise broodiness in its earliest symptoms.

The first indication is when a bird remains on the nest longer than is necessary to lay an egg. Nearly always birds found on the nest in the afternoon during spring and early summer are broody.

A broody hen will ruffle her plumage when touched, she will cluck, croon to herself, shuffle her wings and appear " spread out." A bird that is on the nest for the legitimate purpose of laying may fly off at your approach, but in any case she does not cluck, ruffle her feathers, nor does she return to the nest when you have gone, if she has finished laying.

Also, if a broody hen is taken off the nest, she will open her wings, cluck furiously and may either squat down in a heap on the floor and scream or hurry round in a circle. Excessively broody hens are bare on the breast-bone, and some feathers may even be found loose in the nest bottom.

To Cure Broodiness. The chief step in curing broodiness is to confine the offending bird at once in what is known as a broody coop (which see). In the event of there being an excess of broodies and the proper coop space being full up, the usual thing is to make use of out-of-use slatted-floor pens, or empty brooder houses or rearing coops fitted with a loose, slatted floor—so long as no nest-box accommodation or litter is allowed.

A broody having been discovered, she should be taken to the broody coop and given water, shell and a wet-mash feed straight away. It would be an advantage to give a dusting of insect powder.

Shell and water hoppers should be kept always full. For the morning feed give mixed grain as for the layers, but in smaller quantity ; a half handful will be ample. It is important not to give too much morning grain. At noon give some green food. In the evening give mash—just as much as the bird can eat.

After 3 days the hen should be more active, standing up and having lost the cluck. She should then be returned to the laying house, but watch should be kept on her that she does not return to the nest at night.

BROODY-BREAKING COOP. A broody coop is a special coop in which to " break " or cure broody hens. The coop consists of a compartment in which the floor and front are of slats.

The handiest broody coop is one on legs. Its construction is as follows. The legs are merely continuations of the corner battens, which should be of 2 by 2-in. timber. The height of the legs should be 24 in. and the height of the front of the coop 28 in., making a total height of 52 in. The height of the back is 22 in., allowing a fall in the roof of 6 in. The depth is 15 in., and the length of each compartment 15 to 16 in.

Use tongued and grooved matching for the roof and two ends only, making the rest of the coop of slats. If a storm-board is fitted to the front, so much the better.

The slats are of 2 in. by 1¼ in. lath nailed upright in the front, leaving a feeding space of 2 in. between two. In the partitions only 1 in. should be allowed between the slats to prevent fighting. The slats of the floor should be 1½ in. apart and run from end to end to ensure more comfortable standing. The floor is best made of the bevelled slats used for adult slatted floor houses as they make for better cleanliness.

It is best to make several compartments rather than to have one large one so as to prevent birds from different pens from fighting. Along the front should be a trough or rack with vessels for mash, grit and water.

CAPONS AND CAPONIZING. On the propriety of caponizing, opinions differ. Some persons are definitely opposed to it ; others hold that, as the operation is practically painless and the law permits it, and as the advantages of caponizing are very definite, it is perfectly legitimate.

Caponizing, then, is the operation of removing a male bird's reproductive organs (the testicles). A bird that has been so operated upon is termed a capon, and it bears the same relation to a cockerel as a steer does to a bull. Capons grow much larger than ordinary cockerels, putting on more than one-third extra weight ; and the flesh is far more tender and juicy. There is always a market for capons, good prices being realised even for caponized light-breed cockerels.

In regard to the extra weight obtained by caponizing, whereas an ordinary Plymouth Rock cockerel at 8 months old weighs between 7 and 8 lb., a capon will turn the scale at 10 lb. and over. A pair of Black Giant capons have been known to weigh 26 lb.

The age at which caponizing should be done is round about 8 weeks, when the birds weigh from 1 to 1½ lb.

CAPSULE. Part of the heat regulating apparatus in an incubator. See Incubators.

CARRIAGE. Term applied to the bearing or style of a bird.

CARRYING POULTRY. The correct way to carry or hold a fowl is with the breast lying on the palm of the hand, the birds two legs passing the fingers, and its head tucked under the arm. To hold by the legs or wings may result in internal damage and the possible loss of a valuable bird.

To carry a duck, it should be picked up by holding its neck fairly close to the head and carried by resting the body on the palm of the hand.

CARUNCLES. Name given to the fleshy protuberances on the head and neck of a male turkey ; also seen at the base of the bill of Muscovy drakes and ducks.

CASTOR OIL. Used at the rate of ¼ teaspoonful per bird, is a useful aperient for all kinds of adult hens or ducks. Proportionately larger doses can be used for turkeys and geese, but the dose should never exceed ½ teaspoonful.

CATCHING FOWLS. The best ways to catch fowls for one purpose or another are as follows : If the birds are in laying houses, shut them up at night and drive them out into a catching crate next day ; if the birds sleep in night arks, shut up over-night and secure the birds required by reaching in for them ; if in intensive houses, catch with a net during day ; if in a back-garden run, shut up over-night and catch in the house, or else gently corner them in the run. Fowls that are properly treated are normally very tame, and any odd birds can be picked up readily when being fed with a little corn.

CHAFF. The outside husks of oats and wheat obtained when threshing, or the straw of these cereals cut up in short lengths. Very useful as litter for young chicks. Barley or awned wheat chaff should never be used.

CHICKS, COLOURS OF WHEN HATCHED. Misunderstandings often occur when buyers of day-old chicks, or of sittings of eggs, find that the chicks delivered to them or hatched by them, are not of the colour it was thought the particular breed concerned should be. It is immediately supposed that the chicks or eggs are either not pure-bred or are of some other breed than that which was ordered. The fact is that with many breeds, the chicks are quite different in colour from the adults, and have markings quite different from those they will normally assume when their proper plumage has replaced the baby fluff.

The following are the correct colours and markings for newly-hatched chicks :

Black Breeds. A mixture of black and yellowish white, with sometimes a brownish shade. The white is found on the breast, under parts, wings and sometimes on the head. There may often, indeed, be more white than black.

Barred Breeds. A dark grey, often with a light patch on the head. The chests of some are yellowish. The wings take on the barring first. In many barred breeds, some of the chickens come out all black.

Brown Breeds (which include Brown Leghorns, Brown Sussex, Old English Game, Barnevelders, etc.). Broad brown stripe down the centre of the back edged with a thinner border of creamy yellow ; the rest of the body a pale fawn-brown. The head is marked alternately with fawn or dark brown.

Buff Breeds. These have a creamy or fawn tinge, sometimes speckled sparsely with dark grey or dark brown.

Columbian Breeds, such as Light Sussex, Columbian Wyandottes and other white breeds with black markings of hackle, etc., come out a creamy white with a dark mark on the head. Sometimes the back has a pale greyish marking down it and sometimes the chick has hardly any dark mark at all.

White Breeds. These come yellowish white in Leghorns and Wyandottes. Some may be found to be pure white. In some strains, one may find a trace of smutty colour, but this generally grows out. White Orpingtons and White Rock chicks usually come either a real white or a slightly pale grey.

Red Breeds (including Rhode Island Reds, Red Sussex, etc.). Normal chicks are brownish buff on the head and down the back, with a paler fawn on the under parts.

Mottled Breeds (Ancona and Houdan) come black or brownish along the back and behind the head, with under parts yellow or yellow-white. Andalusian chicks are bluish black at back, but both black and white will appear as well.

Sex-Linked. See Sex-linkage.

CHICKS, DAY-OLD. Numbers of poultry-keepers who have not the facilities for hatching, or wish to introduce a new strain into their flocks, purchase newly-hatched chicks from some near or far hatchery. The chicks are delivered to them by rail or carrier packed in boxes. They stand the journey perfectly well and, provided proper rearing arrangements have been made, and also that, immediately they are received, they can be given the normal management for young chicks, they will make every bit as good adult birds as those which were hatched on the premises.

The reason newly-hatched chicks can stand the journey is that chicks require no food whatever for the first forty-eight hours after emerging from the egg (see Feeding, Poultry).

How Many Chicks to Buy. This depends on how many pullets are wanted. There will be rearing losses, say, on the average, 20 per cent, and half the remainder can be reckoned as cockerels. For every pullet wanted, then, must be allowed as a minimim, 2½ chicks or, to put it another way, 100 chicks should provide 40 pullets.

How the Chicks Should Look on Arrival. On first opening the box it may be difficult to detect exactly the physical fitness of the chicks. If the journey has been a long one and the weather frosty, the chicks may be huddling together for warmth and comfort. They may even be tired. They should, however, brisk up when—still in their box—they are warmed in front of a fire, and should soon begin to chirp, move about and maybe jump out of the box.

Each chick should be handled, examined and perhaps set down on a table. It should be able to stand up quite easily and firmly and run about just as happily as the proverbial cricket.

In no circumstances should there be any pasted vents, soiled fluff, dull, watery eyes, flat, thin legs or full, hard abdomens. The aperture in the abdomen through which the yolk-sac was taken at hatching should be dry and quite unnoticeable when the fluff is parted. The fluff should be loose, full and not stuck down or scanty in any part. No deformity of beak, back, legs or feet should be accepted. Birds that are dead should be kept and the sender notified. The sender should also be notified about any birds that are ailing or deformed.

Guaranteed live delivery of sound and healthy chicks should be pre-arranged.

What to Give for the First Feed. The first feed should be given after the chicks are warmed and recovered from their journey. It should consist of "saturated" chick corn (see Feeding).

Rearing With a Broody. If the chicks are to be reared naturally the chosen hen must be prepared to receive them by sitting her on some china eggs for at least a week before her "family" arrives. The chicks should be given to the hen in the evening of the day of their arrival, being put under one by one, the eggs being removed at the same time (see Rearing Chicks).

Rearing With a Foster-Mother. If the chicks are to be reared by a mechanical foster-mother, brooder or hover, the machine should be run for 24 hours before their arrival. Put in the chicks at night (see Rearing Chicks).

Rearing With a Fireless Brooder. Those who buy day-old chicks and have no broodies and also have no artificial rearing appliances, may rear a few chicks in what is called a fireless brooder. This is a box brooder lined with hay and in which the chicks create sufficient warmth for themselves. See that the "nest" will just comfortably hold the number of chicks you have bought and yet have room to spare. (See also Rearing Chicks.)

CLEANING HOUSES AND RUNS. See Houses and Runs.

COCK. A male bird not less than one year old. See under breed headings for variety distinctions, breeds, weights, etc. A cock has no influence on the number of eggs laid by pullets or hens, only on the usefulness of eggs for hatching. Thus it is a mistake to keep a cock where the eggs are not used for hatching.

COCKEREL. Term applied to a male bird up to one year old.

COD LIVER OIL. See Foods.

COLONY SYSTEM. Method of keeping poultry in which the birds are divided into flocks of anything from 20 to 100, housed in movable houses spaced widely apart about a meadow, each house and its flock thereby forming a "Colony." No runs are provided, the birds being taught to know their own house by confining them to it for a few days before they are released.

Colony houses are much used with birds that are turned into wheat-field stubbles in order to glean the fallen grain.

The Colony system is undoubtedly a very healthy one, excellent for those who have a large amount of spare ground and sufficient assistance to attend to the needs of the birds. For the commercial poultry-farmer the system is usually too wasteful of space and too heavy on labour.

COMB. The fleshy growth on the top of the head. It varies in type and size, according to the breed. The commonest types are the single comb (as in Leghorns) and the rose comb (as in Wyandottes). Other types are the cushion comb (Silkies) ; horn (La Flèche) ; leaf (Houdan) ; pea (Brahma) ; raspberry (Orloff) ; and the strawberry (Malay).

COMB PECKING. An outbreak of comb pecking is a not uncommon occurrence. Not to be confused with the nipping of a hen's comb by the rooster during the act of mating. The comb bleeds and birds peck at it.

Due to lack of greenfood, animal matter or exercise, overheating and laziness.

Remedy conditions at once. Isolate. Bathe comb in disinfectant solution and anoint with sulphur or boracic ointment.

COOPS. Term applied to the small "hutches" in which broody-hens are confined, or in which hens rear their chicks in the natural fashion.

There are several forms of chick coop, the simplest of them being a three-sided box fitted with a sloping roof and with a front of wooden bars. This form of coop is not to be recommended. With it the chicks' food must be placed outside even if it is raining, for, put down inside, the hen would trample on and waste it, leaving her charges to go hungry. Another disadvantage of this type of coop is that the chicks, when they feel like a scamper away from the hen, have no choice but to go out into the cold and damp.

Much preferable is the double coop, which consists of two coops, one with a barred front and one with a closed front.

Each coop should measure as a minimum, 18 in. wide, 18 in. deep, 18 in. high at the back and 24 in. high at the front— all being built under one roof.

Communication is provided between the two by means of a small door in the centre partition.

It is desirable that both compartments be provided with front shutters. In the early part of the season it is vitally necessary to close up coops at night ; otherwise chills and other ills are bound to result. The shutters, however, should have a small opening at the top, to admit fresh air, this opening being covered with wire-netting, to keep out mice and birds.

Each compartment should have a board floor. It doesn't do chicks good to spend their early life on damp soil. There is no harm whatever in rearing the birds on boards, provided these are littered with straw chaff to a depth of ½ in. to begin with and later 1 in.

On wet, damp, foggy or very cold days, the chicks should always be fed in the closed compartment ; on fine, warm days, they can have their meals in the open. During driving rain, the front shutters can be raised to prevent damp, with all its attendant ills, from entering the coop.

COVERTS. Term applied to the part of a fowl's plumage. The wing coverts are the feathers covering the tops of the flight-feathers (which see). The tail coverts are the small feathers at the root of the tail.

CRAMMING. See Fattening.

CREST. Term applied to the crown or tuft of feathers on the head, as in the Houdan.

CROP. The sac at the base of a fowl's neck wherein the food is stored and ground up before passing into the stomach and gizzard for digestion.

CROSS-BREEDING. See under Breeding.

CULLING. Culling means the weeding out of birds that are certain to be worthless if kept.

The breeder of show birds culls out chicks that possess bodily defects; the breeder of utility birds culls out chicks that cannot grow into good layers or good table-birds.

The First Culling. Usually the first culling takes place when the chicks are from 5-8 weeks old.

All permanently unhealthy birds must go. Any chick that has had a disease and recovered, but has been seriously checked is permanently unhealthy. The sign of a bad chick in this respect is over-

long or drooping wings. Permanently soiled abdominal fluff indicates serious disturbance of the digestive tract and that again is a bad sign.

Signs of good health are—bold, dry, bright eye; clean, ample feathers and good appetite.

The next thing is to consider vigour. Signs of lack of vigour are dullness, inactivity, slow growth (when compared with the rest of the batch), and too great a liking for warmth and protection.

The birds are now examined for quality of feathering. Chicks having (1) a distinct lack of body plumage, (2) an excessive quantity of fluff at a later age than its brood mates of the same age, (3) parts of the body quite bare of both fluff and feather, and (4) distinctly slow feathering, should all be suspect. A chick showing any three of these four signs should be culled without hesitation.

Both thin and heavy bone are bad signs, but the heavily-boned bird should not be culled at this stage. The birds to cull now are the ones with thin, shaky legs, of wizened appearance, with a small body framework with loose bone and paper-fine ribs and back.

A bird may be small (compared with her brood mates) because of any of the following influences, and such a bird must be regarded as a cull : (1) Acquired lack of vigour, (2) serious disease setback and (3) inherited lack of stamina. At maturity such a bird will break down shortly after coming into lay, will probably lay small eggs, and will not only herself contract diseases but will also act as a carrier of diseases.

The " dud " head is long and narrow, the eye, small and oval, set half-way between the top of the skull and the top of the wattle, the nostrils are clogged with matter, the tongue is hard at the tip and the beak refuses to shut properly. A bird with a head like this will never lay profitably, and should be culled.

There is no need to cull every bird that has a deformity. Some deformities, such as bent toes, turned-in feet, blindness of one eye, wry tail, unequal thighs, and slightly roach back, do not in any way interfere with laying.

Deformities that will affect laying are —complete blindness, badly crossed beaks, permanently bent thighs (i.e. unable to straighten one leg), and bent back.

The Second Culling. The second culling takes place when the birds are from 4 to 5 months old.

The best birds have a well-fleshed breast-bone, are well grown, without a too great enlargement of comb and wattle, have medium length of feather and a big appetite.

Birds that ought to be culled will have the following signs: Excessively heavy

B

unthrifty plumage, over-long wings, beaky heads, more or less permanent diarrhœa, thin, "bladey" breastbones, "paper" flat legs and lack of size.

In culling birds of this age it is necessary to follow a process known as Handling, which see.

Culling Adult Birds. It is very desirable to go through the whole flock once a year, preferably in July or August, to weed out undesirable birds before the moulting (that is the idling) period. commences.

Over-yeared Hens (those kept for laying during their second year) may not lay profitably for another year, but some may be worth keeping for breeding. Such specimens will be of sound body-size, of especially good breed-type ; show great depth of abdomen ; carry good feather on the breast (showing lack of broodiness) ; have thoroughly bleached legs (see Pigmentation), and possess an exceptionally beautiful head, "beautiful" meaning clear, open, brilliant and bold.

Old Breeders (the birds used during the current season) should generally be culled, except those of special merit.

Previous Year's Pullets will be just finishing their first laying year. Some of them will probably show a good profit in their second year and, depending upon the amount of laying-house space available, one can keep back from one-half to two-thirds of these for another year.

Any suspicion of ill-health, of weakness, of lack of vigour and stamina, and of lack of body size, should be sufficient to cull the bird. Legs that are far finer than the average of the flock, and a head that is "beaky" and long are signs of the cull at this season.

An excess on the other side, i.e. over-large body, fat, round legs that are far coarser than the average of the flock, and a head that is round, wrinkled, fatty or puffy, with a small eye, should also lead to culling.

The good bird will be worn in feather, exceptionally active even at the back-end of the season, still with a large appetite, full red in head, brilliant in eye, bleached in beak and leg colour (if originally yellow), and fine in bone that is not covered with fat.

Used Cocks. If any breeding male under 23 months of age proved itself the producer of really excellent progeny, then under no circumstances should he be sold. If the eggs were not very fertile, or if fertility fell off as the season advanced, or if he was a male that was used just because one hadn't any better, then it would be well to kill him.

Culling Ducks. Health, vigour, and stamina should be the first test that every bird must pass. A duck that is unhealthy will have a thin breast-bone, excessively flat legs, a drooping tail, a dull eye, loose plumage and probably some slight lameness. Such birds should be the first culls.

To test for individual egg-laying powers, it is well to pick up a duck by its neck and then rest its body in the palm of the hand.

A bird to be culled will have an over-long bill, the top line of which will be curved, uneven or dished and not in line with the top of the head. The head itself will be round, with the eye small, lacking fire and set low. The neck of a dud is thick, coarse, short and joining the body with a definite break in what *should* be a lithe, even, well-curved body line.

Long, coarse, straggly feathering is a bad sign, especially if it occurs round the thighs. The back of a dud is also loosely feathered, narrow and short, indicating inability of the body framework to contain large, active egg-making organs.

A shallow and cut-away chest is bad, and this, coupled with a small, unexpanded and drawn-up abdomen, indicates both a poor layer and one out of lay, too. Under-developed back parts in a duck are the worst sign of laying and this, coupled with narrowly-placed legs and a low tail, indicates a producer of 80 eggs a year or less.

Large, coarse bonework suggests a cull, and this may be checked up by feeling the thickness of pelvics, the diameter of the legs and the width of skull.

In regard to the age of laying ducks, let it be remembered that ducks lay much better in their second or third seasons than do hens, and if one has a strain that excels in this respect by all means cull the worst and then keep on, say, 75 per cent of the 1-year and 25 per cent of the 2-year-olds for future laying.

Drakes are hearty eaters, and not one single one should be kept unless it is fulfilling or will fulfil in the near future a profitable purpose.

Bad drakes lack stamina, are small, long-beaked, dull-eyed and narrow-legged.

CUSHION. The feathers over the back part of a hen close to the tail, equivalent to the "saddle" of a cock.

DAY-OLD CHICKS. See Chicks, day-old.

DEAD-IN-SHELL. In some hatching seasons there are many complaints of chicks dying in the shell. The chicks mature normally until the twentieth or twenty-first day and then, instead of emerging, die ; they may even have started to chip the shell before death overtakes them.

The dead-in-shell trouble may well cause a poultry-keeper very serious losses, for not only is there the wastage of eggs to be considered when numbers of them fail to hatch, but there is also the loss of time. It is a serious matter if one finds oneself greatly short of the number of chicks required at the end of the hatching season when it is too late to put down further sittings.

One cause of dead-in-shell trouble is either insufficient or too much moisture during hatching—particularly when eggs are being hatched by incubator.

If an egg is opened and the contents seem watery, filling three-quarters of the egg, it may be assumed that there has been too much moisture, caused either by leaving on the felts or adding moisture deliberately. (See Hatching, Incubator.) If the chick within the egg occupies only just over half the interior and seems leathery and dry, too little moisture is indicated.

Another cause of dead-in-shell is turning the eggs after the eighteenth day. The chick, having worked its way to the top of the shell, has not enough strength to work back again and consequently dies in the shell.

Faulty breeding stock is another prolific cause, resulting in weak germs and lack of vigorous development.

Dead-in-shell trouble can be almost eliminated by breeding only from first-class vigorous stock, attending to moisture, leaving the eggs severely alone from the eighteenth to the twenty-first day, and on the eighteenth day sprinkling the eggs with warm water.

DEAF-EARS. The term applied to the lobes of loose skin hanging from the ear proper.

DEFORMITIES. It is a very sound rule to destroy at once all chicks born with any serious deformity—crooked legs, blindness in both eyes, hump-back, etc.—for such can never grow into useful maturity. Birds with only slight deformities which will not interfere with their activity or laying ability (see Culling) may well be kept on, though, of course, they should never be used for breeding.

Bent Breastbone or *Crooked Breasts.* Bending or denting or kinking of the breastbone is a complaint that is likely to occur during a chick's growth, spoiling its appearance as a table bird or handicapping it as a layer.

It used formerly to be thought that the crooked breast trouble was due to allowing the chicks to perch at too early an age. Nowadays—although it is still held to be a mistake to allow chicks to perch on *ordinary* perches, it being necessary to supply them with flat ones (see below)—it is recognised that the real causes of the evil may be faulty breeding, or lack of sun rays (because of confinement in a house sunshine cannot enter), or faulty feeding.

Birds that have full access to sunshine, that have cod-liver oil, that are fed on standard lines and are bred from healthy normal parents with no trace of breastbone trouble, will never suffer from crooked breasts.

As to perching, chicks should not be given perches until they are in their growing-quarters. If these quarters are of the slatted-floor type perches will not be provided even then, but in ordinary houses perches may quite safely be fitted and used—provided they are of the right type. The ordinary perch consisting of a thin batten *will* cause trouble. The proper kind of perch for growing birds is one made of two or more wooden laths, each measuring 2 in. wide by $\frac{1}{4}$ in. thick, fixed about $1\frac{1}{2}$ ft. from the ground, set $1\frac{1}{2}$ ins. from one another and covered with hay, tied in position with string.

With such a perch, air can circulate freely all round the bodies of the roosting birds. The width of the perch prevents the necessity for clenched claws, a crouching attitude or any abnormal strain. The hay provides a soft cushion.

The hay must be renewed before it becomes in the slightest degree foul, and should be given a sprinkling of insect powder every other day.

Tail Faults. These are most likely to appear in the heavy tailed breeds, particularly the non-sitting light breeds. A bird affected with squirrel tail has a tail which curls over the back, like a squirrel's. In the case of wry tail the tail is carried to one side; in high tail the tail is carried at too high an angle—almost perpendicular; drooping tail is just the opposite—this being a trouble to which pullets are very susceptible.

Failure to provide proper perching is a common cause of these deformities. Birds so affected should not be used for breeding.

Crooked Toes. Sometimes chicks are born with crumpled-up toes. One's first inclination is to destroy them, but the trouble is often curable. All that has to be done is to cut a pad of strong, thick cardboard to the size and approximate shape of the afflicted bird's foot and sew this on to the foot, using thin tape, not thread. The sewing is a two-handed job ; an assistant must hold the bird and steady the leg whilst one wields the needle. Naturally the toes must be straightened before they are sewn. Gentle pressure whilst the work proceeds, operating on each toe in turn, will accomplish this.

The cardboard pad should have an extension at the heel end. This extension is bent at right angles so that it lies along the lower part of the leg and can be secured there.

The chick may walk about during the cure.

DISINFECTION. Disinfection plays a very prominent part in maintaining health in poultry. Disinfectants should not only be used in times of sickness ; they should be used freely at all times and will then prevent much illness. Thus all appliances, particularly incubators, coops, brooders, etc., should be disinfected at the beginning and end of each season of use. Similarly houses should be disinfected whenever a new hatch of birds is to be installed in them, and runs should be disinfected whenever they show signs of becoming foul.

For general disinfecting purposes, any of the recognised proprietary disinfectants, such as Izal, may safely be employed. It is also a good plan to put a little Izal or other disinfectant into the whitewash used for whitening the interior walls of fowl houses.

The Importance of Disinfection. Much may be done by disinfection to prevent the spread of disease. Whenever there has been a case of contagious or infectious illness, everything with which the patient has come into contact—houses, food utensils, etc.—should either be disinfected or, when possible (litter, for instance) burnt. The attendant himself should also do some disinfecting of his clothes, hands and boots before passing from sick birds to healthy ones.

Often diseases are spread by the droppings. If an attendant's boots carry the droppings from sick birds to healthy ones the latter are quite likely to become infected. It is a good plan to have a tray of strong disinfectant standing outside the door of houses in which sick birds are confined so that the attendant's boots may be dipped in this on leaving.

Disinfecting Perches, Perch-sockets and Nest Boxes. These may well be occasionally sprinkled or watered with disinfectant. In addition they should frequently be given a coat of creosote, which is an excellent germ and pest destroyer. The birds should not be allowed to use the house until the creosote is quite dry.

Disinfecting Poultry Houses. Twice a year—spring and autumn—poultry houses should be given a thorough cleaning. Opportunity should be taken of this cleaning to disinfect the place thoroughly. Walls, roof, windows and all internal fittings should be either sprayed with liquid disinfectant or washed in hot water containing disinfectant. The floors should be watered with disinfectant after the litter and the accumulation of droppings have been removed.

Disinfecting Runs. Usually, runs in which the ground is well-covered with grass need no disinfection. The bare earth or gravel run, however, inevitably becomes foul in due course. A practice should be made of disinfecting it twice a year (or oftener if there have been outbreaks of disease).

The runs may be limed and dug over periodically, or, more effectively, they may be watered with either of the following : Sulphuric acid, 7 oz., water 1 gallon ; or crude carbolic acid, 3 oz., water 1 gallon, the soil in each case being well soaked. Fowls must not run on the ground until one month after treatment. When no alternative run space is available, the run might be treated in sections, the section that has been disinfected being fenced off.

Disinfecting Incubators, Brooders, etc. Before use all movable fitments of an incubator—egg drawer, chick tray, felts, thermometer, the bottom diaphragm, etc.—should be thoroughly treated by soaking them in disinfectant for twenty minutes. Movable parts of brooders, hovers, etc., should be similarly treated. As regards the body of the incubator or brooder, this may either be fumigated (see below) or scrubbed with water containing disinfectant. The outside of the incubator, etc., should be disinfected, too, but varnished parts should not be washed with disinfectant solution (which would remove the varnish) but should be treated with a disinfectant polish.

Disinfecting by Fumigation. Both poultry houses and incubators can be disinfected by fumigation. For the former, industrial formalin should be used as follows, it being remembered that it is dangerous for a person to breathe formalin fumes for more than a moment or two. Get the house ready before opening the formalin. Close up all the openings and stop all crevices with paper. Then hang some sacks from the roof. These should come to within a foot of the ground. When all is ready, soak the sacks with the formalin, get out of the house at once and close the door tightly. Leave the place shut up for at least a day. Then air well. Do not let the house be occupied until all fumes have disappeared.

To fumigate an incubator (especially advised after chicks suffering from B.W.D. or coccidiosis have been hatched out), first spray the inside of the machine with warm water, taking care to reach all parts. Then place in the machine a large saucer containing some permanganate of potash crystals, pour over these a little commercial formalin solution, close up the machine tightly, leave for 4 hours, then thoroughly air the interior.

As regards the quantities of the chemicals named, you will require 1 oz. of permanganate of potash and 2 oz. of formalin solution for every 5 cubic feet of incubator space.

DOMESTIC. Name given to a domestic poultry keeper.

DOMESTIC POULTRY FEEDING. See Feeding Fowls.

DOMESTIC POULTRY KEEPING. For many years numbers of householders with ordinary small gardens have been in the habit of keeping a few fowls to supply the home with eggs and an occasional table bird. This practice received strong Government support after the outbreak of war, as a step towards meeting the shortage of eggs, in particular, consequent on the reduction of commercial flocks because of feeding-stuffs difficulties and loss of imports.

A Domestic Poultry-Keepers' Council was established, under the auspices of the Ministry of Agriculture. Area and district organisers were appointed, and Domestic Poultry-Keepers' Clubs were formed in all parts of the country. In 1942 "Domestics" were estimated to number close on 1,000,000.

From the first, the chief feature of the domestic poultry-keeping plan was that the birds should be fed on waste food—housescraps and the like, and such green food as could be spared from the vegetable plot. Rations were limited to small quantities of balancer meal, or laying meal and corn, for those keeping one to 12 or 13 to 50 laying birds respectively (altered in June, 1942, to "not over 25 birds").

In 1941, when commercial producers were required to sell all their eggs to recognised packing stations, "Domestics" were left free either to sell to friends or neighbours (but not for resale) at a controlled price, or to sell them to packing stations at the higher prices paid to commercial producers under an egg subsidy.

In 1942 a further change imposed on all who would keep poultry to supply eggs for their own family the obligation to surrender one egg registration in return for a food ration for each bird kept. Thus a family of six, able to surrender six egg registrations, could get food for six birds. (See also War-time Feeding.)

DOUBLE MATING. See Breeding.

DOUBLE POUSSINS. Term applied to young milk-fed table chickens weighing from 1-2 lb. Sometimes called Large Petits Poussins (see Marketing). They are fattened in the same way as Petits Poussins (see Fattening).

DOUGLAS MIXTURE. Good general tonic for all laying fowls. It is made as follows : Put a gallon of water into an earthenware vessel and add 8 oz. of sulphate of iron (common copperas). When the iron is dissolved, add ½ oz. of sulphuric acid. When the liquid gets clear the mixture is ready for use. To use, add about a gill to the drinking water for every 25 fowls and give every other day.

DRAKE. A male duck.

DRAUGHTS. It is just as bad for poultry as it is for human beings to live in draughty quarters. Draughts should therefore be prevented.

One may determine if a house is draughty by taking a lighted candle into the house, closing up all doors and windows and watching how the candle flame behaves. Any undue flickering will denote a draught and also locate where the draught comes from.

Draughty walls are the result either of cracks in the woodwork, or of warping of the planks, causing the seams to gape apart. Badly cracked planks should be removed and replaced by new timber. Gaping seams can be stopped by putty or pitch. When a wall is " full of draught-provoking places " a good plan is to tar a piece of brown paper over it, the paper being tarred on the outside afterwards.

Much trouble comes of allowing birds to roost on perches that are in a direct draught. In long houses, in which the perches run from one end to the other, a common practice is to put up board partitions at intervals of ten feet, the partition running from the droppings boards to within a foot or so of the roof. If a draught strikes downwards from the roof onto the birds, as it well may in houses that are ventilated from openings above the perches, the best remedy is to fit a deflecting board. The deflecting board consists of a wide board fixed to the wall and inclining upwards towards the roof.

DRESSING POULTRY. See Table Poultry.

DROPPINGS BOARDS. Wide boards fitted beneath the perches to catch the droppings from the roosting birds and prevent them falling into and soiling the litter.

Droppings boards are best made to slip in and out easily so that they can be taken outside the house and scraped. It is also well to have a raised margin along the front.

The boards should be kept surfaced with dry soil, sawdust, or sifted ashes, preferably the first as the manure is then at its best for use on the land.

DROPPINGS PITS. There is a section of the poultry community which believes that it is better to have pits beneath the perches, rather than boards, to catch the manure from the perching birds. The theory is that the pits obviate the need for so much cleaning. On the other hand it cannot be good to allow the fowls to walk about over the manure in the pits.

Droppings pits are made by partitioning off the area below the perches from the rest of the floor by means of a 12 in. board. The soil within the enclosure is dug out to a depth of 4 in. and the hole so made filled with peat moss. This litter is forked over every day and is claimed to do service thus for six months.

DUAL-PURPOSE BIRDS. Breeds of Poultry that are useful for both laying and table purposes. An example is the White Wyandotte.

DUBBING. Term applied to the "trimming" of a fowl's comb, wattles and earlobes so that the head is left smooth. Used formerly to be practised with Old English Gamecocks. Is not now widely practised, and is held by many to be cruel, though others do not accept this, and consider it useful as preventing frostbite—which may sometimes cause severe trouble.

DUCKLET. A young duck of one of the laying breeds.

DUCKLINGS. Duckling is the term applied to young ducks and drakes.
Ailments, See Ailments.
Fattening Ducklings. See Fattening.
Feeding. Ducklings require no food whatever for the first 48 hours; they absorb part of the egg before breaking out of the shell and this is sufficient to support them for the period mentioned.
For the First Week give warm bread and milk, made fairly sloppy, five times a day, giving as much on each occasion as will be cleared up in 10 minutes. Feed the food in troughs.
From 1 Week to 8 Weeks. Give the following wet mash: Middlings, 4 parts; maize germ meal (or maize meal), 2 parts; Sussex ground oats, 1 part; bran, 2 parts; fish meal, preferably impregnated with cod-liver oil, $\frac{1}{2}$ part. Make crumbly moist with hot water and supply four times a day.
From the Eighth Week to One Month before Maturity. Give the same mash but reduce the maize meal to 1 part. Also give a little poultry grain in the morning as a first feed, tipping it into bowls containing 4 in. of water.
Drinking Water, Grit and Shell. Plenty of clean drinking water is required, given in barred troughs or shallow drinking founts. A supply of chick-size flint grit and oyster shell should be permanently before the birds.
Hatching Ducklings. See Hatching.
Ducklings for Table. See Marketing.
Rearing Ducklings. Ducklings can be reared either by natural or artificial means.
By Broody. The management of a broody hen with a family of ducklings, is the same as though she were rearing a family of chicks. See Rearing.
By Brooder. Any good brooder will rear ducklings, but the type that has a lift-up lid is to be preferred, as it facilitates inspection. Ducklings require less heat than chicks, so run the brooder at 85 degs. F. the first week, 80 degs. F. the second, and 75 degs. F. from the third. Weaning can take place when the birds are a month old.

It is important to keep the litter dry, and this means frequent cleanings-out. Peat moss (fine grade) lasts the longest, and is the best because it is absorbent. The litter beneath the brooder must be changed frequently as it speedily becomes soaked through.

It is necessary to watch the ducklings up to the 3rd day of being under the brooder and see that they know where to go back for warmth, as they are often quite stupid. A shield round the brooder to limit their range for 3 or 4 days is often necessary.

For the rest the management of a brooder containing ducklings is the same as though it contained chicks. See Rearing.
Sexing Ducklings. See Sexing.

DUCKS, BREEDS OF. There is a very large number of different breeds of ducks. The following are the most important:

Aylesbury. Excellent table breed; in fact, *the* ideal table duck, being of huge size and rapid maturity. Lays a large bluish-green egg.

The plumage is a glossy pure white; the bill is pink-white or flesh; the eyes dark; the legs and feet bright orange.

Normal weights. Drake 10 lb.; duck 9 lb.

Cayuga. Fairly popular breed combining good laying and table qualities. Unfortunately, the colour of the eggs—a dark green—will at times spoil their sale.

The plumage is lustrous green-black, which must be free from purple or white. The wings should be more lustrous than the rest of the body plumage and free from a brown or purple tinge.

The bill should be slate-black with dense black saddle in centre; the eyes black; the legs and feet dull orange-brown.

Normal weights: Drake, 8 lb.; duck, 7 lb.

Crested. A good laying and quite useful table breed, but not extensively kept. Lays a large bluish egg.

This variety may be of any colour. Normal weights: Drake 7 lb.; duck, 6 lb.

Indian Runner. Formerly classed as the best of all the laying breeds of ducks, but not so generally kept as it used to be. Lays a medium-sized white egg.

There are five varieties, the Black, Chocolate, Fawn, Fawn-and-White, and White.

The plumage of the *Black* variety is solid black with metallic lustre, no grey under chin or wings, no grey wing ribbons or " chain armour " on the breast. The bill should be black; the legs and feet black or very dark tan.

The plumage of the *Chocolate* variety is an even chocolate throughout, but the drake on assuming adult plumage becomes darker than the duck. The bill, legs and feet should be black.

A *drake* of the *Fawn* variety has the head and upper part of neck dark bronze with metallic sheen, probably showing green tinge; lower neck, breast and shoulders rich brown-red; lower chest, flanks and abdomen French-grey peppered with dark brown or black dots on nearly white ground; scapulars red-brown peppered; back and rump deep brown; tail dark brown; wing-bow fawn; secondaries black-brown with metallic lustre; primaries fairly dark brown. When the drake is in " eclipse " he more closely approaches the duck in colour. The bill should be pure black to olive-green mottled with black and black bean; the legs and feet black or dark tan mottled with black.

The plumage of a *duck* of the Fawn variety is almost uniform speckled ginger-fawn; head and neck feathers ticked with red-brown; scapulars ginger-fawn with red-brown pencilling; wing-bow a shade lighter than the scapulars; secondaries red-brown; primaries a shade lighter; back and rump pencilled darker; tail lighter, and belly lighter than upper parts of body. The bill should be black; the iris golden-brown; the legs and feet black or dark tan.

In a *drake* of the *Fawn-and-White* variety the body is uniformly ginger-fawn; cap and cheek markings dull bronze-green, the cap separated from the cheek markings by a projection from the white of the neck tending up to, or terminating in a narrow line more or less encircling, the eye; bill divided from head markings by a prolongation from $\frac{1}{8}$ in. to $\frac{1}{4}$ in. wide, extending from the white underneath the chin; neck pure white; rump and tail similar to head.

The plumage of a *duck* of the Fawn-and-White variety should be the same shade of fawn as the Fawn duck. The fawn and white should meet on the breast with an even cut about half-way between the point of the breast-bone and legs; the base of the neck, upper part of wings, back and tail as nearly as possible the same colour as the fawn of the breast, and from the fawn of the back an irregular branch on either side extending downwards on the thighs to, or nearly to, the hough; the white of the breast extends downwards between the legs to beyond the vent and may overlap the thighs in part; primaries, secondaries and lower part of wing-bow pure white, giving the appearance of a " heart " laid flat on the bird's back.

The bill should be light orange-yellow in young birds; dull cucumber in adult duck and green-yellow in the adult drake; the legs and feet orange-red.

The plumage of the *White* variety should be white throughout; the bill, legs and feet orange-yellow; the iris light blue or grey-blue.

Normal weights are: Drake, 4 lb.-5½ lb.; duck, 3½ lb.–5 lb. Birds bred and shown in the same year as hatched may be accepted for competition at ½ lb. less.

Khaki Campbell. One of the best of the laying breeds. Frequently lays up to 300 white eggs a year.

A drake of this breed has the head, neck, stern and wing-bar a brown shade of bronze. Remainder of plumage an even shade of warm khaki. The bill should be dark green; the legs and feet dark orange.

A duck of this breed is even khaki-colour all over; back and wings laced with lighter shade; head plain khaki, without streak from the eyes.

The bill should be greenish black; the legs and feet as near the body colour as possible.

Normal weights are: 4½ lb. for birds in laying condition in their prime.

Magpie. Mainly kept for ornamental purposes. The head and neck, breast, thighs and rump are white, the head surmounted by a black cap covering the whole crown to the top of the eyes; back solid black from the points of juncture of wing-bows and body to the tail, and extending over the wings, giving the effect of a heart-shaped black mantle; wings white and tail black. The bill should be pale yellow to deep orange; the eyes dark grey or dark brown; the legs and feet orange.

Normal weights are: Drake, 5½ lb.-7 lb.; duck, 4½ lb.-6 lb.

Muscovy. Was at one time kept as a utility bird but is now seldom seen. One of the reasons why it has lost favour is its bad temper, there always being fierce quarrels in the flock. The eggs are white, but laying is not very prolific.

There are several varieties, Piebald. Black and White, White, Black and Blue,

The legs are pale yellow. Muscovy ducks are of exceptionally large size.

Orpington. A very popular laying breed, excellent for free range, being a good forager. Is freely kept for commercial egg production. Lays white eggs. There are two varieties, the Blue and the Buff.

The plumage of the *Blue* variety is an even shade of blue, free from bronze tint, with white bib extending from centre of the neck about 3 in. on to breast ; the head and upper part of the drake's neck at least two shades darker than his body colour. There must be no eye streak or white on the face.

The bill should be blue or slightly tinged with green, but no trace of orange or yellow ; the eyes deep blue iris and black pupil ; the legs and feet dark blue.

A drake of the *Buff* variety is a rich even shade of deep red-buff, free from pencilling ; head and neck seal-brown.

The plumage of a duck of the *Buff* variety is similar to that of the drake but must be free from blue, brown or white feathers.

The bill should be orange with dark bean ; the eyes brown iris and blue pupil ; the legs and feet bright orange-red.

Normal weights are : Drake not under 5 lb. or over 7½ lb. ; duck not under 5 lb. or over 7 lb.

Pekin. An excellent table breed, second only to the Aylesbury. Scores over the Aylesbury by being a better layer. Aylesburys and Pekins are often crossed. Lays a blue egg.

The plumage should be buff, canary or deep cream : the bill bright orange, free from black marks ; the eyes dark lead-blue ; the legs and feet bright orange.

Normal weights are : Drake, 9 lb. ; duck, 8 lb.

Penguin. Usually kept only as a fancy breed. In the drake the head, neck and back are black with beetle-green sheen ; throat and breast white ; wings black with white primaries.

The plumage of the duck is similar to that of the drake, except that the wing-bow may or may not have slight pencilling.

The bill in both sexes should be slate with black bean ; the eyes dark ; the legs and feet as dark as possible.

Normal weights are : Drake, 7 lb.-9 lb. ; duck, 7 lb.-8 lb.

Rouen. Excellent table bird, being of good size and reasonable prolificacy. Is frequently crossed with the Aylesbury, but does not mature so quickly as the latter. Lays a blue egg.

In the drake the head and neck are rich green to within an inch of the shoulders where the ring appears ; ring white, dividing the neck and breast colours, but leaving a small space at the back ; breast rich claret ; stern and flank grey pencilled with glossy black ; tail coverts black with brown tinge, with two or three green-black curled feathers in the centre ; back and rump rich green-black ; large wing coverts pale grey ; small coverts French-grey, finely pencilled ; pinion coverts dark grey or slate-black, grey bars tipped with black forming a line at the base of the flight coverts, the latter feathers slate-black on the upper side and iridescent blue on the lower, each of these feathers tipped with white at the end of the lower side, forming two distinct white bars with a bold blue ribbon mark between the two ; flights slate-black with brown tinge.

The bill should be bright green-yellow with black bean at the tip ; the eyes dark hazel ; the legs and feet bright brick red.

In the duck the head is rich brown, with a wide brown-black line from the base of the bill to the neck and black lines across the head above and below the eyes ; neck as the head with a wide brown line at the back from the shoulders, shading to black at the head ; wing bars white with blue between as in the drake ; flights slate-black with brown tinge ; remainder of plumage rich brown, each feather pencilled, the markings being rich black or very dark brown.

The bill should be bright orange with black bean at the tip and black saddle extending almost to each side and about two-thirds down towards the tip ; the eyes dark hazel ; the legs and feet dull orange-brown.

Normal weights are : Drake, 10 lb. ; duck, 9 lb.

Stanbridge. Mainly a fancy breed. The plumage should be pure white throughout ; the bill should be pale yellow to deep orange ; the eyes blue ; the legs and feet orange.

Normal weights are : Drake, 5½ lb.-7 lb. ; duck, 4½ lb.-6 lb.

DUCKS, LAYER. Duck - keeping may well be very profitable, more profitable in most circumstances than keeping fowls. Among the advantages of ducks, in comparison with fowls, are (1) They need the smallest outlay in housing of any farm stock, (2) they are far less subject to disease, (3) the eggs are easier to hatch and the ducklings easier to rear than chickens, (4) good laying strains yield a greater number of eggs in a year than a similar quality of fowls, and (5) the adult death rate is only about 2 per cent, bearing no comparison to that of fowls.

Formerly ducks used only to be kept for the production of table-birds ; nowadays more ducks are kept for egg production than for table-bird production.

Ducks, in the ordinary way, need not of necessity have swimming water ; they are quite happy on " dry land," with not even so much as a basin of water to bath in. For breeding purposes swimming water *is* necessary ; ducks normally only mate when on the water.

Ailments of Ducks. See Ailments.

Breeding. The breeding of laying ducks is not advised for those who only require one or two dozen birds ; it is far better to purchase sittings of eggs and let a broody hatch them.

When ducks are required in considerable numbers for laying or for table duckling production, breeding is very desirable.

The breeding stock must be the best available, true specimens of their breed and possessing correct breed points and characteristics. For laying ducks the breeding birds should be of at least a 220-egg strain and preferably a 250-egg strain ; the breeding drakes should be of a 270-egg strain. The breeding stock must further be of a strain that lays large eggs—2½ oz. (See also Breeding.)

Feeding Laying Ducks. It is even more important with ducks than with hens to follow a proper system of feeding and adhere to it closely ; ducks dislike changes.

The following pre-war feeding method has been proved to give the most satisfactory results : Wheat 2, kibbled maize 1 and clipped oats 1—parts by weight.

For mash, if the birds have an extensive grass run (say 200 sq. ft. per bird) give middlings 4, bran 2, yellow maize meal 2, Sussex ground oats 2, fish meal 1½—parts by weight. If the birds have an earth run, replace 1 part of the bran with alfalfa meal of the best leafy quality.

The mash should be mixed with water to a rather sloppy state—just a little wetter than that for fowls but not "runny" like pig-swill.

Where circumstances permit midday feeding, give half the mash allowance at noon and the remainder half-an-hour before sunset. If midday feeding cannot be managed, give the whole allowance of mash half-an-hour before sunset.

The *daily* allowance of mash per bird is about 3 oz. weighed dry. Individual flock appetites vary, however ; some flocks will require more than 3 oz. per bird and some less. To gauge actual quantities required, increase the allowance until just a *little* is left in the troughs when all appetites have been satisfied ; then curtail the quantities *slightly.*

As regards the grain feed the best plan is to give 1 oz. per bird early in the morning, and ½ oz. after the evening mash feed. If a midday mash feed is not possible, give 1½ oz. of grain per bird in the morning. The grain should be tipped into the drinking water.

Water, Grit and Shell. Supplies of these essentials must always be before the birds. The water supply should be sufficient to

B*

satisfy all needs until the next time for filling the bowls. Use bowls, founts, etc., which the birds cannot attempt to bathe in, or they will splash most of it away and may afterwards have to go short.

Ducks, being creatures of the water, bibble constantly and during this bibbling they put much mud into the water troughs. This means that it is well to rinse out the troughs every day so that clean water anyway starts off the day. Otherwise an unwholesome black sludge settles on the bottom, especially on some soils.

Water vessels will keep much cleaner if stood upon a slotted platform, but this means a rather shallow drinking vessel. Again it is the best to use one that has a water level sufficiently high for the birds to get their eyes under.

Feeding in War-time. During wartime, rations for ducks are allowed on the same basis as for fowls. Ducks, however, have much larger appetites, and greater use must be made of other foods.

Seventy-five per cent of the day's ration should consist of vegetable matter. Potatoes, carrots and swedes are the mainstay of duck feeding. These three vegetables, or any one of them, when cooked and mixed up well with meal make an appetising mash.

Two meals should be given each day, 4 oz. of prepared mash for each duck at each meal being the correct allowance.

When the ingredients are available, an excellent mash for ducks can be made from : weatings, 4 parts ; bran, 4 parts ; maize meal, 2 parts ; fish meal, 1 part. To this, vegetable matter in the proportion mentioned above can be added.

Housing. It is said by some that ducks can live an entirely open-air life ; but both with laying ducks and table ducks better results are obtained by housing the birds at night. There is no need for an elaborate house with glass windows and other elaborate fittings ; all that is wanted is adequate shelter.

Each bird requires about 4 sq. ft. of floor space in the house ; hence a house measuring 10 ft. long by 8 ft. from front to back will provide ample accommodation for a flock of about 20.

Ducks do not perch ; therefore the house need not be made so high as one for hens. A house 4½ ft. high at the front and 3 ft. high at the back will be quite large enough.

A slatted wooden floor (which helps to keep the litter dry) should be fitted to the house and should be raised a couple of inches off the ground. Straw litter should be spread over it to a depth of about 4 in. and should be renewed every week.

The upper foot of the front of the house should be filled in with wire-netting, protected by a weather-board fixed at an angle of 45 degrees—as in an ordinary poultry house. The whole of the front of the duck house should be made loose or fitted to hinges so that it can be removed bodily or opened in warm weather, since ducks require more fresh air than ordinary ventilators can give them during the summer. The back, too, can also be made to open, but hinged at the top, in one or two sections.

An important point about a duck house is to provide a run-up into the inside. Ducks are very clumsy and easily injure themselves if they have anything of a jump to get into their quarters. For this reason the entrance should be a slightly sloping board. The usual plan is to have the door as the entrance flap, having the hinges at the bottom.

As regards nesting places, these are made on the floor with straw, this being kept in place either with a row of bricks or some battens of wood. A height of 3 ins. is ample for the sides of the nest.

The house ought to face the south-east, south or south-west, so that it will get all the sun possible, and should have a small run attached. The run is necessary because the birds should be confined in it until about 10 a.m., by which time all eggs will have been laid. This saves the trouble of searching for eggs, as ducks will lay almost anywhere.

Judging Ducks for Quality. It is as important with ducks as with hens that only good-class birds be tolerated, all inferior birds being weeded out as soon as their faults can be detected.

Something may be learnt of a duck's capabilities while she is roaming about the run. The good layer is active, first at the food trough and last away ; has a keen appetite and always a more or less full crop ; is early about in the morning and late to bed at night. An adult will be well expanded behind if she is a good layer ; she will walk with legs wide apart and give the appearance of being ideally proportioned. The good layer is all curves ; an angular bird is a doubtful profit-earner.

Management of Laying Ducks. To lay as well as they *can* do—produce up to 300 eggs average a year—ducks obviously must be well looked after ; yet the " management " they require involves very little labour.

Routine work consists firstly of loosening up the floor litter *every* morning and changing it when its whole texture has become foul and there is no clean litter to become available as a sleeping bed. On a solid floor peat moss litter is by far the best, for, while straw will last only a fortnight, moss will last 6 to 8 weeks if kept raked over. One sq. ft. of sleeping space is the most economical, but there must be no floor draughts and rain must be excluded.

Ducks that have a protected run do the best, as exposure to fierce wind may reduce the egg-supply by as much as 30 per cent. A hedge, a row of trees, a wall or a range of buildings would be very satisfactory, or corrugated iron (lengthways) will break the wind.

Ducks respond to tranquillity in everything, and a move to a different house or yard, or a change in feeding when once they have started winter laying, will put them off, probably putting some into a partial moult and always reducing the eggs laid.

The birds should be subject to no change in position or feeding when once they have settled down.

Management of Stock Ducks. Stock ducks, whether of the laying or table breeds, should have as much liberty as possible ; for the rest, their housing and management follows closely on the lines laid down for laying ducks.

If the birds are on free range, give them a small feed of grain in the morning, make them forage all day and top up with a big feed of mash at night.

The best mash is—4 parts, by weight, of middlings ; 2 of bran, one each of maize meal, Sussex ground oats and fish meal, to be fed rather moist.

Give the grain in a shallow trough of water or else scattered about in good, untrodden grass.

As with laying ducks, the birds should be kept confined until 10.0 a.m., as then all that are going to lay *will* have laid.

Moulting Ducks. See Moulting.

Non-laying, Causes of. One or other of the following causes should be looked for when laying ducks are not producing sufficient eggs : (1) insufficient feeding, (2) incorrect feeding, (3) change, move or shock, (4) exposure to winds, (5) foul night accommodation, (6) sodden litter, and (7) inherited poor laying ability. It may have been only one of these causes or the effects of several of them together.

DUCKS, MATING. See Breeding.

DUCKS, TABLE. Table ducks are an excellent proposition. They attain killing size much earlier than chickens— at 10 weeks old they weigh between 4½ lb. and 5½ lb.—and in normal times command a good price. The housing and general management is much the same as for layer ducks. Like layers, table ducklings require plenty of food, but they do well on the " substitute " fare which must be provided in war-time. For the breeds most suitable for use as table birds, see Ducks, Breeds of. (See also Fattening).

DUST BATH. Term applied to a heap, box, or enclosure of dust or ashes in which fowls delight to " wallow," passing the dust between the feathers to the skin. Dust baths ought always to be provided for all fowls, not only because dust-bathing gives them pleasure but also because it helps to keep them free of body insects.

In its simplest form the dust bath consists of a heap of dry sifted soil, or, better, a mixture of sawdust, dry earth, sand and road grit, the whole being kept in a dry place, held in position by boards. Usually however, the material is placed in a box placed on the floor. If standing in the open the box should, of course, have a roof.

A box measuring 4 ft. by 2 ft. is sufficient for 20 fowls.

The dust bath material should be liberally sprinkled with a good disinfectant insect powder.

EARLOBES. Term applied to the skin hanging from the ear. It is red in some breeds, white in others.

EGG-BOXES. For packing eggs for sending by post or rail, boxes can be obtained of various kinds, shapes and sizes. They are made either of wood or cardboard, and fitted with partitions making separate compartments for each egg. For selling hatching eggs special boxes are obtainable, and eggs may safely be sent by post or rail.

EGG-EATING. Hens sometimes develop a taste for eggs. It is usually due to allowing a broken egg to remain on the floor or in a nest long enough for some bird, becoming inquisitive, to peck at it. Bad nesting-places, insufficient nest boxes, dirty nest boxes, boxes in too light and open a position, cramped boxes, insufficient pot eggs, idleness, lack of green food or animal matter, and lack of shell are also predisposing causes.

One may know that an egg-eater is in the flock when one discovers partly-eaten eggs or pieces of shell in the litter or nest boxes.

It is well to prevent the trouble by rectifying the faults indicated. If, through unfortunate circumstances, egg-eating has broken out, a commonly accepted plan is to provide "disappearing egg" nests, those which have a false canvas bottom. As a deterrent, eggs may be blown of their contents and then refilled with mustard and bitter aloes mixed to a paste with paraffin, and placed on the floor.

If an individual is actually seen egg-eating, the points of its beak should be pared down, the points of each half of the beak being trimmed to make them blunt and more soft.

EGGS, ABNORMAL. Abnormality in the shape and appearance of eggs is due to many causes. *If the shell is crinkled*, this is most likely due to some derangement of the oviduct (the shelling organ), probably because a bird, laying heavily, has been fed on foods of a too forcing nature. *The very small egg* that contains no yolk is due to increased activity of the oviduct, thus encouraging it to shell the white without a yolk dropping into it from the ovary.

The flat-sided egg usually owes its queer shape to over-fatness, the layers of fat in the bird's body pressing on the oviduct and affecting the shell when it is just hardening.

Exceedingly long eggs commonly mean a general over-fatness of the birds, while *eggs with pimples* at the ends, although indicating a passing abnormality of the oviduct, point to the need of more ample feeding of a building nature.

Very thin egg-shells are the result of fowls not eating sufficient shell-forming material—that is, crushed oyster shell, cockle shell or limestone grit. A constant supply of one or other of these materials should always be before the birds. It is sometimes useful to mix a little shell with the mash; or the grain ration may be soaked in strong lime water made by putting lime into water and allowing the water to absorb as much of the lime as it will; or a small quantity of powdered chalk may be added to the wet mash.

Eggs without shells are usually the result of the birds eating insufficient shell-making material, when the remedy is to ensure that the birds eat more crushed oyster shell or other shell-making material. Sometimes, however, they are the result of a layer's oviduct becoming inflamed, thus putting the secreting glands out of order. In such an event the ailing bird should be fed on a light diet, containing no scraps, meat meal or fish meal, but an abundance of green food.

If pullets or hens lay *double-yolked eggs* the usual causes are: (1) an excess of meat meal or fish meal, this causing forced production; (2) an excess of spiced foods; (3) fright or a change of quarters; (4) inherited tendencies; or (5) some internal derangement as a result of an accident.

Usually only one or two double-yolked eggs are laid, the layer then producing normal eggs; if a bird persistently lays these eggs she should be specially marked and on no account used for breeding.

EGGS, ADDLED. Fertile eggs in which the germ has commenced to grow and has then died, making the egg "bad."

EGGS, BLOOD IN. Common trouble of which nearly all poultry-keepers have experience sooner or later. The blood appears as specks or particles in the yolk or white. Due to slight internal bleeding, generally when the yolk leaves the cluster.

The trouble should soon right itself without aid. Where it appears to be peculiar to a certain bird, give steel drops, half a teaspoonful with enough water to fill up the spoon, every morning for a week. If this effects no cure, kill the bird, as it is damaged or malformed internally.

EGGS, CANDLING. Method of ascertaining the degree of freshness of an egg. All eggs passing through the packing stations have to be candled. All poultry-keepers producing eggs for sale should candle every egg before it is packed up for distribution.

Candling consists of holding up the eggs, one at a time, in front of a strong light in such a way that their contents become visible.

Simple Candling Devices. There are various devices to render candling quick and easy. One of the best plans is to use daylight. You cut a board to fit half the window of some convenient room. In the centre of this board an egg-shaped hole (say 1¾ in. by 1¼ in.) is cut. If the room is darkened by drawing down the blind to meet the board and if an egg is then held against the hole in the board its contents will be clearly visible.

In an emergency one can view the contents of an egg with the aid of an electric flash lamp. The lens is removed and the egg held over the bulb.

Another plan is to cut an egg-shaped hole in the centre of a large square of cardboard. An egg placed against this hole whilst the cardboard is held up against a strong light can be readily viewed. For serious work, of course, it is desirable to buy a proper candling lamp. This can be had with either oil or electric light as the illuminant. The proper lamps give the best possible view of the interior of an egg.

What You See When Candling. In a new-laid egg (the only egg which should ever be sold for eating purposes) the contents, as viewed against the lamp, should appear absolutely clear, though at the broad end there will be a small air space, denoted by a faint line. This air space should be no bigger than a threepenny piece; that is, when the egg is held vertically the faint line should be very close to the top.

Eggs that show up in any way different from the above should definitely be excluded from a consignment.

If the air space is noticeably larger than normal, say as big as a sixpence or even a shilling and if there is a shading of a darker colour towards the centre, that will be an indication that some broody

hen, say, has sat on the eggs for some hours and started the hatching process.

EGGS, GRADING. See Grading.

EGGS, MARKETING. See Marketing.

EGGS, PACKING. With all market consignments for eggs and for all classes of table poultry, it is definitely desirable to use the form of package prescribed under the National Mark scheme. See Marketing.

Where only one bird is concerned it may be sent by post wrapped in greaseproof paper and packed in a light wooden box lined with straw and containing ventilation holes in the sides.

Where several birds have to be sent away, a wicker hamper is to be preferred, because of the ventilation allowed. Packing should be such that, although it in no way interferes with ventilation, it, at the same time, prevents any possibility of bruising. For this reason, the best packing is straw.

For choice consignments each carcase should be wrapped in greaseproof paper. Pack each carcase closely, interline with twisted straw and paper and, when the hamper is full, put on sufficient straw to keep the packing tight when the lid is put on.

Show birds are best sent on a journey in canvas-lined open-work-wicker hampers well-lined with straw.

For consignments of birds, use special boxes, taking care to keep within the requirements of the law, which stipulate that the containers shall be such that the heads, legs or wings of the birds cannot protrude through the top, bottom or sides, shall be adequately ventilated, shall be high enough not to cramp the birds and shall not be overcrowded.

Packing Sittings of Eggs. See Hatching.

EGGS, PRESERVING. There are many methods of preserving eggs, which will keep best and longest only if they are (1) under 5 days old, (2) unwashed but perfectly clean, (3) infertile, (4) strongly shelled, and (5) free from such defects as blood specks, freckles and shell cracks.

Waterglass Method. The best method of preserving is to use waterglass or one of the proprietary substitutes.

The method consists of packing the eggs in some suitable vessel and pouring sufficient preservative fluid over them to cover them completely.

Any galvanized metal or china or glazed containers will do, but the most convenient are undoubtedly the preserving pails sold specially for the purpose. These are zinc pails containing a removable wire container, which is taken from the bucket, packed with eggs and is then replaced in the bucket.

To make the proper solution, when using water-glass, boil the required quantity of water, allow to cool, and to every gallon add 1 lb. of waterglass. All vessels must be scalded and washed clean.

The proper way to pack the eggs is big-end up, each layer wedged carefully in with that below it. When the eggs are within 3½ in. of the top, packing should stop ; and there should be at least 2 in. covering of solution above the eggs, plus 1-1½ in. free air space above that to prevent spilling.

All vessels must be covered to keep out dust and reduce evaporation of the fluid.

As each container is packed and filled, it must be labelled with the date in order that each batch may be used or sold in the order of preserving.

Regular examination is necessary. Evaporation may reduce the liquid level and water must be added to keep the solution 2 in. above the top layer of eggs.

If any eggs are seen to be floating, this is an indication that (1) the solution is too strong or (2) the egg itself is stale or bad.

Lime-Water Method. A second method is to use lime-water in place of water-glass, the procedure being otherwise the same. The lime-water is made by adding 1 lb. of slaked lime to 5 gals. of cold water, stirring several times during four days, when 1 lb. of salt is added. After settling for 24 hours the clear liquid is poured off carefully for use, the sediment being discarded.

Proprietary Preparations. There are certain preparations in which the eggs are dipped to close the pores, and when dry they are stored, standing on end, and keep well.

Other Methods. Quite recently it has been claimed that eggs may be stored quite safely over a considerable period if they are first dipped for three seconds in boiling water, this hardening the skin lining within the shell. The dipping may be done by placing the eggs on wire-netting holders.

Other means of preservation, mainly for the wholesaler or very large-scale producer, are by gas storage, cold storage or dipping in heated odourless mineral oils.

The ideal *storage* is in a dry, cool room in which the temperature varies less than 15 degs. F. A cellar is ideal so long as it is not musty, damp, or ill-ventilated, the worst place being on a top shelf of a kitchen or dairy. A room temperature of 33-45 degs. F. is the ideal, and over 70 degs. F. conditions are so bad as to cause the eggs to deteriorate rapidly.

EGGS, SITTINGS OF. See Hatching, Sittings for.

EGGS, TAINTED. When eggs have an objectionable flavour the cause is usually external ; the taints are not present when the eggs are laid. The explanation is that eggs readily absorb odours from their surroundings, and if kept in the vicinity of anything having a pronounced odour (onions, for example) they will acquire it. Hence the importance of storing eggs away from anything having a strong smell.

Eggs may also become tainted in the same way if nests are not kept clean and litter is allowed to become musty. Occasionally a bird in a flock may lay eggs having an unpalatable flavour. This is due to some constitutional defect and is very rare.

ENSILAGE. See Foods, Poultry—Silage.

EPSOM SALTS. A very useful medicine in very many poultry ailments. Is particularly valuable in cases of constipation and other internal disorders, and may also be used to reduce over-fatness.

Epsom salts are best given in the drinking water. For flock treatment add one teaspoonful for each pint. For an individual dose, add ¼ teaspoonful to a little warm water and administer by means of a fountain-pen filler.

EXHIBITING POULTRY. The showing of poultry is undoubtedly profitable, as well as being a pleasurable branch of poultry-keeping, whether for the ordinary utility poultry-keeper or the exhibition man. Profit is derived from the prize money won, but more important is the advertising value.

To Enter a Bird for a Show. The first step should be to write to the secretary of some show asking for a schedule in in order that one may see what classes are available.

There are separate classes for " exhibition" and for "utility" birds, and to put a true utility bird in an exhibition class is risking failure. " Exhibition " birds are those bred solely for showing (without regard to egg production). In the utility classes the birds need only possess the characteristics of a utility bird of their breed—that is of a perfect bird such as may be found in any well-bred laying flock of their breed.

If there is a separate class for the desired breed, then that is where it must be entered, but if there is not, then it goes with the " any other variety," heavy or light class.

The entrance fees must be sent with the entries, and then, 3 or 4 days before the show, the would-be exhibitor will receive show labels and sometimes numbered leg bands for the birds.

In most shows the whole of the arrangements are made and the birds despatched and returned without the exhibitor himself having to visit the show. There are special cheap rates by rail for birds being sent to exhibitions.

Selecting Show Birds. The first thing to do in making a selection from a flock of birds of the ones most suitable for showing is to obtain a copy of the Breed Standards of the breed kept and to study these carefully. It is also a help if a nearby show can be visited, the winners in that breed class carefully studied, and a few notes taken so that the would-be exhibitor can have some idea of the type of bird that wins prizes. The following are the points to which most attention must be paid : (1) Breed colour ; (2) undercolour ; (3) legs and headgear ; (4) shape ; (5) size ; (6) condition ; and (7), if utility class, in full lay. Prime condition, bloom of plumage and health are essentials.

It is often difficult to make up one's mind which is the best bird from a number seen in the run. The best plan is to pick out three for every one you want to exhibit and put them in separate show pens. They look altogether different in these wire show pens than when running about and you can make your final choice more easily.

Tameness is a factor to be considered. The very wild bird will never display herself to advantage in a show pen.

Training the Show Bird. It is essential that every exhibitor have some show pens, for it is impossible to train birds for a show without them. Show pens made are of wire and hold one bird each. The pens should be stood on a light platform or table in a dry and comfortable shed, preferably in view of people but not of poultry. The best litter to use for the floor of the pens is chaff or peat moss.

7 to 10 days before the show, the birds should be taken out of the run and confined in these pens, merely for the purpose of getting them used to the pens and for training them to stand up, to remain quiet when the judge's stick is poked into the pen, not to be afraid of passers-by and not to be afraid of being handled.

A proper judges stick is another essential. Immediately it has been placed in the show pen for training a bird should be spoken to and coaxed with tit-bits. Every day for a period the judge's stick should be pressed upwards under the wattles to make the bird hold its head up, then under the chest to make the bird stand up. Then the tail should be moved down and up with the stick. It is just possible that a judge will want to push a bird's legs about with his stick in order to get a better view from the side. The bird must be made quite accustomed to all this.

The bird must further be trained to stand handling, so every day take it out of the pen. Be careful when doing this, however, not to break any feathers. Fetch it out of the pen head first and turn it over sideways and put it back head first, feet down.

Washing a Show Bird. Dark coloured birds need only have the legs scrubbed, the face and headgear polished and the plumage rubbed over with a silk handkerchief to give it a gloss, but a white or light coloured fowl must definitely be washed in water or it will stand no chance of winning whatever.

Three days before the show is the time to wash a bird. Provision must be made for three changes of warm water. A scrubbing brush, sponge, soap flakes, and a couple of towels must be at hand. The bowl containing the water should be deep enough to cover the bird with water.

First the bird must be placed in nicely warm water and thoroughly soaped. The legs must be scrubbed and the tail and wings, too, if they are dirty. The feathers should be sponged down the way they grow. After 5 minutes or so, the dirt will have been removed. Then most of the suds should be squeezed out and the bird placed into a rinsing bath, again of warm water. Ample clean water should be available to pour over the plumage until no soap suds are left. This last point is most important.

After thus rinsing the bird should be given a final bath for the purpose of further rinsing and whitening by the addition of sufficient washing " blue " to make the water a medium tint. After this press the tail and lower feathers to squeeze out surplus moisture and then envelop the bird in towels to dry up as much of the wetness as possible.

This done, place the wet bird on clean straw in a show hamper and dry fairly quickly either in the sun or in front of a warm fire. When dry, return to the training pen and be sure to prevent the bird from becoming dirty. Use short litter such as dust-free chaff and give the bird a low perch at night.

Feeding up Show Birds. Whilst in the training pen feed show birds on wheat and kibbled maize in the morning, greenfood at noon, and a good spiced laying mash (mixed crumbly wet) at night. Supply water. Be sure not to overfeed, and do not leave food in front of the bird all day long.

Despatching Birds to the Show. The best way to send birds to a show is in a special wicker travelling hamper lined with hessian. The show labels will arrive 3-4 days before the show, and these must be tied on to the hampers. The latter should be given a 6-in. bottom of clean, dust-free straw.

Allow ample time for the railway to convey and deliver the birds to the show ground and, before placing them in the hampers, a few last-minute touches are necessary.

Last-Minute Finishing Touches. The legs should be rubbed thoroughly with a rag which has just a little olive oil upon it, and dirty legs should be scrubbed and dried even though this was done previously at washing time.

Any dirty patches on the feathers may be removed with a rag dipped in hot water. If only a little water is used, the feathers will have dried by the time the bird reaches the show.

Comb and wattles should be rubbed with a rag upon which is a little olive oil and vinegar mixed together in the proportions of 4 and 1 respectively. White earlobes can be improved by rubbing with zinc oxide ointment and then dusting with a fine talcum powder.

Just before the hamper goes off give each bird two tonic pills and a really good feed of grain and water to drink.

FACE. Name given to the bare skin around the eyes—generally red, but occasionally white, as in Spanish, purple-black as in Silkies.

FATTENING. Practically all classes of poultry are put through a period of fattening before they are killed for the table. The best methods to adopt are as follows :

Asparagus Chickens. These are surplus cockerels. They are fattened as spring chicken below.

Chickens, Spring. These are young chickens that have attained a weight of 1½ lb. or slightly over but are not, of course, adults. The season when they are most in demand is from February to June, but the demand slackens after March (see Marketing).

To obtain birds in the best condition, start the fattening process when the birds are about 6 oz. short of the required weight, that is when they are just over 1 lb. The trough-fattening method only should be allowed. This consists in restricting the liberty of the birds by caging them in fattening coops ; keeping them warm by protecting the coops or placing them in a shed ; and feeding on mashes specially selected for the purpose of rounding the carcase and lightening the flesh.

Easily the best normal-time spring chicken fattening menu is Sussex ground oats and sour skim milk mixed to a rather moist consistency. If you cannot get milk, use warm water and add fat at the rate of ¼ oz. per day. A cheaper menu is boiled mashed potatoes and barley meal. Feed three times a day, each time giving as much as will be cleared up. ·

After 10-14 days of either of these methods, the birds will be ready. As soon as they seem to go off their feed the time for their killing draws near.

Cockerels of Three Months and Over. Pen singly in coops and feed for 10 days on Sussex ground oats mixed to consistency of cream with sour skim milk or, if no milk is available, with water, in which case add ¼ oz. fat (rendered down) per bird per day. Supply food in troughs. After 10 days birds can be disposed of as " half-fattened " or finished by cramming for a further 10 days, the food for the purpose being Sussex ground oats as before, but in rather more liquid form, with addition of ¼ oz. rendered fat per bird.

Double Poussins. These are what may be termed overgrown Petits Poussins, that is birds weighing just a little more than the 12 oz. of the Petits Poussins (see below), the best selling weight being from 16 oz. to 24 oz.

Fattening is as for Petits Poussins.

Ducklings. To fatten ducklings, confine them in sheds or runs at about 9 weeks of age and feed them on as cheap a mixture as possible. Here is a good one : Potatoes, 1 part ; Sussex ground oats or barley meal, 1 part ; and middlings, 1 part. All parts are by weight and the ingredients should be mixed with sour milk or buttermilk. It is an advantage, though not essential, to add melted fat to the mash at the rate of ½ oz. per bird per day for the first week, increasing gradually to 1 oz. per day at the end of the third week.

Give drinking water only after meals.

The fattening process is completed within 2 to 3 weeks, the birds having to be killed when they are 12-14 weeks old.

Ducks, for Christmas, etc. It is not advisable to attempt to fatten old birds or any of the strictly laying breeds ; the only possible breeds are Aylesbury, Rouen and Pekin.

Ducks that are being fattened should have their liberty restricted by cooping to a small run, but an occasional wash serves to keep them healthy about the head and maintains an appetite. It pays, therefore, to give the birds access to a stream, or deepish bath, say, once a week.

An economical fattening formula is equal parts by weight of middlings, barley meal (or white maize meal) and Sussex ground oats, or small potatoes, if available, may well be substituted for the barley meal.

Feed three times a day, but give at each meal only as much as the birds will clean up eagerly. Take away any food that is left over after each feed.

This regime will broaden the body and hide the breast-bone in ample fat.

Fowls, Adult. The essentials are to confine the birds in a sheltered, quiet coop and feed on special foods for 10 days. The coops should preferably be in a shed or outhouse, and should be thoroughly clean, being creosoted inside and out 5 days previously if any parasites are feared. If the birds are in sight of other stock and they cannot be moved away, old sacks should be hung in front to prevent the inmates seeing their friends running about at liberty. There must be no fretting or the finish you obtain will be only medium.

Collect the birds in the early morning, dust each bird thoroughly to kill parasites, and place in the coops, dividing them out so that the birds have companions in each compartment and yet are not overcrowded. One sq. ft. should be allowed per bird.

Starve the birds for the first 24 hours, after which should be three meals a day, the first as early as possible, the last as late as possible and the other one midway between. These must be at the same times each day, for regularity is essential.

All meals should be of wet mash, no hard grain being given at all, and the most successful food is Sussex ground oats mixed to a creamy mass with sour milk. If milk is not obtainable, some sort of fat should be given frequently at the rate of 1 oz. to every four birds per day. For a cheaper mixture you may use Sussex ground oats and boiled mashed potatoes, half and half, or cheaper still, barley meal and potatoes.

Each meal should be mixed fresh to maintain palatability, and only so much given that the birds will clean up eagerly. It is definitely dangerous to allow any food to be left over in the troughs.

It is best to mix and give the mash just warm for the sake of palatability.

If the birds seem to be losing their appetites after 8 or 9 days' feeding, divide each meal into two lots, giving one immediately after the other. An occasional feed of boiled wheat and white maize in place of the evening meal will serve to keep up the appetite.

No drinking water is given.

After 10-12 days of this treatment, the birds will give signs of being " fed up " and losing their appetite. This means that fattening is completed and the best condition is reached.

Geese, for Christmas. There are two classes of geese sent to the Christmas markets, Runners and Fattened Birds. The former are birds that have just been fed rather well and allowed their liberty up to the time of killing. They make very poor prices compared with a properly fattened goose, and it pays every time to give the birds a proper course of feeding before they are sold.

Severe reduction of liberty is the first important point. Where small numbers of geese are concerned, a disused poultry run or pig-sty can be used ; where the numbers are larger they should be folded behind sheep hurdles. Never attempt to fatten geese singly ; they need company.

Begin feeding about 3 weeks before killing is intended, giving a mixture of two parts each boiled potatoes and middlings, and one of white maize meal or barley meal, twice a day.

Water and shell should be given always so that the meal mixture need not be too sloppy. If appetites flag after 12 days, change the evening feed to boiled wheat and white maize. See that the geese eat heartily and have enough but, at the same time, be sure that none is left over after each feed.

Geese, Michaelmas. For Michaelmas, geese are allowed to have good grazing right up to killing age, but for the last fortnight are given a small feed of wet mash in the morning and also a grain feed in the evening.

For the mash, give boiled mashed potatoes dried off with barley meal. For the grain give either barley or wheat. The grain must be par-boiled and left in its liquor until cool, the grain and the liquor then being tipped into a trough; 2 oz. per bird is the correct allowance.

Petits Poussins. These are young chickens weighing from 12-14 oz. each. To get them into the correct condition, confine them for 10 days in a roomy shed and feed three times a day on the following mash, made rather wet : 4 parts by weight of Sussex ground oats ; 3 parts potatoes or maize meal (former preferred) ; 2 parts middlings and 1 part meat meal. If you can get skim milk, use it to mix the mash and leave out the meat meal.

Spring Chickens. See Chickens above.

Turkeys, for Christmas. Start the fattening process 3 weeks before killing date.

Allow them just a trifle of liberty to keep up their spirits, and give them three meals a day, the first two being of mash and the last of grain. The same fattening mashes are used as for fowls (see above) the best being Sussex ground oats and sour milk. Let the grain be soaked wheat.

If a small run cannot be given up to the turkeys alone, they may be housed in a roomy shed which has been made partially dark by bags hung over the windows— this to prevent fretting for range.

Turkeys often swarm with lice, and you must make quite sure they are free of these parasites before fattening.

Cramming. Cramming is a method of fattening table birds. Practically all first-quality fowls have been crammed in some way.

Cramming may be done either by machine or by hand. A cramming machine is only worth buying when considerable numbers of birds are to be regularly treated.

When cramming by hand, pellets are made and forced down the bird's throat with the finger, till they pass the entrance to the windpipe. They are then worked down from outside. These pellets should be just over one inch long, and about half an inch across. About fifteen pellets will be needed at each cramming. The food should not be made too wet in the main but just sticky enough to hold together well. The ingredients are Sussex ground oats and sour milk.

When machine cramming, the mixture should be of the consistency of thick cream. Cramming should be performed twice a day. In a few days the birds should be perfect.

Occasionally a bird falls sick during this cramming. In such a case, put it by itself for 24 hours and give it nothing to eat. That will bring it round all right.

Cramming is only part of the fattening process. For a fortnight before cramming the birds are fattened by the ordinary method.

Fattening Coops. For fowls in particular it is desirable to use fattening coops. Ordinary chick coops *will* serve, but they are not really satisfactory—one of their disadvantages is that they only hold one or two birds, and fowls fatten best in company.

Slatted pens are the best fattening coops. For indoor use every side and end may be of wooden slats, preferably of half-inch square section battens, placed 2½ in. apart. The bottom must be of rather stouter stuff, say 1 by 1 in. battens, running parallel to the front and about 1½ in. apart.

The pens should be raised on legs to stand at a comfortable height for feeding, say 3 ft. from the floor, and a wooden V-trough should be placed in front to hold the mash.

The full-length coop must be partitioned off into several smaller pens to prevent overcrowding. A good size for each individual coop would be about 2 ft. wide.

If the pens are to stand outdoors they must have a solid roof and back and the extreme ends should also be solid. To prevent smell spread a 2-in. layer of peat moss or dried earth beneath the coops. This layer should be raked over every day to cover up the wet droppings and absorb the liquid part.

FEATHERS. See under separate headings (e.g. Flights, Hackles, Primaries, etc.) for descriptions.

Feathers, Marketing. Certain kinds of poultry feathers are readily saleable at a good price. They must be graded into the following divisions : (1) The down from geese, (2) the soft body feathers of all poultry, (3) coloured body feathers, and (4) quills (that is, tail and wing feathers). Pin-feathers should be discarded altogether. White feathers should always be kept separate from coloured feathers.

The best time to sell feathers is from June until October.

Whilst waiting for sale the feathers should be stored in clean bags or boxes in a dry place.

If feathers are kept for some time it is not unusual to " bake " them to ensure dryness and destruction of insects, and then store in paper bags, tied at the mouth.

FEEDING CHICKS. Chicks require no food whatever for the first 48 hrs. after hatching ; they have absorbed part of the egg before hatching, and that is sufficient to nourish them for the period named. After 48 hrs., however, feeding must be regular and the foods given of the correct kind and of good quality.

If ever there be any thought of stinting chicks or " making do " with any foods on hand, let it be remembered that the way in which chicks are fed determines, to a very large extent, how they will turn out as adult birds—that is to say, the age at which they reach maturity, their ultimate body-size, health, stamina and ability to stand the strain of laying, the size of their eggs, and, in fact, the future profit to be made from them.

Use Proper Appliances. For feeding, both artificially and naturally reared chicks, proper appliances *must* be used. These can either be purchased or made.

Small galvanized iron troughs, divided into compartments by wire hoops, are best, but their size and capacity must be increased with the increasing size of the birds.

For the first fortnight the troughs should be 1¼ to 1½ in. high and not more than 2½ in. wide. A length of 12 in. will be correct for 36 chicks at this age. For fortnight to month-old chicks, as an economy, the same troughs can be used, but it will be necessary to increase their number, and this, of course, will mean more labour than if larger ones are used.

For chicks of a month old and upwards, home-made wooden mash troughs 4 in. high and 4 in. across are excellent.

For water, a drinking fount holding 1-2 pints serves for the first month. For chicks of a month old and upwards the ordinary gallon water founts are suitable. Do *not* follow the old system of putting down saucers of water; these are not hygienic.

Rations. Many proprietary chick mashes and grain mixtures are available, and those sent out by the well-known firms and others with sufficient knowledge of chick requirements to be able to blend the correct ingredients in the correct proportions will give every satisfaction. For those who desire to make up their own mixtures the following are recommended as accomplishing everything required of them.

For First Feeds give saturated chick grain, allowing as much as will go on a shilling as a day's ration for a chick. Saturated chick grain is made by putting some of the dry chick mixture in a saucepan, pouring on sufficient water (or preferably skim milk) to cover, and then heating gradually to simmering but never to boiling point. In 8 to 12 hours the grain will have swelled enough for feeding.

Feed 4 or 5 times a day, and also give dry grain as a scratch after 2 days' feeding. Keep dry bran, skim milk and shell always before the chicks.

The best grain mixture is: kibbled wheat, 6; fine flaked maize, 2; cut groats, 3—parts by weight.

After 5 Days and up to 10 Weeks Old give the following dry mash in troughs: Best middlings, 5; Sussex ground oats, 2; maize germ meal, 3; broad bran, 3; dried skim milk, 1; fish meal, 1½—parts by weight, adding mineral mixture (see Mineral Mixtures) at the rate of 2½ lb. to every cwt. of the above. The troughs containing this mash should be kept filled and available to the chicks.

In addition, give a small morning grain feed (say one handful to 20 or 30 chicks), and a second feed of grain last thing in the evening, this being a large; crop-filling one.

After 10 Weeks readjust the mixture as follows: Middlings, 3½; Sussex ground oats, 1½; maize germ meal, 1; bran, 2; fish meal, ½—parts by weight, with mineral mixture as before. The grain mixture can also be cheapened to wheat and clipped oats.

If feathering is backward add 5 per cent linseed meal (unextracted) to the mash and if the appetite flags, sprinkle the mash with flowers of sulphur (one teaspoonful to 12 birds) and mix well in.

To Speed Up Late-hatched Chicks. The following meal is specially designed to encourage rapid growth and is suitable as a dry or wet mash up to 14 weeks of age : Middlings, 5; bran, 2; maize germ meal, 3; Sussex ground oats, 2; dried milk, 1; Soya bean meal, ½; linseed meal, ½; and fish meal, ½—parts by weight. One pint of tested cod-liver oil should be mixed with every 100 lb. of mash if the birds seem poorly feathered or weak in bone,

and a good mineral mixture should also be added.

For backward growers over 14 weeks old, the best mash mixture is middlings, 3½; maize germ meal, 2; Sussex ground oats, 1; bran, 2; pea meal, 1; fish meal, 1—parts by weight.

The Chicks' Drink. If at all possible let the chicks have skim milk or separated milk or half whole-milk and half water to drink during the first few weeks of their life. It is a great aid to sturdy growth. Where such milk is not available (or permissible as in war-time) water must suffice, but let it be clean pure water, the drinking vessels being frequently cleansed and supplies renewed.

Grit and Shell. Chicks should have fine shell and flint grit always before them from their first week onward. These are best supplied in small boxes or hoppers, convenient for the size of the particular birds.

Teaching Chicks to Feed and Drink. Let the chicks' first few meals be given them on the feeding board. At no meal give more than the chicks will clean up eagerly.

Artificially-reared chicks, having no mother hen to teach them, may not know how to start feeding. In that event, just tap the food board with the fingers. That will encourage them to peck.

If chicks that are kept in a hover or brooder do not come out to feed they must be brought out by hand.

Often artificially-reared chicks also have to be taught how to drink. This is best done by choosing a few of the strongest chicks and just dipping their beaks in the liquid. Get six or so drinking and the others will be quick to imitate them.

When chicks are feeding properly (that is, after 3 days) their grain mixture should be scattered in their scratching litter so that it must be hunted for. Chicks cannot learn to scratch too early. See also Greenfood, under Foods.

Chick-feeding in War-time. Under war-time conditions the foregoing suggestions can only be followed in part, for some foods are prohibited (milk, dried milk) and others may not be obtainable. Further, the rationing regulations do not allow food for rearing chicks by domestic poultry-keepers—save in very special circumstances. Those who are keeping poultry for profit may obtain an allowance of chick grain as well as meal for rearing, and may use a national mixture of meal or be able to mix their own from the limited ingredients available separately.

Domestic poultry-keepers (25 birds or under) must, in general, rear chicks, if they wish to do so, by making full use of waste of various kinds, such unrationed foods (e.g. biscuit meal) as they can obtain, and by using a small portion of their adult poultry ration for "drying off" waste made into a wet mash. As for laying birds, it may be possible to obtain waste fish, waste meat and bones, biscuit meal; small potatoes or other materials, that will enable them to make up good mash, by way of boiling, chopping or mincing, and drying off with biscuit meal, rationed " balancer " or other meal. It is obvious, however, that any adult rationed meal that is given to chicks must be made good for the adults by increased quantities of waste, small potatoes, and unrationed materials obtainable.

On the whole, it will generally be wiser for the small or " domestic " poultry-keeper to purchase pullets of 8 weeks old or more.

FEEDING DUCKLINGS. See Ducklings.

FEEDING DUCKS. See Ducks.

FEEDING FOWLS. Much of what follows applies to normal times, when grain and cereal foods are freely available. In .war-time our methods have to be modified to suit a situation in which many foods formerly considered indispensable cease to be obtainable. War-time feeding is dealt with separately. Taking normal times, then, we have a choice of two main systems, the wet-mash system and the dry-mash system, or combinations of these. With wet-mash feeding the various mashes are given in a moist state; with dry-mash they are given dry.

General Feeding Principles. Opinions differ as to which method is preferable. There are plenty of poultry-keepers normally feeding wet-mash and, on the other hand, plenty who have gone over to the newer dry-mash method. The latter certainly serves in the way of saving labour, and a fowl will lay just as many eggs when on dry-mash as on wet-mash. All things considered, dry-mash feeding is the system recommended, except in winter, when a combination of the two (see below) is to be preferred.

Whether the mash is given wet or dry, the *principles* of feeding fowls remain the same.

To be at the peak of health and production every fowl every day needs grain, mash and greenfood. She cannot thrive and be productive on mash one day and grain another, with greenfood whenever there happens to be a supply available. All three classes of food are needed daily in order to maintain health, vigour and productivity at their best.

With the dry-mash system it is usual to feed grain in the morning, grain (if possible) at midday and more grain just before sunset ; greenfood at midday (unless the birds range over a meadow and can gather their own supplies), and mash provided in hoppers always available for the birds to feed.

With the wet-mash system, sometimes the mash is supplied in the morning, sometimes in the evening. An evening mash gives the best results. The mash should be provided in troughs. Grain is given either as the first or last meal of the day, according to whether the birds have had morning or evening mash and, if possible, also at midday. The greenfood is placed in wire racks or bags, hung up by cords or perhaps by a long nail to a post.

Grain is supplied by scattering it in the litter of the scratching shed, or, in fine weather, in long grass in the run so that the birds must scratch to find it and hence get enforced exercise.

In addition to grain, mash and greenfood, fowls need ample drinking water, grit and shell.

Here now are the rations recommended and the best methods of supplying them :

Breeding Fowls. See Breeding.

Layers. In Autumn. Dry Mash: Middlings, 3 ; bran, 1 ; alfalfa meal, 1 ; Sussex ground oats, 1½ ; yellow maize meal, 1½ ; meat-and-bone meal, 1—parts by weight.

Wet Mash ; Middlings, 5 ; bran, 1 ; alfalfa meal, ½ ; Sussex ground oats, 2½ ; yellow maize meal, 2 ; meat-and-bone meal, 1½—parts by weight.

For grain use wheat and kibbled maize, half and half.

Layers. In Spring. If following the dry-mash system : Morning. One handful grain to each 2 birds, buried in litter, one handful to 5 or 6 at noon for exercise and, half-an-hour before sunset, as much grain as they want to eat. Keep the dry-mash hoppers always open. A good dry-mash mixture for this season is : Middlings, 5½ ; broad bran, 1 ; malt culms, 1 ; Sussex ground oats, 1 ; yellow maize meal, 2 ; Soya bean meal, ½ ; meat-and-bone meal, 1—parts by weight. For grain use this mixture : wheat, 2 ; oats, 1 ; kibbled maize, ½—parts by weight.

If following the wet-mash system : Morning, one handful of grain per bird, buried in litter ; another handful between 7 or 8 birds at noon, and half-an-hour before sunset as much wet mash as the birds will eat. A good mixture for the mash is : Middlings, 5½ ; broad bran, 1 ; malt culms, ½ ; Sussex ground oats, 1½ ; yellow maize meal, 3 ; Soya bean meal, ½ ; meat-and-bone meal, 1½—parts by weight.

Layers. In Summer (May to September). For Dry Mash. Middlings, 5½ ; broad bran, 1 ; malt culms, 1 ; Sussex ground oats, 1 ; yellow maize meal, 2 ;

Soya bean meal, ½; meat-and-bone meal, 1—parts by weight.

For Wet Mash. Middlings, 5½; broad bran, 1; malt culms, ½; Sussex ground oats, 1½; yellow maize meal, 3; Soya bean meal, ½; meat-and-bone meal, 1½—parts by weight.

The grain mixture to use in conjunction with either of these mashes is: Wheat, 2; oats, 1; and kibbled maize, ½—parts by weight.

Layers. In Winter. Have a hopper of dry mash always open and available to the birds.

Early Morning. Scratch feed of wheat and kibbled maize, half and half in deep, loose litter at rate of one handful (2 oz.) to every 2 birds.

Noon. A limited amount of greenfood or, if fine, the birds could be let out on to a grass run.

One Hour before Dusk. A big feed of wet mash, preferably made with warm water. When the birds have eaten this, a mixture of kibbled maize and wheat (half and half) should be put into the troughs for the birds to eat at will until they go to perch, when it is cleared away.

A suitable mixture for both the wet and dry mash is: Middlings, 5; bran, 2; malt culms, ½; Sussex ground oats, 1; yellow maize meal, 1½; maize germ meal, 1; and fish meal, 1¼—parts by weight.

An Alternative Scheme for Layers in Winter. Under the dry-mash system have the following mash always before the birds: Middlings, 3; Sussex ground oats, 1; clover or alfalfa meal, 1; bran, 1; maize meal, 2; fish meal, 1; linseed meal, ½—parts by weight. In the morning give 3 handfuls of grain (kibbled maize, 2; wheat, 1; oats, 1—parts) to every 8 birds; at noon one handful of the same grain mixture to every 8 birds and also a supply of greenfood; half-an-hour before sunset give as much grain as the birds will clear up, supplying it in troughs.

Under the wet-mash system give 3 handfuls of grain (as above) to every 4 birds in the early morning, greenfood and one handful of grain per 4 birds at noon, and, half-an-hour before sunset, the following as a warm mash: Middlings, 5; Sussex ground oats, 2; clover or alfalfa meal, ¾; bran, ¾; maize meal, 2; fish meal, 1½; linseed meal, ½—parts by weight.

Moulting Fowls. See Moulting.

Pullets ready to lay but not yet Laying. Under the wet-mash system the mash should consist of middlings, 5; yellow maize meal, 2; Sussex ground oats, 2½; bran, ¾; clover meal, ¼; fish meal, 1¼; —parts. For a dry mash use middlings, 3; yellow maize meal, 1½; Sussex ground oats, 1½; bran, 1; clover meal, 1; fish meal, ¾—parts by weight.

For grain use: Wheat, 1; oats, 1; kibbled maize, 2—parts by weight.

Pullets Threatening to Lay before Time. The birds should have a mash that enables them to continue gradually to make body-growth yet does not stimulate the egg organs. For this purpose the best dry mash is: Middlings, 5; Sussex ground oats, 1½; maize germ meal, ½; bran, 3; Soya bean meal, ¼—parts by weight. For wet mash reduce the bran to 2 parts. The grain should be wheat and clipped oats, equal parts.

Meals that should be excluded from the mashes given to too-forward birds are: Maize meal, barley meal, pea or bean meal, clover or alfalfa meal, malt culms, maize gluten meal, fish or meat-and-bone meals, blood meal and whale meal.

Mashes, How to Mix. The mashes prepared for poultry consist of a number of ingredients, some of which are more palatable than the others. Unless the meals are thoroughly blended the birds will pick out the portions they like best and leave the rest, with the result that they do not get a properly balanced ration.

The following methods of ensuring this thorough mixing are recommended:

For pens of from 12 to 20 fowls, mixing can be done in any large receptacle such as a bath, old copper, or even a large wooden box.

Put in the ingredients, starting with the lightest one—bran, for instance. Preliminary stirring can be done with a spade, but it is wise to finish off with the hands, having first of all bared the arms. At intervals during the mixing, the receptacle should be tilted over a little to loosen and move that part of the meals which will always persist in collecting at the bottom and in the corners. If you have another vessel available, perfect mixing is ensured if the meals are tipped from one to the other several times.

When larger quantities have to be mixed, the several meals should be tipped out on to the floor of a clean shed and the mixing done with a shovel. During mixing, you should examine for evenness from time to time and when no particle of any one meal can be readily distinguished, the job is finished.

Ensuring Correct Quantities. It is important that the quantities of each ingredient should be correct according to formula. This means that each ingredient should be weighed up in the correct quantity. It is never safe to use a measure holding "*approximately*" the desired quantity; it is definitely dangerous to guess at quantities.

Mixing Wet Mash. Wet mashes should always be supplied hot, or at least warm. This means using hot water (or milk if that is available). The following is the best procedure for ensuring that all the ingredients are uniformly wetted:

The mixture having been prepared and thoroughly blended in a dry state, it is put into a water-tight vessel and then the hot liquid is slowly poured over it, the whole being well stirred the while.

On the question of how much liquid to add, wet mash should be fed in a state known as crumbly moist—a condition in which it is possible to squeeze a handful of the mixed mash into an apparently cohesive mass, but when the ball is thrown on to the ground it breaks to pieces.

"Pudding." See Tottenham Pudding.

Quantities of Food Required by Poultry. It is impossible to lay down really hard and fast rules as to the quantity of food to supply to the different classes of poultry. The only safe plan is to supply as much as the birds will clear up eagerly. If, after the meal, the birds look round for more, you are underfeeding ; if they leave mash in the trough or much grain is found beneath the scratching litter then you are over-feeding.

Very approximately the quantities of food required daily by the different classes of poultry may be set out as follows :

Ducks : 2½ oz. grain ; 2½ oz. mash.
Fowls : 2 oz. ,, ; 2 oz. ,,
Geese : 3½ oz. ,, ; 4 oz. ,,
Turkeys : 3¼ oz. ,, ; 3½ oz. ,,

(With war-time mashes the consumption is greater, amounting to an increase of approximately 100 per cent, which means that fowls, for instance, may consume 8 oz. of war-time mash per day.)

Swill. See Tottenham Pudding.

Tottenham Pudding. Name given to a poultry food prepared originally by Tottenham Council and subsequently by many other local authorities. Made from waste foods collected from households, canteens, etc., roughly sorted and fed into machinery which minces and mixes the mass, then cooks it. The pudding emerges as a brown or greenish-brown material, slightly oily and easily broken up by hand. The material keeps well for about a week, longer if immersed in water.

A typical analysis is : Dry matter 31·9, protein 4·7, oil 3·8, soluble carbohydrates 18·5, fibre 1·8. It keeps fresh for about a fortnight.

Up to 6 ozs. a day can be fed to birds, with meal added, as rations permit. In use the pudding is well broken up, then the meal is mixed in thoroughly. Birds fed on the pudding keep in good condition and production is satisfactory.

Chaff waste is sometimes added to the pudding. The chaff makes the pudding more palatable and improves its texture, also ensuring that it dries hard, thereby being easier to transport. In addition, the chaff materially improves feeding value.

A certain amount of unprocessed waste, in the form of swill, is also supplied for poultry feeding. The swill has to be boiled before use. After boiling a little meal is added.

Swill is not so nourishing as the pudding. Although birds will consume as much as 14 ozs. per day, egg production is very indifferent. Raw swill should be considered only a "maintenance ration," serving to keep birds alive.

War-time Feeding. The advent of the world war called for most revolutionary changes in all preconceived poultry-feeding ideas. Restriction of foodstuffs was further complicated by various restrictive Government measures, which changed from time to time with the progress of the war.

It was considered by the Government that the situation could most easily be met by dividing poultry-keepers into two classes—those who produced eggs for their own home needs, known as "domestic" poultry-keepers and/or "small" poultry-keepers, and those who produced eggs, etc., commercially. Different methods of rationing were devised for each.

Domestics. The only ration issued to domestics was "balancer meal" (which see). This, added to household scraps and other waste foods, formed a "balanced" diet for layers. Balancer was issued only for a stipulated number of pullets over two months old; there was no allowance for chicks or cockerels.

The basis of the "domestics" mashes was household scraps, (leavings of cooked vegetables, meat, fish, pudding, porridge, etc.), chat potatoes, some greenfood and roots. Where raw vegetables were used with the scraps the whole was cooked and thoroughly minced and mixed. Then the balancer was added, usually at the rate of 2 oz. to 6 oz. of other ingredients.

Where it was possible to obtain Tottenham Pudding (which see) in small quantities some domestics made use of this.

The rest of the diet for domestics' birds consisted of spare greenfood and roots served raw.

Commercial Flocks. Rationing for commercial flocks was introduced in February 1941. Allowances were based on the number of birds kept in pre-war days. For a period the basis of allowance was for one-sixth of the pre-war stock. Thus a commercial man with 1,200 birds was allowed rations for 200 birds, on the basis of 1 unit (1 cwt.) of food per twenty birds per month.

Later an acreage reduction was made. For each acre of land in the poultry farm, food for 1 bird, later for 1½ birds, was withheld, on the grounds that a certain amount of poultry food was home-grown.

In 1942 there was a further reduction in

the basis of allowances—from one-sixth to one-eighth of pre-war flocks.

Poultry farmers were thus compelled to rely to an increasing extent on home-grown greens and roots, etc., and waste foods such as Tottenham Pudding. Considerable use was also made of potatoes (up to 90 per cent of the total rations), silage and dried grass.

Waste for Poultry. This may consist of a wide variety of materials—potato parings, vegetable waste of all sorts, fish heads, bones, blood, odd pieces of bread, table scraps. These will be of great value, while additions in season may be acorns, beechmast, hips and haws, elderberries, berries of mountain ash, small potatoes, turnips, swedes, mangolds, and green vegetables of almost every kind.

FIRST CROSS. The progeny of two pure-breeds of different varieties. Some first crosses are especially useful for table poultry production (see Breeding). In addition first-crossing is the main principle of sex-linkage (which see).

FLIGHT FEATHERS, FLIGHTS. The long feathers of the wing that are not seen when the wing is closed. Sometimes called " primaries."

FLUFF. Term applied to the downy feathers on the thighs, around the vent, etc.

FOODS, POULTRY. Following is a description of the many foods suitable for poultry feeding, including those brought into use during war-time.

Acorns. The search for substitute foods during war-time brought acorns into prominence. They are a good poultry food provided they are properly prepared and fed in the right porportions. They must be shelled and well boiled, and then mashed before being mixed with other foods. For the first three or four weeks no more than 5 per cent of acorns should be added to a mash. The proportion may then be increased to 10 per cent, but no more. A greater proportion would lead to discoloration of the yolks and perhaps digestive trouble. Where other nut foods are used (such as horse chestnuts, which see) the total proportion of nuts must not exceed 10 per cent of any mash.

Alfalfa Meal. A poultry food consisting of dried and ground alfalfa, or lucerne as it is called in England. When buying alfalfa meal the indication of quality is the amount of actual dried leaf it contains and its freedom from dust and stalk. The colour should be a good bright green.

Balancer Meal. A mixed meal for domestic poultry keepers, especially designed to form a balanced ration with a variety of household scraps or concentrated waste ("Tottenham Pudding"). The following examples of daily rations for 6 birds may serve as illustrations of what is possible:

1. Balancer Meal 12 oz., "Tottenham Pudding" 36 oz.
2. Balancer Meal 12 oz., household waste 18 oz., fresh greenfood 7 oz., pulses 3 oz., potatoes 8 oz.
3. Balancer Meal 12 oz., household waste 4 oz., fresh greenfood 4 oz., roots 2 oz., potatoes 15 oz., coupon-free meals 5 oz., coupon-free proteins (wet) 7 oz.

Barley. One of the most useful grains for feeding to poultry of all kinds. Good samples will be full and large yet dry and hard. The best grain will carry only a small quantity of husk.

Barley Meal. One of the chief ingredients of most poultry mashes. It should be finely ground but not too powdery. The ideal state is when the husks are broken down, but the particles of barley meal can easily be distinguished. An even creamy colour is required.

Beechmast. The " fruit " of Beech trees. It is a useful protein poultry food, very much liked by turkeys and ducks. If readily obtainable, it may form a small part of the ration—not more than 10 per cent. The beechmast must be prepared in the same way as acorns. See Acorns.

Biscuit Meal. Form of poultry food much used nowadays in mashes. Though it is much dearer than other meals it is considered fair value for money as it is completely digestible so that there is no waste. It is nourishing, as it contains all the valuable proteins and carbohydrates in balanced proportions. As a body-builder it is unsurpassed, the growth being both speedy and strong. It is also palatable, and the birds eat it with relish. It makes what might be otherwise a rather stodgy wet mash light and appetising.

Blood as Food. Fresh blood is excellent food for fowls. The normal method of using it is to take one-half fresh blood and one-half skim milk for mixing the wet mash, but there is no reason why all blood should not be used when skim milk cannot be bought at a reasonable price, or is not available—as during war-time.

When blood is used it must be supplemented by adding a mineral mixture. If alfalfa is one of the ingredients in the mash, $1\frac{1}{2}$ lb. of mineral mixture should be added to every 100 lb. of the meals weighed in their dry state. If alfalfa meal is not used, $2\frac{1}{2}$ lb. of mineral mixture should be added to the same quantity of the meals.

Bone Meal. A cream-coloured meal largely used in poultry mashes. Made from powdered bones. Up to 5 per cent in a mash, is useful for growing chicks, but is not so good as a combination meat-and-bone meal. As some grades of bone meal are used for manurial purposes it is essential that a feeding variety be selected for inclusion in mashes.

A good sample of bone meal should smell sweet and contain no large, unground pieces of bone.

Bones, Green. Fresh bones may be purchased from a butcher, ground in a special bone-crusher and fed to all classes of poultry. It is advisable, however, not to use more than 1 part of ground bone to every 20 parts of other ingredients, by weight.

Bran. A very valuable, if not indispensable, item in most poultry mashes. Its value lies in the fact that it contains minerals ; the digestible parts are easily and quickly digested ; it is a mild laxative ; and it can be given either wet or dry. It maintains health, condition and stamina.

The signs of quality in bran are softness to the touch, largeness of the particles, brightness of colour and sweetness of taste and smell. It should contain a certain proportion of flour adhering to the particles.

Brewers' Grains. By-product of brewing, which, if obtainable cheaply, are suitable for mixing with all kinds of poultry rations. May be wet or dry. If purchased in the wet form must be used within a few days (or in quantity may be stored in a silo), but if dried may be stored for a considerable period, like grain.

Buckwheat. Grain sometimes fed to poultry. A good sample is clean, hard, sharp and unbroken.

May be grown on soils not in rich condition, and though normally best grown on the extensive scale, is well worth considering by any poultry-keeper who has land available.

Buttermilk. A by-product of buttermaking and very useful in poultry-feeding if available regularly and does not contain salt. Can be given separately as a drink or can be used for mixing wet mashes.

Charcoal. Stated by some poultry-keepers to be an aid to digestion in poultry of any age. Should be used in finely-ground form for growing chicks and in pea-sized pieces for adult birds. It should be placed in a separate hopper so that the birds can help themselves.

Clover. Clover is fed to poultry usually in the dried form known as clover-meal. The important test of quality of any sample is the actual amount of dried leaf it contains—its freedom, that is, from mincing, and drying off with biscuit meal, dust and stalk.

Cockle Shell. See Oyster Shell.

Cod-liver Oil. Latterly the inclusion of cod-liver oil in poultry rations has become very popular. There are now comparatively few poultrymen of importance who do not take advantage of its strengthening, vitalising properties. Cod-liver oil has been called "bottled sunshine " —because it has much the same effect on birds that receive it as sunshine, has ; it is, in effect, a substitute for the ultra-violet rays of the sun, providing vitamins. It increases egg production and has a great influence on chicks, increasing their vitality and resistance to disease.

The simplest method of supplying cod-liver oil is to mix the oil with the mash. This is best done by incorporating the allowance of oil with the smallest possible amount of some suitable meal. Afterwards this impregnated meal can be blended with the whole. Bran is a good thing for use for this purpose.

Another way in which cod-liver oil can be supplied is with the grain ration in the later afternoon. When birds are dry-mash fed, it is customary to give the grain in the afternoon in troughs. When this plan is adopted, the oil can be mixed very easily with the grain. Put the grain into a bucket, pour on the necessary amount of oil, and stir for a few minutes until every grain is smeared with the oil.

A teaspoonful daily is quite enough for a dozen small chicks, increasing the dose very gradually until adult birds are receiving 2 oz. to every $6\frac{1}{4}$ lb. of mash.

Dari. Small grain frequently included in chick corn-mixtures. In a good sample the grains are clean, even in size, free from dust and rubbish, and sweet smelling. Dirty samples should be avoided.

Fish Meal. Useful poultry food. Prepared from fish that has been boiled and dried. Good samples are prepared from white fish—cod, etc.—and not herrings. Although possessing a marked smell, it should not be too strong and fishy. The colour is greyish yellow and it should be finely ground, with pieces of bone showing. If, when laid on paper, it imparts an oily patch, there is probably an excess of oil. The oil percentage should not exceed 5, the salt percentage 4.

Food, Digestible Constituents. Correct food mixtures for different classes of poultry are given elsewhere in this Encyclopædia. Some poultry-keepers, however, desire to blend special rations. In such cases it is essential to have the analyses of the foods to be used, for without knowing the exact nature of the foods it is impossible to blend a ration having the correct quantities of albuminoids, carbohydrates, fats, etc.

The following table shows the percentage content of digestible protein, oil and carbohydrates, of foods likely to be employed for poultry :

| | *Digestible Nutrients Per Cent.* | | | | |
| | Crude Protein. | Pure Protein. | Oil. | Carbohydrates. | |
				Soluble.	Fibre.
Roots.					
Artichoke, Jerusalem	1·0	0·4	—	15·8	0·2
Carrots	0·8	0·4	0·1	8·9	0·7
Kohl Rabi	0·7	0·3	—	7·4	0·6
Mangolds, intermediate ..	0·7	0·1	—	8·5	0·3
Parsnips	1·0	0·6	0·1	10·9	0·7
Potatoes	1·1	0·6	—	17·7	—
Sugar Beet	0·8	0·3	—	19·3	0·4
Turnip	0·6	0·2	—	5·2	0·3
Other Green Foods.					
Cabbage, open-leaved ..	1·8	1·2	0·4	6·5	1·7
Kale, Thousand head	1·7	1·2	0·17	7·5	1·8
,, Marrow-stem (singled-out)..	1·6	1·0	0·18	6·4	1·5
Grasses.					
Pasture grass, close grazing : Non-					
rotational	4·5	3·8	0·7	7·8	2·1
Rotational with monthly intervals	2·6	2·3	0·5	7·6	3·7
Pasture grass, extensive grazing ;					
spring value, running off during					
summer	2·5	1·7	0·4	7·3	2·6
Winter pasturage (after close-					
grazing, allowing free growth					
from end of July to December)	2·0	1·5	0·15	7·9	2·6
Green Legumes.					
Red clover, beginning to flower ..	2·5	1·7	0·5	6·3	3·0
White clover, ,, ,, ,, ..	2·8	1·9	0·5	4·7	2·6
Lucerne, in bud	3·6	2·4	0·1	6·8	3·1
Tares, in flower	2·2	1·4	0·3	4·9	2·3
Miscellaneous.					
Artichoke tops (dried)	7·6	6·1	1·1	33·6	4·1
Nettles (dried)	12·8	9·3	4·9	30·0	6·0
Silage.					
Grass	1·4	0·9	0·4	4·7	3·8
Vetch and oats (acid brown) ..	3·8	1·9	1·2	6·4	5·5
Sweet Silage.					
Grass	1·9	0·7	1·3	7·5	5·9
Hay.					
Meadow hay, very good	9·2	6·5	1·5	30·1	12·7
Grains and Seeds.					
Barley	7·6	7·0	1·2	60·9	2·5
Dari	7·7	6·7	3·0	60·5	1·0
Maize	7·9	7·4	2·7	63·7	0·8
Oats	8·0	7·2	4·0	44·8	2·6
Rye	9·6	8·7	1·1	63·9	1·0
Wheat	10·2	9·0	1·2	63·5	0·9
Legumes.					
Beans	20·1	10·3	1·2	44·1	4·1
Peas	19·4	16·9	1·0	49·9	2·5
Vetches	22·9	20·0	1·5	45·8	3·9
Oil Seeds.					
Beechmast	10·7	10·1	24·1	16·8	7·4
Hemp seed	13·7	12·8	29·3	16·8	9·0

	Digestible Nutrients Per Cent.				
	Crude Protein.	Pure Protein.	Oil.	Carbohydrates.	
				Soluble.	Fibre.
Oil Seeds.					
Linseed	19·4	18·1	34·7	18·3	1·8
Soya bean ..	29·5	26·2	15·8	20·8	1·7
Sunflower seed	12·8	11·1	30·7	10·3	9·4
Miscellaneous Seeds.					
Acorns, fresh ·	2·7	2·2	1·9	32·6	4·1
,, dried	4·6	3·8	3·3	55·5	7·0
Buckwheat	8·5	7·5	1·9	42·3	3·5
Horse chestnuts, fresh ..	2·6	1·5	1·2	33·3	0·8
,, ,, dry	4·1	2·4	2·0	48·4	1·2
Maize germ meal	10·4	10·2	11·5	47·0	2·5
Maize, flaked	9·4	9·0	2·0	70·4	0·5
Soya bean meal, extracted ..	40·3	36·3	1·4	24·7	3·6
By-Products.					
Barley brewers' grains fresh ..	5·5	5·2	2·4	9·1	2·4
,, ,, ,, dried ..	13·0	12·1	5·6	27·6	7·3
,, distillers' ,, fresh ..	6·2	5·8	2·6	6·4	1·7
,, ,, ,, dried ..	19·6	18·7	10·2	25·3	4·8
Fish meal, white	55·0	51·0	3·3	1·2	—
Lucerne meal, English (from crop just coming into flower bud) ..	15·9	11·2	1·3	28·4	9·5
Maize gluten feed	20·0	18·4	2·7	49·3	2·5
Meat and bone meal	39·2	29·2	14·3	—	—
Pure meat meal	67·2	63·6	12·5	—	—
Milk, separated	3·3	3·3	0·1	5·0	—
Sugar beet pulp, wet , ...	1·0	1·0	—	8·7	2·8
,, ,, ,, dried	5·3	5·0	—	54·0	16·3
Wheat Offals. (Pure Grades.)					
Finest grade, fine middlings ..	12·6	11·6	3·7	51·1	—
Second grade, coarse middlings or 'sharps (fine wheat feed)* ..	11·6	10·1	3·9	45·9	1·4
Broad bran..	11·0	9·0	2·8	36·9	2·2
Yeast, dried	41·6	39·9	0·2	29·2	—

* Home-produced middlings are now marketed under the names of Weatings (with a guarantee of not more than 5·75 per cent of fibre) and Superfine Weatings (with a guarantee of not more than 4·5 per cent of fibre).

Grain. Term applied to the various cereals—wheat, maize, oats, etc.—fed to poultry. For the quantity of each cereal to use in poultry foods, see Feeding ; for hints on determining whether a grain is of good quality, see under separate headings.

Grain, Substitute for. Apart from its food value, grain scattered among the litter induces the birds to scratch and so gain valuable exercise. The war-time shortage of grain resulted in the ûse of several substitutes which could similarly be thrown down among the litter. Among these substitutes may be mentioned roots chopped into small cubes and dried in the oven ; thin slices of potato similarly treated and then broken up ; and waste bread also dried in the oven, and crushed.

Grass. Excellent food for all poultry· if free from tough or poisonous weeds.

The only times when grass is dangerous is when it is very wet (and then only to chicks ranging over it, for, getting wet, they are apt to get soaked and chilled), when it is frosted (and then all birds should be kept off it, for frosted grass, if eaten, may set up internal troubles), and when, in the summer, it consists of a number of dry hard indigestible stems (when it is liable to form a solid mass in the crop).

Young, short grass has a very high feeding value, as it contains much readily-digestible protein, and every endeavour should be made to keep grass runs fresh and short. When dried, such young grass may have a value roughly equivalent to linseed cake, and it may be used for mixing with other foods to improve the protein content.

To dry grass mowings spread them out thinly in a sunny place and turn them frequently until the cuttings are crisp and dark green, or the grass can be spread out thinly on wire frames placed in a brooder house or in empty night arks. Store in a dry airy shed : the grass must not be allowed to go mouldy. See also Runs.

Green Bone. See Bones, Green.

Greenfood. Greenfood of one kind or another, or a substitute for it, is an essential item in the feeding of all poultry, and a supply should be given every day in the year. Without greenfood health suffers and egg supplies diminish.

Greenfoods Proper. True greenfoods include fresh grass (first and foremost), fresh lawn-mowings, lettuce, the waste leaves of all the cabbage family, pea pods (to a lesser degree), carrot and turnip tops, kohl rabi, fresh cut clover, lucerne, vetches etc. Birds may have as much as they will eat of these.

Roots. When greenfood proper is definitely unavailable certain roots may be substituted, namely carrots (first and foremost), swedes, kohl rabi and beetroot.

Roots should be served raw, *not* cooked. They may either be split into sections and stuck onto nails for the birds to peck at or they may be chopped up into small dice and given in a trough. The birds may have as much as they will eat.

Sprouted Oats. A useful substitute for greenfood proper. To sprout oats, spread a thick layer over the bottom of a tray, water with tepid water and keep in a warm place. The result will be a " forest " of green shoots. The oats may be fed to the birds when the shoots are 3 in. high and still quite green. The matted mass should be pulled into " chunks," these being given to the birds as they are.

Substitutes. When greenfood is short (and when roots are being given) its lack may be made up by including a little extra clover meal or alfalfa meal (up to 7 per cent in the mash).

Frozen Greenfood should never be supplied to poultry, for it may start digestive trouble. The safest plan in frosty weather is to gather the required greenfood on the previous day and keep it for the night in some frost-proof place. In emergency greenfood may be thawed by placing it in *cold* water ; if hot water is used the food will afterwards be flabby and unpalatable.

Greenfood for Chicks. Greenfood is just as essential to chicks as to adult fowls and they should have a supply every day, preferably at midday. The best greenfood is young grass in a run, but this, of course, is only available for chicks that are being reared outdoors. For confined chicks use cut-up young grass, lettuce, spinach, chickweed, dandelions or onion (stems). Cabbage and the coarser greens

may be fed when the chicks are six weeks old, but not before.

Greenfood for confined chicks should be chopped up and fed in the same sized troughs as for the mash. It must never be thrown in the litter. A 10-in. trough of greenfood, level to the top, is ample for 36 chicks.

Growing Food. In some circumstances even domestic poultry-keepers can grow much food for their poultry—green and root crops, buckwheat, rye, sunflowers. Poultry farmers who have an extensive area, as well as very many smallholders, may under war-time conditions find it an important side of their business to devote a part of their land to producing crops for the use of their birds.

Grit. Grit is essential to all poultry. Poultry food has to be ground up in the gizzard (which see) and grit is necessary to do the grinding. There should, therefore, always be a supply of grit before the birds in a position where they can readily find it.

The grit usually supplied is crushed flints, but in recent years limestone grit has been widely used. If the limestone grit is of first quality (the best samples contain 99·8 per cent of carbonate of calcium and, in appearance, are hard, dry, clean and light-grey or white in colour) it is preferable, for it provides the fowls with the lime necessary to their well-being in addition to grinding their food.

Shell-making materials may also be classed as grit, though quite distinct from flint grit. The most commonly used of these are crushed oyster shell and crushed cockle shell. A supply of one or the other of these (or limestone grit) should always be before the birds.

Grit is best supplied in automatic hoppers consisting of a container and a trough, the container replenishing the trough as its contents are consumed.

The dust left in the hoppers should be emptied out every few weeks and the supplies of shell renewed.

Horse Chestnuts. Like acorns (which see) horse chestnuts must be fed to poultry only in limited quantities, 5 per cent in any mash being the maximum. The nuts must be prepared for feeding by being shelled, broken up, soaked for 24 hours, then boiled.

Housescraps as food for poultry. Amateur poultry-keepers can considerably reduce the cost of feeding their birds by making judicious use of the food scraps from their own table; as *must* be done during war-time.

A good mash formula for mixing with the house scraps produced in an average household is : Scraps 5 parts by weight, maize meal 1, alfalfa meal 1, middlings 4, and fish meal 1 ; or scraps 5, " balancer meal " 7, under war conditions.

Pass all the scraps through a chopping or mincing machine before mixing in order to make the resulting mash even, so that the birds can not pick out single whole scraps. See also Feeding Poultry in War-time.

Kale. One of the most useful of the " greens " to grow specially for poultry.

Lettuce. Excellent greenfood for fowls; particularly useful when chopped small, for feeding to chicks.

Lime. All fowls require lime for the production of egg-shell. It is best supplied to them in the form of limestone grit, crushed oyster shell or crushed cockle shell, which see.

Linseed. Useful food for poultry, particularly at the completion of the moult, as it greatly helps feather-formation. See Moulting.

Lucerne. See Alfalfa.

Maize. One of the most valuable of all poultry foods. Is supplied in the form of whole grain, kibbled grain, maize germ meal, maize gluten feed and maize meal.

Malt Culms. Occasionally used for feeding to poultry. See Feeding for mashes in which they may be used.

Meat. Much used in normal times for feeding poultry, in the form of meal. No meat, either raw or cooked, should ever be given alone to poultry. When supplied at all—as when fowls are fed on housescraps—it must be mixed with meals (see Feeding Poultry).

Meat Meal. The best test for this is an examination for hair, gristle and hide, which should be absent. There should not be an excess of bone in meat-and-bone meal. It should smell sweet ; any suspicion of rancidity will make it useless for poultry.

Middlings. See Weatings.

Milk. An excellent food for fowls of all ages and particularly for chicks and fattening birds, and has been extensively used in the past. On account of expense, skim-milk is usually preferred to whole milk. Dried milk is freely used as a substitute. There are many who hold that any from of milk should be reserved for human consumption. See Feeding for quantities and methods of use.

During the war the use of milk for feeding poultry is prohibited.

Mustard. A good stimulant for fowls. Stated to encourage egg production. The special poultry mustard should be obobtained. The correct dose is 1 teaspoonful mixed with the mash for every 12 birds. Mustard should not be given to breeding birds.

Oats. One of the most valuable foods or all poultry. Good quality oats are poken of as " fat," which means that hey are thick and plump, therefore possessing a minimum quantity of indigestible husk. Forty pounds should be the minimum weight of a bushel, usually 42 lb.

Clipped oats are best for poultry. The skin should be thin and wafery ; some oats look plump because of a thick, leathery skin. There should be no dust and dirt in the crease.

There is a simple test for quality in oats. A tablespoonful should be placed in a tumblerful of water and left for 24 hours. At the end of that time, with a good sample, there will only be a dozen or so floating at the top, the rest having sunk.

Oats, Sussex Ground. Another valuable food for poultry. Quality is important. The husk should hardly be seen and the colour should be creamy grey, *not* white.

As a test for quality, a handful may be taken up, the hand closed, and then opened. In a good sample the finger marks will be plainly seen and the ridges caused by them will remain stiffly together. In the bag Sussex ground oats should be solid and fluffy, and loose, should feel soft and smooth to the touch.

Oyster Shell. Crushed oyster and cockle shell are supplied to fowls to provide material from which they may make their egg shells. A hopper of crushed oyster shell (or one of the alternatives, crushed cockle shell or limestone grit) should be always available to all layers.

Potatoes. See Feeding in War-time.

Saturated Corn. See Chick Feeding.

Salt. Salt in excess is poisonous to fowls, and when housescraps are used in the mash care should be taken that. they do not contain much salt. Salt in small quantities is beneficial to adult fowls—especially those not receiving a mineral mixture. The correct allowance is 10 oz. salt to each hundredweight of mash ingredients.

Sharps. Another name for Middlings, or Weatings.

Silage. This is produced from many kinds of green crops by packing and compressing them in a container termed a *silo*, the process itself being named *ensilage*. On a small scale silage may be made in a barrel or tank, but for the poultry farm silos of various types are available—the cheapest and most popular at present being of wooden staves lined on the inside with stout proofed paper.

Crops used are short young grass, clovers, oat and vetch mixture, green maize and many forms of succulent herbage. The materials are usually spread evenly, and compressed by trampling, especially necessary towards the wall of the container. The crops are ensiled in the green state, and if properly trampled may be preserved well. Crop containing a high percentage of protein are commonly the better for the addition of a proportion of molasses.

If standing in the open the container should be well filled after compression, be weighted with soil on top of old sacking and topped in such a way that rain runs off and not down the inside of the silo.

For feeding poultry silage needs to be chopped up fine, and may form up to 10 per cent of the mash.

Spice. There are several excellent proprietary spices. Given to fowls in accordance with the maker's instructions, these have a most beneficial influence on egg production.

Sunflower Seeds. These are a valuable food for poultry, containing seven times as much fat as oats, and more protein than any cereal grains. They also help to keep the fowl's digestive organs in good order. The seed heads should be gathered just before the seeds are ripe and ready to fall. The heads are dried, the seeds are separated from the heads and given a further period of drying. Finally the seeds should be crushed by being passed through a kibbling or mincing machine. The resulting " meal " may be used in mashes in the proportion of 7 per cent. Sunflower seeds may still be added in this proportion even when acorns or chestnuts are included in mashes.

Sussex Ground Oats. See Oats, Sussex Ground.

Swill. See Tottenham Pudding under Feeding Fowls.

Thirds. Term applied in some districts to middlings or weatings.

Waste Food. As a war-time measure (though it may become permanent) household food waste (scraps of all kinds, and vegetable refuse, etc.) is collected by local authorities in a large number of areas, and converted into concentrated waste in the form of "pudding" (first named "Tottenham pudding" after the area in which it originated). This material is of great value for poultry, adults being able to take 5 or 6 oz. daily when admixed with 2 oz. of "balancer" or other meals. See also Tottenham Pudding under Feeding Fowls.

Weatings. Name now commonly used to denote the graded wheat by-product meal previously known as middlings, sharps, thirds, dan, etc.

Wheat. Most useful of all poultry grains. The hard, small, red wheat is the best. It should be dry and should give a hard, metallic sound when dropped on to a hard surface. Often used kibbled for chicks.

Wheatfeed. One of the many names for middlings or weatings.

FOODS, STORING. The small poultry-keeper is likely to store his foods on any handy shelf or in a corner of some out-building. Where a number of fowls are kept, however, proper arrangements have to be made for the storage of foods; otherwise vermin will take heavy toll of them and, further, quantities will be spoilt by deterioration. It is an advantage if the food storage facilities are ample, because then foods can be bought in larger bulk, with the consequent saving in cost.

The store should be close to save labour in carrying the foods about.

Adapting an Existing Building as a Food Store. Any small building will serve for the store so long as it has a dry floor, a damp-proof roof and is vermin proof. Small-mesh wire netting sunk two feet into the ground and carried one foot up the sides of the building, all round, will exclude rats and mice.

When an old building is used as a food store take the precaution of spraying the walls, roof and floor with a recognised good disinfectant solution. This will go far to keep the shed sweet for a considerable time, an important point in a food store.

Building a New Store. If, having no existing shed available, it is proposed to put up a special food store, it should be high enough to walk about in it comfortably. A span roof, 7 ft. high at the ridge and 5 ft. at the eaves, is useful ; the timber should be sound and of a thickness that is not likely to warp.

Inside the Store. The best arrangement for storing the different kinds of foods is to divide the foods under three different headings—(1) the " grits," this including cockle shell, oyster shell, flint and limestone grit ; (2) the grains, meaning wheat, oats, kibbled maize, etc. ; (3) the meals, separate and mixed—and to employ the end of the store for ready-to-use meals, the side for the grains and the part farthest from the door for grits.

Galvanized bins are the best and safest with reference to vermin, but, failing them, grape barrels and treacle barrels, obtainable from grocers, come in quite handy. A scrubbing out of the inside and a coat of creosote on the outside will fit them for long service. A light lid made of $\frac{3}{8}$ in. tongued and grooved boards on 2 in. x $\frac{3}{4}$ in. battens, is a useful addition, helping to keep the contents clean. Never store meals or corn in sacks.

A point in regard to any kind of meal bin: always brush it out, clearing out completely any old meals, before adding the new purchases, or you may get the latter souring quickly.

Grains are best kept in proper wooden bins, especially during the summer, as they are apt to become " sweaty " and to heat if incorrectly stored for any length of time. If corn is to be kept in sacks for over a month it is a good thing to turn the sacks upside down every fortnight.

Food Store Fittings. In every food store there should be an assortment of scoops, wooden spoons (or, better, strong trowels) and buckets. These should be kept tidily together all in one place. Scoops of at least two different sizes are wanted, a large one for grain and meals and a smaller one for mustard, spice and so on.

If you make up your own food mixtures you will find a galvanised bath or wooden tub handy.

Large-scale Storing. Given a vermin-proof store, cubicles are a better proposition than bins. Cubicles cost little, keep the food clean and the food store tidy, make food-handling less irksome and prevent moulding and sweating.

Rough-sawn boards will be excellent for framing the cubicles, say 7 in. by 1 in.

The partitions should only be made as high as needed to keep the foods separated; 4 ft. is ample. The uprights must be firmly fixed both at the bottom and top and should be of 2 in. by 2 in. or 3 in. by 2 in. quartering.

The front should merely consist of separate boards sliding in upright grooves, the groves being made by nailing on to the uprights two pieces of 2 in. by 1 in. batten 1½ ins. apart. This scheme enables the height of the front to be adjusted accordingly to the height of the food inside, so that it may be reached readily.

A cubicle 3 ft. by 2 ft. will hold 2 to 3 cwt. of meal, 3 ft. by 3 ft., 4 to 5 cwt., and almost double that of corn.

FOOT FEATHERS. The feathers to be found on the feet of certain breeds of poultry.

FOSTER MOTHER. Name given to an appliance for rearing chicks artificially. The appliance consists of two compartments under one roof. The inner compartment, or nursery, is heated by a lamp; the outer one is unheated, has a wire-netting front and is used as an exercising and scratching place. The chicks must be confined to the inner compartment for the first day and, when first turned into the outer run, must be driven back into the nursery several times until they learn the way to get back to the heat of their own accord.

FOWLS, CHOICE OF BREED. The various breeds of fowls are described under their respective headings.

The table on page 54 will serve as a useful guide to a choice of breed.

Ancona. One of the most prolific of the light (non-sitting) breeds of fowl. Was highly popular some few years ago, but is not now kept so extensively. Has one disadvantage that it is very nervous and seldom becomes really tame. Lays large white eggs.

The normal weights are : hen, 5 lb. ; pullet, 4½ lb. ; cock, 6-6½ lb. ; cockerel, 5½ lb.

Andalusian. A light-breed (non-sitting) fowl that is not very extensively kept. Lays specially fine large white eggs.

Normal weights are : hen, 5-6 lb. ; pullet, 4½-5 lb. ; cock, 7-8 lb. ; cockerel, 7 lb.

Aseel. A heavy (sitting) breed of fowl popular at one time for cock-fighting. Not now kept, except in very isolated instances. The colours were black, grey, red, black-spangled, red-spangled, white and yellow. A peculiarity of the breed was that the birds were almost bare of plumage at the breast-bone and first wing joints.

Australorp. Popular heavy (sitting) breed of fowls. Excellent table bird. In shape closely resembles its Orpington ancestor, though the body is not quite so deep.

Normal weights are : hen, 7 lb. ; pullet, 5 lb. ; cock, 8 lb. ; cockerel, 7 lb.

Bantams. These miniature fowls are not kept to anything like the extent they should be, probably because the eggs, from a commercial and ordinary domestic point of view, are too small. Bantams are, however, excellent for confined places and where pleasure rather than profit is the aim in view.

Very few complete records are available regarding the number of eggs a Bantam hen will lay in a year. Bantams are usually kept by fanciers who make no effort to obtain eggs except when eggs for hatching are desired, and practically no attention has been given to breeding for egg production. That they are prolific layers, however, there is not the slightest doubt. As a general statement 100 eggs per year is as high a production as most people get, but individual Bantam hens have laid as many as 150 or even 175 eggs in a year.

Bantam eggs vary in size, according to breed but they range from about 12 oz. to 20 oz. a dozen.

The following are the recognised Bantam breeds : Modern and Old English Game, Andalusian, Brahma, Cochin, Silkies, Hamburgh, Indian Game, Japanese, Leghorn, Malay, Minorca, Orpington, Plymouth Rock, Polish, Scotch Grey, Sebrights, Wyandotte, Yokohama, Spanish, Mille Fleur, Booted Rosecomb.

The game Bantams, Sebrights, Rosecombs, Japanese, Polish, Mille Fleur, Booted and Silkies produce particularly good-sized eggs that are classed as white, or may be a little creamy or ivory in colour, though occasional eggs are light brown. The Cochin and Brahma Bantams, on the other hand, produce eggs that vary from deep brown to very light brown. The Brahmas, Cochins and Silkies are

Class	Purpose		Recommended Breed
Smallholder	For Eggs	..	White Leghorn
	For General Purposes	..	White Wyandotte
	For Table ..	..	Rhode Island Red
General Farmer	For Eggs	..	White Leghorn
	For General Purposes	..	Rhode Island Red
	,, ,,	..	Buff Rock
	,, ,,	..	White Wyandotte
	For Table ..	..	Light Sussex
Backyarder	For Eggs	..	White Leghorn
	,, ..	..	Black Leghorn
	,, ..	..	Ancona
	For General Purposes	..	White Wyandotte and Rhode Island Red
	For Table ..	..	Light Sussex
	,, ..	..	Houdan (Good for crossing)
	,, ..	..	Dorking (for size)
Poultry Farmer	For Eggs	..	White Leghorn
	,, ..	..	Brown Leghorn
	For General Purposes	..	White Wyandotte and Rhode Island Red
	,, ,,	..	Light Sussex
	,, ,,	..	Barred and Buff Rocks
	For Table ..	..	Light Sussex (or any table breed)
Commercial Egg Man	For Eggs	..	White Leghorn
	,, ...	..	White Wyandotte
	,, ...	..	Rhode Island Red
Exhibitor	For Eggs	..	Ancona
	,, ...	..	Rhode Island Red
	,, ...	..	La Bresse
	,, ..	..	Light Sussex
	For Table ..	..	White Wyandotte
	,, ..	..	Light Sussex
	,, ..	..	Buff Orpington
	,, ..	..	Barred Rocks

generally classed as sitting breeds, and make good sitters and mothers. The Cochin Bantam is the breed most generally kept for utility purposes in this country. Hens of these breeds are sometimes used to hatch pheasant eggs. The Sebrights, the Mille Fleurs and Booted Bantams show a considerable amount of broodiness but are not so persistent in this respect as the breeds already mentioned. The Game Bantams, Rosecombs, Japanese and Polish are generally classed as non-sitters, although occasional individuals of these breeds will sit and hatch eggs.

Bantams, except for their small size, make desirable table birds. Bantam chickens are just as tender eating as any of the larger breeds.

In most of the Bantam breeds the colour and general characteristics are much the same as in the larger birds of the same name. There are, however, slight differences. Thus the head is usually rounder and larger in proportion to the body, the beak is rather short and more curved, the breast more prominent, the back shorter,

the wings are carried rather lower and the tail higher. Bantams have also a different action from ordinary fowls when walking, being somewhat jaunty in movement.

Barnevelder. Popular heavy-breed (sitting) of fowls, much used for egg production. Is a heavy layer of rich-brown, very large eggs.

Normal weights are : cock, 7-8 lb. ; cockerel, 6-7 lb.; hen, 6-7 lb.; pullet, 3-6 lb.

Barred Plymouth Rock. The most popular variety of the Plymouth Rock breed. See Plymouth Rocks for colour and characteristics.

Black Leghorn. A popular variety of Leghorn fowl. See Leghorn.

Black Orloff. Variety of Orloff. See Orloff.

Black Orpington. Variety of Orpington. See Orpington.

Black Wyandotte. Uncommon variety of Wyandotte, with black plumage showing beetle-green sheen. See Wyandotte.

Blue Leghorn. Variety of Leghorn,

with plumage of an even, medium shade of blue. See Leghorn.

Blue Orpington. Variety of Orpington with slate-blue plumage. See Orpington.

Blue Wyandotte. Variety of Wyandotte with, preferably, plumage of a medium, solid blue. See Wyandotte.

Brahma. Heavy (sitting) breed of fowl, formerly kept very extensively.

A peculiarity of the breed is that the legs are well-feathered down to the feet.

Normal weights are: cock, 10-12 lb.; hen, 7-9 lb.

Bresse. A very useful light (non-sitting) breed of fowl. Grows very quickly and is therefore popular with poultry keepers producing small table birds. Lays a good sized white egg. There are two varieties, the Black and the White.

Normal weights are : cock, 5½-6 lb. ; cockerel, 5-5½ lb. ; hen, 4½-5 lb. ; pullet, 4-4½ lb.

Brown Leghorn. Variety of Leghorn with orange and crimson-red plumage in the cock and golden-yellow and salmon-red plumage in the hen. See Leghorn.

Brown Sussex. Variety of Sussex, the plumage being mainly dark mahogany in the cock, clear to dark brown in the hen. See Sussex.

Buff Leghorn. Variety of Leghorn, with plumage of a uniform shade of buff. See Leghorn.

Buff Orpington. Variety of Orpington fowl with plumage of a clear buff colour. See Orpington.

Buff Rock. Variety of Plymouth Rock with clear, even buff plumage. See Plymouth Rock.

Campines. One-time popular light breed of fowl (non-sitting), laying a large white egg. Is not so prolific as the more widely kept breeds. There are two varieties, the Gold and the Silver.

The normal weight of the cock is 6 lb., and of the hen 5 lb.

Cochins. One-time popular heavy breed of fowls (sitting), laying a large brown egg. Is not so prolific as the more popular breeds. There are five varieties, the Black, the Buff, the Cuckoo, the Partridge and the White.

That of the Buff variety may be any shade of lemon, orange, gold, silver or cinnamon, but must be free from mottling ; variety should have dark blue-grey bars or pencilling on a blue-grey ground.

The characteristic of the Cochin breed is that the legs and feet are well-feathered, the feathering starting out well from the hock and continuing to the ends of the middle.

The normal weights are : 10-13 lb. for the cock and 9-11 lb. for the hen.

Coveney Whites. Seldom-seen light breed of fowls (non-sitting). Lays fairly well, the colour of the eggs being white.

The colour of the plumage is pure white.

Normal weights are 4½-5½ lb. for the cock and 3½-4½ lb. for the hen.

Creve-Cœurs. Seldom-kept heavy breed of fowls (sitting). Is a reasonable layer of large white eggs.

The plumage should be lustrous green-black all over but there may be a few white feathers in the crests of adults.

Normal weights are: cock, 9 lb.; hen, 7 lb.

Croad Langshans. One-time popular heavy breed of fowls (sitting), but not much kept now. Is a moderate layer of good-sized brown eggs.

Normal weights are 9 lb. for the cock and 7 lb. for the hen.

Dorking. Popular heavy-breed (sitting) laying a good-sized white egg. There are five varieties, the Cuckoo, Dark (or coloured), Red, Silver-grey, and White.

Normal weights are : cock, 10-14 lb. ; cockerel, 8-11 lb. ; hen, 8-10 lb. ; pullet, 7-8 lb.

Dumpie. See Scots Dumpie.

Faverolle. Heavy (sitting) breed of fowl, laying a brown egg. There are four varieties, the Blue, Buff, Salmon, and White.

Normal weights of all varieties are : cock, 8-10 lb. ; cockerel, 6½-9 lb: ; hen, 6½-8½ lb. ; pullet, 6-8 lb.

Frizzle. Quaint breed of fowls whose feathers stand out as though "frizzled". The breed has some utility merits, the hens being fair layers of good-sized white eggs.

The plumage is black, blue, buff or white, a pure even shade throughout in the " self-coloured " varieties ; Columbians as in Wyandottes (which see) ; Duckwing, Black-red, Brown-red, Cuckoo, Pile and Spangle as in Old English Game (which see).

Normal weights are : cock, 8 lb. ; cockerel, 7 lb. ; hen, 6 lb. ; pullet, 5 lb.

Game, Indian. Popular breed of fowls much used for crossing with other breeds for the production of table birds. Is a moderate layer of brown eggs of good size.

Game, Indian Jubilee. Seldom-seen breed of fowl with plumage resembling that of the Indian Game.

Game, Modern. Popular breed of fowl, of which there are six varieties—Birchen, Black-red, Brown-red, Golden Duckwing, Silver Duckwing and Pile.

Game, Old English. Breed of fowl formerly extensively kept and associated with cock-fighting. There are several varieties, Black-breasted Red, Dun, Pile, Dun, Duckwing and other types among them.

Hamburgh. Fairly popular breed of fowl laying a good-sized white egg. There are five varieties, the Black, Gold Pencilled, Silver Pencilled, Gold Spangled and Silver Spangled. All are most attractive as birds kept for pleasure.
Normal weights are : cock, about 5 lb. ; hen, about 4 lb.

Houdan. Useful breed of fowl laying a white-shelled egg.
Normal weights are : cock, about 7 lb. ; hen, about 6 lb.

Jersey Black Giant. Popular fowl laying a brown-shelled egg.
Normal weights are : cock, 13 lb. ; cockerel, 11 lb. ; hen, 10 lb. ; pullet, 8 lb.

Lakenfelder. Not very common breed of fowl laying a white-shelled egg.
Normal weights are : cock, 5-6 lb. ; hen, 4½ lb.

Langshan. One-time very popular breed of fowl laying a good brown-shelled egg. There are three varieties, the Black, Blue, and White.
Normal weights for all varieties : cock, 10 lb. ; cockerel, 8 lb. ; hen, 8 lb. ; pullet, 6 lb.

Leghorn. The most popular of all breeds of fowl for egg production. Is a most prolific layer of large white-shelled eggs. Records of 200 eggs a year are quite common and there are plenty of 250-egg records.
There are several varieties, the White, Black, Brown, Blue, Buff, Cuckoo, Duckwing, Exchequer, Mottled and Pile, the first three being the most popular and widely kept, the White outstandingly so. All are small for table purposes.
Normal weights are : cock, 6 lb. ; hen, 5 lb.

Macay. Somewhat uncommon breed of fowl laying a good brown egg and also possessing fair table qualities. There are several varieties but only the Black-red, Pile, Spangled and White are kept in any considerable numbers.
Normal weights are : cock, about 11 lb. ; hen, about 9 lb.

Malines. Breed of fowl laying a brown egg and possessing good table qualities. Of the many varieties only the Blue and the Cuckoo are kept on anything like a general scale.
Normal weights are : cock, 9 lb. ; hen, 7 lb.

Marsh Daisy. Breed of fowl laying a brown-shelled egg. There are five varieties, the Black, Brown, Buff, Wheaten and White.
Normal weights are : cock, 5½-6½ lb. ; hen, 4½-5½ lb.

Minorca. Popular breed of fowl laying large white-shelled eggs. There are two varieties, the Black and the White.
Normal weights are : cock, 6-8 lb. ; hen, 5-7 lb.

Norfolk Grey. Uncommon breed of fowl.
Normal weights are : cock, 7-8 lb. ; hen, 5-6 lb.

Orloff. Not very common breed of fowl laying a white-shelled egg. There are four varieties, the Black, Mahogany, Spangled, and White.
Normal weights are : cock, 8 lb. ; cockerel, 7 lb. ; hen, 6 lb. ; pullet, 5 lb.

Orpington. Formerly one of the most popular breeds of fowl, but now somewhat out of favour. Is a good layer and possesses fine table qualities. Is apt to go broody with annoying frequence. There are four varieties, the Black, Blue, Buff, and White.
Normal weights are : cock, 9 lb. ; hen, 7 lb.

Plymouth Rock. Popular breed of fowl laying a brown-shelled egg. There are four varieties, the Barred, Black, Buff, and White.
Normal weights are : cock, 10-12 lb. ; cockerel, 8-12 lb.

Poland or Polish. Seldom kept breed of fowl laying a white-shelled egg. There are six varieties, the Chamois or White-laced Buff, Gold, Silver, White, White-crested Black and White-crested Blue.
Normal weights are, cock, 6 lb. ; hen, 5 lb.

Redcap. Breed of fowl laying a white-shelled egg.
Normal weights are : cock, 6-6½ lb. ; cockerel, 5-6 lb. ; hen, 4½-5 lb. ; pullet, 4-4½ lb.

Rhode Island Red. A highly popular and widely-kept double-purpose breed of fowl laying a brown-shelled egg.
Normal weights are : cock, 8½ lb. ; cockerel, 7½ lb. ; hen, 6½ lb. ; pullet, 5 lb.

Scots Dumpie. Breed of fowl laying a brown-shelled egg.
Normal weights are : cock, 7 lb. ; hen, 6 lb.

Scots Grey. Breed of fowl laying a white-shelled egg.
Normal weights are : cock, 7 lb. ; hen, 5 lb.

Sicilian Buttercup. Breed of fowl laying a white-shelled egg. There are five varieties, the Brown, Golden, Golden Duckwing, Silver, and White.
Normal weights are : cock, 6½ lb. ; cockerel, 5½ lb. ; hen, 5½ lb. ; pullet, 4 lb.

Sicilian Flower-bird. Uncommon breed of fowl. There are two varieties, the Mahogany and the Spangled.
Normal weights are : cock, 4-4½ lb. ; hen, 3½-4 lb.

Silkie. Curious type of fowl with long plumage resembling silky hair. Lays a brown-shelled egg. There are three varieties, the Black, Blue, and White.
Normal weights are : cock, 3 lb. ; hen, 2 lb.

Spanish. Breed of fowl laying a white-shelled egg. Plumage is black with a beetle-green sheen.

Normal weights are : cock, 7 lb. ; hen, 6 lb.

Sultan. Uncommon breed of fowl with snow white plumage, and the face almost invisible, being covered by the whiskers.

Sumatra Game. Uncommon breed of fowl, laying a brown egg. The plumage is glossy bottle-green or green-black.

Normal weights are : cock, 5-6 lb. ; hen, 4-5 lb.

Sussex. One of the most popular double-purpose breeds of fowls. Is an excellent table bird and also a fine layer of large brown eggs. There are six varieties, the Brown, Buff, Light, Red, Speckled, and White.

Normal weights are : cock, 9 lb. ; hen, 7 lb.

Welsummer. One of the " new " breeds of fowl becoming very popular.

Normal weights are : cock, 7 lb. ; hen, 6 lb.

White Leghorn. See Leghorn.

Wyandotte. This breed of fowl, particularly the White variety, is one of the most useful of all utility poultry breeds. It possesses good table qualities and is a magnificent layer, second only to the White Leghorn. Though a heavy-breed, it does not give much trouble so far as broodiness is concerned. The egg laid is of fine size and a good shade of brown.

There are numerous varieties, the Barred, Black, Blue, Blue Laced, Buff, Buff Laced, Columbian, Gold Laced, Partridge, Silver Laced, Silver Pencilled and White.

Normal weights are : cock, 8½ lb. ; cockerel, 7 lb. ; hen, 7 lb. ; pullet, 5½ lb.

Yokohama. An uncommon breed of fowl, laying a brown egg. Carries a long sweeping tail, in the cock sometimes reaching a length of 4 or 5 feet. The birds are particularly graceful in carriage and are therefore esteemed as a fancy breed.

FRACTURES. See Accidents.

FREE-RANGE. Term applied to birds that may range over a field or meadow rather than remain confined to an enclosed run.

Free-range birds always keep remarkably healthy because they spend so much time in the fresh air, have unlimited fresh greenfood, are on untainted ground and have access to a quantity of natural food —insects, etc.

In general, free range is uneconomical of space for the commercial poultry farmer ; far less land is required when all birds are confined to runs. Free range, however, is practicable and economical

C

where poultry keeping forms one side of the business on large farms and small-holdings.

———

GALL-BLADDER. A small dark green sac attached to the liver of a fowl. Care must be taken not to break this sac before removal, when trussing and dressing a bird for table, or the flesh will be tainted.

GANDER. A male goose. How to determine the sex of young geese is dealt with under Sexing.

GEESE. Provided there is suitable land over which geese may be allowed to roam, it is most unlikely that the breeder will lose money on them.

The profitable life of a breeding goose is five or six times as long as that of the average fowl, and three or four times that of an average duck. Replacements are therefore very small.

Unlike ducks or hens, geese are able to live without concentrated food. For the greater part of the year a goose will manage excellently upon grass, weeds, and unwanted garden refuse.

No special housing accommodation is necessary. A rough three-sided structure, without perches and nest-boxes, affords ample protection. A shed constructed of sod walls and a wooden roof serves as well as any other form of house.

Farmers and smallholders in particular should always keep a few breeding geese growing on for the Michaelmas or Christmas markets.

Ailments of Goslings and Geese. Geese at all ages are particularly healthy birds and very seldom ill. See Ailments.

Breeds of Geese. There are only eight breeds of geese kept on anything like an extensive scale. These are :

African. Good table bird and excellent layer—up to 50 or 60 eggs per year. Much used for crossing with Embdens and other breeds. Colour, shades of brown with whitish underparts. Also greyish and white varieties. Legs and feet, orange. Average weight about 20 lb.

Canadian. Ornamental breed, good for table. Colour, mainly dark brown. Legs and feet, black. Average weight about 14 lb.

Chinese. Very similar in all ways to the African—to which it is closely related —but slightly smaller.

Embden. Most generally useful breed —good layer, good for table, easy to hatch and rear. Colour, pure glossy white. Legs and feet, bright orange. Average weight 18 lb.

English. The White English Goose resembles the Embden but is lighter—

weighing about 14 lb. The Grey English Goose resembles the Toulouse, but is also lighter, weighing at most about 16 lb.

Roman. A smallish but very popular table breed ; also a good layer. Colour, pure white. Legs and feet, orange pink. Average weight about 13 lb.

Sebastopol. Mainly an ornamental breed, but good for table. Colour, snow white. Very light, weighing only 10 lb. on the average.

Toulouse. Shares the lead in popularity with the Embden. Excellent layer—up to 60 eggs per year—and good for table. Colour, shades of grey and white. Legs and feet orange. Average weight up to 30 lb.

Fattening Geese. See Fattening.

Feeding Geese. Geese derive most of their food by grazing and, given a good meadow or common to roam over, will need little supplementary food.

Goslings from 1 day to 8 weeks old. Wet mash of middlings, 5 parts ; barley meal, 2 parts ; Sussex ground oats, 1 part ; bran, 1 part ; linseed meal, ½ part ; meat-and-bone meal, ¼ part. Feed 6 times a day for first week, and 4 times until 8 weeks old.

Stock Goslings from 8 weeks onwards. Wet mash of middlings, 4 parts ; maize meal or barley meal, 3 parts ; bran, 1 part ; meat-and-bone meal, ½ part. Provide above first thing in morning ; in evening provide soaked grain.

Table Goslings from 8 weeks until fattening commences. Wet mash of middlings, 4 parts ; barley meal, 3 parts ; cooked " pig " potatoes, 4 parts ; bran, 1 part ; meat-and-bone meal, ½ part. Provide mash in morning and soaked grain as above.

Laying Geese. Wet mash of middlings, 4 parts ; Sussex ground oats, 2 parts ; maize meal, 1 part ; bran, 1 part ; meat-and-bone meal, 1 part. Give this mash first thing in the morning ; give soaked grain at night.

Breeding Geese. Feed as for laying geese.

Quantities to Feed. Young geese should be allowed to have as much food at each meal as they can clear up without paying frequent visits to the troughs.

Adult geese should be allowed to have as much mash as they can eat in 15 minutes ; they should have as much grain as they can eat.

Grazing. Where mature geese have free range and access to plenty of good grass, they will need much less food than is suggested above—commonly no more than half feeds. At times of flush herbage, as between April and September, one feed a day may suffice—say a little grain at night. Full use of stubbles should be made after harvest.

Grit, Shell and Water. These must always be provided in generous quantities and in positions where the birds can readily find them.

Hatching Goslings. See Hatching.

Housing Geese. Goslings that are destined for the table at an early age should be confined to a run and housed warmly at night. Other geese can be out in the open all day all the year round, merely having a rough shanty of boards, or a framework thatched with straw or bracken in which to sleep at nights.

Mating Geese. See Breeding.

Marketing Geese. See Marketing.

Sexing Goslings. See Sexing.

GIZZARD. That part of the anatomy of poultry into which the food passes, to be ground up by the grit eaten by the birds and prepared ready for assimilation and digestion.

GRADING. Grading is an essential part of the business of every poultry-keeper who desires to make headway. The term grading means the sorting of eggs according to weight, colour, freshness, etc., and the sorting of table poultry according to quality.

Under the National Mark Scheme, grading is imperative ; produce that has not been graded must not be sold under the National Mark. See National Mark.

Eggs. With hen eggs the four grades specified under the National Mark scheme (and which *all* poultry-keepers should adopt irrespective of whether they market under the National Mark or not) are as follows : Special (weighing 2½ oz.) ; Standard (weighing 2 oz.) ; Medium (weighing 1¾ oz.) ; Pullet (weighing 1½ oz.).

There are also four grades of duck eggs, viz. : Special (weighing 2¾ oz.) ; Standard (weighing 2½ oz.) ; Medium (weighing 2¼ oz.) ; Small (weighing 2 oz.).

Under war-time regulations poultry-farmers must sell their eggs to packing stations at controlled prices ; but if they so desire, domestic poultry-keepers (1 to 25 laying birds) may sell direct to friends and neighbours at a fixed price—but such eggs must not be re-sold.

Table Poultry. Just as with eggs it is also imperative to grade Table Poultry under the National Mark Scheme, and all poultry-keepers are advised to follow the prescribed grades whether they market under the National Mark or not. (Under war conditions the price of table poultry is controlled.)

GUINEA FOWLS. These are quaint-looking birds occasionally kept for table purposes, plump Guinea fowls commanding a good price, though they are not in very great demand. In appearance they have only a slight resemblance to ordinary fowls. Their plumage has a slate-coloured ground evenly covered with round white spots. The head somewhat resembles that of a turkey, having red wattles and a bristly mane extending from the back of the head down the neck. The neck is bare of feathers where it joins the head and at this point has the appearance of white kid.

Rearing. The chicks may be slow to get onto their legs for the first few days, but once they are strong they move rapidly. They should, however, be kept off damp ground and out of the rain, dampness being fatal to them.

Their crops are small. They want at least six meals a day until they are a month old, four meals until they are six weeks old and three meals until they are three months old.

They may be fed in the same way as ordinary chicks, except that the animal food—fish meal, meat and bone meal or a milk product—should be increased from one-tenth to one-fifth of the total mash.

It is the nature of the Guinea chick to wander far afield. This is quite all right when the chicks are being reared by a Guinea hen, but is not satisfactory with hen-reared birds. For the first two or three weeks, according to the weather, a small wire-netted run should be provided in front of the coop. Then all the coops may go into large grass runs for another fortnight. After that the birds and the hens can be given full liberty.

Feeding Guinea Fowls. Adult Guinea fowls *must* have wide range; on this they pick up much of their own living. They do, however, need supplementary feeding, the best system being to give them mash in the morning and grain half-an-hour before sunset. A suitable mash is : middlings, 5 parts by weight ; maize meal, 2 parts; Sussex ground oats, 2 parts; bran, 1 part ; meat and bone meal, 1 part. For grain give half wheat and half kibbled maize.

War-time feeding must be based on a proposition of housescraps and such unrationed foods as may be available, plus the small ration officially allowed.

Preparing Guinea Fowls for Table. The best time to kill Guinea fowls is when they are about 3 lb. in weight. The demand is greatest from February—April.

The flavour of the Guinea fowl is its main asset. It is somewhat similar to game, and it is for this reason that the birds are esteemed.

There is no need to fatten the birds by the methods applicable to fowls. In fact, if they are fed exclusively upon a fattening diet, the amount of flesh may be increased, but the unique flavour is destroyed.

All that is necessary is, a fortnight before killing, to give the morning mash with a slightly heavier hand than usual and allow the birds to eat as much maize—in its kibbled state—as they like before going to roost.

HACKLES. The narrow feathers on the neck of a fowl. The feathers on the saddle, or tail end of a bird—pointed at the end in a male, rounded in a female—are known as Tail Hackles.

HANDLING. This term means the examination of fowls, etc., by taking them in the hands for the purpose of noting various utility points and thereby forming an opinion of their capabilities.

All birds should be handled frequently as chicks, as growers, as year-old hens and as 2-year-olds, for the purpose of weeding-out all the poorer specimens and retaining only the very best.

For the weeding-out of wasters, see Culling.

With regard to the formation of an opinion of a pullet's qualities, the best time to make the examination with spring-hatched birds is in early October.

Catch the birds quietly and, for the examination, hold them beneath your arm, tail-end to the front, the breast-bone resting on the palm of the hand, the index and second fingers between the legs, so that the legs can be held firmly if there is any struggling. In this way a bird's body, bones, pelvics, legs, plumage, etc., can be comfortably examined. To look at the head, still hold the bird beneath the arm but turn her front forward.

Bone-framework should first be examined. It should be fine and thin, meaning that the pelvic bones will be thin and flat. the legs should be wide apart, fine scaled and flat, the beak short and the head narrow.

The body must show space for efficient organs. Thus, the back should be wide at the shoulders and long enough to allow space for large egg-making and egg-shelling organs.

The distance between the pelvics and breast-bone should be quite large, say four fingers width.

Feather quality offers helpful guidance as to future possibilities. The feathers on the good pullet will be short and silky and by no means profuse, not even round the abdomen and thighs. One will not find the tight plumage of the bird already in lay. As yet, it will be loose, but after fourteen days of laying it will tighten up remarkably. Quantity is a better test before laying has commenced than tightness and you must cull the bird which seems to be " all feather ".

The flesh should be fine and cover the breast-bone and pelvics quite perceptibly.

The breast-bone itself should not be too long, or it may limit the capacity of the abdomen; neither should it be too short because of the danger of a break-down behind, owing to the internals not having sufficient support. The best advice, therefore, is to choose pullets with medium length of breast-bones.

The head is the next important test-piece. It should be narrow, carrying a medium size, but neat, comb and wattles which hang close together. The face should be flat, with bright, bold, round eyes set high up and protruding on either side. The nostrils should be large and dry and the fowl should breathe without opening its beak.

The wattles should be thin, shiny and of fine quality, must feel waxy, but in non-laying bird should not be fully expanded.

Ducks. Just as the qualities of a utility fowl can be gauged by handling so also can those of a laying duck.

The bird is held so that she faces the back of the operator, the breast resting along the palm of the left hand and the thighs lying in between the fingers, the body resting in the " crutch " made between the arm and the side of the operator. While the front, head, etc., is being examined, the bird is reversed.

The bones of the pelvics should be at least three fingers apart, and quite thin and straight.

In a good bird the abdomen will be big, deep, wide, and pliable. In a bird in lay it will be full, and well expanded.

Next the legs : one should be able to pass a hand between and barely touch them. The legs themselves should be fine-scaled, flat and, if laying has been prolonged and heavy, pale in tint. The bones should be neither too coarse nor too fine. They should be flat, not spindly, round or thick.

The back should be flat, not hollow or rounded, the tail quite " square " to the body and not leaning to either side. The back must be long, as measured from the base of the neck to the front of the tail. Width of back is also necessary. The body itself should be heart-shaped, the top of the " heart " being the back of the bird, the idea being that the bird must *not* have flat sides.

As with fowls, the plumage of a good duck is always tightly carried. It is silky, small, fine in texture and lies quite close to the body. The feathering round the abdomen is bunchy and held well together like a woolly ball. The poor layer has harsh and sticky feathering.

The feathers on the back of a good bird are tight and cannot easily be ruffled up the wrong way—if they are held up

they return quickly to their place. Excessive feathering suggests a poor layer.

The neck, covered tightly with feathering, should be snaky and thin, and fit very gradually to the body. The head, together with the beak, should form something like a wedge. The bill must not be too long or thick. The top of the skull should be flat, not rounded, and the eyes should bulge out from the head when viewed from the front. The tiny feathering round the face should be extremely tight, small and not in heavy profusion. The eye should be large, round, bright, and set high in the skull ; and the nostrils should be large and clean.

HANGERS. Term applied to the feathers springing from the root of the cock's tail.

HATCHING. Care in hatching can make a very great difference in the number of chicks produced from a sitting or hatch of eggs. A careless hatcher will be lucky if he secures one chick from every 2 eggs incubated ; a careful hatcher may reckon on an 80 to 90 per cent hatch with an incubator and more than that with a broody hen—provided, of course, the eggs used are strongly fertile.

Hatching with Broody Hens. In spite of the extensive use of incubators, broody hens are still largely employed for hatching—chiefly on account of their reliability, and the little trouble they involve.

Any box that has a floor space of 14 ins. by 12 in. and a height of 12 in. will do well to sit a broody. Two-compartment onion boxes, grape barrels, candle boxes and tea-chests come in very useful. Three-compartment orange boxes are too small for big hens.

The boxes are best used in a warm sheltered shed ; it does matter if it is open-fronted so long as it backs the prevailing winds. If, for any reason, they *must* be placed outdoors, they must be covered with thick roofing felt and have all the various cracks, etc., covered. They should also be in a sheltered place to which other poultry, cats, dogs and live-stock have no access, and that is not near a busy, noisy road.

Whatever box is used, a detachable light-proof cover must be provided for the front. This may consist merely of a sack weighted down with bricks but is better if of wood.

Put into the box three spadefuls of loose, damp earth, shaping out the centre in the form of a saucer about 1 in. deep and 8 to 10 in. in diameter. The thickness of the earth part of the nest at the centre should be about 1 in.

Upon this shaped earth place a good wisp of well crushed straw or hay. It must be flattened down and given a sprinkling of disinfectant powder. Fourteen pot eggs (or old addled eggs) are put in the nest, and then all is ready for the broody.

Always place a broody on her nest after nightfall. When she is installed close the shutter and leave her undisturbed until the following morning. If some of the pot eggs remain uncovered you will know that she has too many and that the number of eggs proper will have to be reduced. If she is sitting awkwardly and uncomfortably put her in a better position.

Give her her sitting two days after she seems well settled. Take out the pot eggs one by one and as they are removed put in the proper eggs. Do this after nightfall.

A good feed should be given once a day, the ration consisting of whole yellow maize and wheat, half and half, clean water and some small. Every other day give a little crumbly laying mash as well. Supply the food in small tins about 3 in. in diameter.

Give the feed about ¾ hour before sunset. Take the bird off the nest, carefully lifting the wings up to see that no eggs are caught up with them, cover the nest and place the bird on the ground in front of her grain. Allow 10 to 15 minutes for feeding. During this time the bird should relieve herself. If she shows no natural inclination to do so, throw her some 3 feet up in the air and let her flap down.

When she wants to return to the nest allow her to walk gently on to the eggs. Examine all sitters 10 minutes after the feeding interval to see that they are sitting comfortably and all eggs are well covered.

On the 19th day, feed the hen on the nest and disturb as little as possible until all the chicks are out.

Hatching with Incubators. An essential part of the process of hatching by incubator is to have the machines scrupulously clean and properly housed.

A disused or little-used room in the dwelling house will do, or the cellar, or any shed or outbuilding so long as it is capable of maintaining a temperature of between 60 and 65 degrees, is free from draughts and quite weatherproof.

All machines must stand at least 8 in. from the wall, away from draughts between window and window, or door and window, and out of the way of the sun's direct rays at any time of the day.

Wash your hands—especially if they are tainted with paraffin from the lamp—before handling the eggs. Take the egg drawer right out for filling. Mark each egg with an X on one side and an O on the other before placing it gently in the drawer.

Start laying the eggs, X side up, in the centre of the tray but when three-quarters of the tray is full, the eggs should be placed in rows. This makes for easy turning. The drawer should not be so full that the eggs are pressed hard one against another or some may crack during the first heating. The correct filling is such that every egg can be turned easily and without jarring its neighbours.

Put the filled egg drawer back into the machine, sliding it in carefully, for the least jar may do damage.

It is a mistake to shut the door immediately the cold eggs are introduced; they should be warmed up gradually so leave the door open for the first two hours; then shut it and *leave it shut for the first two days.*

The best results are usually obtained if you maintain a heat of 103 degs. F. in hot air machines and 104 degs. F. in hot water machines, but there are circumstances which alter cases. If the air temperature of the incubator room is only 40 degs., or if your incubator is an old one and has a thin case, or if the weather you are experiencing has a wide range between day and night temperatures—then, in all cases, run the machine at one degree higher temperature.

After the first two days the eggs have to be turned three times daily, to prevent the embryo sticking to the shell and to ensure proper heating of the lower part of the egg. Take out the egg drawer, stand it on a firm table and, wetting the fingers on a damp sponge, " rub " the eggs over so that the X side comes on top and the O side goes to the bottom or *vice versa* if the X side is uppermost.

The turning must be done quickly and very gently, and the drawer returned to the machine at once. The cooling of eggs has been proved to be a fallacy, and they should not be left outside the machine to air, as is so commonly done. With self-turning machines, the eggs can often be turned without taking them out of the incubator.

All incubating eggs should be tested on the 7th and 14th days so that faulty ones may be removed and also to ascertain whether the interior atmosphere of the incubator is sufficiently moist or not. See Testing.

If a number of eggs in a hatch are " duds," the contents of, say, two incubators can be amalgamated and the freed incubator refilled with a fresh supply.

Whether moisture is sufficient or not can be ascertained by examining the eggs in front of a testing lamp and ascertaining the size of the air-space at the top of the egg. More simply, an instrument called an hygrometer can be installed in the incubator.

If the hygrometer registers 60 per cent for the first week and afterwards 55 per cent things are going well.

If moisture is deficient it may easily be added to.

One method is to increase moisture in the atmosphere of the incubator room, by standing pans of sand about the room beneath the machines, and keeping these continuously wet, or by spraying the walls of the room occasionally or by hanging up sacks with their ends dipping in trays of water.

If preferred moisture may be added to the interior of the incubator in one or other of the following ways : (1) By keeping a wet sponge in a saucer placed in the centre of the egg tray ; (2) by keeping a pad of wet cotton wool in each corner of the egg tray ; (3) by laying a flannel wrung out in warm water over the eggs for one minute at a time only ; (4) by fitting a wick with one end resting in water at the side of the heater.

The properly-made incubator can never provide too much moisture. The hygrometer or the air-space test (see Hatching Eggs) will only show excess moisture when artificial means of adding moisture have been adopted and then the remedy is obvious.

On the eve of the 18th day open the door, pull the drawer half out so that it rests in position safely, and spray the eggs with warm water so that each is given an even covering of what looks like "dew". Do this quickly, put back the drawer gently and close up at once.

Do not turn the eggs any more after this, and do not open the incubator door any more until at least three-quarters of the chicks are out, which will probably not be until the middle of the 21st day. If the machine has no glass front so that the progress of hatching cannot be inspected, do not open the machine until the middle of the 21st day.

When the chicks are fully dry and fluffed up, they can be moved safely to the brooder.

Winter Hatching. Incubators will bring off just as successful hatches from November to January as from February to April provided the special requirements for cold-weather hatching are studied.

These requirements are : (1) The provision of extra moisture; (2) the adequate heating of the incubator room and proper manipulation of the heating apparatus.

The minimum temperature of the incubator room must be 50 degrees. An oil stove can be used ; or the walls covered with plywood and packed with straw, chaff, peat-moss, shavings or sawdust; lining may be done with sacking similarly packed ; or the room may be partitioned with a double thickness hessian or blanket screen so limiting the space in which the machines are running.

As to manipulation of the heating apparatus, when you expect the night to be

extra cold, the lamp-wick should be turned just a trifle higher to counteract an undue rise of the damper.

Instead of adjusting the damper to swing ¼ in. clear of the chimney top as in warmer weather, it must now swing ½ in. clear. When the weather is settled cold, run the machine at 104 degs. instead of the usual 103 degs. F.

Instead of taking off one felt on the 7th day and the second and last felt on the 14th day, take off the first at the 10th and the last at the 18th.

Hatching Ducklings. No one who wishes to do so need hesitate to hatch ducklings. Their incubation involves no special difficulties, and the young birds are particularly easy to rear. They may be hatched by incubator, by broody hen or by broody duck.

The first essential is to obtain first-grade eggs, clean (best unwashed), firmly shelled, carefully stored, from unrelated, vigorous, trapnested parents, and under 8 days old if hatching with a machine, under 10 days if hatching with a hen or duck. The sittings should be from dams with 220-250 egg records and sires with 270 egg records.

Hatching by broody hen. Treat the broody hen as if she were hatching hen eggs (see Hatching), but with the following differences : Make the nest of earth as usual, but after shaping out, sprinkle with a watering can (using the rose). Then put in the straw and sprinkle with disinfectant powder. An average-sized broody can only cover 10-11 eggs.

Duck eggs take 28 days to hatch, but may chip on the 25th or 26th day.

After the first week, sprinkle the eggs with warm water every other day ; the last week, every day. Thick-shelled eggs, such as those of the Aylesbury, require rather more moisture even than that.

Hatching by broody ducks. Ducks like to make their own nests, but if they do so in too open a spot, it should be screened. They will line the nest with down and little extra attention will be necessary. A duck of the laying breed can cover 9-11 eggs, one of the general purposes breeds 12-13. The management is as with broody hens.

Hatching by Incubator. Any ordinary incubator will serve for hatching duck eggs, though it may be necessary to raise the thermometer slightly so that it is clear of the eggs. The room in which the incubator stands should be kept uniformly warm.

Duck eggs being larger than hen eggs, remember that a 30-egg machine will only take 22 eggs, a 50-machine 39-40, a 60, 50 eggs, and a 100, 85-87 eggs.

Pack the eggs in tight rows so that they can be turned quickly by running a damped finger down the rows.

If the thermometer comes right on top of the eggs run the machine at a temperature of 102 degrees ; if the thermometer is $\frac{1}{4}$ in. above the eggs the hatching temperature should be 103 degrees.

Let the eggs warm up gradually by leaving the incubator door open for 2 hours ; then close it and leave closed for 48 hours. After this, turning three times a day can commence—turning quickly but carefully. No cooling is necessary ; the eggs are sufficiently cooled during the turning.

The supply of moisture must be greater than for hen's eggs, and here are some ways of supplying it : (1) placing a sponge in a saucer of warm water and putting this in the centre of the eggs, (2) filling the troughs attached to each side wall of the machine, (3) keeping a 3-in. false floor of wet sand beneath the machine.

In addition to moisture in the machine, duck eggs need " contact " moisture, and the easiest way to give this is to spray or sprinkle the eggs with warm water (110 degrees F.) once a day for the first fortnight and twice a day for the remaining fortnight.

One felt should be taken out of the machine after 14 days, and the other after 3 weeks. The eggs should begin to chip on the 26th day. Ducklings are extremely slow breaking through, but by the end of the 28th day the hatch should be completed.

As addled duck eggs give off an odour harmful to nearby eggs, testing should be done on the 7th day to take out infertiles and on the 16th day to clear away " addles."

HATCHING GEESE. Goose eggs may be hatched by hens, by the geese themselves or by incubator. The incubation period is from 28 to 30 days.

In using a hen there is a risk that the eggs, being so large, will raise the hen so far off the ground that she cannot keep them properly warm. However, a large hen can often manage to incubate 4 goose eggs.

The hatching procedure is as for ducks.

When using a goose it is always advisable to allow a goose to sit when she has laid two or three dozen eggs. She will cover from 8 to 10 eggs, according to her size.

Let her choose her own nesting place, which will probably be in some hedge, but place over the nest a large packing case, with one side removed, as protection from the weather.

The sitting birds will come off their nests of their own accord and this will generally be at a time when they know that food is waiting for them. If you see, however, that some of the sitters appear to be missing their meals, a good plan is to throw down a little hard grain every afternoon a few yards away from the nest, but in front, so that the bird will see what you are doing. Some birds do not like to wander too far from their nests.

Sprinkle the eggs daily with warm water. In the last week every day dip each one bodily in water heated to 100 degrees.

An ordinary hot-water incubator is easily adapted for goose eggs by removing the wood laths beneath the egg drawer, the eggs then being in the proper position in regard to the thermometer. Most hot-air incubators cannot take goose eggs because they foul the capsule and stirrup. One or two makers fit, if required, a special short stirrup for these large eggs at a small cost.

To conserve heat, it is advisable to place a piece of hessian (not too thick and yet not with an open mesh) on the egg drawer wire bottom and stand the eggs on that.

It is necessary to run the machine at 102 degrees the whole time.

Turn the eggs—exactly half-way over —as speedily as possible, as often as possible, as early in the morning as possible, and as late at night as possible. Cooling is not necessary. Every two days move the outside eggs into the centre. For the rest, incubation procedure is as for hen eggs.

The eggs should be sprinkled with warm water once a day for the first 3 weeks, then twice a day.

HATCHING TURKEYS. It is not advised to hatch turkey eggs with hens, only by turkey hens or incubators. Hatching takes 28-30 days.

Follow the same procedure as described for geese. A turkey can cover from 10-14 eggs, according to her size.

In incubator hatching adjust the incubator as for goose eggs. The eggs must have even heating, and to ensure this lay a piece of very coarse hessian over the wire of the egg tray.

The bulb of the thermometer should be hung so that it is $\frac{1}{4}$ in. from the top of the eggs, and for the first 4 hours of heating the door of the egg compartment should be left a quarter open to ensure gradual heating.

103 deg. F. is the best temperature.

Turn the eggs 3 to 5 times a day quickly, without cooling.

After the first fortnight sprinkle the eggs on alternate days with warm water— during the last week, every day. Take off one felt after the first fortnight and leave the other on until the 25th day.

HATCHING, DATES FOR. For the production of ordinary laying stock (that is for pullets to come into lay in October or November) heavy-breed fowls should be hatched in March, light-breed fowls in April. For pullets to come into lay in August (when adult birds will be going into moult and there will consequently be a shortage of eggs) hatch heavy-breeds in January, light-breeds in February. For birds to be exhibited in the pullet classes of the current year, hatch in early January. For stock cockerels, hatch in January. For table birds hatch all the year round, but particularly in December—January (for the production of spring chicken).

Hatch turkeys for the Christmas markets in April.

Hatch geese as soon as sufficient eggs can be collected, which will be from February—March onwards.

Hatch laying ducks in March—April ; table ducks all the year round.

HATCHING, SITTINGS FOR.

Suitable Eggs for Sittings. All eggs to be incubated must be : (1) Between 2 and $2\frac{1}{4}$ oz. in weight, (2) firmly, evenly and smoothly shelled, (3) free from cracks, and (4) perfectly clean.

Sending Sittings through the Post. The best method of packing is in the type of box in which each egg is held in a cushion of felt or corrugated cardboard. These boxes are expensive, and it is customary to stipulate that they be returned, carriage paid, as soon as they are empty.

The boxes are made of cardboard and are quite efficient in use. In the bottom of each compartment put about $\frac{1}{4}$ in. of chaffed straw, and fill up with more above the eggs.

For dispatch in quantity by rail or lorry, wooden egg boxes, each holding several dozen in separate pockets, are very efficient.

Storing Sittings. A storage room should be a room with a temperature which keeps steady at about 50 degrees. Fluctuating temperatures injure the " germ " in the eggs. A drop below 32 degrees will cause the germs to freeze ; a rise above 72 will start incubation.

The eggs should be placed in trays containing a layer of peat-moss or sweet chaff about 1 in. deep.

HATCHING, TESTING EGGS FOR.

Eggs are tested for three purposes: (1) to ascertain whether they are new-laid and fit for eating ; (2) to check hatching progress and ascertain whether or not a hatching egg is addled or incapable in some other way of producing a chick ; (3) to ascertain whether or not hatching eggs are receiving the right quantity of moisture.

It must be regarded as of the utmost importance to test eggs during the hatching period, whether they are being hatched by incubator or broody hen.

Knowing, as a result of the test, which of the eggs—if any—are unhatchable, one can avoid loss of time, eggs and money.

The eggs are tested by means of a testing lamp or other device (see Eggs, Candling) which renders the interior clearly visible.

The usual testing times are on the 7th day of incubation and again on the 14th day. It may be useful on the 18th day when all eggs except those containing a live chick are taken out. Experts are able to learn something on the 4th day.

If it is desired to test once only during the 21 days of incubation, then this should be the 10th day.

What to Look for When Testing. The chart opposite explains the meaning of all the symptoms that can be noted in an egg when viewed through the testing lamp.

Testing for Moisture. The atmosphere surrounding hatching eggs must contain the correct amount of moisture (see Hatching) ; otherwise hatching failures will occur. With the aid of a testing lamp it is possible to ascertain whether moisture is correct, excessive or insufficient. The size of the air space in the egg is the guide.

Tested on the 7th day, the air space *should* extend one-seventh the way down the egg ; in other words it should occupy one-seventh of the egg. Tested on the 14th day the air space *should* reach nearly a quarter the way down the egg. Tested on the 18th day one-third of the egg *should* be occupied by the air space.

A larger air space than that indicated denotes the need for additional moisture (see Hatching) ; a smaller air space denotes the need for less moisture.

HAY-BOX—for day-old chicks, see Chicks, Day-old (Fireless Brooder).

HAY-BOX—for keeping food warm. A box tightly packed with hay, in the centre of which a vessel of cooked food may be placed to complete the process of cooking and to keep hot until required.

HEN. In the common view a hen is a female fowl that has passed through her first laying season. It is now generally agreed that a hen is a female fowl above 2 years old. (See also Pullet, Yearling.)

HOCK. Term applied to the knee-joint of the leg. Sometimes referred to as the elbow.

HOPPER. Name of the appliance for supplying dry mash, grit and shell, to which the birds help themselves freely.

Hoppers consist of a container to hold a quantity of food, etc., with a trough attached, the contents of the container automatically keeping the trough full.

WHAT MAY BE SEEN WHEN TESTING HATCHING EGGS AND WHAT THE SIGNS MEAN

THE SIGNS	THE MEANINGS
ON THE 4TH DAY	
Evenly clear right through. Possibly yolk may be seen as a faintly dark moving object if the light is strong enough.	This is an *infertile* egg.
A small, faintly dark speck about the size of a small pea, which moves as the egg is turned.	This is a *fertile* egg and the speck is the tiny embryo which has just started to grow.
ON THE 7TH DAY	
Evenly clear right through, larger air space than normal.	Quite *infertile*, as no sign of growing germ.
Distinct pea-size dark speck at top end of egg from which radiate streaks (blood vessels) and which move when the egg is turned. Air space larger than normal but immovable and clearly defined, now being paler than the rest of the egg.	*A strongly fertile* egg. The more this speck moves (of its own accord) and the more pronounced it appears, the more likely is it to grow into a vigorous chick.
A small, confused movable dark mass in the egg, showing no blood vessels. The air space line moves when the egg is moved.	An *addled egg*—i.e. one in which the germ started to grow, but has since died.
A confused, movable dark mass floating at will about the egg. No air space is seen.	An *addled egg* with *broken air space* due to rough handling before incubation.
Small movable dark spot with no radiating blood vessels, but a blood vessel or line along the shell.	*Broken yolk,* due to rough handling, weak stock, or too rapid heating.
Small dark spot which does not move. May have few odd blood vessels.	*Stuck germ* which has become adhered to shell owing to insufficient turning.
ON THE 10TH DAY	
Egg quite clear, air space becoming larger.	*Infertile.*
A dark movable mass (which itself moves) and from which there are moving blood vessels in greater number than are seen before. Air space clear, well defined.	A *fertile* egg doing well.
Confused movable dark mass with few defined blood vessels and with the air space hazy.	*An addled egg.* When the germ died can be told by the size of the confused mass.
ON THE 14TH DAY	
Egg quite clear, air space still larger.	*Infertile.*
The growing chick inside the shell now appears *as a mass the size of a florin,* with many clearly defined blood vessels. Much self-movement is seen. Air space size $\frac{1}{3}$th–$\frac{1}{4}$th down the egg and clearly-defined separation.	A *fully-fertile* egg with strong chick inside.
Confused dark mass with few or no blood vessels filling a large part of the egg.	*An addled egg* with no chance of life inside it now.
ON THE 18TH DAY	
Egg quite clear, air space occupying $\frac{1}{4}$ to $\frac{1}{2}$ the shell.	*Infertile.*
One-third of the egg is clear and the rest filled *with a black mass,* movable in itself, but too compact to permit of moving when turned round. Often the beak of the chick can be seen as a dark triangular protuberance near the shell.	A *fine chick* almost ready to hatch.
Dark movable, confused mass filling most of the shell with hazy air space. The egg feels cold to the touch.	An *addled egg* in which the germ died about the 15th day.

C*

Hoppers in various designs can be purchased, or wooden ones can be homemade. They have a hinged lid for shutting down over the mash at night.

If greater capacity is required, increase the *length* of the hopper, not the height or width.

HORN COMB. A comb somewhat resembling horns. From the base, the comb branches into two spikes, in shape something like the letter V inverted. La Fléche is the typical horn-comb breed.

HOUSING POULTRY. Housing plays a very prominent part in the well-being of fowls. Correctly housed birds will not only thrive better but show a much more profitable return than when in the rough shanties that were formerly considered good enough for them.

For the Garden Poultry-keeper. A 12-fowl house is recommended. It should be at least 5 ft. high at the back. If an outside run will be used then it need only be 7 ft. long and 5 ft. deep, but if the birds are only to have the scratching floor available for exercise then 8 ft. by 5 ft. will give better results.

Intensive Poultry Houses. With these choose a pattern conforming to the following requirements :

The front to give the maximum of sunlight and yet keep out rain. The sashes to be fitted with one of the sunray-passing glasses. Not more than two-thirds of the front area to be glass (or it will make the house too cold in winter and too hot in summer).

If over 7 ft. to 8 ft. deep to be apex-roofed to permit of ridge ventilation. Four square feet of floor space to be allowed to each heavy-breed hen and 3½ square feet for each light-breed.

Semi-intensive Poultry Houses. These to conform broadly to the same specification as above. Even though they will have a run attached, still the floor must be amply lighted.

Slatted-floor Houses. Requirements to insist upon are : Easy closing of nest-boxes, well-made foundation, easy feeding and filling of hoppers (if fitted), efficient slats and easily cleaned dropping boards (if used). Economical houses of good design are 8 ft. by 7 ft. by 5 ft. high for 65 layers.

Fold Houses. These are largely to be judged by their portability and the quality of the timber used. Only timber known as " selected red deal " should be used, and it should be free from knots. Window frames should be mortised and tenoned for strength. Planed framing is advisable (although costing 5 to 10 per cent more). The walls should be of stout timber. For small houses of 25 sq. ft. floor area and under, ⅝ in. matching is ample for a cheap

job, but for large houses ¾ in. stuff is desirable.

Units measuring 16 ft. by 6 ft. overall, having a house part 6 ft. by 4 ft., accommodate, say, 30 layers.

Housing Ducks, Geese and Turkeys. These classes of poultry need only the simplest type of house. See under respective headings for details.

HOUSES, CLEANING. Scrupulous cleanliness is essential with fowls as with all other live stock ; the penalty of dirty conditions is disease, perpetual ill-health of the birds and loss of eggs.

In the ordinary way, so far as the house is concerned, the only daily job of cleaning is to scour the droppings boards in the roosts and scrape out and wash food and drink vessels. Preferably every week, and certainly not less than once a month, walls and roofs should be brushed down with a stiff brush, all windows cleaned (for it is all-important to admit as much light as possible) and the nest-boxes cleared of litter, cleaned out, disinfected and relittered.

In the event of an outbreak of infectious disease still more thorough cleansing and disinfection are essential. See Ailments and Disinfection.

Whenever fresh floor litter is put down (see Litter) the floor should be thoroughly scraped to remove *all* caked droppings, swept perfectly clean and then disinfected.

HOVER. An appliance for rearing artificially-hatched chicks, differing only from a brooder in slight details (see Brooder).

HYGROMETER. An appliance for use in an incubator for determining the moisture contents of the interior atmosphere (see Hatching).

IN-BREEDING. See Breeding.

INCUBATORS. The present-day incubator is a beautiful machine capable of hatching every hatchable egg entrusted to it. The well-known makes are entirely fool-proof, and if the directions supplied with them are adhered to they will continue to give complete satisfaction for many years.

There are two standard types of incubator, the hot-air type and the hot-water type. Which shall be chosen is entirely a matter of personal preference ; neither can be said to be more efficient, simpler or more reliable than the other.

Management of Incubators. See Hatching.

Cleaning and Preparing Incubators. Every incubator requires a thorough cleaning at the commencement of each hatching season, this being one of the secrets of successful hatching.

Take the machine out into the open and there divest it of its fitments such as egg and chick drawers, diaphragms, oil tank, burner, damper, etc.

Have a tub half full of hot water containing disinfectant, several scrubbing brushes and some carbolic soap. Brush each part while dry first to remove dust, etc., and push each part under the water, then scrub thoroughly and allow to dry really well.

Accessories made of material such as diaphragms, chick-cloths (if used) and felts should be taken and soaked in the disinfectant, dried well and then *shaken again* to loosen the fabric.

The burner should be boiled in washing soda solution and a new wick fitted. New micas should be fitted all round and the oil tank swilled out with clean paraffin before filling up to start.

New felts should be provided if the old ones are clogged up with chick fluff, and new canvas for the diaphragms if the old sags. The outside should also be scrubbed down, and given a coat of varnish to prevent undue warping.

The capsule should be tested in warm water (103 degs. F.), and if it expands readily it should be given a coat of dead black paint. The heater itself should be painted dead black, too.

The double panes of glass must be uncracked, and they should be taken out and cleaned inside.

With hot water machines the water run should be scrubbed hard to remove deposits and a new piece of hessian bent over the perforated zinc.

Disinfecting Incubators. If at any time there has been disease among chicks hatched by an incubator, that machine must be properly disinfected before it is used again. Disinfection is particularly necessary after an outbreak of B.W.D., Coccodiosis or Fowl Paralysis.

Fumigation is best and the following method is recommended. Into a saucer put 2 teaspoonfuls of permanganate of potash crystals and, placing this in the incubator, pour on 4 teaspoonfuls of formalin and shut up at once for 4 hours. Then air *freely*. This will efficiently kill all the germs in a machine of 250-egg size.

IN-KNEED. Term applied to a bird whose hocks are close together instead of being well apart. Sometimes such a bird is referred to as knock-kneed.

INSECT PESTS. Lice. These are flat insects and have six legs at the end of which are hooked claws. The eggs, commonly called "nits," are like whitish specks usually to be found fixed to the lower parts of the feathers.

Lousy birds scratch, preen constantly, "bathe" in any pile of dust they can find, and if left without treatment lose appetite and may even die. Growing birds are slowed down and the feathering is roughened. They mope and lose their normal weight.

If a fowl or chicken exhibits any of the symptoms described above, pick it up and move apart the feathers. First of all, part the neck feathers, for here may be found the *Neck Louse*, which keeps close to the skin. If you can see nothing, then work down to the wings, for along the quill feathers may be found the *Wing Louse*. Then go round to the abdomen, which is the commonest place for lice. Here the commonest lice are found—the *Large Body Louse* and the *Small Body Louse*.

These lice are quite small, from 1/30 to 1/12 inch long, and will be found running quickly and aimlessly over the skin.

With all these lice, the remedy is to dust the birds at regular intervals with insect powders.

Fleas. The female fleas lay their eggs in a warm sheltered place, such as in dirty nest-box litter or behind crevices of hoppers. The eggs hatch in a week and in a month they are mature fleas, which plague the fowls day and night.

It is not common for fleas actually to be found on the birds, because of their great activity in jumping off, but they may be found in places such as nest-boxes. If the eggs are seen to be speckled with tiny reddish spots, that is a sure sign of infestation by fleas.

The treatment consists of cleansing the nest litter, treating the whole house with creosote and dusting the hens with insect powder.

Mites. These are minute pests some of them invisible to the naked eye.

Of all the Mite family, the *Red Mite* is probably the most terrible, because of its great numbers. The mites live in perch sockets, or crevices near the perch and, at the fall of night, troop out, crowd on to the body of the fowl, suck the life-blood until gorged, and then go back to their retreat when day breaks. The normal colour is bright grey, but when gorged with blood they are blood-red.

Following an attack of red mite, all perch sockets should be thoroughly cleaned out and painted with perch paint (nicotine sulphate) or with creosote or paraffin, or a mixture of the two. Also all cracks and crevices about the house, especially those near the perches, should be liberally treated with paraffin, creosote or the mixture at frequent intervals as a preventive against further attack by these pests.

Another parasite that frequently gives trouble is the *Depluming Mite.* The symptoms are that parts of the bird's body become bare of feathers, most of them breaking off just under the skin. The usual part affected is the abdomen, but the trouble may spread to the back and thighs. The bare skin is smooth and reddish.

The best remedy is to rub the affected parts with oil of Carraway ointment.

The *Scaly Leg Mite* is the commonest of all the mites, next to the Red Mite. The legs of birds attacked by it are roughened, the scales are shed, and in a few instances the bird loses the use of its legs and the toes drop off. The mites pass readily from fowl to fowl, or from an adult to the chicks it is brooding. In bad cases the legs are encrusted with a black grey scab.

As soon as any roughening of the scales appears, hold the legs in a mixture of paraffin and linseed oil, half and half, for one-half to one minute. Wash the legs clean the next day. One application is generally enough.

Insecticides. Sodium fluoride is probably the best and Pyrethrum powder is good, too. Izal disinfectant powder is excellent, and so are all of the specially made insect powders, such as Keating's, Secto, etc.

Whatever is used should be applied either with a powder bellows or a sprinkler tin. You must be sure to work the powder well in beneath the bird's feathers with the hand.

The parts to pay special attention to when dusting fowls with insecticide are the underside of the wings, the body fluff and round the vent.

INTENSIVE SYSTEM. Term used to denote that method of poultry-keeping in which the birds spend all their time confined to their houses, with no grass run for exercise. It answers perfectly well and is freely practised, especially by town poultry-keepers. The one essential is that the birds be persuaded to take a normal amount of exercise, which is assured by littering the floor of the house with deep, loose litter and sprinkling all grain food amongst this so that the birds must scratch to find what they would eat.

IODINE. Iodine—the ordinary drug in almost daily use in every household—has a great influence on the well-being of poultry. When administered to breeding birds, it is stated to increase the hatchability of eggs and ensures for the chicks from these eggs, better stamina and greater disease-resisting powers. It enables growing birds to make steady, sturdy growth ; helps birds greatly during feathering stages; increases egg production; and gives adult birds greater resistance to disease.

The best plan is to include iodine in a mineral mixture, which all fowls should have. A good recipe for such a mixture is: Calcium carbonate, 4 parts by weight; calcium phosphate, 33 parts ; magnesium sulphate, 5 parts ; flowers of sulphur, 5 parts ; ferrous-sulphate, 2 parts ; sodium chloride, 10 parts ; potassium sulphate, 4 parts ; and potassium iodide, 1/10 part. The potassium iodide is the ingredient that supplies iodine.

Add 4 lb. of this mixture to every 100 lb. of mash, weighed in its dry state.

KEEL. Term applied to the ridge on the breast-bone.

KIBBLED. Term used to denote the cracking of corn (particularly maize) into a number of pieces, fine or coarse, in order to make it more easily digestible by chicks or adult birds.

KILLING POULTRY. Always starve all table birds for 24 hours, before killing. If a bird is killed soon after it has had a good feed, its intestines and crop will be full of food. This is bad, because the full crop spoils the appearance of the carcase, and the full intestines spoil flesh colour and prevent the bird " keeping ".

To Kill a Fowl. Hold the bird by the two feet in the left hand and gather up the tips of the wings in the same hand. Take hold of the head with the right hand, so that the bird's comb lies in the palm of the hand, and extend the neck as far as it will go. When it is straight out, the neck is pulled still further, the head bent back, and pulled again until it gives a jerk.

To ensure ample space for blood drainage, pull the head well away, so that there is a 2 in. space left between the severed ends. Keep the carcase hanging head downwards while plucking, to ensure as much as possible of the blood leaving the carcase, and thereby whitening it.

To Kill a Duck. With both ducklings and ducks follow the same method as above. Large ducks should be held well up in the left hand, so that the right hand has ample room to pull downwards. Make the action decisive, and use a little more strength than with fowls.

To Kill a Turkey. Professionally, turkeys are killed with blocks and tackle fitted in the roof or across a beam, but if you are not killing enough birds to make this worth while, the " ground bar " method is excellent. You get a piece of wood (say 2 in. by 1 in.), about 2½ ft. long, and cut out in the centre a shallow depression about 6 in. to 8 in. long. This bar is laid on the ground, and the turkey's neck put on the ground beneath the depression in the bar, with its comb upwards. Its body is held by the feet so that its breast faces you. Your own feet are placed one on each side of the bar which holds the turkey's head securely on the ground. A steady pull dislocates the neck. One can easily feel when the dislocation has taken place.

To Kill a Goose. The method of killing turkeys is applicable for killing geese.

LACED, or LACING. A stripe or edging of different colour around a feather.

LEADER. Term applied to the single spike of a rose comb. See Rose Comb.

LEAF COMB. A comb somewhat resembling a broad leaf, or a butterfly with outspread wings. Typical of the leaf combed breeds is the Houdan.

LEG BANDS. Bands of celluloid in different colours, or soft metal (these latter bearing different numbers) for affixing to the legs of fowls for the purpose of identifying individuals. All fowls whose egg yields are recorded are leg-banded so that each may be credited with the actual number of eggs it lays. Poultry-keepers also leg-band their birds so that they may know the year when they were hatched, to indicate their strain, to denote any bird that for some reason must be kept in mind (one, for instance, that remains valuable for laying but is unfit for breeding) and so on.

The leg-bands are often fitted to birds when they are in the chick stage. It must then be remembered that a chick grows quickly and that any ring as originally fitted will be too tight after a while. A too-tight ring causes a trouble known as Ring-sloughing, which see.

For methods of fitting rings see Marking Poultry.

LEG FEATHERS. Term applied to the feathers on the outer sides of the shanks, as with Cochins and Langshans.

LIGHT BREEDS. A general name given to the non-sitting class of fowl, such as Leghorns, Anconas and Minorcas.

LIGHTING OF LAYING HOUSES AT NIGHT.* The lighting of fowl houses for a short period night and morning, during the long dark hours of winter, has long been a recognised means

* When night-lighting is done during war-time full regard must of course be paid to the black-out regulations.

of increasing egg-production. In the winter months the hours of daylight are too few (only 9 in comparison with the 16 of summer) to permit a laying hen to eat enough food to produce the number of eggs required of them to-day. With artificial lighting they are able to have a more normal working day, can thus eat more and so produce more eggs. Obviously the feeding costs must be considered, but these are compensated by the extra number of eggs laid.

Three Good Methods. There are three night-lighting methods in use, namely : (1) to provide artificial light for an hour or two in the evening ; (2) to provide artificial light in the evening and also in the early morning ; (3) to provide artificial light in the early morning only.

Advantages of Lighting. In official tests the income per bird from No. 1 was £1, from No. 2 £1 2s., from No. 3 £1. No lights gave an income of 17s. In the same tests the margin over food costs plus lighting expenses per bird was as follows : No. 1 pen, 13s. 4d. ; No. 2 pen, 14s. 8d. ; No. 3 pen 13s. 1d. ; no lights, 11s. The average egg yield per bird was : No. 1 pen, 168 eggs ; No. 2 pen, 183 eggs ; No. 3 pen, 168 eggs ; no lights, 154 eggs.

It will be seen that the best results came from method No. 2.

No. 2 method is not recommended in all instances, however, but only when electric light is available, when the lights can be on from dark to 7 p.m. in the evening and from 5.30 a.m. to daybreak in the morning, being switched on and off automatically.

Without electric light (that is, with the use of oil lamps) No. 1 method is recommended for small poultry-keepers, the lights being put on at 9 p.m., dimmed at 9.45 p.m. and turned off at 10 p.m. On commercial farms No. 3 system will probably be the most convenient, the lights being put on at 5.30 a.m. and turned off at daybreak.

Lighting, Good Forms of. The following are the forms of lighting usually adopted :

For House Containing 6 Birds. An ordinary household paraffin lamp, suspended on the wall about 2½ ft. from the ground so that it shines along between the birds as they are feeding over the trough.

For House Containing 12 Fowls. Hurricane paraffin lamp arranged as above.

For 20-50 Fowl House. Acetylene lamp. An old motor-cycle gas-generator fixed outside the house will do well, the gas being led by a tube to a burner on the house wall. This burner should be sufficiently high off the floor to illuminate the proper area and should be provided with a reflector. An illuminated area of

4½ to 5 ft. will be found enough for 14 birds.

For 100-500 *Fowl House.* Petrol or paraffin vapour lamps. One hung 7 feet from the ground will provide sufficient light for 50-70 birds.

For Commerical Egg Farms. Electric lighting. Reflection should be provided in the form of a conical shade measuring 15 in. diameter and 4 in. high for each bulb. The lights must be 5½ to 6 ft. from the floor level. For preference use 40-watt gas-filled bulbs, the light from one of these being sufficient to illuminate 200 square feet of floor space.

Whatever form of lighting is used arrangements must be contrived so that the lights are dimmed before being finally extinguished, so that the birds may be encouraged to go back to roost and not left stranded in the middle of the house by the sudden putting-out of the light. Hurricane lanterns can be turned half-low by hand and left there until the birds are back on the perches.

Acetylene lanterns with a drip feed are absolutely ideal in this respect as the light may be turned full on, the water supply then cut off, this automatically causing the light to die out without further attention.

Paraffin-vapour lamps are dimmed by slipping over the globe a dimming curtain.

Electric lamps are dimmed either by using another bulb in a separate circuit or by putting a resistance in the current. For these the switching on and off can be done automatically at definite times by clock.

The food given during the lighting period should be that which would normally be given were the day longer. Thus with the morning-and-evening plan, the morning feed should be grain and the evening feed grain or wet mash, according to whether dry or wet mash respectively is fed.

For morning lighting the feed should be grain in the scratching litter. For the " evening lunch " the food on the wet mash system should be trough-fed grain, the wet mash being given 1 hour before sunset as usual. On the dry mash system the lighted feed should be wet mash or grain in troughs.

Night-lighted birds want grit and water as well as food during this extra period, and hence the grit and water containers should also be in the area covered by the lights.

LIME-WASH. See Whitewash.

LINE BREEDING. See Breeding.

LITTER. All fowls that live, or spend some hours of each day, in confinement must have the floor of their quarters well littered. Littering is necessary, for the following reasons : (1) It induces the birds to exercise ; (2) it ensures absorption of excreta and this keeps the house sanitary ; (3) acting as a cushion, it lessens the risk of foot-troubles ; (4) it adds warmth ; (5) it prevents floor-draughts.

It has been definitely proved that fowls living in a house with a well littered floor will lay 20 per cent more eggs than fowls having no litter to scratch in.

Materials to Use for Litter. There are many materials available for littering poultry houses, the more important ones being :

Whole Straw. This is easy to obtain, cheap, well liked by the fowls, bright and cheerful, and easily moved. It is not so absorbent and will not last so long as peat moss. Oat straw, being soft, is least hard-wearing, but is the brightest ; barley straw is hard, but the "ears" are prickly; and wheat straw wears well, is most popular, but is slightly darker in colour. It is not suitable for chicks.

Cut Straw is whole straw passed through a cutting machine, and is generally referred to as chaff. It can be obtained in varying lengths of ¼, ½, ¾ in. It lasts longer than whole straw, is easy to scratch about and quickly covers up the grain scattered in it. It is more absorbent than whole straw, but requires founts to be well out of the way when being vigorously scratched about. It is bulky to store. It should be clean, bright, dust-free, and sweet smelling. Dusty chaff may cause Aspergillosis.

Flights, or fannings, or cavings. These are the outer husks that cover the grain in the ear and are removed and generally left under the drum after threshing. They are often obtainable free for the gathering, but then may be dusty, " spiky " (if of barley), and dirty. A clean sample is well worth having and is then equal to cut straw.

Shavings. These are quite good poultry litter, and are cheap, but are very inconvenient from the point of view of using the poultry manure.

Sawdust is a valuable litter for a laying shed. Is excellent for sprinkling on droppings boards, and can be used well as a "bottom" under other litters.

Sawdust can often be had for the gathering, and anyhow will cost very little. When it must be removed it may usefully go into the compost heap for a few months, when it should be useful as a manure.

Hay in its whole state is least suitable of all materials for litter as it mats so quickly, and is too valuable as a stack food to use it as litter. Poor, rough hay from waste spots may, however, be used.

Cut Waste Hay is similar to cut chaff, and may be used with success.

Bracken is excellent litter, and can·

often be had for the cutting. Should be cut when 8 in. to 10 in. high, so that the thin leafy tops are mainly used and very little of the thick woody stems lower down. With the poultry droppings it forms a manure little inferior to that made with straw.

Peat Moss is most popular because of its highly absorbent nature. It is not costly, considering that a bale will litter 32 square yards, and that the material will last three to four times as long as any other litter. Being absorbent, it is particularly sanitary, and is not liked by parasites. Its faults are that it darkens the house, does not give so much " fun " as the longer litter, and is often so dusty as to coat everything with a thick layer.

It is important to obtain peat moss that is as dust-free as possible, of the correct grade (fine for chicks, coarser for adults) and even in texture, i.e., easy to break up.

Dead Leaves are thoroughly good, are liked by the fowls, provide keen exercise and are free for the gathering. They are sometimes hard to dry because of inclement weather, are bulky to store and are not particularly long lived as litter, but that sums up their disadvantages.

The Correct Litter Depths. The following table shows the depth in inches at which each of the popular litters should be spread on the floor :

weeks in summer, these periods being for fowls with an outside run and droppings boards.

Feeding Grain in Litter. The grain part of the fowls' ration is always buried in the litter so that the birds must scratch over the litter to find it. It is not enough to throw the grain on the top ; it must be raked over as soon as it is thrown down.

MANDIBLE. The " movable " part of the beak of poultry.

MANURE, POULTRY. Poultry manure is highly valuable as a fertiliser for vegetables, mushrooms, flowers, fruits and farm crops. It has to be stored under proper conditions, however, before it can be used.

A common plan is to store it under cover in alternate layers with dry soil or sand, the heap being turned anew and well-mixed before use.

There has grown up a very good trade in dried poultry manure and many hundreds of tons are being despatched yearly to gardeners and farmers all over the country. The fresh manure is put through a special drying plant and is converted into a fine, dry powder, odourless, clean and convenient to handle.

Any poultry-farmer having a large

Litter.	For Layers.	For Chicks.	For Growers.
Whole Straw	6	not used	2
Cut Straw	2½	¼—¾	2
Flights	2½	¼—¾	1—1½
Shavings	3	not used	1—2
Sawdust	2	¼—½	1½—2
Hay	4	not used	2
Cut Hay	2	¼—¼	1½
Bracken	4—5	not used	2—3
Peat Moss	2—2½	¼—¼	½—1
Dead Leaves	3 in. small leaves 5—6 in. large leaves	not used	2—3

Use chick litter for birds up to 6 weeks old, grower's litter for birds from 6 to 12 weeks old.

When to Renew Litter. It is quite impossible to give a stated, definite time for cleaning out used litter, this being essentially a matter for personal observation. Immediately it becomes (1) dampish, (2) smelly, (3) heavy, (4) foul with droppings, or (5) " cakey," then is the time to clear it away.

As a guide, however, allowing 3 sq. ft. per bird, whole straw will want clearing out in the winter once in 6 weeks and in the summer once in 8 to 10 weeks. Peat moss will last 8 weeks in winter and 12

number of birds would do well to consider the installation of a manure-drying plant. The dried manure brings a good price, which, with the rapidly growing use of the dried manure, will tend to increase.

MARBLED. Term used to denote spotting on a fowl's plumage.

MARKETING. Many improvements in the marketing of poultry and eggs were introduced before the war. Different methods had to serve after 1939, but as peace brings more normal times the old rules will apply. If top prices are to be realised—and they alone pay—all classes of table poultry and eggs will have to be sent to market in the most attractive form possible.

Table poultry must be finished in proper fashion, prepared after killing in accordance with market requirements, graded into definite and prescribed grades and packed neatly in the form of package preferred by the market.

Eggs must be candled to make *sure* of their freshness, graded according to size and packed in the type of package prescribed by the market.

A National Mark scheme for eggs and all kinds of poultry was introduced some years ago, and full particulars were published by the Ministry of Agriculture (55, Whitehall, London, S.W.1).

As a war measure in 1941 it was ordered that all eggs must be sold to packing stations at controlled prices, but " domestic " poultry-keepers were given the option of selling to packing stations or of supplying their surplus eggs direct to friends and neighbours at the prevailing retail price.

Candling is done by means of a testing lamp, and enables the contents of an egg to be seen clearly and its age, degree of freshness, etc., judged. See Eggs, Candling, for methods and signs.

Preparing Poultry. Usually poultry need to be " drawn " before being marketed. The operations, in the order in which they are carried out, are as follows: (1) Lay the carcase on to its breast, (2) cut the neck off where it joins the body, (3) take out the windpipe and crop, (4) put the carcase on to its back, inserting the fingers into the body as far as possible and work round and loosen all the forward " internals," (5) make a cut 1 in. long across between the parson's nose and vent, (6) hook out the large intestine and cut away the vent with this attached to it, (7) loosen the rear " internals " through this hole, (8) grasp the gizzard and pull gently, when the whole of the internal organs, including the lungs, should come out together.

Do not cut the neck and neck-skin off close, but leave a 3-in. flap of neck skin to cover over the hole left when the neck is removed.

Be sure not to break the large intestine (near the vent end), nor to squeeze out any excreta that may taint the carcase. Another warning : When grasping the internal organs to pull them out do not squeeze them too tightly or you may break the gall bladder and give a bitter taste to the carcase.

The legs are left on, but to clean them up, dip them into boiling water, scrape them and finally cut off the big toe and all claws.

It pays to draw the sinews of fowls as follows : Make a slit 2 ins. long down and

along the outside of the leg just through the scales and the skin. Push a skewer into the hole thus made, feeling for and working under all the white cord-like tendons, and then bringing the end of the skewer out through the same hole on the other side of the tendons.

Hold the leg straight out, twist the skewer round three times to tighten the sinews, and then bend the leg down on to the thigh. This snaps the sinews, which can then be pulled out and cut away at the ankle joint.

Dust the carcases lightly with flour before packing them.

Marketing Poussins. These are baby chickens weighing from 12-14 oz. each. The best time to sell is late February or early March, and the most attractive birds have been those *not* exceeding 12 oz. in weight, plump, with round bodies and of the white-fleshed table breeds. They sell far better killed and cleanly plucked—every tiny feather off.

Marketing Double Poussins. These are overgrown Petits Poussins. Weight for weight, they do not sell so well as the smaller birds.

After February and early March, when the Poussins proper have been sold, the markets want a slightly bigger bird and then these may be turned into cash. It is wise to get in touch with the salesman with whom one deals before sending on a consignment.

Marketing Spring Chickens. Birds go by this name after they have attained the weight of 1½ lb., and in the past they made good money every year. The season is from the end of February until June, but the climax of the market is usually reached in March. It is advisable not to sell spring chickens until they turn the scale at from 2½ to 3 lb., although up to 4 lb. weight will be accepted.

Spring chickens may be sold dead or alive. The best prices are paid for birds that have been killed and plucked.

Marketing Ducklings. Weight and age are the chief factors in securing a good market for ducklings. The market is open for ducks weighing over 4 lb. but they should normally be fed so that they attain that weight between 9 and 12 weeks old. If not marketed in good time they start to moult and then plucking becomes a terrible business and even when done most carefully the young, growing stubs that are left give the bird a very untidy appearance.

Marketing Geese. Careful plucking and shaping are all-important with Geese. See under respective headings for instructions.

Marketing Turkeys. Care in plucking is very important. The birds must essentially look well. A further attention that is well rewarded is the drawing of the leg sinews. This is done as follows :

First break the bones at the top of the foot with sharp blows from a hammer. Then tie a piece of strong cord round the foot and the ankle (where it is broken) and suspend from a hook about a foot above your head. Grasp the thighs and pull till the feet, complete with sinews, come right away. Grasp the legs below the hock-joint, otherwise you may pull the skin off the thighs.

Packing Eggs. See Eggs, Packing.

Packing Poultry. See Packing.

Shaping Poultry. See Shaping.

Trussing Poultry. See Trussing.

MARKING POULTRY. Poultry-farmers and other poultry-keepers who mark their birds provide each of their birds with an identification mark so that individuals may be recognised and any facts about them turned up in a moment.

Marking Chicks. Chicks may be marked by toe-punching, wing-banding or leg-banding.

(1) Toe-punching consists of punching holes in the webbing of the feet by means of a special instrument. The operation is not painful and causes no inconvenience to the chick. As one or two holes may be drilled in each web on each foot, a large number of different identity marks can thus be provided. One lot of chicks may have one hole on the left foot, another one hole on both feet, another two in the inner web of the left foot and so on.

The special instrument sold for the purpose, called a toe-punch, is made in various patterns.

The operation is fairly simple, the chick being held comfortably in the hand, whilst the fingers of the same hand grip the leg just above the foot and extend it. The punch should be held along, not across, the toes, with the pin-side uppermost. That way it will make a clean hole, which should be well away from the edge of the web but not too near the toe.

(2) The wing-banding method is preferred by many for marking, as it allows an unlimited number of variations. For the purpose you require a supply of wing-bands. These are made of aluminium and bear consecutive embossed numbers. They are made in two sizes, 1¼ in. long for day-old chicks and 2¼ in. long for month-old chicks.

Some poultry-keepers fit the larger size band to the day-old chick, but it is rather cumbersome, and the better plan is to fit the smaller first and change to the larger ones at one month old.

The band is placed through the flap of skin across the wing. It must not be too close to the bone or there may be chafing.

Nor must it be too near to the outside, or it may break away. Nor must it be fitted behind one of the smaller wing bones.

The skin is punctured with a sharp pen-knife, the thin end of the wing-band inserted, then taken through the hole at the end of the wing-band and then bent back over itself.

(3) Some poultry-keepers fit miniature, numbered leg rings. They are affixed in a moment. The disadvantage is that constant watch has to be kept on them and a change to a larger size made directly they are becoming tight. Tight bands cause Ring-sloughing, which see.

Marking Adult Birds. The invariable method of marking adult birds is by affixing to their legs rings of celluloid or bands of aluminium. The celluloid rings are in a variety of colours ; the aluminium bands bear numbers. With celluloid rings a great number of different identification combinations are possible, for the birds can be ringed on either or both legs and can have up to three rings on each leg.

The celluloid rings are usually in the form of a spiral ring. To affix them, the pointed end is pushed upon and slipped onto the leg. A twisting motion to the ring then " screws " it on to the leg.

MASHES. See Feeding.

MEALY. Term applied to plumage that is dotted or stippled with a lighter shade. It counts as a defect in a buff-coloured bird at a show.

MINERAL MIXTURES. Mineral mixtures may play an important part in maintaining fowls in good health and providing the wherewithal for egg production.

One good mixture, which includes a proportion of iodine, is given under Iodine. Excellent proprietary mixtures are also obtainable, and these, being carefully blended by experts, may be recommended.

Another good recipe for a home-made mixture consists of :

Oyster dust or ground limestone, 3 stones ; bone meal, 6 stones ; common salt, 2 stones ; potassium iodide, 4 oz. It is important that all the ingredients are finely powdered, not merely crushed to grit. This formula is good for adults, chicks or growers, and the proportion of the mixture to use is 3 oz. of mineral to every 10 lb. of mash, or 2½ lb. per cwt.

MOSSY. Term applied to indistinct marking of a fowl's plumage—a defect in show birds.

MOULTING. Poultry normally moult any time from July to March.

The later in the season the moult starts, the longer the moulting process lasts and the longer the birds are unproductive and unprofitable. It is therefore an advantage to encourage early and rapid moulting. The right time for laying fowls to moult, in the ordinary way, is July.

Encouraging Fowls to Moult. It is not difficult to induce fowls to start their moult in July.

The procedure is to make drastic changes in the bird's diet without curtailing the quantity of food consumed. This means cutting out mash entirely and substituting grain—clipped oats and wheat, half and half.

In the morning give a scratch feed in the litter of three handfuls to four birds. At noon, give one handful to two birds, and at night give a fairly large feed—as much as they can eat in a quarter of an hour.

Fresh greenfood must be given at midday—in fairly liberal quantities.

As further aids to feather-casting, the drinking water should be medicated with Epsom salts. Make up a stock solution of ¼ lb. of salts to the pint of water, and of this give one dessertspoonful to the pint of drinking water every other day for 10 days ; and the birds should be confined to a warm house. If you can arrange to change the birds about to different houses, so much the better.

Usually 14-18 days of this treatment will see laying almost stopped, comb and wattles smaller and pale, and feathers falling out freely.

To Encourage Quick Re-feathering. As soon as moulting really has started—as witness the litter on the floor and the bedraggled appearance of the birds' plumage—the birds should be put onto a re-feathering diet.

The most important change is to restore the daily allowance of mash—a special feather-forming mixture. For dry mash the following is recommended :

Middlings 3, yellow maize meal 2, Sussex ground oats 1, linseed meal ½, pea meal ½, bran 2½, soya bean meal ½, fish meal ½—parts by weight, with cod-liver oil at the rate of 1 pint to 112 lb.

For wet mash use : Middlings 4, yellow maize meal 2, Sussex ground oats 2, linseed meal ½, pea meal ½, bran 1, soya bean meal ½, fish meal ½—parts by weight, with cod-liver oil as before.

When using either of these mashes give 1 teaspoonful of tincture of iodine in each gallon of drinking water.

The best grain mixture to go with these mashes is : small wheat 2, clipped oats 1, kibbled yellow maize 1—parts by weight.

If you use a proprietary mash it can be made suitable for moulters by adding 5 per cent *each* of linseed meal, pea meal,

soya bean meal and bran to it, together with the cod-liver oil as above.

On the dry-mash system let the morning feed be 1 handful of grain per 4 birds. At noon give 1 handful to 8 birds just as a scratch feed, and at night give as much grain—fed in troughs—as they can eat in 15 minutes. Let the birds have access to their dry mash all day.

When feeding wet mash, give 1 handful of mixed grain to each 2 birds in the morning, 1 handful to each 4 birds at noon, and then a heavy feed of mash last thing at night.

Supply fresh greenfood at noon.

A supply of pea-charcoal given in empty shell hoppers or boxes will help. So also will flowers of sulphur, 1 level tablespoonful to 12 fowls in the wet mash before mixing, or ½ lb. to each 2 stone of dry mash.

To Bring Moulting Fowls Back to Lay. The mash again plays a leading part. The best possible egg-encouraging mash for use wet is : 5 parts, by weight, of middlings, 2½ of Sussex ground oats, 2 of maize germ meal, 1½ of meat-and-bone meal or fish meal, ½ of pea meal and 1½ of broad bran.

For a dry mash use : 3 parts, by weight, of middlings, 1½ each of Sussex ground oats and maize germ meal, 2 of bran, 1 of fish meal and ½ of pea meal.

Equal parts wheat and oats is the best grain mixture.

As a fillip to get them to lay their first egg use a mixture of the following : Carbonate of iron 2 oz. ; ginger 4 oz. ; cinnamon 1 oz. ; iron sulphate ½ oz. ; gentian ½ oz. ; aniseed ¼ oz. Each ingredient should be finely powdered. The proper proportion is 1 tablespoonful to 24 hens if fed on wet mash and, for dry mash, use the same quantity to every 3 lb. of mash, weighed when still in the dry state.

To Treat Moult-fast in Fowls. Sometimes there is a perceptive halt during the moult. The fowl appears ill, with pale, shrunken comb. Examination reveals young feathers and bare parts at the same time.

Give Epsom salts every day for a week. Supplement the special refeathering mash already mentioned with ½ part sunflower seeds among 25 birds per day in troughs.

Premature Moulting in Fowls. It is not unusual for early-hatched pullets (those hatched in January) to commence moulting in the autumn. It is usually the neck-feathers that fall. The causes are an overheated, sun-scorched house ; lack of sufficient food to maintain body strength and stimulate egg organs to produce eggs.

Move the birds to cool, shady quarters by day and allow plenty of air at night. Encourage birds to eat as much as possible, going to the length of providing an additional meal of grain, in a trough last thing at night.

Encouraging Ducks to Moult Early. The best time for ducks and geese to start their moult is during August.

Ducks can be encouraged to moult by moving them to a fresh house and run and changing the feeding method. The run should be more limited than that they have been accustomed to, and there should be no swimming water.

In regard to feeding, the mash should be cut down in quantity, gradually decreasing day by day until hardly any is fed. The loss of food, however, must be made up for by grain, which should be given at both feeds.

The usual feeding scheme should be altered, too. If the birds are ordinarily fed mash in the morning, alter it and feed at night, and *vice versa*.

After about a week or 10 days of this treatment, the egg supply will have fallen off, feathers will be strewing the run and, if birds are handled, new feathers will be seen making their appearance. This indicates that a change to a feather-growing ration should be made.

Refeathering Ducks. The following mash will quickly re-clothe moulted ducks with new plumage : Middlings 5, maize meal ½, maize germ meal 2, Sussex ground oats 1½, bran 1½ and fish meal 1½—parts by weight.

If the weather happens to be bad or the ducks appear to feather with difficulty, add ½ part linseed meal to the mash.

For grain feed equal parts kibbled maize, oats and wheat. Feed it in the drinking water in the early morning and give only a small supply. Give the mash in the evening.

When the ducks are practically re-feathered, the maize should gradually be left out of the grain ration, and the linseed meal withdrawn from the mash.

To Bring Moulted Ducks Back to Lay. When feathering is complete, put the birds on this return-to-lay mash : middlings 5, Sussex ground oats 1½, yellow maize meal 2, bran 1½ and fish meal 1½—parts by weight.

At the same time return the birds to the quarters that will be their home through the winter.

Moulting Geese. Geese, though easy moulters, must be cared for properly or they may " go light " and die.

Some strains moult early and require no special encouragement beyond keeping them decently fed and providing facilities for them to reach a protected shed. Otherwise, it will be necessary, to limit their range by erecting a light netting fence. This should also keep them away from swimming water.

If grain can be purchased cheaper than mash, feed grain, soaking it occasionally for a change. Here is a cheap mash : Middlings 4, maize germ meal 1, maize meal ½, barley meal ½ and fish meal ½—parts by weight.

When the tail feathers fall out, it is an indication of a successful moult being in progress and the growing of the wing feathers points to the moult nearing completion.

(*Note*.—The heavy feeding of grain and of mash suggested above is, of course, impracticable during war-time, and must be modified accordingly.)

MUFF. Term applied to the tufts of feathers found on each side of the face of some breeds, for instance, Faverolles,

MUFFLING. Term applied to the whole of the feathering of the head with the exception of the crest.

NEST BOXES. In every laying house for fowls there must be an adequate supply of nest boxes. The correct allowance is 1 nest box to every 4 birds. The correct size for each nest box is 12 in. wide and deep and 15 in. high. The boxes should be situated in the darkest part of the house. They may either stand on legs, 18 in. above the ground, on a shelf fixed to the same height, or may be fixed to the outside of the house, the entrance to them, of course, being from the inside.

With outside nest boxes it is important to have them weather-proof and warm. For small houses it is convenient to have the back, or top, on hinges, so that eggs can be removed without the need for entering the house. Care must be taken to ensure that no rain water or snow can leak through the joint between the hinges and fixed part.

Nest Box Litter. Straw—just enough to prevent eggs coming in contact with the hard floor—is the best litter for nests. Chaff, flights, coffee husks and peat moss can also be used but are not so good. Whatever is used must be frequently changed or dirty eggs will result. The litter should be freely dusted with insect powder.

Creosoting Nest Boxes. Twice a year—oftener if there has been an attack of fleas or other pests—nest boxes should be creosoted, and kept closed for 3 days before the birds are again allowed to use them.

NEST EGGS. The presence of a dummy egg in the nest does something to encourage fowls to use the nest. The " pot " eggs that have disinfectant properties are the best to use.

NESTS. See Hatching.

NIGHT ARK. Term applied to a special type of poultry house, which is portable and is provided with a slatted floor and usually an apex roof. Is freely used for growing chicks and occasionally for adult birds. Also termed Sussex Ark.

Night arks must be moved to a fresh position every 3 days; otherwise the accumulation of droppings that have fallen through the slatted floor may cause trouble. The droppings should be cleared from the sites after the house is moved. At intervals the floor slats should themselves be scraped clean.

Better still is it to have boards or a sliding tray below the slatted floor, so that the droppings may easily be removed to a wheelbarrow for disposal. The ark need not then be moved frequently. See also Housing.

NIGHT-LIGHTING. Term applied to the installation of lights in laying houses so that these may be illuminated early in the morning and/or late in the evening during the winter months. See Lighting.

NON-SITTER. Term applied to the light breeds of fowls, such as the Leghorn, that rarely desire to sit.

PEA COMB. A triple comb, consisting of three small combs joined at the base, the middle section being the highest. A typical pea combed breed is the Brahma.

PENCILLING. Term applied to the markings or stripes on feathers. The markings may run straight across the feathers, as in Hamburghs; may be V-shaped, as in Campines; or follow the outline of the feather, as in Partridge Wyandottes.

PERCHES. For adult fowls the best perches are lengths of 2 in. by 3 in. timber, with the upper edges bevelled. If there is more than one row of perches, there should be a space of 15 in. between perch and perch. There should be a gap of 8 in. between the back wall and the perch, which should be about 8 inches above the droppings board, in its turn not more than 21 in. from the floor. The perch should be 24 in. from the nearest point of the roof, there should be enough perches to allow each bird 7 to 9 in. of perch room.

Perches should never be fixtures; they should always be easily removable, so that they can be regularly and properly cleaned and also so that proper measures can be taken against insect pests.

The best method is to have slots—or brackets—fixed to the walls in which the perches can fit loosely. These slots can be continually painted with paraffin and then no red-mite or other pest will collect there.

Perches for Chicks. Not until chicks are almost full grown should they be allowed to perch on ordinary perches. Instead they should be provided with special perches consisting of two or more wooden laths, each measuring 2 in. wide by $\frac{1}{2}$ in. thick, fixed $1\frac{1}{2}$ in. from one another and set $1\frac{1}{2}$ feet from the ground. These perches are better if covered with hay, kept in position by tying it on with string.

PERCH PAINT. Preparation for the treatment of perches when red mites are present. See Insect Pests.

PERMANGANATE OF POTASH. Useful preventive of colds and other ailments. Enough crystals to colour the drinking water light pink should be used at any threat of trouble, when one or two birds are heard to sneeze and it is required to prevent an outbreak of colds from contagion.

PETITS POUSSINS. Term applied to young chickens weighing 12-14 oz. and sold for table at this size. See Fattening, Marketing.

PIGMENTATION. Term used to denote a test of a laying fowl's abilities, founded on the amount of pigment, or colour, in different parts of the body.

In accordance with the number of eggs a fowl has laid so the yellow parts of the body become bleached. Thus a bird that is a good layer is more bleached than one that is a poor layer. By noting the amount of bleaching in the individual members of a flock it may be possible to tell which of the birds are first-class layers and which are " duds ".

The various parts of the body bleach in a definite order—the vent first, then the eye-ring, then the earlobes, then the beak, and finally the legs. The vent loses its colour quickly, so that when 4 or 5 eggs have been laid it is quite pink.

The eye-ring is the inner edge of the eyelid, and this gets pale after 6 to 7 eggs have been laid. The test for the earlobes can only be carried out on a bird that has white lobes, or if red, a few patches of white. After 10 to 14 eggs have been laid these lobes will be white.

The beak bleaches in a regular order, beginning at the base and working to the top part, the last spot to bleach out on the beak being a patch just on the curve of the top half of the beak.

The legs pale far more slowly, and they,

too, do so in a definite order, the part at the back being the last of the leg to bleach.

The very last part of the body to bleach is the point at the back of the hock, and this serves as an index to the natural depth of colour tint.

When the bird stops laying the colour comes back, in the same order as it left.

PLUCKING TABLE POULTRY.

The way plucking is done can make or mar the appearance of a table bird.

Plucking should be done quickly but carefully. One tear in the skin, especially on the breast or side of the crop, will often mean a reduction in the value. Keep to a regular order in plucking. Pull out large wing feathers and large tail feathers first, because they come out so much easier then. Then pluck the breast from the top down to the vent, then the back in the same direction, and finish up with the legs and wings.

The flesh tears by far the easiest on the shoulders, just in front of the parson's nose and each side of the crop, so exercise particular care over these parts. If the worst happens, hide the tear by stretching the flesh *and* the tear down so as to be covered by the thigh when the bird is tied.

Turkeys are plucked in the same way except that it is customary to leave on a few feathers over the hips.

Geese and ducks take slightly longer to pluck than fowls—say another 1½ to 2 minutes per bird—because under the feathers there is a coat of down. They do not tear so easily. The short feathers on the wing-ends need not be plucked off with ducks, geese or turkeys.

Neck feathers are left on all poultry.

Stubbing and Singeing. After plucking comes stubbing. For this use the dull blade of a small knife, such as a pocket-knife, grasping the stubs between it and the thumb.

After stubbing, singe the carcase (to get rid of hairs)—either over a spirit lamp or a Primus stove.

PRIMARIES.

The primary feathers of the wings used in flying, also termed " flights ".

PULLET.

A female fowl from birth until she reaches the age of one year and becomes known as a yearling.

PURE BREED.

Birds that have no blood of another breed in them, as distinct from cross breeds, produced by the mating of two distinct breeds.

RASPBERRY COMB. Comb resembling a raspberry cut in half. The Orloff is typical of the breeds of fowls wearing this comb.

REARER.

General name given to the appliances (brooders, hovers, fostermothers, etc.), in which artificially hatched chicks are reared.

REARING CHICKS.

When chicks are to be reared by broody hen the rearing quarters should be ready well before the date of hatching.

The first requirement is a sound, weatherproof coop. It should be 2 ft. wide by 2 ft. deep by 22 in. high in front, made of tongued and grooved matching, the roof being either of weather-boards or wood well covered with tarred felt. There should be a movable floor that pulls out for cleaning, night shutters and a wire-netting run 3 ft. long, 2 ft. wide and 1 ft. high. For the run use ½ in. mesh netting.

The ideal position for the run to stand is on a patch of short, fresh, untainted grass, sheltered on the north by a hedge or range of buildings. If the site is exposed, it is well to put up a wind-break of galvanized iron sheets or wattle hurdles.

To prepare the coop for occupation, sprinkle the floor with disinfectant powder and litter with clean, yellow, dustfree, dry chaff.

The equipment necessary is a feeding board 10 in. by 4 in., a saucer, a pint water fount and a trough 1 in. high, 8 in. long.

Installing the Broody and Chicks. Do not move the chicks from the nest to the rearing coop until they are thoroughly fluffed up and dry. When they *are* ready to be moved, carry the chicks to their new quarters in a haylined box or basket, with a double-thickness of flannel placed over them.

Let the broody be comfortably installed before giving her the chicks. Dust her with insect powder before putting her in the coop, but rub off surplus powder with the hand.

Darken the coop for an hour after the broody and her family have been installed. Then give the first feed, putting the feeding board *inside* the coop and a tablespoonful of good chick corn. The hen will generally teach the chicks to eat, but if she does not, you can do so by tapping the board lightly. Put some water, or, better still, in normal times, milk, in the saucer, and set it in the coop. If the hen drinks a little, so much the better for teaching the chicks.

After a few minutes shut the coop up again and see that the hen broods the chicks. In 2 hours' time repeat the feeding, and also the shutting-up until you know that the hen will definitely sit down to brood the chicks.

The broody is fed grain mixture and water in tins, so placed as to prevent the baby chicks from eating the too-large grains.

As the chicks grow bigger, they will not go to bed at night very early, and will be roaming about on nice evenings quite late. It is best to shut up the coop at dusk, for even at this age chicks can easily be shut out all night and will be killed by the night cold.

When the chicks are 3 to 4 weeks old runs should be done away with and larger dry-mash hoppers, say 2 ft. long, 8 in. wide and 2 ft. to 3 ft. high, should be used, and the gallon-size galvanised adult water founts. These should be stood close to the coop.

Separate the chicks from the mother bird after 5 to 7 weeks, according to the weather.

Artificial Rearing. Incubator-hatched chicks may be reared by brooders, hovers or the mechanical appliance known as a foster-mother (see separate headings). The most satisfactory method, in the view of most poultry men, is an indoor hover housed in a well-lighted, draught-proof, dry shed, to which is attached a 10 in. high sun parlour, made of wire netting (½ in. mesh for the floor and 1 in. for the sides and top) on 2 by 1 battens. This should be fixed to the pophole of the house and raised at least 3 in. off the ground.

For 100 chicks the brooder house should have a 7 ft. by 5 ft. floor area and the sun parlour should have an area of at least 10 sq. ft. The chicks are kept in this house and not let run on the ground for the first 6 to 8 weeks of their lives.

Hovers, brooder-house, sun parlour, etc., should be thoroughly cleaned and disinfected before use.

Dealing first with the brooder-house sun-parlour scheme, litter the house with bright, clean, dry and dust-free chaff to a depth of ½ in. only. Place the hover in the darkest, warmest corner; and at least 6 in. from the walls. Scrape out a hole in the litter for the lamp, fit up the hover (or brooder), start the lamp going and put in the thermometer, the bulb being 2 in. from the litter and midway between the outside wall of the hover and the lamp.

Provide a " first-feed " board measuring 24 in. by 12 in. for every 70 chicks, 2 1-pint water founts and a dry mash trough 24 in. long and 2 in. wide, barred or holed to protect the mash. Get a piece of ½ in. netting, 3 in. longer than the circumference of the hover and put this round for the first 3 days to keep the chicks inside overnight.

If, instead of a sun parlour the chicks are to run on the ground, provide a fresh grass or earth run upon which no chicks (and, if possible, no adult birds) have been for at least a year.

If a foster-mother is to be used stand it

in a sheltered spot on clean, short grass. Litter the hover compartment with fine, dry peat-moss and the scratching compartment with chaff as above.

With hovers, brooders and foster-mothers the temperature beneath the hover should be 85 to 90 degs. F. for the the first week, 80 to 85 degs. for the second, 75 to 80 degs. for the third, and 70 degs. from then onwards to weaning time.

Do not, however, be guided entirely by the thermometer. The appearance of the chicks themselves will provide the surest guide whether temperatures are correct or not.

If the chicks are huddled close to the lamp, increase the heat a little ; if they are scattered around the edges, reduce it a little. If there is moisture on the inside of the brooder in the early morning, it is due to overcrowding or lack of ventilation, and if the latter you can reduce the litter around the hover, make extra holes in the metal or wooden sides or pin up parts of the curtain.

The chicks should be transferred to the brooding appliance as soon as they are dry and fluffed up in the incubator. When doing this, make sure that they are well wrapped up and protected from the cold. It is best to instal them at night. Place them well under the brooder or hover, bank up the chaff round the curtains, put round the netting guard, see that the lamp is working properly, and leave them thus.

In the morning (and let it be early), pin up a 4-in. strip of curtain and entice the chicks out by tapping gently on the feeding board, on which some chick grain has been scattered. After 5 minutes put them all back.

Management of Brooders, etc. Examine the litter *beneath the hover* and renew very frequently. Look at the chicks after dark and see if they are comfortable. Permit of sun bathing by allowing the chicks to squat down in the litter on nice days in full, unstopped sunlight ; on nice days, too, put one dry mash hopper and one fount out in the sun parlour.

Move foster-mother the very day after the chicks are 6 days old and choose a hedge-protected spot if at all possible.

Remember that as chicks grow they need more space. A 100-size brooder will only accommodate 40 birds at a month old. Further, when the chicks are a month old more air must be admitted beneath the brooder : (1) by raising the brooder so that the curtains are lifted 1-2 in. away from the litter, (2) by pinning up ¼ of the curtain space, (3) by taking away one or more of the curtains if the hover is a square one, or (4) by fitting a ½-in. mesh netting or wire cloth floor.

If there is overcrowding, stock can be reduced by culling wasters (see Culling) and sorting the birds into sexes (see Sexing).

The Care of Growing Chicks. Chicks should be weaned from the hover, brooder or foster-mother when they are 6-7 weeks old. The weaning process should be gradual, reducing the lamp flame in the brooders somewhat day by day, and then turning it out altogether, still leaving the birds in the brooder.

Night arks provide the best accommodation for the chicks after weaning and during the growing-on stage. If the weather is cold when the chicks are installed in the arks the slotted floors can be littered with straw and a hurricane lantern can be hung up in the house.

The arks should be placed in a well-sheltered situation, protection from wind being particularly necessary.

RINGING. See Marking.

RING SLOUGHING. A common trouble where leg rings are used. Tight rings may cause swellings that, if not noticed, may enclose the ring and cause great agony.

The cause is usually growth of the leg or too tight-fitting ring, or gouty tendencies.

Take ring off, bathe carefully, and paint with tincture of iodine once a day.

ROSE COMB. A comb that lies nearly flat on the top of the head, and is covered with small spikes. The longer spike, at the back, is called the leader. This is one of the commonest types of comb. The Wyandottes are an example of the rose-combed breeds.

RUNS. When runs are fitted to poultry-houses, there is certain to be a higher egg production, because (1) the direct sun rays can reach the birds ; (2) shade that is far cooler and fresher than that inside a house can be provided ; (3) a change of outlook and a freshening of spirits is given the birds ; (4) increased accommodation is ensured ; (5) extra exercise is provided by scratching and ranging ; (6) animal matter in the shape of insects is free for the gathering ; (7) fresh, live greenfood is at hand.

Earthen Runs. Runs on the bare soil provide all the above conditions except a growth of greenfood. If the run has no roof, allow a minimum of 20 sq. ft. per bird, but a great economy can be effected if it is roofed over, for then you can allow double the number of birds to use it.

Earth runs go foul quicker than grass runs. Occasionally, therefore, they should be purified by slaking some fresh lump quicklime by sprinkling water on it in a shallow heap until it breaks down into a powder. The slaked lime is spread about ¼ in. thick all over the run and then dig

in. Keep the birds shut up till digging is completed.

Grass Runs. Grass runs may be either of the free-range or semi-intensive type. Laying is increased if the grass in these is always short, freshly-growing, green (as opposed to dry and brown), and with a good, thick "bottom". To ensure good grass scythe regularly throughout the year as soon as the grass reaches a height of 4 in. Dress with basic slag in the autumn—dark, blue-green grass will do better with a liming. Cut all coarse, tufty grass hard back with a hook.

With large grass runs, it is difficult to do any cleaning proper ; the best plan is to shut the birds away from them (or from a section of them) so that the grass may have an opportunity to absorb the manure, recover its strength and again grow strongly.

RUMP. Term used to denote the cushion ; sometimes spoken of as the " parson's nose ".

RUST. Name given to the red-brown patch sometimes seen on the wings of dark plumaged breeds.

RYE. Occasionally used as grain food for fowls. Other grains are better, however.

SADDLE. The posterior part of the back of a male bird—equivalent to the " cushion " of a hen.

SADDLE HACKLES. The long, narrow feathers hanging from the saddle.

SCRATCHING SHED. A covered, open-fronted shed or house in which intensively kept fowls spend most of their time. The floor is littered with loose litter and the grain food is distributed in the litter.

SECONDARIES. The quill feathers of the wings. The secondaries, unlike the primaries, are visible when the wings are folded.

SELF-COLOUR. Plumage of the same colour throughout.

SEXING. *Chicks.* It is very desirable that, except when mated for breeding, the sexes should be kept apart. The means of distinguishing the sexes from a day-old and onwards are discussed below.

There are very definite good reasons for sorting out growing chicks into sexes at the very earliest opportunity, and thereafter housing and running the sexes apart. Among these reasons are:

(1) Excessively rapid maturity in the pullets is prevented when the attention of the males is removed.

(2) A better profit is provided by the sale of the unwanted males at an early age.

(3) The pullet chicks can be given a special mash for the maintenance of growth, bone and feather without wasting it on male chicks that do not require it.

(4) The overcrowded growing-on houses will be eased and more room will be available for the pullets.

(5) Growth will be stunted if the males and females are allowed to run together.

How Sexing is Done. There are certain rules to be followed if sexing is to produce reliable results.

(*a*) The first rule is to sex together chicks *of the same age.* This means picking out chicks that were hatched within about 3 days of one another, sexing these first and then dealing with older ones.

(*b*) The second rule is to sex chicks of the same breed division, i.e. either heavy breeds or light breeds, together. Light breed male chicks **feather** up much more speedily than do heavy **breed** males of the same age ; in fact, **they** feather almost as quickly as the pullet chicks. Tail-length in light breeds is by **no** means such a good guide as in heavy **breeds**, for whereas with the heavies **the** pullet definitely grows a larger tail (age for age) than the male, in the light breeds tail length is often the same in cockerels and pullets.

(*c*) It is well to **compare** chicks of the same strain, because **most** chicks of the show strain (i.e. bred for the chief purpose of exhibiting) grow **more** profuse feathering, but grow it less quickly than chicks of utility strains, i.e. from bred-to-lay parents. So a slow-feathering show pullet may wrongly be classed as a cockerel when compared with a utility pullet.

Show chicks, too, are slower to show signs of an increasing headgear than utility chicks.

Bone in exhibition chicks is also generally far stouter than in utility chicks.

(*d*) One should recognise the difference in breeds, as it may cause difficulty to compare, say, White Wyandottes and Rhode Island Reds. Thus Rhode chicks are longer and stouter on the leg, so that it is quite easy to mistake a Rhode pullet chick for a male, because of its length of leg, if it were compared with a Wyandotte of the same age.

Some breeds at maturity are from 1 to 2 lb. less in weight throughout their life. Thus, to compare at 6 weeks a Jersey Giant chick with a Wyandotte may result in false conclusions because of the smaller size wrongly suggesting that the Wyandotte is a pullet rather than a cockerel.

(*e*) The fifth rule is that of management, which includes feeding and housing.

A chick reared intensively and warmly has a greater wealth of feather than one reared on mother earth from a week onwards. Battery-reared chicks are often larger (though softer and more abnormal) than those reared under harder conditions.

Means of Sexing. In recent years it has become possible to distinguish the sex of chicks at a day old, by inspection of the sexual organs. The practice is now generally followed, and guarantees of 95 per cent accuracy are sometimes given. The difference in the organs at that age is very small, but quite definite, though those who practice day-old sexing must be capable, experienced and rapid at the work to make it an economic proposition. When a chick is held lightly in the hand, and the organ gently pressed upwards from the body, a cockerel chick shows a distinct but minute protuberance, whereas a pullet chick does not.

Apart from the sexual-organ test, the sexes can only be distinguished, at one day old, in sex-linked birds (see Sex-Linkage).

Other Signs of Sex. Some breeders believe that at a week old the comb of male chicks is a shade higher than in female chicks. It cannot be trusted.

At two weeks females of heavy breeds show a little tail feathering—about $\frac{1}{8}$-$\frac{1}{4}$ in. Males of heavy breeds show no tail whatever, but of light breeds the tail will begin to show quite plainly.

At three weeks females of the heavy breed are better feathered than the males. The pullets show a tail up to $\frac{3}{4}$ in. long and the shoulder plumage is $\frac{3}{4}$ in. long. Males of the heavy breeds have much less feather, show more fluff and carry hardly any tail, very often none.

(Where chicks of an exhibition or exhibition-utility strain are concerned, the pullets often have hardly any tail at this age, and it should not be regarded as too definite).

Light breed chicks are difficult to distinguish by the size of tail, and in some strains there is hardly any difference between the male and female tail. Generally, however, the male shows the strongest and perkiest tail and the males feather almost as fast as the pullets.

At four weeks females of the heavy breeds will show feathering down the back, down the chest and along the thighs. Tail length in the utility breed will be approximately 1 to 1$\frac{1}{4}$ in.

Males of the heavy breeds will show much more fluff. The tail will be appearing, but will be quite small, and there will be hardly any back plumage.

Light breeds may be sexed by comparing the combs carefully, those of the males being larger and redder than of the females. This comb-test was formerly generally relied upon.

At five weeks females of the heavy breeds will have shorter (by ½ in. or so) and thinner legs, longer bodies, longer tails and better feather.

Males of the heavy breeds will have longer and stouter legs, short, cobbier bodies, stumpy tails (1 in. or so long) and not much feathering along the back and wing-bows.

Light breed males are rather longer on the leg at this age, but the comb will be racing in size away from that of the female and offers the best sign.

At six weeks females of the heavy breeds will now show hardly any fluff at all and the apparent length of the body will be more easily noted.

Males of the heavy breeds will be much better feathered than at an earlier age, but they will still show bareness at the shoulders, wing-bow and back.

The way the birds carry themselves is a sign now in both heavy and light breeds. The males will appear taller and with definitely stouter legs, while the pullets will appear much closer to the ground owing to their small legs and thighs.

At eight weeks, females of the heavy breeds offer the following guides : smallness of comb, abundance of plumage, shortness and thinness of leg, length of body, length of tail (1 to 2 in.) and bold, dashing manner.

Light breed pullets are distinguishable by the neatness and shortness of leg, perfection of feather and smallness of comb, the latter being the best sign by far.

At ten weeks pullets of the heavy breeds will appear definitely of the " weaker " sex, being neat and smaller in body than the males, while the wattles will still be small.

Males of the heavy breeds will be taller and larger, with increased expansion and reddening of comb and wattles, and almost complete feathering.

Light breed males show a definite difference in comb and wattles.

(Note that with many utility heavy breeds, at three months the sexes appear about the same, and it may even be difficult to distinguish a backward male as compared with a forward pullet, so that it is a wise plan to ring or separate the pullets when the sex is easier of determination, say at eight weeks).

At twelve weeks, bone continues to increase, so that the males seem to stand higher and on much stouter legs. The male headgear increases more rapidly, too, and becomes definitely light red.

At this age the hackle test can be applied. The feathers on the saddle-hackle of the male will be pointed at the end and those on the pullet rounded, but be sure it is the saddle-hackle plumage you are looking at and not that growing higher along the back.

At this age, the spur of the male is generally larger than that of the female, but this is not a really easy feature and not to be relied upon.

Sexing Ducklings. It is important that ducklings be separated according to their sex at 10 weeks of age at the latest, not only because it is harmful for the sexes to run together but so that unwanted drakes and culls can be sold off at once and no money wasted on feeding them for a further period.

In ducklings you cannot judge sex by (1) body-shape, (2) length of leg, (3) rate of feathering, (4) head gear, (5) stance, and (6) sizes. The signs you *do* have to go by are : (*a*) the sexual organ, (*b*) plumage-colour, (*c*) voice, (*d*) feathers just above the tail.

(*a*) The sexual organ is easily seen, even in a day-old bird, by gently pressing each side of the vent, whilst holding the duckling gently. If there is a pointed organ which protrudes as pressing continues, the bird is a male, but if the vent remains as a depression, the test shows a female.

(*b*) The plumage test can only be applied to ducks of the coloured breeds, such as Khaki Campbells, Fawn or Indian Runners and Rouens. At 6 to 8 weeks of age a properly fed duckling is getting its feathers well, i.e. losing its baby fluff. The duckling's wing feathers are more advanced, far brighter, more defined and more contrasting in colour than those of the drake.

(*c*) At 8 weeks and sometimes before (in highly-bred-to-lay strains) the ducklings get their voice. Each duckling should be picked up and held by the neck, to make it call out.

The female will have a harsh quack, something like a youth when his voice begins to " break," and it will be much louder and plainer. The male youngster will given an empty, voiceless sort of hiss like a blowing out of breath and there will not be any hoarseness.

At 10 to 12 weeks the voice of the female develops into a definite quack, and as this is the proper age for selling the surplus drakerels, so the voice test is the best one to apply.

(*d*) When the birds are 20 to 22 weeks old, they will be getting their adult plumage and with it the curled tail sexing feathers. These only appear on the male. Two in number, they grow upwards and forwards over the body just in front of the tail and on the back.

Ducks. In adult birds sex can be told by the feathers just above the tail. In the duck, these lie flat. In the drake two of the feathers curl upwards towards the back. There is also a distinct difference in carriage, the hind parts of the ducks being much lower and heavier; while the voice of the duck is deep and the quack of the drake much softer.

Geese. The most obvious sex sign in adult geese is the behaviour of the ganders when cornered. They will "form-up" on the outside of the geese, put their heads down and hiss, the geese meanwhile remaining quiet. They are also usually taller, longer necked, lighter and closer in the body.

Goslings. The best way to ascertain the sex of goslings is to examine the sexual organ of each bird in turn.

Take up a bird and hold it down on a table. Then gently force the vent open by pressing backwards and downwards with the fingers, which are on top and pulling downwards and outwards with the lower fingers. The organ will then appear. In baby gosling males the penis is small, worm-like, quite pale and only comes out about a quarter to half-inch when pressed. In a female, you will see quite plainly that there is nothing to be expelled by continued pressure.

When the gosling is 6 to 7 months old, the same procedure to determine sex is followed, but in a rather different manner. It is best then to hold the growing gosling between your legs so that both hands are free. All the organs will be correspondingly larger, but the penis itself will not be appreciably larger until the gosling has become an adult goose and is mated. At this stage the organ may be from 2 in. to 3 in. long (of a deeper tint, pinkish yellow), and of a corkscrew shape.

When the goslings are from 6 to 8 weeks old, the heads of the males are rounder, stouter, and carried higher than those of the females. The upper half of the beak is a trifle thicker and coarser, also the neck of the male is stouter than that of the female.

Guinea Fowls. At 10 weeks of age, the voice is the only decent means of telling the sex. The females utter a cry of two syllables, which has been likened to "come-back," or "buck-wheat". The male's usual cry is faint and of one syllable—the "wheat" part. It is important to note, however, that when very excited, both sexes utter practically the same cry, so this test should be applied when they are not over-disturbed.

At a later age the male has a larger "helmet," more pronounced wattles, and the head is coarser. The bird itself is more pugnacious and often walks in a forward position on its toes, as it were. The back is sometimes arched.

Turkeys. The male chicks are longer on the leg than the females and the width of body is rather less. They are also more prone to assert themselves.

At about 8 weeks turkey chicks "shoot the red," i.e. the head appendages begin to grow. In birds of the same breeding, hatch, uniformity of growth and age, the male chicks will "shoot the red" first.

At about 5 to 6 months the male has definitely more length of leg with a greater amount of thigh showing below the side feathers, and his body is carried higher on the leg.

Many turkeys have a tassel growing in front of the wish-bone on the chest. In some breeds and some strains, both males and females have this tassel, but it is larger in the male and begins to grow at an earlier age. In the coloured breeds, the male shows a more defined colour from the shoulders to the tail.

The wattles, or caruncles as they should be called, are always larger in the male, but more especially so is the spike at the top of the head, in a breeding bird reaching perhaps 4 or 5 in. in length and hanging down over the beak.

SEX-LINKAGE. For some years now it has been known that the chicks resulting from mating certain birds of distinct colours are distinguishable so far as sex is concerned at birth, the male chicks coming a very different colour from the female chicks. The male chicks, indeed, take the colour of the mother hen and the female chicks take the colour of the cock. This only happens with certain matings—not with all matings in which the father and mother are of different colours. Chicks produced as a result of the specified matings are known as Sex-linked chicks.

The advantage of knowing at birth which of one's chicks are cockerels and which pullets are, of course, obvious (see Sexing).

The most usual practice is to mate a "gold" male on to "silver" females, this resulting in male chicks that are silver and females that are gold. An example of this is the Brown Leghorn male and Light Sussex females.

The leg-colour of certain breeds is also sex-linked. If we mate a Gold Pencilled Hamburg cock (dark legs) and a Brown Leghorn hen (light legs), we get male chicks with light-coloured legs and females with dark legs.

The "barred" and "black" breeds behave as sex-links as well. If a Black Rock male is mated to Barred Rock females, we get black female chicks and barred male chicks.

"Cuckoo" breeds act in the same way as barred.

There is, again, sex-linkage in the eye colour. If a dark-eyed Langshan cock is mated to a light-eyed Brown Leghorn, we get light-eyed male chicks and dark-eyed female chicks.

It must be clearly understood that the

dark factor must always be possessed by the male and the light by the female : the opposites do not hold true.

Included in the term "gold" males would be males of tints of black-red, brown, partridge, buff and gold ; and "silver" females include the lights (e.g. Light Sussex), the silver greys, the barred, the duckwings and the speckled.

A White Leghorn is *not* one of the silver breeds and will not respond to the laws of sex-linkage.

The following is a list of many sex-linked crosses that are available ; it will be noticed that, unfortunately, the majority are with birds possessing little or no utility value :

Mate any of these "gold" males: Barnevelder, Brown Leghorn, Brown Sussex, Brown Buttercup, Mahogany Flower Bird, Golden Duckwing Leghorn, Black Red Game, Indian Game, Partridge Cochin, Black Red Malay, Partridge Wyandotte, Wheaten Marsh Daisy, Mahogany Orloff, Gold Campine, Gold Wyandotte, Gold Hamburg, Golden Buttercup, Buff Leghorn, Buff Rock, Buff Orpington, Rhode Island Red, Red Sussex, Red Dorking.

With any of these "silver" females: Light Brahma, Light Sussex, Columbian Wyandotte, Silver Grey Dorking, Duckwing Game, Silver Duckwing Leghorn, Barred Rock, Silver Campine, Silver Hamburg, Silver Buttercup, Silver Wyandotte, White Wyandotte, Ancona, Exchequer Leghorn,

And the chicks will be buff-coloured females, and light-coloured males.

Mate any of these "Black" males: Black Rock, Croad Langshan, Black Leghorn, Black Langshan, Black Minorca, Black Orpington, Black Silkie, Black Spanish, Black Wyandotte, Black Cochin, Crève Cœur, Black Scots Dumpy, Black Hamburg, Black Orloff,

With any of these Barred females; Cuckoo Leghorn, Coucou de Malines, Cuckoo Dumpy, Barred Rock, Scots Grey,

And the chicks will be black females ; and black males having a light patch on the head and often also on the rump.

Mate either of these dark-legged males: White Bresse, Brown Leghorn,

With either of these light-legged females: Light Sussex, Gold Hamburg,

And the chicks will be dark-legged females and light-legged males.

Mate this dark-eyed male: Langshan,

With this light-eyed female: Brown Leghorn,

And the chicks will be dark-eyed females and light-eyed males.

Of the many crosses above mentioned, those set out in the table below are the most worth considering from a utility point of view.

SHAFT. The quill part of a feather.

SHAPING POULTRY. Most classes of table poultry are greatly improved in appearance by a process known as shaping. It consists of leaving the birds for a time in an appliance called a shaping press. This consists of a V-shaped trough, a clean board and some bricks. The birds are packed tightly together in the trough, side by side, breast down, legs tucked into the sides. The board is placed along on top of the birds and on the board the bricks to weigh it down. Choose the coolest room available for the shaping.

Always shape ducks, too, as it gives their breast the desired flat appearance. They can be shaped on a shelf, and weighted just so that the breast-bone doesn't stick up like a ridge.

Geese are not usually shaped ; neither are turkeys—in the ordinary sense of the word, but turkeys are given an appearance of width by hanging them by their legs with a piece of wood about 7 in. long wedged in between the thighs to keep them apart.

SHOWING POULTRY. See Exhibition Poultry.

SICKLES. The long curved feathers of a cock's tail.

SILAGE. See Foods.

PARENTS.		OFFSPRING.	
MALE	FEMALE	MALE	FEMALE
Brown Leghorn	Light Sussex	Excellent for table	Good layers
R.I.R.	White Wyandotte	Good table	Excellent layers
Brown Leghorn	do.	Average table	do.
Buff Rock	do.	Good table	do.
Buff Orpington	do.	do.	do.
Indian Game	Light Sussex	Very excellent table	Fair layers
R.I.R.	do.	Excellent table	Good layers
Black Leghorn	Barred Rock	Good table	Good layers
White Bresse	Light Sussex	Excellent table	Very good layers

SINGEING. See Plucking Table Poultry.

SINGLE COMB. A narrow, flat comb surmounted by a number of spikes or serrations.

"SMALL" POULTRY KEEPING. Term applied in war-time to class of poultry-keeper keeping over 25 and not more than 50 birds. (See also Domestic Poultry-keeping.)

SMUTTY. Term applied to the dark indistinct patches sometimes seen on the plumage of buff and other coloured birds.

SPANGLING. Term applied to the spot of colour at the end of each feather, the spots being of a different hue to the ground colour of the feathers. Spangled Hamburghs are one of the breeds having this marking.

SPRING CHICKEN. Class of young fowl sold for table. See Fattening, and Marketing.

SPURS. The horny spike on the shanks of cocks and some hens. Sometimes a cock's spurs cause serious wounds on hens, especially in the breeding pen, and if so they should be shortened, or the sharp points filed down until blunt.

SQUIRREL TAIL. A tail that is carried very high, reminding one of the manner in which a squirrel carries its tail. See Deformities.

STAG. Name given to a turkey cock.

STRAIN. This is the name given to variations within a breed of fowl, the variations being specially evolved for some definite purpose. Thus with the Light Sussex there has been evolved a special utility strain, table strain, dual-purpose strain and exhibition strain.

STRAWBERRY COMB. A comb somewhat resembling half of a strawberry, as in the Malay breed.

STUBBING. Term applied to the removal of feather stubs from a table bird after plucking. See Plucking.

SUN PARLOUR. Term applied to a wire-netting enclosure fixed to a brooder house or foster-mother, in which artificially-reared chicks may exercise and rest in the sunshine. Sun parlours are usually raised off the ground, being provided with a floor of small-mesh wire-netting, preferably with a tray below to receive droppings. Thus birds using it do not come into direct contact with the soil and can-

not therefore pick up harmful disease germs.

For a brood of about 100 baby chicks a sun parlour would need to be 3-4 ft. wide and long and 6-8 in. high.

Sun parlours are also used very extensively in connection with intensive poultry houses both for growers and for adult birds. They are neatly wired-in enclosures, entrance to which is gained through ordinary openings like pop-holes. They may extend for a considerable part of the length of large intensive houses.

SUSSEX ARK. Small slatted floor poultry house. See Night Ark.

TABLE BREEDS. Among fowls, the Light Sussex is usually considered to be the best breed for table. Dorkings, La Bresse, Faverolles, Orpington and White Wyandottes are also fairly extensively used. Often the table poultry producer prefers a cross-breed bird. The following are the most popular crosses : Indian Game—Dorking ; Indian Game—Buff Orpington ; Indian Game—Light Sussex ; Faverolles-Sussex.

Of ducks, the breed kept most extensively for table is the Aylesbury. The Pekin and the Rouen are also good table breeds.

The most popular table breed of turkey is the American Bronze, but all breeds are serviceable. Similarly all utility breeds of geese serve for table purposes.

TABLE POULTRY. Term applied to poultry raised for killing for food. See also Fattening, Killing, Marketing.

Old Birds, To Make Them Tender. Give the birds 1 dessert-spoonful of vinegar 1 hour before killing. Give the dose in two lots ; a bird cannot swallow as much as a dessert-spoonful at once.

It is worth remembering that if a bird is killed, plucked and cooked at once while still warm after killing, it will be tender. If left only 2 days it will be tough, but if allowed to hang head down in a cold pantry for 6 or 7 days until limp, it will be found to cook fine and tender.

Old birds may also be made tender by using a pressure cooker ; or by steaming for three hours before roasting.

Plucking. See Plucking, Stubbing.

Plumping Thin-breasted Birds. The bird having been drawn, take a pair of strong scissors and, guiding them inside the empty carcase, feel for and cut through the ribs, taking care not to damage the under skin. Both sides are so cut and then the breast-bone is tapped down with the handle of a knife, so changing the blade-like appearance of the breast to a nicely rounded form.

An alternative method—but not so good—is to break the breast-bone with a wooden striker.

It is also possible to camouflage a scraggy thigh. First, the middle finger is placed inside the carcase and pushed through the flesh so as to reach the inside of the thigh skin. The skin is worked loose from the thigh by pushing the finger about. It will now be found that, if the leg (shank) is grasped, the body skin can be pushed right over the thigh so as to half-hide it from view and in all cases make it more attractive.

Singeing. After a table bird of any kind has been plucked clean, the hairs on the carcase should be removed by burning. The simplest way to do this is to use a piece of lighted brown paper and pass the flames quickly over the carcase.

Other methods are to singe over the blue flame of a gas stove, or the flame of methylated spirit lighted in a tin lid.

Shaping. See Shaping.

Trussing. See Trussing.

Table Ducklings, Geese and Turkeys. See under Marketing.

TAIL COVERTS. Term applied to the curved feathers at the root of the tail.

TAIL FEATHERS. The straight, stiff feathers of the tail.

TOE PUNCHING. See Marking Poultry.

TONICS. See under Ailments, Health Tonics.

TRAP-NESTING. Trap-nests are laying nests that hold prisoner the birds using them, so enabling the poultry-keeper to know precisely how many eggs a year each one of his birds is laying. The birds must remain on the nest until the poultry-keeper releases them.

Trap-nesting is essential for all poultry-keepers who would compete in any of the big laying tests, and also for all who wish to sell sittings of eggs, or chicks, or stock at a good price. It is a definite asset to be able to advertise that eggs, day-old chicks or stock are from trap-nested parents with an egg record of, say, 215 eggs in their pullet year.

There are various types of trap-nest, some of them to be bought very cheaply. Any handyman, however, can make his own trap-nest.

The simplest home-made device consists of a wire flap, bent as here.

The hooks at the top are hooked on to staples at the top of the nest box and the bottom impinges against the bottom flange of the nest box. Thus a bird wanting to use the nest box pushes

against the wire flap, raises it and enters. Immediately she is on the nest, the flap drops back into place. It cannot be pushed forward when the bird desires to leave ; pressing against it will merely push it up against the flange of the nest. Thus the bird is held prisoner until the poultry-keeper himself raises the flap.

Naturally all birds must have an identification mark (a leg ring or leg band) when trap-nests are used. Also there must be a proper egg-recording card hanging on the wall near the nests. When a bird is being released from a nest, her identification number or colour is noted and at once the egg is jotted down to her score on the recording card.

Birds must be released as soon as possible after they have laid : all the time they are prisoners they can neither eat nor exercise. The ideal procedure is to go the round of the nests at fixed times daily —twice during the morning, again at the time of giving the noon feed and once during the afternoon.

TROUGHS. For sizes of food troughs see p. 41.

All food utensils must be kept scrupulously clean and should frequently be scrubbed with boiling water and soap, disinfectant being added to the water.

TRUSSING POULTRY. Trussing means tying-up of table poultry carcases so that they may look shapely and attractive in appearance.

Fowls. For this there will be required a special 10-in. trussing needle, and, for each bird, 24 in. of good, non-fluffy string in two 12-in. lengths. The trussing operation is as follows :

Having drawn the bird (see under Marketing) place it on a table. Start by bending the wings over on themselves so that they lie on the back. Now take the trussing needle threaded with one of the lengths of string and pressing back the thigh, drive the needle in at the angle of the thigh-bend, pushing it through so that it comes out in the same position on the other side of the carcase, the thigh on that side being pressed back in turn to allow this.

Now carry the needle—still threaded with the same piece of string—to the wing on the side where the string came out. Pass the needle through the centre of the middle joint and out through the centre of the end joint. Unthread the needle and, threading it with the other end of the string, treat the second wing in the same way.

Unthread the needle and, taking the two ends of the string, tie them tightly together, at the same time tucking in the neck-skin flap.

Now pass to the tail end of the bird.

Threading the needle with the second piece of string, push it through the flesh just behind the "parson's nose," driving it in deeply enough so that it just skims the back-bone. In order that the string may be concealed take the needle through the flesh along the side of the back.

As the needle emerges, let it continue through the hock joint, round the leg, through the flesh at the end of the breast-bone, on through the hock joint of the opposite leg, round this leg, up through the side flesh and out at the same point behind the "parson's nose" as that at which the string entered, the string then being knotted once to prevent it slipping.

The "parson's nose" is now tucked down into the opening below, the string is drawn tight to pull the legs and carcase close together and hold the "parson's nose" in position, and is then knotted and trimmed.

The result of this method of tying is a carcase that looks attractive and plump and is easily carved.

Ducks. The best way is to follow the same procedure as with fowls. Although many carcases are skewered, a much neater effect is obtained by tying with string.

Where the skewer method is preferred three skewers should be employed. The first one is pushed through one wing and out through the other wing, so that both are held low down and close to the body. The second one goes through the legs to hold these in position (also low down) and a third one can be used to hold the second joint of the leg in a shapely position.

The flap of skin over the neck should be kept in place by a short skewer.

Turkeys. For turkeys skewers are used, metal ones being considered better than wood.

Place the carcase on its back, push the thighs close into the body and low down, and push a skewer through the wings, taking up a little of the lower part of the thigh flesh to hold it down. The second skewer is pushed through just behind the hock joint (holding the hocks low down) and through the rear end of the back.

The third skewer holds the shanks in place, being forced through the leg about 1 in. from the ankle, the feet having been cut off, of course.

A small skewer is used to hold the flap of skin in position over the neck entrance.

A second operation, known as stringing, follows the skewering. It consists of winding string backwards and forwards from one skewer to the other, starting at the end leg-skewer and tying off at the wing skewer the other end.

Geese. Follow the skewer method as advised for ducks.

TURKEYS. Turkeys are a most profitable class of poultry for anyone having a reasonable amount of ground. They are an ideal side-line for commercial egg and table-chicken producers, for fruit-growers, market-gardeners, small-holders and farmers.

One of the chief essentials to success with turkeys is the possession of plenty of good, fresh grass land. With up-to-date rearing methods it is not really necessary to have unlimited space, but the ground used must have been clear of all other poultry for at least 6 months.

Ailments. The chief ailments from which turkeys may suffer are : Blackhead, Diarrhœa, Emphysema, Gapes, Going-Light, Scaley Leg and Swelled-head. See under separate headings, Ailments.

Breeds. The four most important breeds are the Black (or Norfolk), Bronze (American Mammoth or Cambridge), the Buff and White (Austrian or Holland).

Normal weights are : Bronze : Cock, 20-30 lb. ; cockerel, 15-25 lb. ; hen, 15-20 lb. ; pullet, 12-18 lb. Other breeds : Cock, 20-25 lb. ; cockerel, 15-20 lb. ; hen, 12-20 lb. ; pullet, 10-18 lb.

Fattening. See Fattening.

Feeding Turkey Chicks. *From 1 day to 10 weeks old:* Wet mash of best middlings, 4 parts ; maize germ meal (or maize meal), 1 part ; Sussex ground oats, 2 parts ; alfalfa meal (or clover meal), ½ part ; broad bran, ½ part ; unextracted linseed meal, ½ part ; fish meal, impregnated with cod-liver oil, ½ part.

Feed 8 times a day for first week, 6 times during second week, 5 times a day for next fortnight, 4 times thereafter.

Turkeys from 10 weeks onwards: Wet mash of middlings, 5 parts ; Sussex ground oats, 2 parts ; maize meal, 2 parts ; bran, 1 part ; meat and bone meal, ¾ part. Up to 4 months old give 4 meals a day, 2 of above mash and 2 of corn, alternately, starting with corn as first meal. After 4 months give 2 meals only, mash in morning, corn at night.

For Laying Turkeys: Wet mash as for growing turkeys, except that 1 part meat and bone meal instead of ¾ part is given. Give mash morning, grain evening.

Hatching Turkeys. See Hatching.

Marketing Turkeys. See Marketing.

Mating Turkeys. See Breeding.

Plucking Turkeys. See Plucking.

Rearing Turkeys. Rearing follows the same lines as with chicks. See Rearing.

Sexing Turkeys. See Sexing.

Shooting the Red. At 10 weeks of age, turkeys enter upon a period known as "shooting the red". This is not a disease but a normal process in which the birds' wattles grow fully. The strain which is often spoken of as being put on the birds at this period need not be feared so long as they have been bred from healthy stock, are well fed, and warmly and drily housed.

As a tonic during this period give cod-liver oil and Parrish's Chemical Food in equal proportions at the rate of 1 teaspoonful per bird, per day.

Turkeys should always be taught to roost before shooting the red.

Trussing Turkeys. See Trussing.

UNDERCOLOUR. Colour of the body-fluff to be seen when the plumage is brushed aside.

UTILITY POULTRY. Poultry kept for egg or table-bird production as opposed to those kept for exhibition or fancy purposes.

VENTILATION. All poultry-houses must be so ventilated that they do not appear to be stuffy when they are entered of a morning after being closed up all night. Ventilators must on no account permit a draught (see Draughts). Roof ventilation is usually considered best. Louvre-board ventilators are ideal as they deflect the entering air upwards. The type of ventilator which consists of an opening along the apex of the roof is also to be commended. See also Housing.

Ventilation of Incubator Room. See Hatching.

VERMIN. See Insect Pests.

WATER, DRINKING. All poultry at all ages require a regular and constant supply of pure, clean drinking water. Fowls stinted of water do not lay as well as they should ; fowls compelled to drink impure water are unlikely to remain healthy.

All water vessels (founts, etc.) should be kept scrupulously clean, and be regularly washed out, especially during summer.

For medicated water earthenware vessels are always desirable.

Sun-warmed water is harmful to fowls. All drinking vessels should stand in the shady parts of the house ; or when in the open for the use of birds on free range or in runs, they should be in a shady place or be provided with shade, such as a thatched hurdle or an up-turned three-sided box.

WATER FOUNTS. Many types are available, each being the most suitable for some particular purpose. Thus, for providing water on grass range, the bucket fountains which lie on their side are the best, as they are easy to fill and carry. For chicks, and mature birds galvanised founts of various sizes are available, to suit different numbers and sizes of birds—e.g. ½ gal. to 3 gal. or so.

For hanging in a house the 1 and 2-gallon wall founts, with hood over the drinking trough, are useful.

Top-fill founts holding 3 and 4 gallons are excellent when larger quantities are required. For still larger quantities one can use an empty oil drum or barrel fitted with a rubber tube leading to a valve placed in any shallow drinking vessel.

Earthenware founts are also available in various sizes ; while wine bottles can be used for small poultry outfits by fitting them to a " bottle tray ". Earthenware founts keep water cooler and are more sanitary than galvanized iron, but are more expensive.

Founts, or fountains, are much better than open vessels, for providing the water supply.

Any vessels should always be so placed that the fowls cannot step in them or scratch dust and litter into them.

As a guide to the number of vessels to provide, a hundred fowls will drink about six gallons of water a day, but it would be safer to allow eight gallon capacity founts per hundred birds. Even so the vessels should be filled morning and late afternoon ; otherwise the birds, when they come off the perches early in the morning, may find nothing to drink.

Cleaning Water Founts. The usual way is just to rinse them out every time they are refilled, giving them, perhaps a really good cleansing once every month or so. This amount of cleaning is not nearly sufficient to keep the water pure. Under such conditions the insides get covered with either a greenish slime or a blackish deposit. Slime and deposit are definitely harmful.

Founts that can be taken to pieces should be scalded out at least once a week. Then they should be scrubbed out with a weak disinfectant solution and finally thoroughly rinsed. This is additional to rinsing at each filling.

With founts that are not separable into parts each fount may be quarter-filled with cinder siftings and an equal quantity of water. Then shake the fount about for a few minutes and empty out. Do this several times until the water comes away clear.

Frozen Drinking Water. It is very desirable to prevent the drinking water from being frozen, because otherwise there must be short periods during which the birds go thirsty until the water has been thawed and fresh water supplied. The following are alternative methods of preventing freezing: (1) Hang old sacks on the outside and inside of the house wall near the founts ; (2) move the founts to the warmest part of the house—e.g. close to the roosting birds at night ; (3) collect all the founts into a group over-night and hang a hurricane lamp near them; (4) provide water founts in enclosed boxes on the house wall, for the birds to drink from inside, and have a small heater in the boxes, near the founts.

Open-air water vessels should be emptied at night during severe frost, and refilled with warm water in the morning. When unexpectedly frozen over it is easy to thaw by using hot water.

WATER-GLASS. Preparation used for preserving eggs. See Eggs, Preserving.

WATTLES. The folds of skin or flesh at each side of the beak.

WEB. The skin between the toes; the flat or plume part of a feather ; the skin seen when a wing is extended.

WHITEWASH. Whitewash is the best covering for the interior of poultry houses as it helps to make them lighter. Some poultry-keepers also whitewash the outside of the roofs of their scratching sheds and intensive houses in summer, as the interior then keeps cooler—white not absorbing heat so readily as dark colours.

For obvious reasons the whitewash used must not rub off at a touch. Following is a recipe for a whitewash that will stick on well and will not rub off on contact :

Place the lime lumps in a receptacle and pour enough hot water over them to cover them by about 1 in. Soon the water will come to the boil. Before this happens add a little fat or tallow—$1\frac{1}{2}$ lb. to every 7 lb. of lime. When boiling has subsided, add enough additional hot water to make the whole into a thin wash, the consistency of milk. Apply the liquid whilst hot.

Another good stick-on whitewash is made as follows by slaking $\frac{1}{2}$ bushel of lime with hot water as above, adding 2 lb. of zinc sulphate and 2 lb. common salt that have previously been dissolved in hot water.

Always add a good disinfectant to Whitewash.

WING BARS. The two lines of stiff feathers across the wings.

WRY TAIL. Term applied to a tail carried to one side instead of in a line with the backbone. See Deformities.

YEARLING. Female fowl from the age of 1 year to 2 years, after which it is known as a hen.

VEGETABLES

A CONITE. One of the important medicinal herbs. It is a perennial and grows best in damp, shady places. May be raised from seed sown in August, but propagation by division of the roots is simpler. The roots are the part of the plant required by druggists. Also see Herbs.

ALLOTMENTS. Allotments fall into two main groups—those provided on ground owned or rented by local authorities (by far the larger group) and those on ground owned or rented by allotment societies. At the outbreak of war there were some 900,000 allotments in this country. Under various Cultivation of Land Orders very considerable areas of new land were made available for allotment purposes, and by the beginning of 1942 more than 1,500,000 plots were under cultivation.

The produce from allotments has played an important part in the feeding of our people in wartime, especially since by carefully-planned cropping an average allotment—10 rods or 300 square yards—can produce enough food to feed a family of four for the greater part of the year.

ANGELICA. Biennial herb, the stalks of which are in fair demand by makers of confectioner's supplies. Grows best in moist soil. Raised from seed sown in early summer. Sow in rows 2 ft. apart and thin plants to 2 ft. apart. Also see Herbs.

ANISE. Annual herb, the seeds of which yield the oil known as aniseed. Seed is sown in April in a sheltered corner, in rows 6 in. apart. Thin seedlings to 3 in. apart. Harvest seeds of the crop in September. Also see Herbs.

ARTICHOKES. There are three different types of Artichokes—the Chinese Artichoke, which is a " luxury " vegetable, the Globe Artichoke, and the most-widely grown sort, the Jerusalem Artichoke.

Chinese Artichokes. These are somewhat similar to Jerusalem Artichokes but are smaller and the tubers are queerly curled.* They are very prolific. They are grown from tubers planted in March in well-drained, manured soil, in drills 4 in. deep, the tubers being 10 to 12 in. apart in the drills and the rows 18 in. apart. After planting, keep the land free of weeds and fork between the rows at odd times to keep the soil loose and free. No earthing up is required.

The tubers may be left in the ground until required or stored in a clamp.

Globe Artichokes. It is the globe like heads of fleshy leaves that are eaten. The plants are raised from seed sown in drills made in rich soil in March or April. The seedlings are thinned out to stand 6 in. apart in the rows and left until the following spring, when they can be transplanted to a permanent bed of rich soil, being put in 4 ft. apart.

For quicker results plants may be bought in April. These will yield some " globes " the first summer.

Once the plants are established as many further plants as are required can be obtained by removing the sucker growths which are produced freely in April and planting them 4 ft. apart in rich soil.

In autumn, after all the globes have been gathered, the top-growth of the plants is cut back to within an inch or two of the ground. In winter, the somewhat tender roots are protected from frost by a covering of straw or ashes.

Chards. Chards is the name given to the blanched shoots of Globe Artichokes. These shoots are reckoned as a great delicacy and some growers use their plants more for the production of chards than of globes. Usually, however, the plan is to use young plants for globes and older plants for chards, the latter being destroyed at the end of the season, for the production of shoots will have exhausted them.

Blanching Chards. The growths are produced in spring. When they have reached a height of 2 ft. they are bound round with straw and covered with leaves or other litter. About 6 weeks later the blanched chards are ready for gathering. A couple of handfuls of dried poultry manure per clump in spring is a good stimulant.

Jerusalem Artichokes. These produce underground tubers similar to potatoes. The plants grow 8 to 12 ft. high and thus are often used to form a wind-break screen at the end of a plot, or as a screen for a manure or compost heap. Tubers are planted 6 in. deep and 1 ft. apart in rows 2 to 3 ft. apart—in soil on the light side ; too rich a soil produces foliage at the expense of the tubers. January is a good planting month. During the summer only an occasional hoeing is required.

The stems are cut down in October. The tubers can remain in the ground until required or lifted and stored in a heap of sand. It is desirable to lift all tubers at the end of each season and make a new

D   89

planting for the next crop. The common practice of leaving unwanted tubers in the bed year after year leads to very poor quality crops.

ARTIFICIAL MANURES. See Manures.

ASPARAGUS. Although Asparagus is regarded as a luxury vegetable, it offers no difficulties to the grower. Moreover, it is a real " permanent " crop ; once a bed has been planted it will remain productive for many years. It is one of the most profitable vegetable crops for the commercial grower.

The bed can be 3 or 5 ft. wide and any length desired. It should be made in an open, sunny situation sheltered from north and east winds.

Preparing the Bed. The bed should be prepared in winter, planting usually being done the following April, though in sheltered districts planting may be done in October.

Having marked out the bed, take out the soil two spits deep, keeping the spits separate, and fork over the third spit. Good drainage is essential, so unless the soil is very light it is advisable to put in below the 2 ft. depth a 6 in. deep layer of drainage material—lime rubble, broken bricks or stones.

Now return the second spit, mixing with each barrowful half a barrowful of a compost of equal parts lime rubble, littery manure and leaf-mould. Having returned this spit, sprinkle on the surface ¼ in. bones, if available, 4 ounces per square yard. With each barrowful of the top spit mix half a barrowful of a compost of 2 parts well-rotted manure and 1 part lime rubble. Break up both spits finely.

The additions to the soil will, after moderate firming, raise the bed 9 in. above its surroundings, as is desirable.

Planting Asparagus. The roots are planted in rows 15 in. apart each way, the crowns or root tops being set 4 in. below the surface. The simplest planting plan is to make right along the bed a trench with a shallow ridge down the centre. A bed 5 ft. wide will accommodate three rows of plants. Cover the ridge with a thin layer of sand and set the crowns on the ridge, spreading the roots evenly down both sides. Cover gradually and carefully with fine soil. Finally, mulch the bed with a 3 in. layer of well-rotted manure.

Year-old crowns are cheapest but with these no crop will be forthcoming for two more years. For a crop the season after planting, buy 3-year-old-crowns. Conover's Colossal can be recommended as a leading variety both for market and home use, on account of its reliability and the high quality of its stems.

Cutting Asparagus. Having planted 3-year-old crowns, a few shoots may be taken the first summer after planting, but the less any plants are cut during their first season, the stronger and more productive will they be in subsequent years.

When the bed is established, cut all shoots produced in the early part of the season when they are 6 in. high. It is best to use a special asparagus knife, which has a saw-edge ; a clean cut, as with a table knife, causes the stumps to " bleed " and so weakens the roots. Draw away a little soil from around the shoot and cut an inch below the surface. Cutting should cease entirely about the middle of June, all remaining strong shoots being allowed to grow and form the well-known feathery foliage. This foliage must not be cut for vases, bouquets, etc.

Cultural Hints. During late spring asparagus beds need frequent hoeing to keep down weeds. Should the weather prove dry, especially soon after planting, water must be given—at least ½ gallon per plant on each occasion.

After cutting has finished in mid-June, weakly shoots which were not worth gathering for the kitchen should be cut off. The bed should then be given a dressing of salt at the rate of 2 oz. to the square yard. During the summer—after the first year—the bed should be fed. Monthly sprinklings of nitrate of soda (1 oz. to the square yard), dressings of a mixed fertilizer and weekly waterings with weak liquid manure are all good.

When the foliage has grown tall, as will be the case as summer advances, it should be supported by means of cords tied to stakes set around the bed.

The foliage plays an important part in building up the crowns for the next season.

Each spring asparagus beds require top-dressing or mulching after being lightly pricked over with the fork. Should the soil in which the plants are established be a heavy one mulch with sifted leaf-mould but, before applying it, scatter ¼ lb. of good-grade bone meal, guano, or garden fertilizer over the ground, and round, but not touching, the plants. Rich compost, sandy soil or sifted, rotted-down compost heap may also be used for this mulch. Whatever is used the fertiliser mentioned above must be applied first.

With light soil it is best to mulch with rotted manure. This keeps the soil moist and warm, while at the same time feeding the plants.

Dig the alleys between the rows when the mulching is finished.

Forcing Asparagus. Early crops can be obtained by digging up the roots from the outdoor bed and planting them in a frame over a hotbed, setting them 4 in. deep in the usual way. Forced roots are usually too weak to be worth planting out in the garden again.

Asparagus Pests and Diseases. Asparagus has only three serious enemies, as follow :

Asparagus Beetles. These are yellow and black beetles which gnaw the stems and foliage of plants, particularly from June to August. Remedy is to dust or spray with derris.

Asparagus Fly. This pest which rarely causes serious trouble in this country, punctures the shoots in June and July, and deposits an egg in the puncture, the egg hatching out into a maggot. The presence of the maggot causes the shoot to yellow and afterwards die down. If it has appeared in previous seasons, give a preventive spraying with derris.

Asparagus Rust. This causes yellow-brown pustules to appear on the stems and foliage, which then turn black. The trouble is most prevalent from June to September.

Spray with Bordeaux mixture (for recipe, see Bordeaux mixture) after cutting is finished and repeat twice at 10-14 days intervals. When foliage is cut down in autumn, promptly burn it.

AUBERGINE, or Egg Plant. Vegetable that has gained in popularity in recent years and is well worth the attention of market growers. The plants produce large " fruits " which are both ornamental and edible. The " fruits " are egg-shaped or similar to small marrows, according to variety, and white or purple in colour. They are delicious when sliced and fried, or they may be stuffed with various mixtures and then cooked.

Plants are usually grown in a greenhouse. They are raised from seeds sown thinly in February in shallow pans filled with light soil and grown on in temperature of 65 to 75 deg. F. When large enough to handle, the seedlings are potted and subsequently potted on until they are in 8 in. pots in which they fruit. Plants may also be grown in sheltered positions in the open and will grow well if summer be hot.

B **ALM.** Perennial herb, the leaves of which are used medicinally, being gathered from May to October. Can be grown in ordinary garden soil, seeds being sown in April. in rows 1 ft. apart. Thin out to 1 ft. apart. Plants can afterwards be propagated by means of cuttings, or by division of roots. Also see Herbs.

BASIC SLAG. Useful slow-acting manure. See Manures.

BASIL. Annual herbs of two kinds

—Bush Basil and Sweet Basil. Leaves of both are used for flavouring. Sow seeds of the Bush Basil in March, in rows 1 ft. apart, thinning to 8 in. apart ; sow seeds of Sweet Basil in May, in rows 15 in. apart, thinning to 8 in. apart. Leaves can be gathered from Bush Basil from July to September, from Sweet Basil from September to April. Also see Herbs.

BASTARD TRENCHING. Another name for mock trenching. See Digging.

BEANS. Beans rank among the most popular and profitable of all vegetables, whether grown for home use or market. This applies equally to French beans, runners and broad beans. Haricot beans and butter beans are two other useful kinds.

Broad Beans. These will grow readily in any well dug ground in good heart. Rake the soil fine just before sowing. Make drills 9 in. wide, 2 in. deep in heavy soil, 3 in. in light soil. Sow in a double row alternately at 6 in. apart, laying the seeds flat on their sides. Half a pint of seeds will sow a 30-ft. row. Fill in firmly with fine soil. If there are several drills leave 2½ ft. between them.

The first sowing can be made in late February with successional sowings in March and early April.

Good varieties for general use are the longpods Bunyard's Exhibition and Seville Longpod, and the Windsors Homestead and Broad Windsor.

Cropping is earlier and more generous when the side branches that form on the main stem are rubbed off as soon as they appear. The tips of the plants should be nipped out when flowers form as a preventive of black fly, the worst pest of Broad Beans.

Half an ounce of nitrate of soda to the yard of double row is a useful spring fillip. A week later apply superphosphate at the rate of 2 oz. to the yard run. Each time hoe in the dressing.

Butter Beans. The golden pods borne by these beans are cooked whole. They have a very pleasing flavour and are not stringy. The plants are grown like dwarf French Beans, being sown indoors in boxes in early April, or outdoors at the end of April.

They like a sunny spot and when planted out, should stand 1 ft. apart in rows 18 in. apart.

Dwarf French, or Kidney Beans. One of the most profitable crops. Not only are supplies wanted by the markets but also many tons are required by the canning factories.

For ordinary outdoor crops, make successional sowings from April to July— at least three sowings during this period. Sow on good rich soil 6 in. apart in drills

4 in. deep. With more than one row, let the drills be 2 ft. apart. Half a pint of seed will sow 60-ft. row. Many growers sow a single row of French Beans along the top of the celery and leek ridges.

For Early Supplies. Earlier pickings of beans can be obtained by sowing seeds in pots or boxes, in a greenhouse or frame, in early March and planting out the plants in the garden as soon as the weather becomes settled.

A further means of obtaining early supplies is to sow in early March in a warm, sun-bathed border, sheltered from the north and east by a hedge, fence or wattle hurdles.

Where there is greenhouse space, profitable out-of-season crops can be grown, sowings being made in succession from August to March. The first sowings provide pickings in October and November, when prices are very high.

For this purpose use clean 8 in. pots, drained by placing over the hole at the bottom a large inverted crock and a layer of small ones round, covered with 1 in. layer of rough loam or compost riddlings. Half fill the pots with the following compost : Loam 4 parts, leaf-mould, manure, nut charcoal, and sand 1 part each, with 4 oz. of bone meal to the barrowful of compost.

Dibble 7 seeds into the compost at an equal distance apart, and ½ in. deep. Stand the pots in a draught-free corner of the greenhouse or frame. When growth is 1 in. long, remove the beans to a light stage. Spray the plants lightly on bright mornings when there is reasonable prospect of evaporation and a dry air before night. Don't spray at all on dull, rainy days.

Reduce the plants to 5 when they are big enough to tell which are the strongest. Top-dress as soon as the first flowers show, filling up the pots to within 1 in. of the rim with a compost of loam 3 parts, manure and sand 1 part each.

Feed weekly with very dilute liquid manure after the first pods are 1 in. long until the crop is finished. Outdoor crops should be fed as recommended for Runner Beans.

Choose either Lightning or Osborne's Early Forcing for the early sowings, Canadian Wonder or Masterpiece for the main sowings.

Climbing French Beans. These are sometimes grown in place of runners outdoors or as a greenhouse crop. The beans, smaller than runners, are stringless and very tender. For greenhouse culture the beans are sown in the same way as the Dwarf French Beans, being grown on in prepared beds or large pots.

Outdoors the seed is sown in May, the plants being trained and treated generally in the same way as runners.

Haricot Beans. These have received much more attention from gardeners during recent years because they are so useful for winter food. The sowing is made at the end of April or early in May, when weather conditions appear to be settled. Sow in drills 2 in. deep and treat generally in the same way as French Beans.

The pods are left to hang until they begin to turn brown. Then the plants are pulled up whole, tied in bundles and hung up for the beans to ripen properly. When quite dry the beans are shelled out and stored.

Brown Dutch and White Rice are good varieties.

An up-to-date idea is to grow white runner beans, some of which can be eaten green, like ordinary runners, while the rest can be left for their seeds. These seeds, when dried, are a perfect haricot substitute. The plants are grown in the same way as other runner beans. To provide seeds, the beans are left to dry as described above for haricot beans.

Runner Beans. To obtain good crops of Runners, proper trenches must be prepared as early in the winter as possible.

Having selected the position—one in full sun but protected from the east— mark out the site for the trenches, which should run north to south, not east to west. Make the trenches 2 ft. 6 in. wide on ordinary soil, 3 ft. wide on heavy soil.

Start by taking out the top soil to the depth of 1 ft., depositing it in a ridge on the *right* of the trench. Into the trench put a good supply of manure, decayed garden rubbish and the like and dig this into the sub-soil to the depth of 1 ft. Leave the trench thus for a while. In 3 weeks' time sprinkle the newly-turned sub-soil with basic slag, 4 oz. to the yard run and prick it in 2 in. deep. Also sprinkle the top-soil—which was piled up at the side of the trench—with 2 oz. basic slag per yard run.

Leave the trench alone again until February ; then return the top-soil to the trench, aiming to leave the surface a couple of inches below the surrounding level.

Sowing Runner Beans. A pint of seed is required for each 15 yards run or row. Streamline, Prizewinner and Scarlet Emperor are reliable varieties.

Tread the soil in the trench firm, then take out drill 2 in. deep and 12 in. wide. Sow, 9 in. apart each way, two rows of seed in each drill. Sow the seed alternately, thus ensuring to every plant the maximum space. Where beans of show standard are required, sow one row of seeds 12 in. apart down the middle of each drill.

Before filling up the drill (if the plants

are to be grown on stakes) fix, 1 in. away from each bean, a substantial stake rising 8 ft. out of the ground, and going 2 ft. down. Staking should not be deferred until after growth starts or some vital roots will be pierced with the stake end. Tie firmly, half-way up the row and extending the whole length, a line of crosspieces to give additional stability.

Sow a few extra beans at the end of the row to provide a spare supply of plants to replace any " casualties " in the rows proper.

If gaps have to be filled do this before the running stems begin to bush (at the 4 in. high stage).

Sowing Against Fences. Runner Beans can be grown well against a fence. A south-facing fence is rather too hot except in northern or exposed districts.

To prepare a site at the base of a fence mark out a strip 2 ft. to 2 ft. 6 in. wide, put on a layer 3 in. deep of rotted manure, or, failing that, $\frac{1}{2}$ lb. of hop manure per yard run of row and dig spit deep.

Put the beans in in a single row—4 in. away from the wall, 3 in. deep and 9 in. apart. As soon as the young central runner begins to push, fix up a stout string for it to climb to—one to each plant. Fasten one end of the string to a peg driven into the ground, and the other end to the wall or fence near the top.

Dwarfing Runner Beans. In market gardens it is too expensive a business to stake runners ; they are therefore dwarfed and grown without sticks. It is also customary to dwarf the runners in exposed gardens. Dwarf runners do not crop so heavily as those grown on sticks, but all the same they yield well enough.

Dwarfing is done as follows : When the plants are 12 in. high, nip off the growing point. This will cause branching. The tips of the branches are nipped off in turn and thereafter every trail that is produced is pinched back. The result is bushy growth, like that of a dwarf bean.

Aids to Heavy Cropping with Runner Beans. For really prolific crops of particularly good quality beans, " stop " each plant when 1 ft. high, take the three best of the resultant side-growths and grow them up separate stakes.

Always syringe beans frequently during fine weather. This helps the flowers to set—preventing that usual trouble of a number of flowers on each spray failing to produce beans.

Early in summer spread a thick layer, 1 ft. wide, of rotted manure or lawn mowings or peat on each side of the rows.

Feeding Runner Beans. As soon as the runners have begun to twine round the supports, give a dose of superphosphate at the rate of 2 oz. per yard run of row. Hoe and water it in. Repeat the dose in 6 weeks' time.

From the beginning of July liquid farmyard manure once a week, or liquid cow and soot manure, should be given, these applications being kept up until the end of August, when the manuring season is over and growth stops. It is important, when doing the feeding, to keep the liquid well away from the leaves and stems of the plants.

Saving Runner Bean Roots: If the roots of Runner Beans are dug up in the autumn, stored in a frost-proof shed, under a covering of dry soil, and planted out in the garden again in early May, they will yield a good and early crop.

Diseases and Enemies of Beans. Beans are not subject to many troubles. Their worst worries are the following :

Black Fly. These are sometimes found in Runners but mainly attack Broad Beans, on which they are often a positive plague. They can be kept down by pinching off the tip of each plant just when the flowers have set. As a further precaution it is advisable to spray the plants with quassia solution (see Quassia Solution for recipe) several times from the time flowers begin to appear until the pods are forming.

Red Spider. This is the worst pest trouble of Runner Beans, and also affects Dwarf Beans. The symptoms of attack are brown or coppery foliage with moving red specks on it. The remedy is to spray with paraffin emulsion (see Paraffin Emulsion for recipe) followed with clear water.

Canker. Both French Beans and Runners are subject to a canker which shows, first on the stems and then on the pods, as dark irregularly shaped patches edged with red, forming depressions. The plants should be sprayed with half strength Bordeaux mixture at 10-14 days intervals.

BEETROOT. The round, or globe, type should be sown in April to provide early roots to pull in summer. Sow the long type in early May and June. A July sowing of the globe type will provide roots to pull in winter. Crimson Globe and Cheltenham Green Top are reliable varieties.

Beet grows best in a rich, loose, deeply-worked, light loam. The site should be trenched two spits deep in autumn or winter and should be one which, although not freshly manured, was manured well for the previous crop. Freshly manured ground causes the roots to fork and become unshapely.

The seeds should be sown in drills 1 in. deep, being covered with fine soil. Rows should be 12-18 in. apart. $\frac{1}{2}$ oz. of seed will sow a row 50 ft. long. Germination is quicker when the seeds are soaked in water for two hours before sowing.

The seedlings should be thinned out as

early as possible. Thin Round Beet to
4-5 in. ; Long Beet 5-6 in.

Feeding Beetroot. Immediately after
thinning, apply a stimulant in the shape of
1 part sulphate of ammonia, 1 part sulphate of potash and 2 parts superphosphate. Apply at the rate of 3 oz. per
square yard of bed.

Diseases and Pests of Beet. *Rust* is
the only serious disease. It shows itself
in the form of raised pustules on the leaves,
the pustules being covered with a rusty-
red powder. The symptoms show up any
time from mid-June to mid-September.
As a remedy, spray with Bordeaux mixture (which see) at half-strength twice,
with a 14-day interval between the
sprayings. After harvesting burn all
tops and trimmings.

Beet Fly. This is a common pest. It
causes pale blisters or blotches on the
foliage in June and July. As a remedy,
spray with paraffin emulsion (which see)
on the first sign of attack.

Harvesting and Storing Beet. Mid-
October is the normal time for harvesting.
Even so, progressive lifting from early
September on is far better procedure.
There are no outward signs of maturation
on the foliage ; the size of the roots is
the only guide. If these are large enough,
they are also good enough, and they will
keep fairly well.

Beet are stored in a heap, with dry
sifted soil or sand between the layers.
With long beet spread a layer of soil 2 in.
deep on the floor of a shed, cellar, or other
suitable place, in the form of a semi-
circle backing against the wall and 18 in.
to 2 ft. in diameter (according to the size
of the roots). Then lay the roots down
crowns outward—after twisting, not cutting off the tops—tips inwards on this
semi-circle. The tips will just about meet
in the centre. Cover to hide from view
with dry sifted soil or sand.

Proceed in similar fashion with the next
layer of roots, and the next, until the stack
is complete. Then cover with 3 in. depth
of soil to finish. In this way the crowns
of the roots are all peeping out round the
stack.

With round beet the centre of the stack
is filled up as you go to complete each
layer of roots, but these interior ones do
not keep so well as those on the outside.

BEET, SEAKALE. Also known as
Swiss Chard and Silver Beet. Provides
a change from the ordinary run of vegetables. The leaves are cooked like spinach, the stalks and mid-ribs in the same
way as Seakale. Sow in April, in good
soil, in drills 1 in. deep and 18 in. apart.
Thin to 15 in. apart. Hoe frequently in
summer and work during dry spells.

BELLADONNA (Deadly Nightshade).

Valuable medicinal herb. It is a perennial and, owing to its poisonous nature,
precautions must be observed in growing
and harvesting the plants. The roots,
leaves and seeds are all required by wholesale druggists. See Herbs, Medicinal.

BIRD DAMAGE. Some birds, especially sparrows and wood pigeons, are a
nuisance to the vegetable grower, robbing
seed and seedling rows. For protecting
seed beds there is nothing to beat fishing-
netting. On small-scale beds black cotton criss-crossed over the soil and fixed to
12 in. high pegs inserted here and there
will serve. In recent years an excellent
cotton spinner has been introduced.

Birds have a dislike for anything of a
glittering nature, and pieces of mirror
and bright tin suspended over seed-beds
will scare them.

Movable guards for rows of seedlings
can be bought, or made up out of wire-
netting. These guards are particularly
necessary over young green peas.

BLANCHING. The term " blanching " means that the natural colour of a
plant is removed so that it turns white.
In the process the " strong " flavour of
such vegetables as celery, cos lettuce,
leeks, endive and cardoons is replaced by
a much milder and improved flavour.

In the main, blanching consists of excluding light from the plant for a period
before harvesting. For the methods of
blanching to be adopted with the different
plants which are so treated, see under
separate headings.

Blanching also indicates a special
treatment in the preservation (canning
and bottling) of fruit and vegetables.

BLIGHT. Term applied to sap-
sucking insects that smother the shoots
and foliage of various plants. The Blight
which attacks beans, for example, is the
Black Fly, which sucks out sap from
young shoots, causing loss of bloom and
crop. The treatment is to spray early in
the season with quassia mixture, nicotine
emulsion or paraffin emulsion (see under
separate headings for recipes).

For special forms of Blight attacking
specific vegetables and the special methods
of dealing with the pests, see under vegetable concerned.

BOLTING. Term applied to the
premature running to seed of vegetables.
The usual causes are too early sowing,
too thick sowing and checks in early
growth. A hot, dry, shallow, and poor
soil is also likely to lead to bolting. Plants
that have bolted are mostly useless and
should be pulled up straight away. With
young cabbage plants which are bolting
a plan which may be tried is to make a

small slit in the stem and insert a slip of wood in this.

BONE-MEAL. See Manures.

BONFIRE. A bonfire can be started with little trouble if the precaution has been taken to store a small heap of the more inflammable rubbish under cover—where it will keep dry—for use at the outset. Even when rain is almost continuous, given a few dry sticks and some very inflammable material like straw, the fire may still be lighted and kept burning by adopting the following procedure : Dig a shallow trench and over it lay an iron hurdle or some bars that will keep the rubbish from falling through into the trench. On this hurdle make a good fire of the *dry* stuff. As soon as this is burning well, put on some of the looser materials from the damp rubbish pile, such as hedge trimmings and stubs. When they have caught, put on a thin layer of the wetter material.

As this material begins to smoulder, pile on more rubbish, but never much at a time or the bonfire will be choked. Pack some of the damp stuff round the bottom of the burning pile to be drying. As it dries, it will gradually consume ; then more of the wet material can be disposed of in the same way. Whenever there is sign of a flame bursting out, put a forkful of damp material in that particular spot.

If the fire seems to be almost dead, thrust a stick through it, to allow the air to penetrate. This will revive it. When the centre is a mass of glowing ashes, pack up the fire well and leave it. It will continue for hours slowly burning away. (During war-time care must be taken that the bonfire is extinguished before black-out time.)

The ashes from a garden bonfire are a useful fertilizer. See Wood-ashes.

BORAGE. Annual herb, the flowers of which are used for culinary purposes, although not in very great demand commercially. The young leaves are sometimes used in salad. Sow seeds in March, in ordinary soil, in rows 1 ft. apart. Thin the seedlings to 1 ft. apart. Also see Herbs.

BORDEAUX MIXTURE. One of the best fungicides available and useful as a *preventive* of many diseases. Used freely for spraying potatoes against *the* potato disease. Also effective against rust and other diseases of various vegetables. •

Bordeaux mixture can be bought in ready-prepared form, and this is to be recommended, as it has only to be diluted with water. Where a home-made wash is desired, a simple formula is : Copper

sulphate, 1 lb. ; quicklime, 1 lb. ; water, 10 gallons.

The mixture should always be freshly prepared when required ; it loses its fungicidal powers if kept in store. Also, it must be prepared in a *wooden*, not metal vessel. The dissolved lime should be added slowly to the dissolved copper sulphate solution—never the other way about. And *cold* water must be used.

BORECOLE. See Kale.

BROCCOLI. Broccoli, when it can be produced early and late, and in good quality, is one of the most useful of all market crops.

There are as many as a hundred varieties of broccoli, but these are divided into four classes, according to the time in which they mature. The first section are ready for cutting in late summer, autumn and early winter. One sowing does not ensure cutting over such a long period ; two or three sowings must be made at intervals of three or four weeks. The same applies, in a less degree however, to the other three sections.

The second section matures in December, January and February. The latest plantings of the first section overlap the first plantings of the second section, and the latest of the second section overlap the third.

The third section matures before and after Easter. The fourth may be called summer broccoli, since they mature in April and later. The very last of this section just meet the first section.

Good varieties are : First section (sow February to April) : Michaelmas White, Early Roscoff, Walcheren and Autumn Protecting.

Second section (sow in April) : Middle Roscoff, Winter Mammoth, Penzance, Early Market, Early White.

Third section (sow April to end of May) : Late Roscoff, Late White, Leamington, Snow White and Eastertide—the latest.

Fourth section (sow in May) : Whitsuntide, Bouquet, Model, Standwell and Late Queen. Those mentioned have splendid heads, difficult to distinguish from cauliflower.

The earliest sowings should be made in the frame, or, if wanted *very* early, in the greenhouse.

The later sowings are made in the open ground in drills ½ in. deep and 10 in. apart. Following thin sowing the plants can be left where they are until they are big enough to go into permanent quarters where they are planted 2-2½ ft. apart.

The soil that suits broccoli best is one which is moderately rich, without fresh manure.

An ounce of seed will produce 1,000 plants.

Perennial Broccoli. In addition to the above there is a distinctive type of perennial broccoli. This produces from nine to twelve heads, each the size of a good cauliflower and of excellent quality. The plants are left in the ground after all the heads are cut, and the soil is well enriched with manure. In the following and subsequent seasons the plants will yield further supplies of good heads. Seed is sown from the end of April to the end of May, and culture generally is as for other broccoli.

Feeding Broccoli. No feeding should be given young broccoli. When the young plants are making headway after planting out they can be greatly helped by the following : 1 part sulphate of ammonia, 1 part sulphate of potash and 3 parts superphosphate ; 3 oz. of the mixture being applied to each square yard of bed. Also give the plants ½ oz. nitrate of soda per square yard, once when the plants are mid-way in their growth and again when they are heading up. Alternatively the plants can be given monthly doses of liquid manure.

Pests and Diseases. As for cabbage. See Cabbage.

Green Sprouting Broccoli. See under Calabresse.

BRUSSELS SPROUTS. These are an excellent market crop when they are of good quality (large and really tight " buttons ") and are produced early and late in the season. Cambridge No. 5, Wroxton and Exhibition are reliable varieties.

Sprouts require a well worked and generously manured soil. Avoid ground on which any member of the cabbage family was grown the previous year.

To secure big plants bearing a profusion of large tight buttons phosphates must be added to the soil. Work 2 oz. of bone meal into each square yard of each spit when digging, or spread basic slag on the surface of the dug bed, 3-4 oz. to the square yard, and leave it to wash down.

Half an ounce of seed will give 700-800 plants.

Seeds are sown in February or March in a cold frame, and outdoors in April or August. Sow in drills ½ in. deep and 1 ft. apart. In June and July the young plants from February to April sowings should be planted out 2 ft. apart each way. The seedlings from the August sowing should be moved into soil which has not recently been manured, to stand the winter. In spring they should be transplanted into rich soil.

Firm planting in well-settled soil does much towards preventing loose buttons. Another aid to the formation of solid sprouts is to allow plenty of space. It is also a good plan, during August and September, to take off here and there some of the large *lower* leaves.

A precaution often neglected is to provide support for the plants. Actually they *must* be kept upright ; if they flop over their roots are loosened in the soil and, as already mentioned, loose roots result in loose sprouts. A simple method of support is to run cords tied between stakes, down the rows of plants.

For feeding, make a mixture of 1 part sulphate of ammonia, 1 part sulphate of potash and 3 parts superphosphate, and apply 3 oz. of the mixture per square yard of bed soon after planting. Give two dressings of ½ oz. nitrate of soda per square yard, one in mid-July and the other in mid-August.

When gathering sprouts take only a few from each plant, rather than many from one and none from another. Work up the stem from the bottom in each case. Do not cut off the " cabbage " at the top until the last sprouts have been gathered.

Pests and Diseases. In common with the rest of the cabbage family, sprouts are subject to attack by the Cabbage Root Maggot and Club Root. For methods of dealing with these troubles, see Cabbages.

BURDOCK. Useful biennial medicinal herb raised from seed sown in spring. The dried roots from plants in the first year of growth are required by wholesale druggists. Also see Herbs.

BURGUNDY MIXTURE. Useful for the same purpose as Bordeaux Mixture and preferred by some gardeners to it. The recipe is : Copper sulphate, 2 lb. ; washing soda, 2½ lb. ; water, 17 gallons. Made in same way as Bordeaux Mixture (which see).

BUTTERFLIES. Almost all the brightly-coloured butterflies which flit about the garden are harmless to plants. The greatest danger is to be feared from the plentiful white cabbage butterflies which are the parents of the hordes of yellow and green caterpillars that attack crops of " greens " and convert the leaves into a network of holes.

From the end of May the eggs (like little pearl drops) should be hunted for on the leaves and rubbed off. Later the greens may be cleared of caterpillars by putting a handful of common salt into 2 gallons of water and syringing the solution into the plants during the evening. The caterpillars that will be driven out should be dropped into a jar of salt water.

CABBAGE. Under this heading is included only cabbage proper. For details concerning other members of the cabbage family—broccoli, cauliflower, savoys, brussels sprouts, etc.—see under respective headings.

For an All-the-year-round Supply.

It is possible to have a supply of cabbage almost the whole year round. The first sowing is made in a frame or on a warm border in early March, this giving supplies for July and August. The second sowing is made at the end of March or early in April, in the open, for supplies from August to November. The third sowing is made in May to provide supplies from November to February. The fourth, and most important, sowing is made from mid-July to mid-August—usually two sowings are made, one at the beginning and one at the end of the period named, to be on the safe side—to provide supplies for spring and early summer.

A Selection of Varieties.

There are numerous varieties to choose from for each sowing. Usually local custom dictates which of these varieties shall be selected. Also the nurserymen's catalogues provide good guidance. With the above reservations, a useful range of varieties would be :

First Sowing. Ellam's Early, Flower of Spring, Earliest of All.

Second Sowing. Carter's Velocity, Tender and True and also as for first sowing.

Third Sowing. Nonpareil, Carter's Model, Carter's Pioneer, Sutton's All-heart, Sutton's Favourite, Christmas Drumhead.

Fourth Sowing. Harbinger, Winningstadt, Imperial, Sutton's Pride of the Market, Carter's Beefheart, Wheeler's Imperial, Enfield Market.

Sowing Cabbage.

For frame sowings, the preparation of the frame bed follows the usual lines. The soil used should not be rich and the surface of the bed should be within 8 in. of the light. For warm border sowings, the bed should be raised above the surrounding level and consist of well-dug and fine, but not rich soil.

An ounce of seed will give 1,200-1,500 plants.

For the outdoor sowings the seed-bed should be in an open situation and consist of well dug soil rather on the poor side. The ground should not have carried a crop of any member of the cabbage family, or of turnips or wallflowers, in the past three years.

To prepare the seed-bed (which of course, should have been well dug—but not manured—in winter) fork over the soil 6 in. to 7 in. deep and break it up finely so as to secure a fine surface tilth. Do not add any manure, either natural or chemical. If the soil is thought to be lacking in lime, give enough slack lime to whiten the surface and stir that in before sowing. If you limed that particular site

D*

the previous spring, give a dressing of weathered soot—enough to blacken the soil—before sowing and stir that in.

Sowing may be in $\frac{1}{2}$ in. deep drills, 6 in. apart, or by broadcasting. For quite clean soil broadcasting is the better way, in which case sprinkle thinly and make $\frac{1}{2}$ oz. of seed do 2 square yards; then rake the seed in.

Immediately after sowing net the bed to protect both seed and young seedlings from birds, keeping the nets stretched over wooden hoops at least 1 ft. high. Peg the nets down at the side.

If the ground is weedy, adopt the drill method and net over as before.

Planting Out Cabbage.

The seedlings are ready for planting out in their finishing quarters when they have made strong little plants. They should be lifted carefully, with all roots intact, and replanted as soon as possible afterwards. If the soil is dry give a good watering an hour or two before lifting.

In regard to situation, the sites on which peas, broad beans and onions were previously grown will do splendidly for cabbage without further manuring. If other ground is close—a potato plot, for instance—it is advisable to dig in a little rotted manure.

For the earlier plantings choose a situation sheltered from east winds.

In the garden trowel-planting is advised but in the market-garden dibber-planting has usually to be practised.

For the larger varieties rows 1 ft. 9 in. to 2 ft. apart, with plants 14 in. asunder in the rows will be right. For the smaller sorts, 1 ft. 6 in. between the rows and 12 in. between the plants will be enough room to last the crop until hearting time.

Firm the soil well around the plants. Loose planting causes loose hearts.

If the ground is dry, water in after planting.

In most lots of seedlings there are some specimens which are " blind "—have no growing point—and will not therefore heart up. These should be discarded. So also should plants which have very sunken growing points.

Market growers are adopting a new planting method with spring cabbage which cuts out winter losses—which are caused mainly by the wet state of the soil rather than by frost. The cabbages are planted in blocks of five rows at 18 in. apart. Between each two blocks a strip of soil 3 ft. wide is left.

When the plants get into their stride after planting out a channel about 1 ft. wide and deep is cut in the bare strip, the dug-out soil being tucked around the plants. The plants are thus made cosy while the channel collects and carries away the surplus moisture. When the plants are set out in the usual way small channels cut between the rows are a great help in winter.

Puddling is a safeguard adopted as a precautionary measure against Root Maggot. When the plants are lifted, and before they are planted out their roots are swilled in a mixture of clay, soot and lime, with sufficient water added to make it the consistency of paint.

An alternative precautionary measure against Root Maggot is to provide each plant with a protective disc of tarred felt. This encircles the stem close against the soil. The disc should measure 3 in. across. If a slit is cut in one side of the disc, running to the centre hole (through which the stem passes) the disc can easily be slipped on the plant, and pressed flat on the soil, so preventing the flies from depositing their eggs.

The discs require occasionally pressing down so that they may remain in contact with the soil.

Bolting Cabbage. Bolting is of frequent occurrence with spring cabbage.

Generally, the way to prevent it is to make growing conditions as favourable as possible. If the ground between the plants is firm in March, fork it 3 or 4 in. deep, and after giving each plant ¼ teaspoonful of nitrate of soda, earth up the cabbages as far as just below the bottom leaves.

Long-jointed growth is the first sign of bolting. If this is observed in spite of the above precautions—or because of their neglect—pierce the stem of the offending plants with a pocket-knife an inch or two above the soil level and place in the slit a small pebble or match stalk.

Feeding Cabbage. Cabbages are gross feeders. Start by applying basic slag, 4 oz. per square yard, six weeks after planting. This is particularly useful to the autumn-sown plants. Stimulate these spring cabbage with fortnightly dressings, from March onwards, of nitrate of soda. Either give this as a liquid (½ oz. per gallon of water and 1 pint of this per plant) or in dry form, at the rate of ½ oz. per square yard.

Stimulate summer cabbage with a mixture of 1 part sulphate of ammonia, 1 part sulphate of potash and 3 parts superphosphate. Apply three weeks after planting at the rate of 3 óz. per square yard. Also give two dressings of ½ oz. nitrate of soda per square yard, one in mid-July and the other in mid-August.

Cabbage Pests and Diseases. The commonest pests of the cabbage family are the caterpillars of the *large and small white butterfly* and of the *cabbage moth.* These attack the plants at the hearting stage. The plants should be hand-picked where possible. Spraying with weak salt water, 1 teaspoonful of salt per gallon —the spray being directed down into the hearts of the plants—will dislodge the pests.

Another pest is a short fat maggot—the grub of the *cabbage root fly*—which feeds on the roots, causing the plants to become stunted and to flag quickly in dry weather. Badly crippled plants should be pulled up and burnt. Other plants should be watered each week with a weak ammoniacal solution. The plants should also be earthed-up to induce the formation of cotton roots. Puddling and the fitting of felt discs as already described, are valuable safeguards against this pest.

Black Rot, which cripples the attacked plants at any stage in their growth, cannot be cured. Diseased plants must be lifted and destroyed, the ground being dressed with a solution of formalin in autumn.

White Rust shows itself in crippled, stunted seedlings with swollen white polished patches on the leaves, which presently become powdery. The seedlings should be sprayed with Cheshunt Compound (which see) and the ground treated with formalin as for black rot.

The commonest trouble with all members of the cabbage family is *Club Root.* For methods of combating this ailment, see Club Root.

Blight is also apt to attack cabbages. See Blight.

CABBAGE, CHINESE. Also known as **PE-TSAI.** Uncommon vegetable, resembling a cos lettuce, which can be grown outdoors only in the milder part of the country. May be used as a salad or cooked as a vegetable. It is the leaves and mid-ribs which are eaten. Raised from seed sown in July, seedlings being transferred to a nursery bed, then into good soil in a sunny position.

CABBAGE, COLEWORT. Colewort is the name applied to a small quick-hearting cabbage. Sown in the autumn, it matures during the winter and comes into use well before spring cabbage. It may also be sown at other times to provide a quick crop. Sowing and cultivation is as for Cabbage, which see.

CABBAGE, PORTUGAL, or **COUVE TRONCHUDA.** Member of the cabbage family. The leaves of the plant have very large, thick, white midribs. These can be trimmed clean, tied in bundles and used just like seakale, making a very attractive and easily procured " luxury " dish.

In addition, Couve Tronchuda hearts up just like an ordinary cabbage, and can be cooked in the same way. Seeds are sown in the open, in a prepared seed-bed, early in April. The seedlings are transplanted to other quarters in May and June, being placed a yard apart, all ways.

CABBAGE, RED. Variety much used for pickling and occasionally for boiling. Is in good market demand if well-hearted and of good size. The chief sowing is made in July or August, the seedlings being planted 4 in. apart into a sheltered bed in early autumn, wintering there and being transferred to an open, sunny well dug and manured bed in March-April. The heads attain excellent size by the following autumn. Sowings are also made in March-April for use the following autumn.

General culture is as for Cabbage, which see.

CABBAGE, SAVOY. Variety of cabbage which does well anywhere and will stand up to the most severe winter and heaviest soil. Ormskirk, Early, Medium and Late will provide a succession of good heads. Sow in February, in frame for earliest supplies, in March and mid-April outdoors for main-crop. Plant out 18 in. to 24 in. apart in June or July. The rest of the culture is as for cabbage, which see.

CALABRESSE. A distinct type of green sprouting broccoli that is gaining popularity in this country. It produces a large central head in summer. After this is cut sprouts are produced in profusion for two to three months. Head and sprouts should be cut with 6 in. of stem, which is peeled before cooking, and has a delicious flavour. Seed is sown in the open in March. The plants are set out 2 ft. apart in row 3 ft. apart.

CAPSICUM. Plant bearing pods used for salads and pickles. Sow seeds in pans plunged in a hot-bed in February or early March. Prick off plants into 3 in. pots when 2 in. high and from thence pot on into 6 to 8 in. pots in which they will fruit. Plants are usually grown from first to last in greenhouse but in warm gardens will thrive in the open if planted out in rich soil in June. Fruit ripens in September.

CARAWAY. Annual herb, the seeds of which are used in cakes and for medicinal purposes. Raised from seed sown in April in drills 1 ft. apart. Seedlings are thinned to 8 in. apart. Also see Herbs.

CARDOON. Uncommon vegetable that deserves to be more extensively cultivated. It closely resembles celery. After blanching, the inner stems are useful for salads and soups. Should be grown in deeply-dug and well-manured soil. The seed is sown in boxes in a frame in February or outdoors in April in rows 3 ft. apart, and 3 to 4 in. deep. Plant out the plants in " celery " trenches

when 6 in. high. In autumn the stems are blanched by wrapping thick bands of hay around them to within 1 ft. of their tops, then earthing up the stems with soil. The plants are ready for use about four weeks after being covered up.

CARROTS. Carrots are an indispensable crop in all food-gardens, whether grown for home use or market.

For early work: Early Gem (excellent variety for frames, maturing for use eight weeks after sowing), Carter's Delicatesse, Early Horn, Nantes, Early Market. For main crop: Scarlet Intermediate, James' Intermediate, Long Red Surrey.

½ oz. of seed will sow 60-ft. row.

Choose stump-rooted varieties for early work, long or intermediate varieties for main-crops—unless the soil is shallow, when a stump-rooted variety is again to be preferred.

Sowing Dates. Carrots are frequently grown in frames for early crops, sowings on a warm border providing the next succession, outdoor sowings then carrying on supplies for summer use and for storage for winter.

For first sowings sow in January in a frame. Sow again in February on a warm border. Sow in the open in March and April, in June and July (a stump-rooted variety for pulling in autumn) and in September for standing the winter and coming into use in spring. If supplies are short, or " new " carrots are in good demand, sowings may be made any time from March to August.

Preparing the Seed Bed. Frame and warm border culture follows on usual lines.

For the ordinary outdoor sowings select an open position in full sunshine where the soil is well-drained. Good carrots cannot be grown in heavy soil, sandy soil, or sour, water-logged soil. The ideal soil is one which is loamy and free, tending to the light side rather than to the clayey side and of practically the same consistence as that right down to the depth the carrots will grow, or deeper.

Where the soil is not all it might be, it is worth while to make up a special bed for the crop. Let the soil be good to a depth of 15 or 18 in. ; mix with the loam plenty of gritty sand and, if peat is available add up to one-third of that.

All carrot beds should be moderately rich, but this richness must not be secured by adding fresh manure to the soil. Fresh manure causes forking of the roots. The best carrots grow on soil that was well manured six or more months in advance. If the ground is poor, one may add manure, to the bottom spit only, but the best plan is to choose ground that was heavily manured for the previous crop.

Assistance is given to all beds by raking a dressing of fertilizer into the ground when it is being prepared for sowing—any general fertilizer will do.

Sowing Carrots. Sow in drills $\frac{1}{4}$-in. deep. Sow shorthorns in drills 6 in. apart and sow thinly, so that no thinning out is required ; simply pull up roots here and there as they become of size for kitchen use. Sow intermediates in drills 1 ft. apart and the longs in drills 15 in. apart. ·

Rub the seed first in the hands with a little dry sand ; it can then be sown more conveniently and thinly. Germination is stronger when the drills are filled in after sowing with an equal part mixture of leaf mould and old potting soil.

Half an ounce of seed sows a drill 50 ft. long.

Carrots are not now thinned so severely as used to be the practice. It is sufficient if the rows are so thinned that the retained roots just have enough space in which to swell. With the later thinnings the withdrawn roots will be of usable size.

Feeding Carrots. A fine stimulant for frame carrots is a teaspoonful of dried poultry manure per foot of row, applied when colour is beginning to show.

For other carrots, apply a mixture of 1 part sulphate ·of ammonia, 1 part sulphate of potash and 3 parts superphosphate, at the rate of 1 oz. per yard of row, a fortnight after thinning. Also give $\frac{1}{2}$ oz. sulphate of ammonia per yard of row just as crop is reaching mature size.

Harvesting and Storing Carrots. The signs of approaching maturation in carrots are as follows ; First the foliage takes on a yellow or reddish-bronze colouring. Then the outer leaves fall and lie on the ground, only the central tuft of leaves on each root remaining erect.

To make sure, lift a test root or two. If they come out cleanly (small side roots having dried off), if the root is firm and solid and the skin is set, lift at once, whatever the date. A green colouring at the crown of the root just under the tuft of leaves often accompanies full maturation.

The roots are stored in dry soil or sand in a cool shed or cellar after the tops have been cut off close to the crowns. Put down a 2-in. layer of soil, then a layer of carrots, another layer of soil and so on, finishing with a topping of soil 3 in. deep.

For large quantities, carrots can be clamped as potatoes. See Clamping.

Pests of Carrots. The chief pest enemy of carrots is the maggot laid by the *Carrot Fly.* The fly deposits its eggs on the plants just below the soil surface and the resulting maggots infest the roots, causing them to rot.

The risk of attack is lessened when thinning is done only on dull days or after the sun has gone down and the seedlings are watered freely after thinning. Watering the seedlings with a solution of $\frac{1}{4}$ pint paraffin in 1 gallon water makes them distasteful to the pests without interfering with the plants' growth. Another remedy against carrot fly is to moisten some sand with paraffin and distribute this in a little ridge on each side of the newly-thinned seedlings. This must be renewed after rain, or naphthalene can be applied weekly to the rows. 1 oz. per yard run—from the time the seedlings appear till late June.

Carrot Fly seldom gives trouble after August.

Wireworm are also a trouble to some carrot growers. The presence of the pests is denoted by stunted plants and premature bronzing of the foliage. For methods of trapping these pests see Wireworm.

CATERPILLARS. Many good proprietary remedies are available for gardens that are plagued with caterpillars. Some of these, such as Derris (which see) are non-poisonous, and can be sprayed on vegetables without harm accruing. Dustings of lime or soot over infested crops are also effective. Caterpillars that conceal themselves during the day, feeding at night, can be destroyed by sprinkling a mild soil-fumigant, such as naphthalene, on the soil between the plants. A night hunt with a flash-lamp is another good method of attack, the pests being picked off and dropped into a jar of paraffin. Also see Butterflies.

CAULIFLOWER. Cauliflowers are a crop for every vegetable plot. They are also a good market crop. The demand for quality heads is invariably in excess of supply and prices are usually good.

When to Sow. It is usual to make a first sowing in a heated frame or greenhouse in January, another greenhouse sowing in March, a third sowing out of doors in April or May, a fourth sowing outdoors in August:

The January-sown plants are put out in a warm border in April and provide heads in June and July. The March-sown plants are put out in the open in May and provide heads in August. The April-sown plants are put out in the open when 6 in. high and provide heads in autumn. The August-sown plants are grown on in the greenhouse or frame and provide heads for May.

Recommended Cauliflower Varieties. For under-glass sowings and for the August-sown crops choose Magnum Bonum, Early London, Carter's Forerunner, Carter's Defiance Forcing, Snowball or First Crop.

For the first outdoor sowing, choose from Early Autumn Giant, All-the-year-round, Mont Blanc and Early White London.

For main crops choose from Autumn Giant, Eclipse, Walcheren.

Half an ounce of seed will give 600-800 plants.

Sowing Cauliflowers. For the under-glass sowings sow in boxes placed over a hot-bed in a greenhouse or frame. Prick off the plants into other boxes, when they have been up three weeks and plant out, in due course, after hardening off.

For the spring outdoor sowings choose an open situation where the soil is rich and friable but not freshly manured. No other member of the cabbage family, or turnips, should have been grown on the site for the past three years.

Sow in drills ½ in. deep, and 1 ft. apart. If the weather is hot and dry cover the seed bed with a mat to conserve moisture. Remove the mat when the seedlings show. If no mat is used, net the seed bed as a protection against birds. If the bed is overcrowded it is an excellent plan to prick out the plants into a nursery bed, spacing them 4 in. apart in rows 1 ft. apart, leaving them here until they are planted out in their final quarters.

For the August sowing—a most valuable one and well worth the trouble involved —special treatment is required.

Sow the seeds as advised above. By the middle of September the seedlings will have their first pair of true leaves. Now fill a frame with sandy soil. Trans-plant the seedlings into the frame, at least 2 in. apart. Plant them down to the lowest leaves and make firm. Water them in and keep the sash tightly closed for a few days till they get a grip of the soil and no longer show signs of flagging.

During the winter, ventilate fully when-ever the weather allows. At all costs keep the soil dry, yet not dust dry so that the plants flag. Damp is the chief enemy of cauliflowers in winter. The plants will form a nice root system in the dryish conditions and will be ready to go outside in early February.

Alternatively the plants can winter in an unheated greenhouse. In this case put them in boxes of sandy soil and allow full ventilation.

The plants need a fortnight's hardening, first in a cold frame, then in a sheltered corner, before being planted out.

Planting out Cauliflowers. The earliest plantings should go in a sunny sheltered border. Put them in groups of three so that each group can be covered with a cloche or hand-glass in times of frost.

For the later plantings choose an open, sunny situation where the soil is rich and well dug but which has not grown any member of the cabbage family, turnips or wallflowers for three years. Plant out 18 to 24 in. apart in rows 18 to 24 in. apart, the greater distances being for the large summer main-crop varieties.

Keep the plants growing without a check, watering freely in dry weather; otherwise they will produce small heads little bigger than a tea-cup.

Feeding Cauliflowers. A pinch of nitrate of soda dropped alongside young plants once every three weeks until they attain size is a good stimulant. For the rest, feeding is as for Brussels Sprouts, which see.

Protecting Cauliflowers. When the curds have formed summer cauliflower require protection from the sun (for hot sun causes yellowing) and early and late cauliflower protection from frost. Pro-tection in both cases is afforded by break-ing and binding a large leaf so that it covers the curd.

Storing Cauliflower. When cauli-flowers are turning in more quickly than they can be used they may be "held in suspense" by pulling them up and hang-ing them head downwards in a dark, cool shed, with their roots encased in moss. Heads may also be buried in peat, after most of the outer leaves have been re-moved.

Pests and Diseases of Cauliflowers. As for Cabbage, which see.

CELERIAC, or TURNIP-ROOTED CELERY. Uncommon vegetable rap-idly gaining popularity. Produces turnip-like roots which are cooked and served like beet, used for flavouring soups, or grated for inclusion in fresh salads. Sowings should be made in boxes in moderate heat in March. Prick off the seedlings into boxes at an early date and plant out the rows 2 ft. apart in rich, sandy soil in June. Requires large quantities of water. Does not need blanching. Lift the crop in autumn and store as Carrots, which see. Remove all leaves save those in the centre of the crown before storage.

CELERY. One of the most useful vegetables grown, being equally palatable when eaten in the raw state as when cooked. Good quality produce is always in demand at a good price by all markets. Celery varieties are grouped according to colour of the leaf stems into white, pink and red varieties. Also certain varieties are favoured for an early crop and others for the main or general crop. For a full supply such varieties as the following should be grown:

Early. White: Mont Blanc and Earli-est White; Pink: Sulham Prize Pink.

Maincrop. White, Champion Solid, White Plume, Solid White and White Queen; Pink: Clayworth Prize, Selected Favourite Pink; Red: Standard Bearer and Giant Red.

There is also a self-blanching celery which can be grown on the flat, like lettuce.

An ounce of seed produces 1,000 plants.

The seed of the early sort should be sown in mid-February and the main crop in early March. Sow in the greenhouse or hot-bed frame, maintaining a steady heat of 60 degrees. Sow in boxes of light soil and leaf-mould mixed and passed through a fine sieve. Sprinkle the seeds over the surface very thinly (otherwise damping-off will take toll of the seedlings); then cover them from sight with a thin covering of more fine soil. Transfer the seedlings to other boxes when ½ in. high. Place them in a cold frame when they are well up, and so gradually harden them off for planting out in June—when they will be 4 or 5 in. high.

All but the self-blanching variety of celery are usually grown in trenches, which should be made in April. Late varieties *can*, if desired, be grown on the flat.

Preparing Celery Trenches. The trenches should run from north to south and be prepared as early in the spring as possible. They should be 18 in. wide to take two rows of plants—the usual planting plan. If several trenches are to be made side by side there should be at least 4 ft. from centre to centre. Start by digging the trenches 2 ft. deep. Fill in 1 ft. depth of rotted manure and after that put 4 in. of fine soil. In this the plants will be planted. If manure is scarce, dig out only 1 ft. depth and then dig into the lower foot a generous dressing of rotted manure and garden rubbish mixed.

In either event the soil removed is piled up in a ridge beside the trench, the ridge being given a flat top. This ridge may be planted with various crops (see Intercropping) as the soil forming it will not be required for earthing-up the plants until late summer.

Planting Out Celery. The young plants, having been lifted with a good ball of soil and denuded of blistered and yellowing leaves and any suckers (small out-growths) are planted with a trowel, alternately, thus .˙.˙.˙. 6 in. apart, there being 9 in. between the two rows which go into each trench.

Before planting the trench will need some preparation. Fork the " floor " and at the same time work in 1 oz. kainit and 4 oz. soot to each yard run. This should be done some time in March. A week before planting, firm the sides of the trench, tidy up and firm the ridges and again fork the floor to the depth of an inch or two.

After planting, water regularly as required, keep down weeds, remove suckers as they appear, hoe between the plants occasionally and check pests.

Feeding Celery. The most useful stimulants are nitrate of soda—1 oz. dis-

solved in a gallon of water ; potash—1 oz. dissolved in a gallon of water.

Give a dose of nitrate of soda one week and of potash the following week, and so on right up to the time when manuring has to stop, which will be when earthing-up has progressed to about half-way up the sticks—some time after the middle of September, probably—a little earlier for the early pink varieties, such as Sulham Prize Pink.

Blanching Celery. Start by going along the rows and removing all damaged leaves, and suckers. Then loop the sticks of each plant loosely together with raffia to keep them together whilst earthing proceeds. Give the rows a heavy soaking with water.

Blanching occupies from 6 to 8 weeks. There are two methods—the use of " collars " of corrugated paper or stiff cartridge paper and earthing-up. With both methods the weather must be fine and dry at the time.

The collar method—which is usually only employed with exhibition celery and celery grown for home use—consists of fitting over the plants collars of a sufficient height to come up to the leaves and tight enough to enclose the stems comfortably and yet exclude light from them. The collars are held together by two string ties, one at the bottom and one at the top.

As the plants grow, additional collars are added or, alternatively, the collar is pushed up to the top of the plant and the section uncovered is banked up with soil.

The earthing-up method consists of banking up soil—taken from the ridges—against the stems of the plants. The banking up should be a gradual process. From 4 to 6 in. of soil are applied first ; then additional soil is added every week or ten days until the plants show nothing but the top of the leaves, and are banked up like ridges of potatoes, the edges being some 12 in. high.

The soil must be broken up finely before each earthing and care must be taken not to allow any earth to fall in between the sticks. The ridges should be patted firm each time.

In severe weather, when celery is standing the winter, cover the tops of the plants with straw.

Self-blanching Celery. There is a variety of celery, known as Golden Self-blanching, which does not require any earthing up. It is early and compact in habit. The heads are shorter than those of other varieties, but they are solid, very nutty in flavour, and exceptionally crisp. The plants are grown in flat beds and without any blanching treatment, and are ready for eating at the end of July. Apart from the earthing-up the culture is the same as for other varieties of celery.

Celery Troubles, Cause and Remedy:
Bolting. Caused by dryness at the root, particularly at earthing time. Remedy : Keep plants well watered.

Pithy Leaf Stems. Due to over-feeding and over-watering. Remedy : Obvious.

Poor Hearts. Due to neglect in taking out suckers formed round base of plants. To get a good stick of celery—and with it a good heart—each plant must be restricted to a single central growth, with one central rosette of leaves.

Crooked and Distorted Hearts. Caused by bad earthing-up. Always chop up the earthing-up soil finely—not draw it up in lumps. Also caused by unequal pressure on the sides of the stick. Let the drawn-up soil exercise pressure evenly on all sides by reason of its weight.

Rotting-heart. Caused by fine soil getting into the hearts when earthing-up. To prevent that, before drawing up any soil and after having taken off the suckers and short lower leaves, tie the longer ones together in a bunch, by a single tie placed immediately under the top tuft of green leaves. Use raffia for this—not string.

Rot at the hearts is also caused by allowing soil to lie on the top tuft of green leaves at the final earthing-up. That top tuft of green leaves must always be clear—all through the winter.

Gnawed and Disfigured Sticks. Slugs, as well as earth-worms, may get into the centres and disfigure the central leaf stems. *Remedy :* Keep soil from centre of plants.

A pinch of chloride of lime sprinkled round each plant at each earthing-up will make things unpleasant for both slugs and worms and will not harm the celery.

Celery Pests and Diseases. **Leaf-Miner.** This is the only serious pest of celery. It causes whitish blotches and streaks on the leaves, which afterwards turn brown and rot. The blotches and streaks are caused by a maggot which tunnels about the leaf between the upper and lower surfaces. The maggots are hatched from eggs deposited in the leaves by the parent fly. The trouble is most prevalent from June to September.

Spray the plants every ten days from late May with quassia solution (which see), thus deterring the fly from egg-laying. Pick off and burn badly affected leaves. With leaves showing only a mark or two, hunt for the maggot—which will be found as a slight lump—and squash it.

Leaf Scorch. This is a disease causing brown patches on the foliage, which spread and lead to wilting. Attacks are to be anticipated in all stages of growth. As a preventive steep the seeds, before sowing, in 1 part formalin, 600 parts water for three hours. In the event of attack spray young plants with liver of sulphur solution (which see), older plants with Bordeaux mixture (which see).

Rust. Another common disease causing small pale green spots on leaves, the spots changing to brown and then to grey. Leaves turn yellow, droop and die. Usually attacks occur from mid-June to the end of August.

Cut off and burn affected leaves. Spray in mid-June with Bordeaux mixture (which see) at half strength and repeat every fortnight till mid-August.

Leaf Spot. Common disease occurring from mid-June to the end of August and causing dark brown patches dulled with black. Dealt with similarly to Rust.

CHALKY SOILS. For methods of dealing with chalky soils to increase their fertility see Digging and Manuring.

CHAMOMILE. Perennial herb, the flowers of which are used as medicine. Raised from seed sown in April in rows $2\frac{1}{2}$ ft. apart. Thin to 2 ft. Also see Herbs.

CHARDS. Name given to the sucker growths of globe artichokes, which are eaten as a vegetable. See Artichokes.

CHERVIL. Fragrant annual herb useful for salads and flavouring. A curled variety is also available for garnishing. Sow seeds in May, in a shady position, in rows 9 in. apart and thin out to 4 in. apart. Also see Herbs.

CHESHUNT COMPOUND. Useful remedy against stem-rot, fungus and other diseases. Is particularly useful in checking outbreaks of damping-off among tomato and other seedlings. To prepare the compound, take 2 parts copper sulphate and 11 parts *fresh* ammonium carbonate. Crush to a fine powder, mix and store in a tightly corked glass or stone jar for twenty-four hours. One ounce of the mixture should be dissolved in warm water and added to 2 gallons of water. Do not put into vessels made of iron, tin or zinc.

CHICORY. A very profitable market crop and also extensively grown in the ordinary amateur's garden. The value of chicory is twofold. Eaten raw, it is a tasty white salad with endive flavouring ; when boiled, it has a flavour resembling that of seakale. It is also served as a green vegetable.

To start with chicory, either roots are bought in early November, or seed is sown outdoors in May, in drills 1 ft. apart. For salad purposes, sowing may be made at intervals during the summer, in a shady position. Cutting commences as soon as the plants have made 3 to 4 in.

Blanching starts in late November. Roots are dug up as required and prepared by cutting off the lower half. The retained halves, each bearing at its crown a terminal bud, are then potted or boxed up and forced in a dark, warmish cellar, or you can have a batch of them under the greenhouse stage on a heap of leaves capped with soil, the whole being covered round with sacks to ensure darkness.

Without shed or greenhouse, there is an easy way to force chicory outdoors. First of all build a hot-bed 3 ft. high, of 1 part strawy stable manure and 3 parts old leaves. Cap it with about 6 in. depth of top garden soil and plant the trimmed chicory roots on this bed. Allow 3 in. clearance between them. Next, shovel up more soil and heap it on top of the chicory until it is 7 in. or 8 in. in depth. Finally heap fallen leaves generously round and over the bed.

The chicory is ready for cutting as soon as the tips of the white shoots begin to push through the top soil.

CHINESE ARTICHOKES. See Artichokes.

CHINESE CABBAGE. See Cabbage, Chinese.

CHIVES. Perennial herb, the leaves of which make a mild substitute for salad onions. The roots are bulbous and sometimes used for pickling. Propagated by dividing up the tufts in autumn or spring.

The bed should be re-made every third year. Lift the clumps with a fork, starting at one end of the bed. The bulbs will be found packed in clusters very much like those of shallots. The mother bulb at the centre is surrounded by several offsets which become smaller towards the outside. Don't replant the smallest, nor yet the mother bulb. There are plenty of medium size to re-make the bed.

Chives succeed best in a partially shaded spot such as beneath a fruit tree or behind a north wall or fence.

CHOU DE BURGHLEY. Uncommon vegetable, part cabbage, part broccoli. Sown in spring, a cabbage-like plant may be cut in autumn or winter. If the plant is left until the following spring a broccoli-like heart develops. Culture is as for cabbage. See Cabbage.

CHOU DE RUSSIE, or Russian Kale. Form of kale with deeply-cut, greyish-green leaves, producing in autumn a close, white heart that is excellent for winter cooking. If left until spring a profusion of tender sprouts can be gathered.

Sow in April-May and grow as ordinary kale. See Kale.

CLAMPING. Method of storing potatoes and other vegetables when the crops are too big for storage in sheds, etc.

The method, briefly, is to pack the vegetables in a heap and cover this with soil.

Make the clamp in a dry situation, if possible sheltered from north and east winds. Fork the site 1 ft. deep to facilitate drainage. Spread on the forked soil a 2-in. covering of ashes as a further guarantee of dryness. Then pile the potatoes, carrots or whatever the crop may be carefully on the ashes, working from a 4-ft. wide base to a pointed ridge at the top.

Examine the tubers, etc., carefully before clamping them. If diseased specimens were stored, they would soon go rotten and involve the sound specimens near them.

With potatoes, as a safeguard against winter rot, sprinkle freshly-slaked lime or flowers of sulphur freely amongst the tubers when arranging them.

As soon as the crop is in position, cover with a 4-in. layer of clean straw, and leave until heavy, continuous rain or keen frost make a 3-in. covering of soil necessary. The latter you take from immediately around the margin of the clamp. Dig a foot-wide channel, and lay the soil carefully over the straw. As soil practically excludes air, and as an airless condition favours rot, provide ventilation by fixing at intervals of 3 ft. along the ridge of the clamp a 3-in. land tile or good wisp of straw.

CLARY. Aromatic herb, the leaves of which are used for flavouring. Although a perennial, it is best treated as a biennial. Sow seeds in April in rows 15 in. apart and thin to 1 ft. apart. Also see Herbs.

CLAY SOIL. For methods of dealing with clay soil to increase its fertility see Digging and Manuring.

CLOCHES. Cloche gardening is now an established part of the routine of a large number of market and private gardeners. The former find it exceedingly profitable ; the latter have proved that it is the best method of raising first-class out-of-season crops.

There are several types of cloches— tent, barn and so on. Some are made of glass, others of strong glass substitute. All are good, trapping every bit of light and retaining heat sufficient to ensure healthy growth during winter.

At all seasons they can be put to good use. The following are some of the many sowings which can be made under cloches. In January: Broad beans, Brussels sprouts, cabbage, kale, cauliflower, onions, parsley, peas, radish, spinach. In February: Carrots, celery, lettuce. In March: French and runner beans, beet, broccoli, turnips. During the summer the cloches can be used for the cultivation of tomatoes, and ridge cucumbers and bush marrows. Then more sowings can be made. In August: Corn salad, endive, lettuce. In September: Carrots, onions, spinach. In October: Cabbage, kale, cabbage lettuce, parsley.

Naturally a sheltered sunny position will be chosen for the cloches. If possible the rows should be run north and south, then there will be no need to close up the ends except during the hardest weather.

Leave a space of 18 in. between each two rows of cloches, to allow for the necessary attentions.

The ground on which cloches are to be used must be prepared thoroughly and generously. Work it 1 ft. deep, incorporating a 3-in. layer of well-rotted stable manure (or one of the accepted substitutes) and 1 oz. of superphosphate with each yard run of row to be covered. Break up the soil finely.

A week before sowing water the ground well, unless it is already moist, and put the cloches in position. Then the soil will be nicely warm at sowing time.

CLUB-ROOT or Finger-and-Toe. Fungus disease which attacks cabbages, cauliflowers, brussels sprouts, turnips, radishes (and certain flowers, such as wallflowers, in addition) causing the roots to become a mass of swollen tissue. Any crop badly attacked by the disease is definitely useless. The germ which causes the disease lives in the soil and can remain there for seven years.

The Lime Cure. An infested crop should always be burned—the first step in eradicating the trouble. Then the germs left in the ground must be killed so that they can do no further harm. These germs succumb readily to quicklime. The time to apply the lime is early autumn, and the quantity 1 lb. per square yard. Place the quicklime in a heap and gently pour over it roughly its own bulk of water. Allow to stand overnight, so that it may crumble into powder. Scatter this over the ground and dig it in as soon as possible afterwards. Some time between the application and cropping time, fork the land to mix the lime and soil together properly.

No member of the cabbage family should be grown on club-root-infested land for two years after the liming ; other vegetables may be grown there, however.

The Corrosive Sublimate Cure. Club-root may be combated even more effectively by the following method :

A start is made by watering the soil to be used for the seed boxes—or the drills if outdoor seed-sowing is practised—with 1 oz. of corrosive sublimate in 20 gallons of water Do the watering two days before sowing.

The next stage comes when, after they have been set out in final quarters, the plants are watered with a much stronger solution—10 oz. in 10 gallons. About ½ pint of liquid is required for each plant. The cost of this operation is but a few pence per 100 plants.

In cases where any site devoted to cabbages and turnips is carrying a particularly heavy load of infection, the dose may be repeated six weeks later.

It should be borne in mind that *corrosive sublimate is poisonous* and must therefore be used with care. The precautions to take are to ensure that the liquid does not come into contact with plants which are to be eaten within a few weeks, to gather up and *burn* dead worms found upon the plot and to wash the hands thoroughly after handling the chemical.

There are also excellent proprietary preparations for use against club-root.

COLCHICUM (Autumn Crocus). Imported medicinal plant. Both the "corms" and the seeds are required by wholesale druggists. Corms are planted in the first place and left for two or three years to increase. Also see Herbs.

COLEWORT. See Cabbage, Colewort.

COMFREY. Medicinal plant, of which both leaves and roots are required by wholesale druggists. Propagated by division of the roots in spring. Also see Herbs.

COMPOST. Valuable manure for the garden, consisting of decayed vegetable matter, annual weeds and turf heaped until the whole has rotted. See Manures.

CORIANDER. Annual herb, the seeds of which are used in curry, the leaves in salad. Raised from seed sown in April in rows 1 ft. apart. Thin out to 1 ft. Also see Herbs.

CORN COBS. See Maize.

CORN SALAD, or **LAMB'S LETTUCE.** Leaves are used in salads and are an invaluable stand-by when lettuce is scarce. Should be grown on dry, well-drained soil. Sow seeds in spring, and at fortnightly intervals through till the end of October, in drills 6 to 9 in. apart, and ½ in. deep. Thin seedlings to 6 in. apart.

COUVE TRONCHUDA. See Cabbage, Portugal.

CRESS. See Mustard and Cress, also Watercress.

CUCUMBERS. Cucumbers are an excellent crop for all. They are a distinct asset in the private garden, while the commercial grower with a good range of glass and the ability to grow them well will find them as profitable as any vegetable.

Sowing Times. Cucumbers can be sown at any time of the year, but the usual sowing times are August-October for autumn and winter crops, early December for early spring crops, early January for April and May crops, March-April for summer crops. It can be reckoned that 10–12 weeks are required from sowing time to harvesting time.

Cucumber Varieties. Select from Improved Telegraph, Lockie's Perfection, Prolific, Tender and True, Every Day, Butcher's Disease Resister. Telegraph is probably the most popular of all varieties.

Raising the Plants. The plants must be raised under glass, with a minimum temperature of 60 degrees (better 65 degrees). For the cold-weather sowings it is desirable to have a brisk bottom heat, the seed vessels either being plunged in a hot-bed or in a propagating case on the hot-water pipes. For quantities of plants, sow in boxes. For smaller supplies sow singly in 3-in. pots, as follows :

Fill the pots with ordinary sandy sowing compost and in the centre of each press in one seed, on edge, about 1 in. deep. Water the pot with tepid water and then put it in a warm propagator, or plunge it in a box of fibre, moss or mould over the pipes. When the seedlings appear water freely with water as warm as the greenhouse.

The seedlings will soon fill the pots with roots. Then give them a move to 6 or 7-in. pots.

The place selected for the plants after potting ought to be near enough the glass to prevent them from getting drawn. In three to five weeks, according to season, the plants are ready to go into their growing quarters.

Planting Out Cucumbers. Cucumbers may either be grown in a heated greenhouse or in a frame on a hot-bed.

In the Greenhouse. For private use cucumbers can be grown either in large deep boxes or large pots. For commercial work the usual plan is to grow them in mounds of soil placed on slates or sheets of corrugated iron on the staging. The soil to use is a mixture of two parts turfy loam, roughly broken up, and one part of leaf-mould and burnt garden refuse, mixed. At least a bushel of soil will be needed for each plant. The mounds of soil should be spaced out 3 ft. apart.

The ball of soil surrounding the roots of the cucumber plant should be sunk into the mound so that the lower leaves of the plant are just clear of the soil. All that remains is to make the plant firm, and then insert against it a cane which will reach the lowest wire on the roof.

There must be roof wires on which the growths can be trained. These wires should be spaced out under the roof from 6 in. to 1 ft. apart. Strong cord or string will answer. The supports should not be nearer than 9 in. to the roof glass.

In the Frame. To grow cucumbers in a frame, a hot-bed must first be made up, this consisting of good stable manure and leaves mixed into a heap and turned several times, at intervals of a few days. This should then be trodden down into a firm bed, 4 ft. deep and 3 ft. wider than the frame.

Put the frame on the bed, and a few days later, when the heat from the manure has subsided a little, place over the hot-bed a layer of from 7 to 9 in. of the soil-mixture advised for greenhouse cucumbers. The small cucumber plants are planted in this soil. Until they are established they must be shaded from the sunshine.

Two cucumber plants are sufficient for a two-light frame, one planted under the centre of each " light," and trained to the four corners of its particular section.

Training Cucumbers. In the Greenhouse. The first step in the training of cucumbers is to pinch out the growing point. The time to do this is when the plants have formed ten to twelve good leaves. As a rule, five side-shoots are produced after the first pinching. The bottom one is invariably so weak that it is better to cut it out with a sharp knife. When the four shoots retained are long enough, tie them out evenly on the wires. As the stems will swell considerably, the tie should be left loose. Make it immediately above a leaf, and wrap the raffia once round the support to prevent slipping.

As soon as a cucumber is swelling, pinch the shoot bearing it, one good leaf beyond the fruit. Soon there will be another set of side-shoots. Cut out all but the two best, and when cucumbers on each of these shoots are swelling, pinch them one good leaf beyond the fruit. By following this policy there is break after break of fertile growth.

At no time allow more than two side-shoots on each branch, and whenever any sideshoot shows no sign of cucumbers at a foot long, pinch it out.

In the Frame. Training follows the same lines as above. It is, however, advisable to peg out the first side-shoots evenly on all sides. Otherwise the growth may cluster at one side of the frame and sterility result through over-crowding. Use hooked wood pegs and lower the shoots by stages.

Cultural Hints. Cucumbers only thrive in a moist heat. It is therefore necessary to syringe them freely both when growing in a frame and a greenhouse. Two syringings a day are not too many in summer.

Frame lights and greenhouse windows should be only slightly opened.

When the plants have been growing awhile their roots will be noticed to be appearing above the soil. When this happens a top-dressing of rich potting mould should be applied. The plants will also benefit from fortnightly waterings of weak liquid manure.

Cucumbers should *not* be pollinated. When they are the fruits are bull-necked and of very little value. If time permits the male flowers—the "plain" flowers, not those with a little cucumber beneath them, which are the female flowers—should be picked off.

There are sometimes complaints about bitterness in cucumbers. Bad drainage, resulting in stagnant water collecting about roots is one common cause; violent fluctuations in the temperature are another. The use of superphosphate as a fertiliser is a third cause. Dilute liquid manure and blood manure are the only fertilisers that should be used for this crop.

To Prolong Cropping. When frame cucumbers have apparently exhausted their fruiting powers they can be given a new lease of life by cutting out all exhausted growths, training out any available new shoots, pegging these into the soil and heaping over them at the pegged-down point little mounds of rich soil. This induces the shoots to root on their own account.

Ridge Cucumbers. Cucumbers can be grown outdoors, a special sort being available for the purpose. This is known as the Ridge Cucumber. Stockwood Ridge and King of the Ridge are reliable varieties.

To prepare the bed for outdoor cucumbers, select a sunny place which, if possible, is sheltered from winds. Dig a trench from 18 in. to 24 in. wide, 9 in. deep and any convenient length. Obtain some partially-decayed manure or hop manure, and fill this into the trench until it forms a bed 9 in. above the surface of the surrounding soil. Cover the whole with 6 in. of good rich soil, leaving it with a flat top. Let the soil removed from the trench be piled up to act as a check to north winds.

The cucumbers can either be obtained by sowing seeds on the bed after a few days, when it has settled, or, by raising them in a greenhouse or frame, or by purchasing young plants.

If you wish to sow seeds, direct in the bed, groups of three should be sown at intervals of 3 ft. down the centre of the bed.

Sow the seeds about 1½ ins. deep—on their sides, *not* flat—covering them with soil and then inverting pots over them until germination takes place. When the seedlings come up, reduce them to one at each station.

In the case of young plants it will be sufficient to plant them at 3 ft. intervals down the bed. Until they have taken hold of the soil these should be covered with handlights or bottomless boxes covered with glass.

When the young cucumber plants reach a height of about 1 ft., the point of the growing shoot should be nipped off. The side-shoots which ultimately appear should be pinned down to the soil in such a manner that they will spread to the corners of the bed or section each occupies.

In northern and exposed districts it is necessary to cover the plants during the early stages with a glass-topped box or cloche or hand-light.

Copious watering is required through the growing season. Cutting the points while they are young encourages further production.

Pests and Diseases of Cucumbers.
Root-knot Eelworm. Bores into stem of plants and causes yellowing of leaves and weakness of plant. Pull up and burn plant. In future sterilise all soil used—that is the only plan. See Sterilising for methods.

Wireworm. See Wireworm for preventive measures and remedies.

Thrips. This only appears when the atmosphere of the growing quarters is kept too dry. Free syringing will drive away the pests.

Blight. To be controlled by spraying with any good insecticide.

Mildew. Causes greyish powder on shoots and subsequent wilting. Dust affected parts with sulphur powder or spray with liver of sulphur solution, which see.

Leaf Spot. Common disease to be found in several forms. In one form the symptoms are pale-green semi-transparent spots on the leaves, the spots increasing in size and turning brown. In another form the spots are red. The usual causes are an excess of moisture and sudden changes of temperature. Cut off and burn all badly affected leaves and spray with liver of sulphur, which see.

Wilt. This results in the collapse of the plant. There is no remedy. Can be prevented by sterilising the soil (see Sterilising) or watering the soil with Cheshunt compound, which see.

CUSTARD MARROW. See Marrows, Custard.

DANDELION. Dandelion shoots are in demand nowadays for salads, having a flavour similar to that of chicory.

Box or pot up the roots in ordinary top garden soil or, better, sifted leaf-mould, and place the filled pots or boxes in a dark cellar, shed, or under a darkened part of the greenhouse stage. Cut the shoots when they are about 9 in. high.

In some gardens sufficient dandelion roots can be obtained for forcing purposes. Where large supplies are required, sow seeds of an improved variety in spring. They go into an outdoor bed, 8 in. apart in rows 1 ft. asunder. No attention is needed once the seed is sown, the roots being lifted, pulled up and blanched as required.

Dandelion roots also have a medicinal value and are required in large quantities by wholesale druggists. Also see Herbs.

DERRIS. An insecticide very efficient against caterpillars, blight and other pests, but non-poisonous to human beings. It can be applied in powder form, with the aid of bellows, or as a liquid spray. The formula is derris, 1 to 2 oz.; soft soap 4 oz., water 5 gals.

Dissolve the soap in one gallon of water, the derris in another gallon. Add the derris to the soap solution, stir vigorously, and add the remaining 3 gallons of water.

DIGGING AND MANURING. Digging is the basis of all good gardening, the work that lays the foundations of success. As digging and manuring are usually done together, autumn and winter being the time for these operations, it will be most helpful to deal with the two subjects under one heading.

All digging comes into one of four general categories: (1) Plain digging; (2) mock trenching ; (3) full trenching ; (4) ridging. As indicated below, variations in the methods are necessary to meet the needs of certain types of soils—very heavy and very light soils, stony and chalky soils, etc. Following are the general details of the four digging methods.

(1) *Plain Digging*. The quickest and simplest method of digging in which the soil is turned over only one spit deep.

A start is made by taking out at the end of the plot a 15-in. wide trench to the depth of the spade blade. The excavated soil is wheeled to the end of the plot where digging will finish ; it will be required to fill the last trench.

With a long plot, such as an allotment, it is convenient to divide the plot in two, lengthways, and to work the two portions separately. Digging will then finish at the end of the plot where it was begun, and the soil excavated from the first trench should be left there, to fill the last trench.

Into the trench turn over the next strip of soil. Thus another trench is opened up. Into this the soil from the next strip is turned. Digging proceeds thus until the whole plot has been treated. To make the digging as easy and systematic as possible work in strips $7\frac{1}{2}$ in. wide. This enables the digger to get a nice spadeful of soil each time. The spadefuls will lift more cleanly if the end of the portion to be lifted is nicked before the spade is driven in. The spade must be driven in vertically to keep the trench a full spit deep.

As each spadeful is thrown forward it should be inverted by a twist of the wrist.

Where stable manure is being used allow a barrowload for each 25 square yards of ground. Spread the appropriate allowance of manure in the bottom of each trench as it is opened up, so that it is covered over by the soil from the next strip. Well rotted compost is a good alternative. Where neither of these manures is available hop manure is a good substitute. It should be spread in the bottom of the trenches at the rate of 6 oz. to the square yard.

(2) *Mock-trenching*. Also known as double digging. This method has the advantage that the soil is loosened to a depth of 2 ft. instead of only 1 ft. as in plain digging. The greater depth gives roots a far better chance to plunge deeply.

Take out a trench 2 ft. wide and spade-blade deep, across one end of the plot, moving the excavated soil to where digging will finish.

The subsoil thus exposed is loosened with the fork, to the full depth of the tines. The manure, applied at the rate mentioned under plain digging, is then spread on the surface of the subsoil and lightly forked in.

The soil from the next strip is turned over on to the manure. To maintain the proper width of trench, it is a good plan to nick the soil with the spade to mark the 2 ft. width. It would not be convenient, however, to work the full 2 ft. width at a time ; the soil should be handled in 8 in. wide strips.

When the soil from the three 8-in. strips has been moved forward the first trench will have been filled in and another 2 ft. trench opened up. Fork the subsoil, put in manure, and proceed with the next 2 ft. strip. Continue this until the plot is finished.

(3) *Full Trenching*. In this method the subsoil is brought to the surface, the top spit being buried beneath it. It is most advantageous in old gardens and allotments where the top soil has got " tired " after years of cultivation, whereas the subsoil is full of humus and in a condition to provide the roots with all they need.

The first step is to take out a trench 2 ft. wide and 1 ft. deep. Next a foot depth of subsoil is taken out, and heaped up separately from the top soil.

The bottom of the trench is well forked over, to open up the soil and improve drainage and aeration.

The next 2 ft. strip is then dug. The top foot of soil is dug out and thrown into the bottom of the first (2 ft. deep) trench. Follow by digging out a foot depth of subsoil and throwing this into the trench. The result is that the first trench is filled, with the former top soil below and the subsoil on top, and another 2 ft. deep trench is opened up. Proceed thus over the whole plot.

Manure is mixed with the old top soil as this is thrown into the bottom of the trench. A large pailful of stable manure or compost heap or 4 oz. of prepared hop manure to the yard run, will be appropriate allowances.

More manure, at the rate mentioned, should be placed on the surface of the old top soil before the trench is filled in with subsoil.

(4) *Ridging.* This is the digging method which should be used to improve soil consisting almost entirely of clay. In wet periods such soil becomes greasy and sodden ; in dry spells it becomes almost rock hard.

The work is divided into three stages —the first in November the second in mid-January, the third at the end of February.

Start by dividing the plot into 5-ft. wide strips, nicking the soil with the spade to mark the various sections.

Next dig out a foot depth of soil from the first strip and spread it over the surface of the adjoining strip (strip No. 2). Because of its nature the soil will be lifted in large lumps. Do not attempt to break them down ; the weather will do that. As digging proceeds mount up the soil in fairly sharply pointed ridges.

Proceed to dig out the soil from strip No. 3, ridging it up on strip No. 4 ; from strip No. 5, on to No. 6 and so on.

Now return to the first strip. Loosen the subsoil to a depth of one foot and then dig out the soil from the sides to form a central ridge in the trench.

Make similar ridges in each of the other trenches, thus exposing the maximum area of soil to the weather.

By mid-January the soil, both in the trenches and on the intermediate strips will be well broken down.

Now comes the second stage. Level out the ridges in the trenches, at the same time working in manure or compost heap (a barrowful to 25 square yards) or hop manure (6 oz. to the square yard). Next the trench is filled in with soil from the adjoining ridge.

When all the trenches have been dealt with similarly the undug intermediate strips will be ready for attention.

The soil is dug out from these strips and ridged on the filled-in trenches in the manner described. The soil in the trenches is also ridged as before.

At the end of February the soil is levelled in the trenches, manure is added and the trenches are filled with the ridged soil. Thus the whole plot is thoroughly dug and its condition improved immeasurably by the long exposure to the weather.

Digging Grass Land. If the grass is long, matted and tousled it should first be cut down with sickle or scythe. Too much top growth turned into the soil would make it spongy and difficult to work during its first season.

After cutting down begin at the end of the plot and mark off a strip 2 ft. wide. Lift the turf from this strip, taking it about 3 in. deep. Stack this turf where digging will finish.

Dig out the soil to a depth of 1 ft. and heap it near the turf.

Break up the subsoil as in mock-trenching to a depth of 1 ft.

Mark off another 2 ft. wide strip, pare off the turf and place it in the bottom of the first trench, grass side downwards. Chop it into pieces 4 to 6 in. square.

Dig out 1 ft. depth of soil from the second trench to fill the first. Break up the bottom of the second trench, pare off the turf from the third strip, and proceed as before until the whole plot is dug. The turf and soil from the first trench will fill in the last trench.

When rotted down the turf should provide all the manure needed for the time being.

As turfed land invariably teems with wireworm the clipped turf should be treated with a wireworm destroyer— equal parts lime and naphthalene sprinkled at the rate of 2 oz. to the yard run of trench, or one of the proprietary soil fumigants.

Digging Heavy Soils. Some soils are heavy and difficult to work without being so stubborn as to make ridging essential. Here is the method of dealing with such soils.

Open up a trench as described for mock trenching. Spread on the surface of the loosened subsoil a good dressing of strawy stable manure or substitute, such as vegetable refuse, or newly fallen leaves. Apply at the rate of a large bucketful to the yard run of trench. Rough straw chopped into 6-in. lengths, two bucketfuls to the yard run, is another useful substitute for stable manure.

Dig over the top soil from the adjoining strip, and work another dressing of manure into that. Treat all the strips similarly.

Sand or sifted coal ashes can be used to lighten heavy soil. The rate of application is three quarters of a bucketful to the yard run, both to sub-soil and top soil. The disadvantage of sand and cinders is that they are not nearly so long-lasting in their effect as is manure.

Soak-away pits are also a help to heavy land. See Drainage.

Digging Light Soil. The object when digging sandy soil is to improve its moisture-retaining capacities. This is achieved by working in plenty of binding material and putting in a kind of " breakwater " over the subsoil.

The general procedure is to mock trench the ground, placing the " breakwater " over the subsoil after this has been broken up and manured. Turves placed grass-side downwards and chopped up make the best " breakwater ".

A layer of leafmould or spent hops will serve, or even a good layer of old grass.

More manure is added to the top soil as this is turned over from the adjoining strip.

Where available, cow and pig manure are ideal for light soil. Alternatively, rotted vegetable refuse, spent hops or leafmould will serve, used at the rate of a large bucketful to the yard run of trench. Prepared hop manure or shoddy, used in accordance with the instructions supplied with the material, are other alternatives.

Digging Stony Ground. The big stones should first be removed from the surface. If the plot is a sharply sloping one—as it may be in hilly districts where most stony plots are found—a good plan is to use them for walls to " shore up " the soil so that a more level surface can be built up.

Then the plot should be plain dug or mock trenched, other large stones encountered being thrown out. It would be tedious work to try to pick out the smaller stones and in any event they are no material drawback to good plant growth.

If the soil is light, as well as stony, a " breakwater " of turf, as prescribed for light soil, will be an advantage.

Digging Shallow Soil. In some plots there is only a 6 to 9 in. depth of soil over a rocky, shaly or concrete-hard yellow clay subsoil. Nothing can be done to such a sub-soil and the only alternative is to deepen the soil from the top.

The best plan where practicable is to import a supply of good soil. Sometimes rotted turf, surplus top soil or leaf-mould can be obtained from the local authority, especially where the authority has a number of open spaces. Such matter should be spread over the original top soil of the plot and be well mixed with it during the process of plain digging. Manure should also be worked in.

Where extra soil cannot be obtained all the available manure, rotted garden refuse, leaf mould and the like should be spread over the hard subsoil as digging proceeds.

Digging Peaty Soil. Sourness and wetness are the disadvantages of peaty soil and some form of drainage is necessary. Land tiles are the complete solution, but faggot or rubble drains will serve. See Drainage.

When some form of drainage has been introduced the ground should be mock trenched, but in this case no manure need be added ; there is already plenty of humus and adequate food in the soil for one season's cropping.

Lime should be used freely in both spits —1 lb. to each 3 yards run of trench.

Old tree branches are often found in peaty land. The larger should be removed, the smaller chopped up with the spade and left in the soil.

Digging Waste and Weedy Land. Where the plot is merely covered with thick grass this should be cut down and the land treated as described under Digging Turfed Land.

A plot filled with deep-rooting weeds— docks, dandelions and the like—must be thoroughly mock trenched, every scrap of weed root being taken out as the work proceeds.

Ground covered with scrub, such as gorse and elder, calls for extensive clearance work. The top growth should be cut down to ground level, then the stumps must be got up. Take out the soil around each stump, chop off the roots and it should then be possible to lever out the stump. All the roots must be dug out; if left in they would impede proper digging and when rotted might convey mould to the crops.

All rough ground of the types described needs mock trenching.

DILL. Aromatic herb, the leaves of which are used for flavouring soups and sauces. Raised from seed sown in March, in rows 1 ft. apart. Thin out to 1 ft. apart. Also see Herbs.

DISEASES OF VEGETABLES. For the various diseases which attack vegetables and the ways of combating them see under separate headings and also under respective vegetables.

DRAINAGE. Crops grown on ground which is constantly wet, such as peaty soil, are severely handicapped. The cropping power of such badly-drained ground can be increased enormously by the introduction of some form of drainage.

The thorough method is to insert land tiles, but at the present time the cost is heavy. There must be a ditch or the like into which the tiles can carry the water.

The tiles should be laid in rows at 12 to 18 feet apart, depending upon the soil. Starting at 18 in. deep they should slope gently to the ditch. They should be covered with ashes before the soil is returned over them.

Except on a very large plot there is no need to herring-bone the drains ; each row can run straight to the ditch.

As an alternative to land tiles tree branches and brushwood can be used. Trenches 9 in. wide and 18 in. deep are taken out. The bottom of each trench is covered with a double layer of branches cut into 18-in. lengths. A layer of old turves are placed grass-side downwards over the branches to prevent soil falling among then and clogging up the "drains".

Still another plan is to make rubble drains. These also are spaced 12 to 18 ft. apart. Trenches 9 in. wide and 2 ft. deep are taken out and the bottom 9 in. filled with rubble—broken bricks, large stones, clinkers and the like. Again a layer of turves is required.

Soak-away Pits. For a small plot or to drain a section of ground on which for some reason water collects. Soak-away pits will serve well.

To make a soak-away mark out a 2-ft. square of ground and remove the soil to a depth of 2 ft. Fill the bottom 1 ft. with rubble, cover with old turf grass-side downwards and return the soil.

Where several soak-aways are to be made 10-12 ft. apart is the appropriate spacing.

DRILL. Term used to denote a depression or shallow trench, cut in the soil by means of the edge of the rake, the hoe, etc., for the reception of seeds. See Seeds and Seed Sowing, and under separate headings for depths, distances apart, etc. Also the implement or machine used for drilling or sowing seed in drills.

DROUGHTS, METHOD OF COMBATING. Mulches of strawy manure, leaf-mould or lawn mowings spread on the soil around plants help to conserve moisture. Crops particularly needing mulching are peas, beans, marrows, globe artichokes and outdoor tomatoes. The hoe, however, is the gardener's best friend during periods of drought, constant stirring of the soil to keep the surface loose preventing the evaporation of moisture.

When once watering is started it should be kept up until rain comes again. Artificial watering results in many young roots being formed close to the surface, and if these are not kept supplied with water they will wither, and the last state of the plants will invariably be considerably worse than the first.

After-effects of Drought. Apart from depriving crops of water, a droughty summer has other effects in certain cases.

With potatoes, if the drought is followed by a wet spell, as often happens, super-tubering is inevitably the result, the potatoes that have formed themselves commencing to grow out and produce little potatoes. The remedy is to lift the crop at the earliest opportunity, even before the haulm has quite died down.

The danger of autumnal growth will be heavy on beet and carrots also, for it will assuredly cause root-splitting, which, in turn, may mean loss of keeping powers. So lift these roots early, too.

Lift salsafy and scorzonera early also ; the roots may not split if left in, but they will almost certainly send out a lot of side roots and become stringy in consequence.

Greens may look blue in late summer after a drought but they will certainly recover. They must not be given stimulants in the hope that this will encourage growth in spite of the dryness of the ground. A sprinkling of salt—just enough to whiten the ground—is permissible, for salt will attract moisture from the atmosphere.

DUTCH LIGHTS. See Frames, Heated and Unheated.

EARTHING-UP. Drawing up soil to the stems of plants, either to protect them, as is done with winter greens, to blanch them, as in the case of celery, or to encourage the production of underground tubers, as with potatoes. For methods see under headings mentioned.

EDIBLE-PODDED PEA, or SUGAR PEA. See Peas.

EELWORMS. Small, destructive creatures that burrow into plant and vegetable stems and tubers. The pests cannot be got at when present in plants so the soil must be cleansed of them before sowing or planting, by applying fresh gas-lime in autumn at the rate of 1 stone per square rod. To prevent eelworm attacks in greenhouse plants, sterilise the potting soil. See Soil Sterilising.

EGG PLANT. See Aubergine.

EGYPTIAN ONION, or Tree Onion. See Onions.

ELECAMPANE. Useful medicinal herb. May be propagated from seeds sown in the spring or by division of the old roots. The roots are the part required by wholesale druggists. Also see Herbs.

ELECTRIC HEATING for hot-beds. See Frames.

ENDIVE. Ranks as one of the most popular salad plants and is also coming into use for cooking. It is an indispensable crop for private gardens and a good crop for market growers, finding a particularly ready sale in autumn, winter and spring.

A glance through any good seed catalogue will reveal that there are now a number of different varieties and it is advised that growers should include at least two of these.

For summer use, Green Curled is a good choice ; Batavian Broad-leafed for autumn and winter.

When to Sow. The following are the usual sowing dates : A small sowing in May for early use, a main-crop sowing in mid-June for early autumn use, two sowings at the middle and end of July for winter use, a mid-August sowing for late winter use, a January-February sowing (in a hot-bed) for spring use.

Cultivation. Any ordinary garden soil will grow good endive, but the most suitable is a medium loam. Plants fail to do well in a heavy, moist border and they demand a sunny site.

Prepare the soil by digging it well and adding, if possible, a moderate dressing of old manure. Avoid the use of fresh, rank manure. Where manure is not available, rake in ¼ lb. per square yard of good garden fertilizer.

Sow the curled type in drills 12 in. apart, the Batavian in drills 15 in. apart. Thin the plants by stages until the curled stand 12 in. and the Batavian 15 in. apart.

In very light soils it is best to sow the seed in drills sunk about 5 or 6 in. deep so that they are the better able to take advantage of soil moisture and to secure the full value of water or liquid manure applied. These drills can be flooded with water, which will flow round the plants and soak in where it is most needed.

Blanching. The flavour of green endive is too " full " for most people ; hence it requires to be blanched. In summer all that need be done with the curled variety is to tie up the plants loosely like a cos lettuce. At other times they can be covered with an inverted flower-pot with a " crock " over the drainage hole ; or a spadeful of leaf-mould or ashes can be spread over each plant.

In summer about fifteen days is required for a full "blanch". In winter the best plan is to lift the plants before the onset of severe frosts, place them in boxes of soil and stand the boxes in a shed or cellar from which light can be excluded. This will give a perfect blanche.

FARMYARD MANURE. See Manures. Also see Digging and Manuring.

FENNEL. Perennial herb, the leaves of which are used in sauces and for garnishing. Raised from seed sown in April, in rows 1 ft. apart. Thin out to 1 ft. apart. Also see Herbs.

FERTILIZATION. The union of the male and female elements that follows pollination, producing fruit and viable or "living" seeds.

FERTILIZERS. See Manures. Also see Digging and Manuring.

FINGER-AND-TOE DISEASE. See Club-Root.

FLEA BEETLE, TURNIP BEETLE, or **HOPPER.** Injurious insect, laying eggs on the first young leaves of turnips, and sometimes attacking other plants. Seedlings are eaten bare to the ground. Watch for yellow spots on leaves, then dust with mixture of equal parts dry soot and lime, puffed through a blower —or, better still, dust in the same way with derris dust. Trenching the turnip site in winter buries the grubs deeply and smothers them.

FORCING VEGETABLES. Term applied to the growing of vegetables out of season in heated frames or greenhouses. The following are the vegetables most usually forced : Asparagus, beans, (broad and French), beet (globe), carrots, chicory, cauliflowers, cucumbers, endive, cabbage, lettuce; mustard and cress, mint, mushrooms, onions, potatoes, peas, parsley, radishes, rhubarb, spinach, spinach beet and seakale.

For methods to follow in forcing, see under separate heads. For details concerning frame preparation and management, see Frames.

In the ordinary way, a minimum greenhouse temperature of 45 to 55 degrees suits most vegetables though with cucumbers and dwarf beans 60 degrees is the minimum.

In the greenhouse, the custom is to prepare beds of rich soil either on the staging or on the floor for vegetable forcing, the cultivation routine then following on normal lines.

FOXGLOVE. One of the leading medicinal plants. Is cultivated for commercial purposes in the same way as when grown in the garden for its showy blooms. The leaves are required by wholesale druggists. Also see Herbs.

FRAMES, HEATED AND UNHEATED. Frames are of various types —the ordinary single-light box frame, the two-light box frame, box frames in ranges up to a dozen or more, brick pits, sunk pits (merely pits dug in the soil, lined with boards and with frame lights over them) and walls of turf over which frame lights are rested.

For ordinary garden purposes, the one- or two-light box frame is best. A good size for a one-light frame is 4 ft. 6 in. from back to front, 3 ft. 6 in. wide, 18 in. high at the back and 9 in. at the front. With two lights the frame should be about 7 ft. long, 4 ft. 6 in. wide. They are easily made at home, however, when timber supplies are normal. Solid, stout timber should be used.

These box frames can be moved from one part of the garden to another as required, for sheltering any desired crop. In winter they can be used for forcing work by placing them on a hot-bed.

For market-gardening work, brick pits are best. These are heated either by manure hot-beds or, better, by a range of pipes from a boiler. A common plan is to have these pits adjoining a greenhouse so that some of the heat from the house passes to the pits.

Sunk pits and turf frames are really only a make-shift for use when extra frame space is needed or in emergency.

Making Hot-beds. The usual material for making hot-beds is fresh strawy horse-manure, sometimes mixed with fresh garden rubbish and leaves also may be employed. The procedure is as follows :

When the manure arrives, tease it out and let it lie in a loose heap, protected from rain, for a few days, so that it may lose its rankness. The making of the hot-bed may then proceed.

First of all, put a layer of coarse, littery stuff in the bottom and, covering this with a thin layer of manure, trample the whole firm. Then put on another layer and again trample or beat into a solid mass. Keep on doing this again and again until there is a solid block of manure at least 18 in. thick—more than that, if possible. Trim the edges and leave everything neat and tidy.

If leaves are used, build them up in the same way, wetting each layer and trampling it until the solid block results. Leaves, however, are liable to be blown about by the wind, therefore bank up a hot-bed of leaves with soil on all sides.

When it is a portable frame that the hot-bed is to accommodate, the bed should be at least 2 ft. longer and broader than the frame itself. Thus, when the frame is put into position on the bed, a margin of 1 ft. all round is secured. Brick frames and deep wooden frames should merely be filled with the manure or leaves from 9 to 18 in. (according to the crop) from the sash.

The hot-bed should be left for three or

four days after it is made because it nearly always heats very violently for a while. Bore a hole in the hot-bed with a dibber, insert a proper hot-bed thermometer and after, say, ten minutes, take the reading. It is necessary to wait until the heat is near 70 or 75 deg. F. before attempting any cultivation.

Another method of testing is to put a stick in the hole, remove it daily, and feel it, trusting to the guidance of your hand to tell you when to start planting or sowing.

If crops are to be raised directly over the manure, cover the hot-bed with 4 to 6 in. of fine, sandy compost.

Electric Heating for Frames. In normal times, and in areas where current is available at a reasonable rate, frames may be heated by electricity, special appliances being available for the purpose.

A pit is dug 1 ft. deep and 1 ft. wider all round than the frame and lined with coke breeze, ashes or similar material covered with 3 in. of sand, this to insulate the frame from the cold soil and prevent wastage of heat. If the frame is to be of the sunk-pit type, the pit is dug correspondingly deeper—to allow 1 ft. of coke breeze, etc., below the required depth for the pit.

The method of heating is to run parallel lines of special cable in the sand bed from back to front of the frame, spacing them out 4 to 9 in. apart, according to the desired temperature. The cable lines which are covered over first with wire netting (to prevent damage when digging) and then with the seed-bed soil—just as manure is covered—are supplied with current from the nearest source. Each frame is provided with its own switch and there is also a fuse box. If desired, an automatic temperature controller, consisting of a thermometer which switches the current on or off as the temperature falls or rises, may be installed.

For a 6-ft. by 4-ft. frame, 50 ft. of heating cable are needed.

The running costs (consumption of electricity) will depend on several considerations, but for beds maintained at a temperature of from 60 to 70 deg. F. in the early part of the year, it has been found that about 1 unit per square yard per day is consumed.

Protecting Frames from Frost. At times of frost much can be done to maintain a good temperature in a frame by covering the lights.

For this purpose Archangel mats measuring 9 ft. by 4 ft. 6 in. are definitely best. As an alternative quilts made of sacking and stuffed with straw or leaves, plain sacking or old carpets may be used. Some growers put on a 4-in. covering of straw or bracken, holding this in place with boards or wire-netting.

Whatever is used must be kept dry and

dried if it becomes wet. Wet coverings do not keep out frost.

Put the coverings on before the temperature falls to freezing point, and do not take them off until the frost has gone. During long spells of uninterrupted frost keep the coverings on altogether, though you should not neglect an opportunity to take them off for an hour or two if the temperature rises above freezing point.

In addition to matting them, box frames can be further safeguarded from frost by piling up thick banks of leaves or manure or ashes, or even soil, around the frame walls.

Frame Crops. See under Glasshouse and Frame Crops.

Dutch Lights. Name given to a special type of light used originally by Dutch smallholders who settled in this country some few years ago mainly to grow tomatoes. Our own growers are now using them, not only for tomatoes, but also for lettuce, cauliflower and occasionally for radishes.

The lights contain a single sheet of glass usually 5 ft. and 2½ ft. Frames and " greenhouses " are built up with these lights, (much as rows of cloches are put together) some of the " houses " covering large areas of land.

Usually the "houses" are cold but occasionally heat is introduced into a small one for forcing work.

FRENCH BEANS. See Beans.

FROST, PROTECTING VEGETABLES AGAINST. It is in the early autumn that the protection of vegetables against frost is often desirable. The first frosts of the season are frequently followed by an "Indian Summer" spell, and crops which are not frost-nipped may carry on some while longer.

Runner and French beans can be protected by throwing tiffany or old lace curtains over the face of the row on the side facing the wind. If any plants happen to be frosted, the foliage should be syringed all over with clear, cold water early in the morning before the sun is on the plants.

To protect cauliflowers against any average frost it is quite enough to go over the bed two or three times a week and bend the leaves over the heart. If a really sharp frost comes along, advanced cauliflowers should be pulled up and hung head downwards in a shed, where they will keep for some time.

As regards marrows, a sprinkling of dry bracken over the fruits and plants will be sufficient protection until early October, when all the marrows should be ripe. If nothing better can be obtained, newspapers will do for covering these plants. A mat, or a little tiffany, hung over

tomatoes against walls or fences will serve to keep them going for another few weeks, to give fruits which are only just showing colour a chance to become fully ripened.

Celery, carrots, beet and salsafy will actually be improved by frost, while potatoes will come to no harm as long as they are well earthed up.

Protecting Frames. See Frames, Heated and Unheated.

FUMIGATION. See Soil Fumigation.

GARLIC. Member of onion family. Should be given same cultural treatment as shallots (see Shallots). February is the planting time. The bulbs consist of a number of separate cloves enclosed in a silvery membrane. The bulbs should be carefully opened up, the cloves separated and planted 12 in. apart and 2 in. deep.

Harvest crop in summer and store.

GHERKIN. Small pickling cucumber, usually grown outdoors. Culture is as described under Cucumbers, Ridge.

GLASSHOUSE AND FRAME CROPS. The cessation of all imports of early produce as the war developed gave an added importance to our own glasshouse production. By the winter of 1942-3 very considerable quantities of food were being produced, not only in commercial glasshouses but also in privately-owned greenhouses. Large supplies of vegetable seedlings were also being raised.

Cropping orders and fuel restrictions regulated the production from commercial houses, the position early in 1943 being that in houses artificially-heated between November 1 and February 28 only tomatoes, lettuces or mustard and cress could be grown, in addition to young tomato and vegetable plants.

Up to that time no such limits had been placed on private greenhouse owners, except that they were allowed fuel for greenhouses on the understanding that the houses were used only for food-growing.

Crops for Heated Houses. Assuming that there is no material alteration in the position of private greenhouses heated with boilers, or where good oil heaters can be worked, the following are suggested as suitable crops for winter: French beans, early cauliflowers, lettuce, radishes, salad onions, cucumbers, shorthorn carrots, six-weeks turnips, round spinach, peas, broad beans, mustard and cress. These should all be grown on the stages in pots and boxes.

Under the stages chicory, rhubarb and seakale can be forced; or mushrooms or turnips for tops can be grown.

Crops for Unheated Houses. A greenhouse can be turned to very good account even if not heated at all during winter, though obviously, the crops will not come along so quickly as in a heated house.

The vegetables recommended for an unheated house are lettuces, salad onions, spring cabbage and radishes.

Succession sowings should be made, at three-weekly intervals, to maintain the supply of lettuce, carrots, turnips, etc. Boxes about 6 in. deep should be used for these crops. For the culture of French beans, and the forcing of rhubarb, see separate entries. For the culture of tomatoes, the main glasshouse crop, also see separate entry.

Management of Greenhouse Crops. The best temperature for winter greenhouse crops is from 55 to 60 deg. F. The temperature must be kept steady at as near that range as possible.

Even in winter the sun's rays passing through the glass can soon run up the temperature of a house. For that reason, and because the air must be kept as fresh and buoyant as possible, the house must be ventilated whenever weather conditions permit.

The correct watering of winter crops is all-important; water should be given only when the crops really need it—when the soil is light brown (compared with the dark brown colour of moist soil).

A tank or tub should be kept in the house, and be re-filled after each watering, so that the next time water is required the chill has been taken off it.

Wash down the glass as often as may be necessary to keep it perfectly clean, so that all light available can pass through. Use a long-handled brush for quick work and rinse down well after washing so that no dirt is left to dry on the panes.

Frame Crops. The most useful and profitable crops for frames are lettuce, carrots, turnips and spinach. Seed should be sown in a 4-in. deep bed of good soil made up in the frame. The seed of carrot should be sown broadcast; the seed of turnips and spinach sown on $\frac{1}{4}$-in. deep drills 6 in. apart; while lettuce may be sown thinly, or seedlings transplanted from a greenhouse.

Other suitable crops for frame culture are salad onions and radishes (sown broadcast), and corn salad sown in $\frac{1}{2}$-in. deep drills 8 in. apart.

Keep the lights closed until the seed germinates, then ventilate whenever the weather allows. Water sparingly, thin as much as is necessary to allow for proper development, and wash the glass frequently to allow all possible light to pass into the frame. (Also see Frames, Heated and Unheated).

GLOBE ARTICHOKES. See Artichokes.

GOAT MANURE. See Manures.

GOOD KING HENRY, or **MERCURY.** Perennial vegetable producing foliage that is used like spinach, while the stalks may be used as a substitute for asparagus. Sow in spring, the seedlings being transplanted in deep, rich soil, in rows 18 in. apart, with 15 in. between the plants in a row. Established plants can be increased by division in spring.

GOURD. Useful vegetable producing large fruits of marrow type, but curiously shaped, in summer. Sow in warm greenhouse in April and plant out in June or sow outdoors in May and grow in same way as marrows. See Marrows.

GREENFLY. Is not so troublesome among vegetables as among flowers, but can sometimes be a great nuisance, as on beans, cabbages, Brussels sprouts. As a remedy, spray with any non-poisonous spray. Derris solution is a useful remedy (see Derris). Pyrethrum wash (see Pyrethrum) another excellent non-poisonous insecticide. Vegetables recently sprayed have merely to be washed to be fit for use. Greenfly-infested crops may alternatively be dusted with derris or pyrethrum powder when the foliage is damp.

GREEN MANURING. See Manures.

GREENS. General name for all members of the cabbage family—cabbage, cauliflower, Brussels sprouts, etc. See separate entries.

GUANO. See Manures.

HANDLIGHTS. Term applied to appliances made of glass which are used for placing over vegetables as a protection against frost and also to hasten maturity. They play a very prominent part in many market gardens, especially those engaged in the profitable trade of producing early vegetables.

Two main types are available—those which are complete in themselves and those which may be " built up " to cover almost any length of row. Also see Cloches.

HARICOT BEANS. See Beans.

HARVESTING VEGETABLES. For signs of maturity and methods of harvesting, see under individual crops.

HENBANE. Useful medicinal plant. There are two varieties, annual and biennial. The latter is grown commercially, the leaves, flowering tops of branches and the seeds all being required by wholesale druggists. Also see Herbs.

HERBS. Not only is a herb border, or even a herb garden, useful in the private garden but herb culture can also be profitable commercially.

Culinary Herbs. For details of culture with these, see under separate headings. The most commonly used culinary herbs are mint, parsley, sage, thyme and tarragon.

Harvesting and Storing Herbs. The harvesting season varies according to locality and the nature of the season, but can be determined upon in individual cases by remembering that the leaves are most full of flavour just before the first flowers are ready to open ; that, therefore, is the time when they should be cut. Cutting should be done with a sharp knife, preferably on a dry and sunny day, a good length of stalk being taken with them. They need to be tied up in small bundles, with as little delay as possible, to dry. It is a common mistake to dry herbs by spreading them out on a sheet of paper in full sunlight. Such rapid drying parches out the oils that contain the flavour. The right plan is to hang up the bunches, head downwards, in a dry, warm current of air where the sun does not reach them and where they can be protected from dust, leaving them thus until each leaf is crisp and brittle. Then the leaves may be easily rubbed from the stalks. A greenhouse is a good place in which to dry herbs. 70 degrees is the best drying temperature.

Pack the dried herbs either in air-tight bottles or tins previously dried thoroughly.

Medicinal Herbs. Up to 1914 practically all the medicinal herbs required by our druggists were imported from abroad. The serious shortage which occurred before the end of the 1914-18 war led the Government to encourage medicinal herb growing in this country, and a number of " herb farms " were started. These helped to meet the shortage created by the cutting off of imported supplies from 1939 onwards. Again Government encouragement has been given to medicinal herb growing. Exactly how profitable the project will be in the post-war years depends on whether cheap supplies are again available from foreign countries.

There is a wide range in the category of medicinal herbs, some of which are required in far greater quantities than others by the wholesale druggists. The best plan for those contemplating medicinal herb culture would be to get into touch with one of the firms specialising in the handling of these herbs and consult them as to their needs.

Proper drying is the crux of success with medicinal herbs. The herbs *must* be dried under the right conditions of temperature and moisture, which means the employment of a proper drying chamber—one specially built, or adapted in the correct manner from a barn, outhouse or the like.

The more important medicinal herbs which are cultivated—many grow wild—are aconite, angelica, belladonna, burdock, chamomile, colchicum, comfrey, dandelion, dill, elecampane, fennel, foxglove, henbane, lily-of-the-valley, male fern, marigold, marsh mallow, mullein, wormwood. See under separate headings.

HOEING. The regular use of the hoe saves a considerable amount of watering and weeding, and its use also definitely encourages growth.

To remove weeds, use the Dutch hoe, not cutting through the weeds just where they enter the soil, but getting the blade well under them and jerking them right out. Gather up the weeds after hoeing and put them on the rubbish heap ; otherwise, if rain comes, the hoed-out weeds would root again.

Hoeing helps to save watering by creating a mulch of fine earth which protects the lower soil from the sun, thus retarding, or even stopping, evaporation. Use a Dutch hoe, just pushing the blade to and fro into the surface soil, and walk backwards, *not* forwards, so as to avoid treading on the ground that has already been treated. Care must be taken not to injure roots growing near the surface or to go too close to the stems of plants.

When time permits, it is a good plan to Dutch hoe vegetable plots regularly once a fortnight throughout the summer.

HOP MANURE. See Manures.

HOREHOUND. Perennial herb, the shoots and leaves of which are used in cough mixtures. Seed is sown in April in rows 10 in. apart. Thin to the same distance. Also see Herbs.

HORSE MANURE. See Manures.

HORSERADISH. Useful crop for private gardens and some growers find it a good crop for market. Both for home use and sale, horseradish must be of good quality, the roots being long and straight. To this end it should not be considered as a permanent crop, but should be treated rather as an annual crop, to be lifted once a year and the bed entirely remade.

Autumn is the time to dig up the bed. Get *all* the roots out ; do not leave any in the soil, for the tiniest piece of root may grow and become a great nuisance. Sort out enough long roots for the winter's needs and bury these in some convenient spot.

From the remaining roots, select the best and cut them into pieces 5 or 6 in. in length. Make a slanting cut at the lower end and a straight cut at the top so as to indicate, later, which way up the root should be planted.

Now dig over the piece of ground which is to be the horseradish bed, remembering that about nine plants can be accommodated in a square yard. It is not necessary to add any manure unless the soil is in very poor condition.

Set the line and dibble holes 9 to 10 in. deep. Into each drop one piece of the prepared root so that its top comes 3 or 4 in. from the surface. Fill in soil loosely to cover.

Plant in rows 1 ft. to 15 in. apart, the plants being 1 ft. apart in the rows.

HOT-BEDS. See Frames, Heated and Unheated.

HYSSOP. Perennial herb, with evergreen aromatic leaves which are used in cough mixtures. Propagated from seed sown in April and division in spring. Grow in rows 18 in. apart, with 18 in. between the plants in a row. See also Herbs.

INSECT PESTS. See under separate headings for insecticides to use and other remedial measures.

INSECTICIDES. The chief insecticides for use in the vegetable garden are quassia solution, derris in solution or powder form, pyrethrum powder and pyrethrum wash. See under these headings for recipes.

INTERCROPPING. Term applied to the common practice of growing a quick-maturing crop in the spaces between the rows of slow-maturing crops and on other idle ground. Intercropping adds greatly to the yield of a given area of land and may even double the total amount of produce taken in the year from the ground.

The following are good examples of intercropping, all of them well worth following :

(1) Sowing spinach, radishes or lettuce between onion or parsnip rows.

(2) Sowing spinach, radishes or lettuce on marrow or mushroom beds.

(3) Sowing the above crops, or French beans, on the celery or leek ridges.

(4) Planting mushroom spawn on marrow beds.

(5) Planting tomatoes between fruit trees growing against sunny walls and fences.

(6) Planting cos or cabbage lettuce between each two rows of runner beans.

(7) Sowing stump-rooted carrots (for drawing young) between rows of Brussels sprouts.

(8) Sowing Windsor broad beans between the rows of main-crop potatoes.

(9) Planting tomato plants between rows of maincrop potatoes.

JERUSALEM ARTICHOKE. See Artichokes.

KAINIT. Slow acting potash manure. See Manures.

KALE, or BORECOLE. There is a tendency for Kale to be neglected in the average garden and by the average grower. There is nothing to beat them from the continuous-production point of view. Their hardiness enables them to resist the severest of winter frosts, thus maintaining a regular supply of greens right into the spring. There are sufficient varieties to provide a distinct and welcome change in the menu.

Asparagus kale yields succulent sideshoots that are eaten like asparagus. Cottager's kale produces endless sideshoots in spring. Hearting kale is dwarf, with curled leaves, and forms hearts like the cabbage. In addition to these there are the Scotch and curly kales, and also ornamental kales, with pretty variegated foliage—white leaves delicately fringed and veined with green, for instance.

Hungry Gap Kale, a newer variety, is exceptionally hardy. It will flourish where other winter kinds are destroyed by hard weather.

Kales like good, well-dug soil and an open position. Sow in April in a seedbed and plant out 2½ ft. apart in rows 3 ft. apart in sites cleared of potatoes, peas, etc., in late summer. Sow Hungry Gap Kale in June to August. The practical details of cultivation, feeding, etc., are as for cabbage, which see. Pests and diseases are also as for cabbage.

KOHL RABI. This vegetable, half turnip, half cabbage, is coming to be extensively grown and there is even some market demand for it in the better localities. Covent Garden, for instance, takes fair supplies. Its flavour is unique and distinctive, a combination of cabbage and turnip. White and purple sorts are obtainable.

Kohl Rabi is raised from seeds sown in April, in rows 12 to 18 in. apart, 1 oz. of seed being sufficient for 8 square yards. Plants are planted out from the seed-bed 1 ft. apart in rows 3 ft. apart.

The season of use is from July to December, bulbs being pulled when they are the size of a turnip ; if left to grow large the flavour is coarse.

LAMBS LETTUCE. See Corn Salad.

LEEKS. Leeks share with onions the distinction of being one of the most health-giving of all vegetables. During recent years they have become a still more popular market crop. Musselburgh, Lyon and Prizetaker are reliable varieties. Half an ounce of seed will yield 500 plants.]

When to Sow Leeks. For long, stout sticks, sowings are made in a warm greenhouse in mid-January or in a cold frame in February. For pot leeks, sow from mid-March to mid-April out-of-doors.

How to Sow Leeks. Under Glass. Sow in boxes 2½ to 3 in. deep, drained with layer of leaves ½ in. deep, and filled with sifted compost from an old cucumber or hot-bed. Make the soil moderately firm, water it with a rosed can and allow it to drain. Then space out the seed ½ in. apart each way on the moist surface and cover to hide them from view with more of the sifted compost.

Alternatively use thumb (2 to 2½ in.) pots and sow two or three seeds in each.

Sowing Outdoors. The best place to sow is in a drill by the side of the spring-sown onion bed ; the rich soil there will be just to the liking of leeks. Draw a drill, ½ in. deep as for onions—12 in. from the last row of the latter—and dust into it a little powdered chalk to show up the seeds as they are sown. Then space out the seeds ½ in. apart in single row, cover them in, tread them in and give a few touches of the rake up and down the row.

The Three Planting Methods. There are three methods of growing leeks:

In Trenches. This method is used where big leeks of the best quality are required. The crops are ready from about the end of September, but cannot be depended upon to last much after the beginning of December.

In making trenches follow similar lines to those recommended for Celery, which see. A trench 18 in. wide with a ridge 2 ft. wide will take one row of leeks 8 in. apart in single row ; a trench 2 ft. wide with a ridge 2 ft. 6 in. wide will take two rows of leeks, the rows being 1 ft. apart and the plants 8 in. to 10 in. asunder in rows.

Get the trenches ready in March, so that the manure you dig in can be on the way to rotting down by planting out time which is in mid-April.

In Pot-Holes. This method is the best for culinary leeks. It consists of making holes 6 in. to 7 in. deep and, say, 3 in. to 4 in. across and dropping a plant in each. You do not fill up the holes at planting time, but do that later on, when the plants are growing strongly. Finally, you complete the blanching by earthing up each individual plant, these mounds of soil being from 4 in. to 5 in. high.

Allow 18 in. between the rows and 12 in. between the " stations " (pot-holes) in each row. Planting time is from the middle to the end of April.

On the Flat. This is the least troublesome method and provides quite good culinary leeks up to the end of March. It consists of transplanting seedlings to a piece of well cultivated ground just as you would cabbages or lettuces.

There is no earthing-up.

For big leeks on the flat, planting time is mid-April. Allow 18 in. between the rows and 8 in. between plants in the row.

For late plantings, as " followers " after early potatoes, broad beans, etc., 12 in. between the rows and 6 in. between the plants.

Planting Out Leeks. It is good practice, a short time before leeks are to be transferred from seed boxes or seed bed, to shorten back the foliage slightly. At transplanting time, any overlong roots may be shortened with advantage.

Always do the planting-out in showery weather if possible. The dibber is the handiest planting tool. Water afterwards if the weather is dry—at least until the plants are established.

With pot-hole leeks, the seedlings may simply be dropped into the hole and then each hole well watered. The water, on subsiding, covers the roots with a fine soil sediment which serves until, later, soil is pushed into the hole as advised above.

Leeks want plenty of good, fat farmyard manure. So any site devoted to the crop must be liberally dressed with nearly rotted stuff or have been well-manured for the immediately preceding crop.

If farmyard manure is short give the site, before planting, 3 oz. of superphosphate and 1 oz. of sulphate of potash per yard run of row. Fork this into the top 6 in. of soil and *mix* it in thoroughly. Then, ten days after planting, sprinkle on enough soot to blacken the ground.

After Care. Leeks require copious watering for a while after planting—if rain does not oblige.

To assist the building-up of the stick two or three bouts of leaf-shortening are to be recommended. The first is given to the trench leeks about mid-June and the second a month later and the last one about the beginning of September. Leeks in pot-holes and on the flat have their leaves shortened about the end of June, the middle of August and the end of September. In all cases only the long *mature* leaves must be cut; the young leaves in the centre of each stick must not be touched.

Feeding Leeks. Leeks are hearty feeders. When they are growing strongly they may have fortnightly doses of liquid manure (2 pints per plant) right throughout their growing season, until mid-November, in fact. The liquid manure may be made from horse, cow, pig or poultry manure and soot. It is a good plan to alternate the manure and soot water applications.

An occasional dressing of agricultural salt, ¼ teaspoonful per plant, sprinkled around each plant is appreciated by leeks and is a preventive of bolting.

Pests and Diseases of Leeks. Leeks are remarkably free from insect pests and fnugoid diseases. The only trouble to be feared is Heart-Rot, and this will not occur unless the plants are grown in very heavy, water-logged soil.

LENTIL. Not very often grown in this country as a vegetable, but easy to grow from seed sown in April. Plenty of space—about 20 in.—should be allowed between the rows. The plants flourish best on a dry soil, and the only cultural requirement is to keep the rows free from weeds by regular hoeings. The crop is mature when the leaves turn yellow and the pods become a darkish hue. It is the seeds that are used.

LETTUCE. Lettuce is a crop which every garden should supply almost every month of the year. It is a good market crop when of the right quality.

Reliable varieties of cabbage lettuce are All the Year Round, Feltham King, and Carter's Holborn Standard. For early spring cutting sow Arctic King or Stanstead Park. Good Cos lettuces are Giant White, Giant Green, and Paris White. There are many other good varieties. An ounce of-seed will yield 2,000 plants.

When to Sow Lettuce. The succession of sowings begins in March. Up to June they should be made in a bed in a sunny position. Through July and August a shady position—at the foot of rows of peas or beans, or a border facing north—should be selected. In September a border facing south is the choice. This border will keep supplies going well into the autumn, and as the colder weather approaches part of the crop can be covered over with a frame. For winter, sowings can be made in a frame on a hot-bed or in boxes in the heated greenhouse. In January or February further sowings can be made in the south-facing border.

How to Sow Lettuce. Sowing lettuce in boxes to be grown on in the greenhouse, or in beds in the heated frame follows on normal lines. For market crops in the greenhouse, it is customary to plant out in beds made up on the floor or on the staging.

With sowings made after early May transplanting is not desirable. The seeds should be sown in drills ½ in. deep and 12 in. apart, the seedlings afterwards being thinned—cabbage lettuce to 9 in., cos to 12 in.

With lettuce sown up to early May transplanting is necessary, otherwise the plants will run to flower prematurely. When transplanting cut off ¼ in. from the end of each tap root. This encourages the formation of surface fibrous roots and promotes early and firm hearting.

Above all things lettuce demands well-dug, rich soil, which holds the moisture well. Lettuce grown in poor dry soil is so long in maturing that it is apt to bolt, besides being " heartless ". All sites where it is to be grown should have plenty of rotted manure, or decayed garden rubbish worked into it, and further, the surface soil between and among the plants must be regularly and freely hoed.

Blanching Cos Lettuce. Cos lettuce yield whiter, crisper hearts after blanching, ties being placed around the outer leaves to blanch the centres. This is done when the plants are almost full-grown.

Using ordinary raffia, damped and twisted (or thin elastic bands), tie the middle of the plant. Make a second and closer tie a week later, this time near the top of the plant.

Feeding Lettuce. With lettuce grown in a greenhouse or frame, it is a good plan to give each plant ¼ teaspoonful of sulphate of magnesia a week after planting-out, the fertilizer being sprinkled around the plant and watered in.

Generally, fortnightly waterings with liquid manure of any kind (or ½ oz. nitrate of soda per gallon of water) will encourage rapid growth and good hearting. A teaspoonful per foot of dried poultry manure is another good stimulant.

Pests and Diseases. The only insect pest which normally troubles lettuce is the Root Aphis—which collects in masses on stem, and stunts the growth of the plants. Young plants which have been attacked may be freed by lifting them, washing the roots in clear water and re-planting. Old plants should be dug up and burnt.

Under glass seedlings are liable to Mildew. The remedy is to water with Cheshunt Compound, which see.

LIGHTS. See Frames, Heated and Unheated.

LILY-OF-THE-VALLEY. All parts of this plant are of medicinal value—leaves gathered in May-July, flowers in June-July, and the roots in autumn. Also see Herbs.

LIME AND LIMING. While lime must be regarded as plant food, it is a necessity to plants for other reasons. It sets loose the stores of natural plant foods locked up in the soil ; it neutralises sourness in the soil ; it checks disease and soil pests ; it makes cold soils warmer; tends to bind very light soils and makes heavy soils lighter ; and it makes possible the activity of the nitrogen-giving organisms that are associated with leguminous crops.

All vegetable ground needs liming once in three years—sometimes even chalky soils in which lime is not always present in the top soil in sufficient quantity to fulfil the many functions performed by lime.

How to Tell if Lime is Needed. There are occasions—as when a new piece of ground is taken over—when it is not known whether liming is required or not. In such cases, the proper procedure is to test the soil as follows :

Take a representative sample of soil from the garden—a handful here and a handful there—and mix well. A tablespoonful of the mixture is all that is required.

Put that tablespoonful into a tumbler, add sufficient water to make a creamy paste, and then drop in 6 drops of hydrochloric acid.

Should the pasty mud fizz markedly, lime is not needed : should it fizz mildly or not fizz at all, lime is definitely required.

In recent years the Government have made grants to enable many food growers (allotment holders as well as smallholders and farmers) to obtain lime at special prices. While this opportunity persists advantage should obviously be taken of it. Lime ready for immediate application is available in some instances ; in others lump or cob lime is supplied. This must be slaked (broken down to a powder) before application.

To slake it, take out a small pit in the garden of a size appropriate to the quantity to be slaked. Place the lumps of lime in the pit and cover with 4 in. of soil. The lumps will absorb moisture from the surrounding soil and fall to a fine powder in two or three days. Then the lime can be applied.

The method of application depends on when the soil was dug and manured. If this was done a month or more previously the method is to spread the lime over the soil at the rate of 12 oz. to the square yard, and fork it lightly in.

The lime becomes mixed with the soil evenly and is all ready to wash down as the rains percolate the soil.

By this method there is no likelihood of the lime coming into direct contact with the manure which was dug in when the soil was turned over. Actually lime and manure are bad mixers : in certain respects one spoils the other if they are allowed to come into close contact.

For that reason, it is unwise to fork the lime in where soil has been dug and manured recently—the lime would easily wash through the " open " soil and soon reach the manure. Instead, the lime should merely be spread over the surface and allowed to lie, when it will then wash down more slowly.

Use a shovel for spreading, depositing the lime with an even, sweeping motion. Thus there is an even spread over the whole surface—by any other means it is impossible to avoid getting a patchy spread, with too much in some places and not enough in others.

The ready slaked lime—hydrated lime, it is called—is used in the same way and at the same rate as slaked cob lime.

Occasionally ground limestone is offered. It is simply pure ground quarried limestone. It is equally good. Its only difference from other limes mentioned is that it is not burned in a kiln. Spread it at the rate of 20 oz. to the square yard, to get a good result.

When liming remember that the ground on which potatoes are to be grown should not be treated ; lime encourages scab. Rhubarb and celery need little if any lime. On the other hand the cabbage family, turnips, swedes and beans all like generously limed soil.

LIQUID MANURE. Excellent stimulant for all vegetables during growth.

For Large Supplies. Where a good quantity of liquid manure is needed, the best plan is to mix a stock solution. Secure a barrel holding 20 to 30 gallons.

Fill the barrel with clear water, and plunge into it a bag containing solid cow, horse, sheep or pig manure. Remove any straw there may be in the manure. For every gallon of water add 1 lb. of manure and cover the barrel top to prevent wastage by air. In four days the liquid will, after diluting to quarter strength, be ready for use.

Immediately after drawing from the barrel always fill up with clear water. Do this two months, when the barrel should be emptied and a fresh supply of manure and water put in. Meanwhile, the liquid withdrawn should be diluted less and less for use.

An excellent liquid manure can be made from fresh poultry manure. It is so much more concentrated, however, that only ½ lb. is needed to a gallon of water. Before using, dilute to quarter strength.

For Small Supplies. These are best

prepared as required. For instance, a supply of dried, powdered poultry manure can be bought. Add 2 oz. of this to a gallon of water, stir until the powder dissolves, and there is a complete plant food at the correct strength. This applies equally to dried blood, though here 1 oz. dissolved in a gallon of water is the correct strength.

Two ounces of prepared hop manure left to dissolve in a gallon of water overnight also forms a splendid fertilizer. Again, any of the patent proprietary manures make a good liquid stimulant, the usual strength being 1 teaspoonful to the gallon of water. Nitrate of soda or sulphate of ammonia, ¼ oz. to the gallon of water, is another good stimulant.

To Use Liquid Manure. Before feeding any plant with liquid manure make sure that the soil is moist or scald will develop. If there is any doubt, water with clear water a few hours before feeding.

Do not allow the liquid to splash the foliage or the parts touched will be permanently disfigured.

LIVER OF SULPHUR SOLUTION. Excellent spraying fluid for use as a remedy against mildew and other diseases. The commonest recipe is 2 oz. liver of sulphur (potassium sulphide) and 8 oz. soft soap dissolved in 10 gallons of water. Another recipe is 1¼ oz. liver of sulphur, 1½ oz. of flour paste (added to give the wash greater adhesive powers) and 2 gallons water. The wash must be made fresh for each occasion. Do not let it come into contact with white paint.

MAIZE, or **SWEET CORN.** Vegetable of delicious flavour which is gaining popularity in this country. Produces large cobs which, boiled for ten minutes to half an hour, according to how tender they are, and served with sauce, have a very definite appeal.

The plants are raised from seeds sown in boxes during April, the boxes being placed in a heated greenhouse or frame. Later sowings can be made in the open during May. Plants raised indoors should be hardened off in cold frames and then planted out from mid-May to June.

MALE FERN. Important medicinal plant raised from spores sown in autumn in damp shady spots. Must be left untouched for a few years to mature and increase. The brown scaly roots are the parts required by wholesale druggists. See Herbs, Medicinal.

MANURES. The various kinds of manures which may be incorporated with

E

the soil at digging time are dealt with under Digging and Manuring; the special manures preferred by different crops are given under the separate entries. Here we will consider the qualities and purposes of the various manures, natural and artificial, the substitutes which may be used for natural manures, methods of storing manures and the like. Here, as with manuring suggestions generally throughout this book, regard must be paid to the fact that many fertilizers are in short supply, and it may be necessary to manure with very small quantities or with substitutes.

Animal Manure. Horse, or stable, manure is the best for all general purposes and particularly for clay or heavy loams that have a tendency to become sticky or run together. Horse manure is most valuable when mixed with straw and rotted, but is worth having when peat moss has been used for bedding the animals.

For light, dry soils, cow and pig manure are the preference. The ideal scheme is to employ a mixture of the two.

Animal manures are better than all other forms for digging into the soil in winter or spring.

The correct amounts of animal manure to apply are:

Minimum Annual Dressing. 10 tons per acre—1¼ cwt. per sq. pole (30¼ sq. yds.) —12½ cwt. for an ordinary plot of 300 sq. yds.

A Good Economical Dressing. 16 tons per acre—2 cwt. per sq. pole—1 ton (a good load) per plot.

Artificial Manures. The three elements essential to the proper development of plant life—nitrogen, potash and phosphates—which are all contained in stable manure, can be supplied to plants in the form of artificial (or chemical) manures.

The most useful nitrogen-supplying manures are nitrate of soda, sulphate of ammonia, nitrate of potash, nitrate of lime and nitro-chalk.

Potash, which improves the general growth of plants, and the colour and substance of the leaves, can be applied in the form of kainit, sulphate of potash, potash salts, or nuriate of potash.

Phosphates, which encourage root development and hasten maturity, are contained in superphosphate of lime (superphosphate, or super, as it is also called), basic slag, bone-meal and steamed bone flour.

Compost. This consists of garden waste, the leaves, grass mowings, the unwanted foliage from cabbages, carrots, beet and other vegetables (provided it will decay rapidly and is not diseased or pest-ridden)—that has been allowed some months in which to rot down.

Compost is commonly used for digging into the soil during winter cultivation.

In conjunction with artificial manures, it is a complete substitute for animal manure, the compost making humus and the artificials supplying quickly soluble plant food.

The best way to rot down garden rubbish into compost is to pack it into a pit dug in an odd corner or stack it up in a heap. As each layer is put down it should be dressed freely with slaked lime or new soot. Certain proprietary preparations applied to the garden waste, cause it to rot more rapidly and also considerably increase its value. Admixture with garden soil and sprinklings of sulphate of ammonia will also be effective in rotting down waste and making a good compost. Where a little manure is available excellent compost can be made by building up the heap in layers of garden rubbish (6 in.), manure (2 in.), and soil dusted with lime (1 in.). Make the heap 3 ft. high, finally covering it with soil. Turn over a month later and again cap with soil. The compost will be ready for use a month or so later.

Green Manure. This can be strongly recommended for all gardens. It is obtained by sowing some quick-growing foliage-crop, allowing it to grow for a while, then digging it into the ground to rot. In that way it performs the same duties as compost. Similar to compost, it forms, in conjunction with artificials, a complete substitute for animal manure.

The time to sow green-manure crops is as early in the summer as the ground becomes vacant—preferably in August. You can either sow the broad-leaved rape, or mustard or a special green-manure mixture as sold by seedsmen. Whatever you choose, sow it thickly, broadcast, and it will soon make tremendously thick top-growth—even though the soil is poor.

In the autumn, or winter, when a good height of top-growth has been made, dig the lot in, burying the stuff not less than a spade-blade deep. It is easier to dig in if beaten down and crushed with the spade.

Guano. This manure—the droppings of sea birds—is similar to poultry manure (which see) in its benefits to plants generally. It is only used as a stimulant to growing crops.

Hop Manure. The prepared hop manure sold under " brand " names, and

THE COMMONLY-USED MANURES

Manure.	Rate per Sq. Yd.	Complete or In-complete.	Supplementary Dressing.	Rate per Sq. Yd.	When to Apply.
Stable or Farm-yard	¾ pailful*	Complete	— —	—	— —
Compost Heap ..	¾ pailful*	Incomplete	Superphosphate of lime	2 oz.	2 weeks before sowing or planting.
Spent Hops ..	1 pailful*	Incomplete	Equal parts superphosphate of lime and sulphate of ammonia	3 oz.	2 weeks before sowing or planting.
Leaf Mould ..	1 pailful*	Incomplete	Equal parts superphosphate of lime and sulphate of ammonia	3 oz.	2 weeks before sowing or planting.
Granulated Peat	½ pailful*	Incomplete	Equal parts superphosphate of lime and sulphate of ammonia	2 oz.	2 weeks before sowing or planting.
Moss Peat ..	½ pailful*	Incomplete	Equal parts superphosphate of lime and sulphate of ammonia	2 oz.	2 weeks before sowing or planting.
Prepared Hop Manure†	6 oz.	Complete	— —	—	— —
Prepared Shoddy†	8 oz.	Complete	— —	—	— —.
Treated Sewage†	10 oz.	Complete	— —	—	— —
Treated Corporation Manure†	Follow Instructions.	Complete	— —	—	— —

*Use a 3-gallon pail.   † Proprietary products.

also the spent hops obtained from breweries are an excellent complete substitute for stable manure. They may be dug into the soil in place of animal manure, for in addition to their fertilising qualities they also form the essential humus, thus adding "body" to the soil. The hop manure is dug into the soil in exactly the same way as stable manure. It is sweet smelling, clean to handle, free from weeds and insects.

Liquid Manure. See Liquid Manure.

Poultry Manure. Poultry manure is an extremely valuable product, good for almost every horticultural purpose. It can be dug into the ground during winter cultivation. Combined with decayed garden refuse, it is excellent for that purpose. However, it is more commonly used as a stimulant for growing crops, either in liquid or powder form.

Poultry manure is now obtainable in air-dried form, that is, in the form of a dry powder, odourless, clean to handle and in every way convenient.

Rabbit and Goat Manure. These manures also are mostly used as a stimulant to growing crops. In any case they should not be applied direct to the soil as they have a souring tendency. If to be used for general manuring, the best plan is to throw the droppings into the compost heap and let them rot down and mix with it.

Seaweed. Where supplies of seaweed are available, the material can be turned to good use as manure, forming very nearly a complete substitute for animal manure. It is rich in potash. Left for six months to rot down, it can be dug into the soil in winter, at the rate of 2 cwt. per rod. Or dried seaweed may be burnt and the ashes sprinkled on the surface of the soil.

Shoddy. This is refuse from woollen and other mills. It is not advisable to use shoddy alone. A mixture of shoddy, spent hops and fallen leaves is a very useful manure for ordinary purposes. Take, say, a couple of 2-bushel bags of shoddy, break it up well, mix with it twice the quantity of spent hops, and then 4 times the bulk of these two combined of leaves. Give it a shaking out and rebuilding once a week for two months ; then it can be dug into the garden.

In no case must shoddy be dug in "raw," especially on the lighter soils, as it is naturally rather slow in action and intractable, although it is rich.

Sewage Manure. A manure used in increasing quantities during the war. As prepared by most local authorities it is an excellent substitute for stable manure. It is, indeed, richer in nitrogen and phosphoric acid than stable manure and

only slightly behind it in its potash content.

Sewage manures vary somewhat, but all can be dug into the ground for any crops.

In some localities sewage manure can be obtained at the sewage works for the cost of carting. In other instances sewage is specially treated to increase its manurial value still more and it is sold by the cwt. In all cases the price is very reasonable.

MARIGOLD. Grown as a medicinal plant, it is the yellow ray florets of the blossoming head that are required by wholesale druggists. Culture is the same as when the plant is grown in the flower garden for its bright blooms. Also see Herbs.

MARJORAM. Perennial and biennial herb, the leaves and shoots of which are used for flavouring. The perennial or Pot marjoram is propagated by seeds sown outdoors in March or April and by root division in spring. The biennial, or sweet marjoram, is usually treated as an annual, the seed being sown in April. The plants need plenty of space and should be thinned or planted to 1 ft. apart in rows of 1 ft. apart. Also see Herbs.

MARKETING VEGETABLES. Before the war a very high standard of quality, grading and packing was being reached by our market growers. The National Mark scheme was in full operation—this being the scheme under which growers bound themselves, with penalty in case of default, to pack only produce conforming to a specified Government standard. The scheme covered many vegetables—notably asparagus, broccoli and cauliflower, Brussels sprouts, cabbage lettuce, bunched carrots, cucumbers, peas and tomatoes—and was being extended to other produce. The growers concerned were in return allowed to use on their packages a special label called a National Mark. The mark was accepted as a guarantee of quality and produce bearing it gained the highest ruling prices.

Although growers have done their best to maintain the quality of their produce during the war, naturally the National Mark Scheme could not operate successfully under war-time conditions and needs.

Considerable control over prices—of potatoes, onions and leeks, tomatoes, cucumbers and other vegetables—has been another war-time development. Further, the method of marketing and sale, both wholesale and retail, has been regulated.

The general food situation in the post-war period, and the extent to which vegetables are imported into the country, are factors that will doubtless help to determine future marketing rules and policy ; but it will be accepted that in any circumstances the highest prices will be received by the growers who pay the greatest possible regard to quality, grading and packing.

MARROWS. Marrows are so useful in the kitchen that every gardener should grow a big crop of them. Apart from their use as a vegetable they make excellent jam, " ginger," " cream," and so on. They are a good market crop, too, especially if they come in early. An objection sometimes raised against marrows is that they take up a considerable amount of space. This objection can be ruled out by growing the bush types which have become very popular during recent years.

In the garden marrows are often grown on the rubbish heap. But they do far better if given a properly-prepared bed, just as in the market garden.

The Choice of Methods. For early supplies—and in cold districts—it is advisable to grow marrows in a frame—usually a hot-bed frame which has been used earlier for forcing vegetables. The bed needs no further preparation. Just clean out weeds and plant about the middle of April—two plants per " light," measuring 6 ft. 6 in. by 4 ft. 6 in. For a square light of 4 ft. 6 in. side or thereabouts one plant per light will do. To push growth along the heat in the frame is maintained by external linings of fermenting manure, keeping the frame quite close and giving no ventilation, and very little, if any, shading. In the summer, the lights are removed and the marrows allowed to grow beyond the limit of the frame.

The common method of growing marrows, however, is on a raised outdoor bed. The most convenient way of making this is to build the walls of old turves and fill in the hollow centre with fermenting leaves and manure. Tread these well down and cap with 6 in. deep of chopped loam in which to plant.

The raised bed should be 3 ft. in diameter and 2 ft. 6 in. to 3 ft. high. The marrows are set out in pairs in the middle and trained in opposite directions. Planting out is done about the middle of May, and for a week or ten days after planting bell-glasses or cloches are used to shield the plants from winds.

A third method is the sunken outdoor bed, to make which you take out the soil on the selected site to a depth of a foot or so. Then you fill up slightly above soil level with fresh manure, treading this down firmly. Finally, you cap 5 in. deep with some of the excavated top-soil.

Here, too, the beds may be squares, each to take two plants, of 3 ft. side, or

long beds 3 ft. wide in their shorter axis to take the required plants in pairs, the pairs being a yard apart.

If marrows are required in quantity, of course, they are grown on the flat—in stations prepared by digging out holes and putting manure in each.

Raising Marrow Plants. Whatever the type of bed in which you intend to grow the marrows the plants should preferably be raised from seed sown in mid-March, in pots in the greenhouse or frame. With the pot method, use 3-in. pots, well drained and filled with turfy loam, to which sufficient sand has been added to make it porous. Sow one or two seeds in each pot—on edge, not flat—pressing them into the soil. Cover with glass and paper until germination takes place. As the seedlings grow, give them more and more air, and harden them off well for planting out.

The plants can also be raised by sowing direct into the outdoor beds in April. It is a help to place a glass jam pot over each seed. The jars provide useful protection and promote germination and early growth.

Cultural Hints. Marrows want plenty of water at all times.

Liquid cow and soot manure can be given weekly from the time that fruiting commences. Guano and dried poultry manure are the best of the concentrated fertilisers.

On the formation of the seventh or eighth leaf remove the growing point of both trailing or bush kinds, to induce the formation of fruit-bearing side-shoots. The bush kinds require no further stopping. In the case of the trailers, when it is seen that the fruits are swelling freely the growing point of the bearing shoots should be removed two leaves beyond the fruit. Follow this course with all subsequent side-shoots and the plants will waste no energy on superfluous growths.

Pollinating Marrows. It is necessary to pollinate the marrow flowers when these appear. Pluck a male flower, strip off its petals and rub the pollen-laden centre against the centre of the female flower. The latter is easily recognisable by having a miniature marrow behind the petals, the male flower having only an ordinary stem.

When marrow plants do not produce female flowers they can be encouraged to do so by nipping off the tips of the trailing shoots.

Pests of Marrows. Marrows are notably free of disease, but they are liable to attack by Red Spider and Black Fly. The rusty-red foliage resulting from a Red Spider infestation should be cut off and burnt and the rest of the plants sprayed forcefully with clear water.

Dirty, crippled and stunted foliage are the symptoms of Black Fly attack. Cut off affected foliage and dust with tobacco powder, spraying with clear water half an hour afterwards.

Custard Marrows. Marrows of the bush types which bear very solid fruits with a distinct and delicious flavour. The plants are ornamental when growing.' Culture is exactly the same as outlined on page 124 for other marrows.

MARSH MALLOW. Useful medicinal herb, grown for its leaves and roots. Raised from seed sown in the spring and from cuttings. Also see Herbs.

MERCURY. See Good King Henry.

MINT. Valuable kitchen herb, much used for flavouring. Indispensable in the garden and provides a useful crop for market growers, especially early in the season and during the winter.

Prefers a cool, semi-shaded position and moist soil. Best grown in an out-of-the way border as it spreads rapidly. Plants can remain undisturbed for several years, merely being cut back closely in autumn and given a top-dressing of rich soil.

Remaking Mint Beds. At intervals, mint beds should be re-made. Do this in autumn or spring. Lift all the old plants with a fork and then break up the clumps. Cut the long stringy roots in pieces, scatter these over the bed and cover with an inch of good soil.

If replanting in the same place, dig up the bed first and mix some leaf-mould or well-decayed manure with the soil. More of this enriching material will be wanted if the soil is light than if it is heavy.

When replanting elsewhere every stray bit of root must be got out of the old bed; otherwise mint will be coming up everywhere.

The best method of starting a new bed is to buy a few roots, cut these up, scatter them and cover over as already described.

Forcing Mint. Supplies of fresh mint can be obtained in winter by lifting some roots in autumn, planting them in boxes, and bringing these into warm or cold greenhouse or frame.

MOCK-TRENCHING. See Digging and Manuring.

MULLEIN. Useful medicinal plant grown for its leaves and flowers. It is a biennial, raised from seed sown in rich soil in a fairly sunny situation. See Herbs, Medicinal.

MUSHROOMS. Mushroom growing received a great fillip during the war owing to the high food value of the crop as well as its tastiness. Now mushroom culture should be part and parcel of every home food grower's programme. As a market crop mushrooms offer wonderful opportunities. Not only are large quantities required for hotels and restaurants as well as for homes, but manufacturers absorb big supplies for the many bottled and tinned specialities in which mushrooms are used.

Until fairly recent times a plentiful supply of good horse manure was essential for mushroom growing. The fact that such manure is now almost unobtainable in many parts is not a bar to the cultivation of mushrooms. Special synthetic composts have been introduced which, used with straw or chaff, make beds capable of producing magnificent crops.

Mushrooms can be grown outdoors during the summer. The method most favoured, however—both by home and commercial growers—is to grow the crop under cover, where the crop can be so much better controlled. Sheds, cellar, unused garages, barns, stables, and all such places can be used for the purpose.

Following are details of preparing the beds where horse manure is available and where the synthetic compost is used.

Preparing Mushroom Beds. For manure beds the manure is stacked under cover but open to the air. Turn it over on alternate days for a fortnight to three weeks, during which period the heat will fall to the region of 80 deg. F.

When turning mix the droppings and straw thoroughly. On each occasion transpose the outside of the heap to the centre. If you come across dry patches, damp them.

Make up the bed by easy stages, treading each complement firm. When the temperature falls to 75 degrees Fahr., the bed is ready for spawning.

Using one of the synthetic materials as an example, a single bag, together with 2 cwt. of clean straw, or straw chaff, will make 60 square feet of bed.

The straw chaff is rather better to use than the whole straw, the cut-up stuff being simpler to compress in a small heap. Poor quality wheat, barley or oat straw is better for the purpose than good straw.

This straw chaff is then prepared, on a site previously disinfected against soil insects and diseases. This preparation consists of watering the straw three times, at intervals of 12 hours.

A layer of damp straw, 6 in. thick, is then spread in a circle about 4 ft. in diameter. The layer of straw is given a dusting of the preparation, and that is washed in by a very light spray of water through a fine rose. These layers are continued until the 2 cwt. of straw and bag of compost is built up into a conical heap.

Four days after the heap generates a temperature of 120 deg. Fahr., it must be rebuilt, turning inside to outside, moistening any dried parts. Turning is done a second and a third time. The mixture will then have attained a dark brown colour, smell quite sweet, and slightly of ammonia, will bind together when squeezed in the hand, with no excessive moisture.

Make the beds flat-topped, firm, and 9 in. deep. The width may be anything up to 5 ft., but don't exceed the 5 ft., for the sake of convenient working.

A dusting with pyrethrum powder when making up the bed is a safeguard against pest attack.

Spawning Mushroom Beds. Always use the best sterilized spawn ; it is never contaminated with disease, and gives the best and heaviest crops. Break up the spawn into pieces the size of a walnut, moistening them with aired water if they are in the least dry. Plant them, either in horse manure or synthetic beds, an inch deep and 9 in. apart, making the holes with the fingers.

The spawn will not long remain inactive. Delicate white threads called hyphæ will issue from it. When they do, cover the beds 1½ in. thick with clean sifted compost. A mixture of 3 parts subsoil and 2 parts granulated peat is ideal.

It is in this soil-cover that the mushrooms are formed. The cover must be kept moderately moist, but never on any occasion give more than 4 pints per square yard, or water will drain through into the compost, and rot the running spawn. The compost itself will remain permanently moist enough. Hundreds of beds are ruined by over-watering.

With under-cover beds there must be darkness or subdued light, a little draught-free ventilation, and a temperature of 50 to 55 deg. F. Between April and August this temperature is obtained by natural means. At other seasons a fume-proof oil stove will probably be adequate to provide the required temperature.

To Ensure a Succession. Beds spawned in September (the usual month for commencing the season's operations) come into bearing early November, and continue for 2 to 3 months. To make perfectly sure of a succession until the following July, when supplies become available from outdoor beds, spawn in early November, early January, early March and early May.

Each square foot of bed spawned should produce up to 2 lb. of mushrooms.

Gathering Mushrooms. When mushrooms appear do not cut them. When the veil covering the young button bursts, and the gills can be seen pulling can start. Take hold of the stem base, give a sharp twist and pull simultaneously, and everything will come out cleanly.

After pulling, fill up the little holes with the compost recommended for soiling.

If any mushrooms break, or there are bits lying about, gather them up and burn them, or they may on decay give rise to serious troubles.

MUSTARD AND CRESS. Can be grown outdoors, but keeps cleaner and is easier to cut if sown in boxes and kept in a greenhouse or frame. Place a layer of leaves on the bottom of the box, cover with rich sandy soil and, using a rammer, make it quite firm.

Sow the seeds thickly, pressing them into the soil but not covering them, and moisten the earth with a little tepid water. Sow the cress seeds four days before the mustard since the latter is quicker to germinate. By making a small weekly sowing throughout the year you will have a regular supply.

Some commercial men make up a bed on a glass-house stage and cut the salad as required for sale.

Some growers place coarse canvas over the soil and sow the seeds on the canvas. This prevents grittiness in the salad.

NAPHTHALENE. A cheap and effective soil fumigant. Spread on freshly dug soil at the rate of 2 oz. per square yard and lightly raked in, it will kill soil pests such as wireworms, chafer grubs, weevils, springtails, and millipedes. It may be used among growing plants, having no harmful effects upon them even though it touches their foliage.

Naphthalene has also been found effective in safeguarding members of the cabbage family against Root Fly, safeguarding onions against Onion Fly and carrots against Carrot Fly.

NEW ZEALAND SPINACH. See Spinach.

NICOTINE EMULSION. A reliable insecticide for use at any time of the year. but should not be employed on vegetables that are shortly to be eaten, especially such as lettuce. Is more potent than either paraffin emulsion or quassia emulsion.

Recipe : 1 quart of tobacco extract, ½ lb. soft soap, 3 gallons of water. Dissolve extract and soap in boiling water and dilute with the 3 gallons of water.

NITRATE OF SODA. See Manures.

ONIONS. Of all crops onions are considered one of the most important for the food garden. Equally they are an important crop for the market grower. They have the advantages that the thinnings can be disposed of in bunches as spring onions whilst the matured crops, properly stored, are saleable over a period of many months. The war gave a tremendous impetus to onion growing, to replace supplies that had hitherto been imported.

It is essential that, for market work, onions be of good size, clean and well-grown. "Scrub" onions, wrongly-shaped, small in size or with thick necks meet with little demand.

Methods of Growing Onions. Onions may be grown from seeds or "sets"— "sets" being small onion bulbs sown the previous year and lifted when immature.

When growing from seeds, sowings may be made outdoors in February-March, outdoors in September (the plants then standing through the winter to grow and mature the following early summer), in boxes, greenhouse or frame, in January-February, the seedlings being hardened off and planted out in late March.

For general crops the March outdoor sowing method serves. For early crops the September sowing is advised. For extra large bulbs the underglass method of raising the plants alone will do.

Sowing Onions. In boxes, sowing follows the usual lines. The boxes are drained with a layer of leaves, and filled with good, light soil. The seeds are sown very thinly and covered with a sprinkling of soil. When large enough to handle they are pricked off into other boxes and planted out in March or April into beds prepared as advised below.

In the open sowings are made very thinly in drills $\frac{1}{4}$ in. deep (deep sowing leads to thick-necked bulbs) and 12 in. apart on beds prepared as advised below. The seedlings are thinned as they become large enough for use as spring or salad onions, the object being ultimately to leave them standing 6 in. to 12 in. apart, according to whether normal or extra-large-sized bulbs are wanted.

The larger onion seeds produce the stronger seedlings. If time permits it is worth while before sowing to pick out and discard the small ones—in normal times. A sprinkling of weathered soot over the seeds helps germination.

Reliable varieties for spring sowing are Bedfordshire Champion, Ailsa Craig, Long Keeping ; for pulling green, White Lisbon; for autumn sowing, Autumn Triumph, Flagon.

Preparing the Onion Bed. Whether for planting seedlings, planting sets or sowing seed, the onion bed must be very rich, porous, well-drained and in a situation where it obtains all the sunshine possible.

When the site is settled, mark off the area necessary for the number of plants desired to be grown. Four square yards

can take 100 plants, although more room should be given when it is intended to grow for exhibition.

This ground is best when dug to a depth of $2\frac{1}{2}$ ft. or 3 ft. Do not bring up the sub-soil to the surface ; it tends to be cold and barren. Keep the good surface spit where it is unless there is reason to believe that it is "sick" through over-cropping. In such a case, dig it into the sub-soil, remove some of the sub-soil and wheel in a suitable amount of nice old turf for the new top spit.

When breaking up the sub-soil, try to improve it. A clayey sub-soil will become better if plenty of rough garden rubbish, littery manure, leaves, and, in extreme cases, some ashes are dug in.

A sandy, gravelly sub-soil will become more spongy and retentive if sewage or cesspool sludge is available, or rotted vegetable matter or green turf is dug in. In both cases a dressing of lime may be beneficial.

The upper spit should be about a foot or more, in depth and consist of a good, loamy, porous soil. Garden soil serves ; rotted turf is excellent, but both must be porous. If not quite "open" enough, add sand and gritty matter and in all cases dig in a light dressing of lime. Ridge it up to let the frost get at it.

This work should be completed, if possible, in December. Then in January the surface spit is turned over again, plenty of good fresh manure then being added. Also fork a good dressing of soot into the surface.

Shortly before sowing tread or roll the surface. A firm seed or planting bed is essential to onions. A loose soil encourages thick necks and flabby bulbs.

Planting Onions. *Onion Seedlings.* Seedlings are planted out in showery weather 6 in. to 12 in. apart in rows 12 in. to 15 in. apart, according to whether normal or large-sized bulbs are the aim. Plant so that the little bulbs are one-third buried.

Onion Sets. Advice is sometimes given to plant sets so that their tips are left just exposed. This is not the better way ; it is very apt to lead to bolting ; the sets are liable to push themselves out of the ground.

The correct method is to plant the bulbs below the surface of the ground—3 in. deep in light soils, 2 in. deep in other soils. Plant at least 9 in. apart in rows 1 ft. apart. Also, the bulbs must be planted very firmly and the soil must be well firmed before planting.

Small onions from the previous year's crop cannot be used as sets. They would run to seed or develop thick neck.

It is, however, a simple matter to raise a supply of sets. A special sowing should be made in early April, Bedfordshire Champion, Up-to-Date, Ailsa Craig being good varieties for the purpose. The secret of success is to broadcast thickly on very poor ground. Bulbs raised under these conditions will, when replanted, continue the growth which the soil poverty interrupted. When growth yellows, lift the bulbs, ripen them in full sun for a fortnight, and store them until planting time.

Feeding Onions. However thoroughly the bed is prepared summer feeding is essential to the production of large or good quality bulbs.

Starting at the beginning of June apply nitrate of soda ($\frac{1}{2}$ oz. per square yard) on three occasions at intervals of five days. Two or three days afterwards a healthier green will be noticeable in the foliage, while the young leaves will be bigger and finer than the old. (Sulphate of ammonia may be used insread of nitrate of soda).

A week after the last nitrate application, follow with superphosphate of lime (1 oz. per square yard), repeating dose a week later still. This fertilizer not only increases the number of long, white roots, but makes all the roots better food foragers.

Afterwards all the crop needs is something to help them fill the bulb. Sulphate of potash ($\frac{1}{2}$ oz. per square yard) is unrivalled for that purpose. Dress the plants with it weekly as long as they are able to accept and make use of food—which will be some time in August, when their leaves will then fall over and the lustre associated with active growth will depart.

Do not allow the fertilisers to touch any part of the onions. Give a watering after each dressing if there is no rain.

Ripening, Harvesting and Storing Onions. When the bulbs are fully formed and the foliage is beginning to yellow, push over the foliage towards the alleyways between the rows, using the back of the rake and bending the foliage down to the ground. Next scrape away loose soil from under the bulbs, so that each bulb is left anchored only by the downward growing roots.

The onions can be lifted when the foliage is generally yellow and dry to the touch.

After lifting spread out the onions in a sunny spot outdoors or in an airy shed, in a single layer, to dry.

The bulbs will store successfully, after drying, if spread out in an open layer on a dry shelf, but a better scheme is to rig up a " hammock " of wire netting a few feet from the floor of a dry, frost-proof shed, cellar, or loft, spreading the onions out on that.

A good way to keep small quantities of onions is to hang them in bunches, or ropes, from the ceiling in a cool, airy shed.

This is the method of roping. Begin by sorting the bulbs into sizes. Take a 3-ft. long stick or length of stout rope and, starting at the bottom, tie the neckend of each bulb to the stick or rope. Place the largest bulb at the bottom, grading to the smallest at the top. For the tying use string or stout raffia.

Pests and Diseases of Onions. One of the commonest causes of trouble with onions is the maggot of the *Onion Fly*. The pest can be kept at bay by dressing the soil between the rows with naphthalene (which see) or by moistening some sand with paraffin and distributing that between the rows.

Where the maggot *does* obtain a hold— as evidenced by stunted plants, and browning and rotting foliage—the only plan is to pull out and burn bad specimens.

Mildew and *Smut* are the two diseases most to be feared. The former shows as yellow patches on the leaves, which are afterwards covered with a powdery mould; the latter as dark spots and streaks on the leaves, giving rise to a black powder.

To combat mildew dig up and burn badly attacked plants and dust remaining plants with sulphur and lime (2 parts sulphur powder, 1 part lime) when leaves are damp with dew. Later, spray with liver of sulphur solution.

When smut occurs all that can be done is to pull up and burn affected plants.

Marketing Onions. See Marketing.

Pickling Onion. For the growing of onions specially for pickling, a piece of poor, dry ground will serve quite well. It must be in full sunshine, though, and must be rolled or trodden until it is very firm. Sow the seed broadcast. White Silver-skin and The Queen are two popular varieties for pickling.

Cover the seeds with soil, or lightly rake them in. Then tread the ground or roll it again. Harvest and store as for ordinary onions.

Potato Onions. Small onions that grow in dense clumps of closely-packed bulbs. They may remain in the ground year after year without disturbance other than the pulling up of such bulbs as are required.

Egyptian, or Tree Onions. Species of perennial onion seldom seen in modern gardens, but none the less quite useful for pickling or cooking. Once planted, they go on bearing year after year, being practically immune from pests and disease. Each bulb throws up a stout stalk on the top of which grows a tuft of little bulbs about the size of marbles. These can be plucked when ripe and, if required, stored as shallots. Plant in March, setting bulbs 1 in. deep and 4 in. apart.

PARAFFIN EMULSION. One of the best sprays for use against aphis, caterpillars, etc., though it should not be used on vegetables shortly to be eaten as it would be " tasted ". Take paraffin, 1 quart ; soft soap, 1 lb. ; soft water, 2 quarts. The soft soap is dissolved over the fire in the 2 quarts of water. Then the solution is taken off the fire and the paraffin added gradually and stirred in briskly. The mixture is then bottled for use as a stock solution. It must be kept closely corked and given a good shaking before use. To use add 1 part of the solution to 49 parts water. If foliage is young, the emulsion should be syringed off with clear water an hour after use.

PARSLEY. An indispensable crop for every garden and there is money to be made by growing it for market. To grow quality sprays, a soil which has been deeply dug is essential. Parsley prefers a medium to heavy but well worked loam ; light soil must be well manured before the crop will do well on it.

Parsley is one of the few vegetables which do well in shade or, rather, semi-shade. It will thrive on a north border.

Make three sowings, one, early, in heat or in a sunny frame to furnish plants which may go outside early in May, the second sowing early in March directly in the open, in rows 15 in. apart, and the third sowing in June-July in the open.

Thin out the plants where they come up in dense patches, leaving them standing 9 in. apart.

In regard to the third sowing, this gives good, sturdy plants that stand the severe frost of winter better than those of the earlier sowings.

A further plan to ensure winter pickings is to cover some of the plants, where they stand, with a frame, or to pot up a few plants in autumn and grow them on in the greenhouse.

PARSNIPS. This useful root crop gives the best results when sown early in the year. If the weather is favourable, early February is a good sowing time. The roots require a long season of growth and they like their first two months in the ground while the weather remains cool.

Lisbonnais, Student and Hollow Crown are reliable varieties.

Half an ounce of seed will sow a row 100 ft. long.

A new idea is to sow parsnips in mid-September for pulling young—such parsnips having a very pleasing flavour. Pulling can start nine or ten weeks after sowing, when the roots are about the thickness of the finger, and continue for several weeks. Any of the above-mentioned varieties can be used for this purpose.

E*

Preparing the Parsnip Bed. Parsnips will grow in any kind of soil provided that it is deeply dug and the seed sown early. However, it prefers a sandy or medium loam and a position in full sunshine.

Do not be content with merely digging the ground ; trench the site. Remember that, in reason, the roots will go down as far as you dig. On no account put any fresh manure into the soil ; otherwise the roots are inclined to branch or " fork ". Select a border which was richly manured for the crop it carried the previous year, such as the old celery ground or the onion bed, or choose ground that had manure dug into it early the previous autumn.

A light dressing of fertiliser can be raked in just before the seed is sown.

Sowing Parsnips. Sow in drills ½ in. deep and 15 in. apart. Press the seed into the soil before covering in ; this helps germination. Many gardeners mix radish seed with the parsnips ; the radishes come up quickly and mark the drills, enabling early hoeing to be carried out. Thin the seedlings later on, to some 8 in. to 12 in. apart.

Where the soil is unsuitable for deep-rooting crops excellent parsnips can be grown in this way : With a crowbar or stout stick make holes 12 in. or so deep, 10 in. apart. Fill these holes to within an inch of the surface with sifted soil. Sow 4 or 5 seeds in each hole and cover with more sifted soil. When the seedlings appear thin to the two strongest and later draw out the weaker of the two.

Feeding Parsnips. Not only does this feeding programme encourage fine rooting ; it also keeps at bay the common parsnip trouble, Rust. As a preliminary, water with a solution made by dissolving ¼ oz. of sulphate of iron crystals in a gallon of water. Do that in late June, giving each parsnip 2 pints. A week later start to feed weekly with soot-water, sulphate of potash and dilute liquid manure in succession. The correct dose of each is : Soot-water and dilute liquid manure 2 pints per parsnip, sulphate of potash ¼ oz. Spread the last-named on the soil around the root, avoiding it and the leaves.

Celery Fly or leaf-miner occasionally attacks the plants. For remedy see Celery—Pests and Diseases.

PEAS. Peas are among the most esteemed of vegetables for the home food grower. With a properly planned succession they can be available for many months, while special varieties can be grown to provide dry peas for winter use. They are an equally valuable crop for market growers, and there is an excellent sale to canning factories. For the latter purpose special varieties have to be chosen and special methods of cultivation followed. Inquiry at any canning factory will bring details of requirements.

When to Sow Peas. Early sowings are made in January under glass, for planting out at the end of March, and in mid-February on a warm border. Following is the programme for outdoor sowings to provide full succession :

First Earlies : Sow mid-March, English Wonder (1¼ ft.) ; sow end March, Thomas Laxton (3 ft.).

Second Earlies : Sow mid-April, Rentpayer (2½ ft.) ; sow end April, Paragon Marrowfat (3½ ft.).

Maincrop : Sow mid-May, Alderman (5 ft.).

Late Maincrop : Sow end May, Gladstone (4 ft.).

In early to mid-June sow the quick-maturing Peter Pan (20 in.) for extra late gathering.

Peas for drying—Lincoln Blue (2½ ft.) and Harrison's Glory (2 ft.)—should be sown in mid-March.

The Best Soil and Position. A rich, deep, friable soil, well stocked with humus and lime is what peas like, and the best crops are only to be obtained when these conditions are given. On the other hand, it must be very poor soil indeed and a very bleak situation in which fair results cannot be obtained.

The best place for peas is the site of last year's celery trench. Next best is a spot of ground that has been mock-trenched two spits deep and well stocked with manure during the winter.

In no case should peas be sown on sites that have been late-dug and then manured with comparatively raw manure. Such ground is too spongy for peas.

Preparing the Ground. For market work, preparation follows on the usual lines, the recommendations given above being duly observed.

For garden purposes the following is the procedure advised :

Take out trenches (in January-February) about 2 ft. deep where the rows are to be, break up the bottom soil with the fork, and put in a layer of decayed vegetable rubbish to which has been added, if possible, some rotted manure. This layer should be about 4 in. deep. Next comes a layer of soil 4 in. deep, to which a little rotted manure and also 4 oz. of basic slag per yard run, have been added, the trench then being filled up to within 4 in. of the top with good soil. In this depression the drills are drawn.

Sowing Peas. The normal procedure is to take out 9 in. wide flat-bottomed drills, 2 in. deep in heavy soil, 2½ in. in light. In these drills the seeds are sown in three rows, with the seeds spaced 3 in. apart.

A different method, however, is now being adopted by some growers. Two ordinary drills are made 2 to 2½ in. deep, and 9 in. apart, with the corner of the draw hoe. In each drill a single row of peas is sown, 2 to 3 in. apart.

The purpose of this method is to permit of a better scheme of sticking. With the wide drill three-row sowing, the pea sticks are put up along each edge of the row. In effect, the peas are boxed in by the sticks.

With the new sowing method the sticks are pressed into the soil *between* the two rows of peas. Thus they can climb up the sticks in full light. Growth is far more vigorous as a result and the crop appreciably heavier.

In either case a ¼-in. layer of silver sand in the bottom of the drill will help germination and early growth.

Another plan adopted by some growers is to make the drills 3 in. deeper than required, though only covering the seed with 2 in. of soil. This means that a depression is left in the soil above the seeds. The plan is advantageous in dry seasons, allowing all water applied to get to the plants. When the peas are for drying, space the drills for Lincoln Blue at 30 in. apart, and for Harrison's Glory at 24 in. Set the seeds at 3 in. apart.

Protecting the Pea Seeds from Mice and Birds. The best method of protecting the seed from mice and birds is to red lead it before sowing. Damp the seeds, put them in a tin, sprinkle red lead powder over them and then shake them up in the tin. In the garden the seeds sown can further be protected from birds by the use of fine-mesh pea guards, stretched netting, or a barrage of criss-crossed black threads.

Staking Peas. All peas do best if supported—even the dwarf varieties. The supports prevent the bending and cracking of the haulm which is common in windy places. Twiggy sticks make the best supports.

For staking tall peas, use bushy sticks 1 ft. taller than the advertised height of the peas. Thus for a 3 ft. variety use 4 ft. sticks. Allow four sticks per yard ; that is, two on each side. When putting in stakes at the sides of the usual 9-in. wide drill set the sticks upright, not inclining inwards toward one another.

Feeding Peas. When the plants are 1 in. high dress them with ½ oz. of sulphate of ammonia per yard run of row, watering this in. When the plants are 3 in. high give ½ oz. of superphosphate per yard run, this ensuring well-filled pods. A month later give ¼ oz. nitrate of soda or sulphate of ammonia per yard run. When flowers are forming repeat the superphosphate dressing.

Pests and Diseases of Peas. Pea Moth.

This causes maggots in the peas. Where there has been trouble with this pest, fumigate the seed bed before making the next sowings with 2 oz. naphthalene per square yard.

Pea Beetle. This is a pest of the seed grower. Its maggots bore into the seeds and prevent them germinating. The only remedy is prevention—as for Pea Moth. Seeds containing holes should never be sown.

Pea Weevil. This pest bites the leaves of the plants, attacks mostly being experienced in May and June. As a remedy spray attacked plants with quassia solution (which see) or dust the plants with equal parts soot and lime.

Mildew. This is the only serious disease. It causes the foliage to become covered with a white-pinkish mould or, in another form, the leaves have yellow-green blotches which are later dotted with yellow, brown or black pustules. Attack usually occurs when the plants are at the half-grown stage. As a remedy, spray with Bordeaux mixture (which see) at half strength. Burn all haulm after crop is gathered.

Edible-podded, or Sugar Peas. These are grown like ordinary green peas, but differ in that they do not need shelling; the pods have no membranous lining and are cooked whole.

PEAT SOIL.

For methods of dealing with peat soil to make it fertile see Digging and Manuring.

PENNYROYAL.

Perennial herb with creeping stems. The aromatic leaves are used for flavouring purposes. Propagated by seeds in April or root division in autumn. Where seedlings are raised they should be planted out in their permanent quarters, preferably in a damp gravelly soil, in September. Also see Herbs.

PERMANGANATE SOLUTION.

Useful fungicide. Made by adding sufficient crystals of permanganate of potash to water to colour it rose-red.

PESTS.

For the various pests which infest vegetable crops, see under separate vegetable headings.

PE-TSAI.

See Cabbage, Chinese.

PICKLING CABBAGE.

See Cabbage, Red.

POLLINATION.

Term used to denote the transference of pollen from one flower to another. Usually, this transference is effected by the wind and by insects, but certain vegetables—notably marrows and tomatoes—are best pollinated by hand. For the method of carrying out the pollination, see under Marrows and Tomatoes.

PORTUGAL CABBAGE.

See Cabbage, Portugal.

POTASH.

See Manures.

POTATOES.

There is no need to enlarge on the value of potatoes in the ordinary kitchen garden, allotment or holding. They are everybody's crop. Where space permits, all sections should be grown to provide for current needs (or sales) and to ensure adequate stocks for winter.

The choice of the seed potatoes is a matter of first importance. It definitely pays to buy the best. The year after the planting of the bought seed, tubers for planting can be saved from the home crop; but new seed must be obtained for the third year.

It will be noticed in the seed potato catalogues that some varieties are marked as "immune"—immune, that is, from wart disease, a very serious disease which hampers cropping capacity and spreads very easily.

The Government views wart disease so seriously that it has made illegal the planting of any seed of a non-immune variety in land where the disease has occurred, while the advisability of planting only immune varieties on all types of land has been stressed. There are quite a number of good, non-immune varieties which may be planted in "clean" land, however, if growers care to select them.

Following are reliable varieties in the different sections :

Among Immunes.

First Early : Arran Pilot, Ballydoon, D. Vernon, Doon Early, Snowdrop (Witch Hill), Ulster Chieftain.

Second Early : Arran Comrade, Arran Luxury, Arran Signet, Ben Lomond, Catriona, Dunbar Rover, Edzell Blue, Great Scot, Ulster Monarch.

Early Maincrop : Arran Banner, Arran Peak, Doon Star, Gladstone, Majestic, Redskin.

Late Maincrop : Arran Consul, Arran Victory, Champion, Dunbar Archer, Dunbar Standard, Golden Wonder, Kerr's Pink.

Among Non-Immunes.

First Early : Sharpe's Express, Duke of York, May Queen.

Second Early : Eclipse, Epicure, British Queen.

Early Maincrop : King Edward, Arran Chief, Up-to-Date.

Late Maincrop : Field Marshal, Red King.

The amount of seed of each section of potatoes required for planting can be calculated accurately. All seed potatoes must conform to regulation sizes and thus it can safely be estimated that for a 30-ft. row 7 lb. of first earlies, 6 lb. of second earlies, and 5 lb. of Maincrops or late maincrops will be needed.

When to Plant Potatoes. For very early supplies of " new " potatoes, sets may be planted in a frame, or in pots in a warm greenhouse, in December or January. For early outdoor supplies plant on a warm border facing full south, in February. Plant first-earlies in early March, second-earlies in mid-March and maincrops from the end of March to mid-April.

Preparing the Ground for Potatoes. Do not grow potatoes on the same strip of land two years in succession, as this entails the risk of eelworm attack or fungus disease.

If the selected site (after peas or members of the cabbage family is good) has already been dug, fork it a foot deep, and break down the lumps finely. Then tread firmly. If the site is still undug, work it a foot deep, and with spade blade, fork tines, and feet, establish a similar degree of fineness and firmness.

Where ground which was not manured during winter digging is used chemical fertilizer should be applied. A good mixture is superphosphate 2 parts, sulphate of ammonia 1 part. This should be distributed over the potato ground before planting, at the rate of 3 oz. to the square yard.

Sprouting Seed Potatoes. All seed potatoes do better when sprouted before planting, growing more quickly and cropping more heavily. To sprout the tubers, set them up in boxes in January. A suitable box is one 3 in. deep, 24 in. long, and 15 in. wide, the bottom being made of slats 2 in. wide with about an inch of space between them. So that room can be saved, the boxes should be made so that they can be stood one on top of the other—by nailing a 6 in. high upright in each corner of each box.

The seed potatoes should be placed in the box so that the end which has two or three eyes close together is uppermost. Then take the boxes and stack them in a light, airy, frost-proof shed, cold greenhouse, or spare frame. Cover with mats during severe frost.

If the potatoes are good, the shoots will be plump and green, with perhaps a purplish tint, and not much more than an inch long, by planting-out time.

Sometimes the shoots become infested with greenfly during sprouting. The best remedy is to syringe off the pests with slightly tepid water. With only a few sets to deal with the individual tubers can be washed clean under the tap.

No more than three or four shoots are required for a normal-sized seed tuber (one that isn't to be cut into sections). Any beyond this should be rubbed off or cut out. One of the shoots to be sacrificed should be the " crown " or central shoot.

If by chance any shoots are over-long, cut them back to just about the second joint.

Discard entirely any tubers which have failed to sprout at all; they will not " grow ".

Cutting Seed Potatoes. It is uneconomic to plant very large tubers, but with earlies don't adopt such a rigorous cutting policy as with second earlies and maincrops. Cut specially large tubers into half, lengthwise, retaining on each cut portion an equal number of sprouts. Cut the tubers 24 hours before planting. Stand the cut tubers in their boxes, in the greenhouse, frame or other warm, moist place and they will heal well. There is no need to dip the cut surfaces in ashes or sand.

The varieties Majestic and The Bishop should never be cut; they would " bleed to death ".

Planting Potatoes. There are various planting methods, but the only one commonly practised now is in drills.

The distance apart for the drills varies as follows : For earlies, 2 ft. ; for second earlies, 28 in. ; for main-crops, 30 in. Similarly the depth of drill varies. Normally it should be 5 in. on ground which has been dug and manured, 8 in. on ground which has not been prepared. On heavy soil 5 in. is sufficient depth and on light soil the depth should be 6 in.

Make the drills with a spade or draw-hoe. In the case of undug soil, when they will be 8 in. deep, spread a 3 in. thick layer of rotted manure in the drills, and plant on that. Where natural manure is not available use prepared hop manure. Make the drills 6 in. deep, spread the hop manure in the drill at the rate of 6 oz. to the yard run, cover with an inch of soil and set the tubers on that.

The distance apart at which to space the tubers is 10 in. in the case of earlies, 14 in. second earlies, and 15 in. for maincrops.

Space the tubers—at the distance required, as above—upright, sprouts uppermost, on the manure or bare soil, as the case may be. When returning the soil heap it up in a ridge above the row. This applies especially to early plantings, the ridges affording extra warmth.

Lazy-bed Planting. Even potatoes, good drainage lovers as they are, can be made to succeed on boggy land by planting them in " lazy beds ". The same method is useful when planting in soil which has not been properly dug. The beds are 6 ft. wide, allowing for three rows.

Divide the ground, already dug 1 ft. deep, into 6 ft. wide strips. Spread on alternate strips a 3 in.-thick layer of manure, plant the potatoes on the manure covering them with 5 in. of finely broken soil taken from a strip 3 ft. wide on each side of each bed.

Planting in Pots and Frames. Potatoes planted in pots and grown on in a heated greenhouse give crops of new potatoes in early spring. Use 9 in. pots and the following soil mixture : 2 parts sifted leaf-mould ; 2 parts sifted loam or good garden soil and 1 part rotted manure.

Drain the pots with crocks and a layer of leaves. Put in a 3 in. depth of soil and place three tubers upright on this and just cover them out of sight with more soil, made firm. When the shoots are 2 in. long, add more compost almost to bury the shoots again. Continue to " earth-up " in this way until the pots are filled to within 1 in. of the top ; after that let the shoots grow ahead uncovered. When the shoots are 6 in. above the pot provide twiggy stakes and tie with raffia.

Feed the plants weekly, from the time the shoots are 3 in. high, with superphosphate ($\frac{1}{2}$ teaspoonful per pot).

Plantings can similarly be made in deep boxes.

In a temperature of 60 degrees minimum crops are ready ten weeks from planting.

For frame work prepare the bed in the frame by digging up the soil at the bottom and taking it out to the depth of 9 in. Fill the 9 in. depression with fallen leaves, or a mixture of leaves and strawy manure. Fork it out evenly, then throw over that a covering of good soil.

Tread down ; then plant the sets 1 ft. apart in rows 18 in. apart.

The frame must be in a sunny spot, and if it backs close against a wall or fence, so much the better. Mats or sacks will have to be placed over the glass in times of frost.

Lack of frame-depth prevents the growth being earthed-up as with an outdoor crop, but in this case it makes little difference. The crop is to be forked out and used as soon as the tubers are of a decent size.

Growing Potatoes from "Eyes."

This is an old method of potato growing, revived in Britain during the war following its adoption on a large scale in Russia. The tubers are set up to sprout. As soon as the sprouts, or eyes, are sturdy and a good $\frac{1}{4}$ in. long they are cut out of the tubers, with a small portion of the flesh of the tuber at the base. Care is taken that the base of the eye is not injured during the cutting. The eyes are then planted in manure-lined drills just as the usual seed tubers are planted. Cutting out the eyes in this way means that the bulk of the tuber is saved and can be used for human consumption or stock feeding.

Earthing Up Potatoes. When the potato shoots are showing 6 in. above ground, fork between the rows, breaking up the soil finely. At the same time dress the soil with a mixture of superphosphate, 3 parts ; sulphate of ammonia, 1 part ; sulphate of potash, 2 parts, applying this at the rate of 1 oz. per yard run and keeping it off the plants. A week later, hoe up the soil in a ridge—that is, " earth-up "—against the plants. Continue to draw up soil to the plants at intervals during the season, but leave 6 in. of the tops exposed.

If any flowers appear on the plants nip them off.

It pays to remove weakly shoots from maincrop potatoes on which the haulm is crowded. The yield is generally better as a result and disease is less likely to be troublesome.

It is always a good sign when the swelling tubers lift and crack the soil lying above them. You can be sure that tuber formation is going on satisfactorily then. But if this goes on the noses of these tubers will be pushed up to the light and will be greened. Draw up, therefore, a little fine soil with the drag-hoe to cover, and do it again in a fortnight's time if the soil-lifting and cracking continues.

Spraying Potatoes. There is a disease, known as *The* Potato Disease, which causes havoc among crops in damp seasons. The risk of an outbreak can be entirely obviated by spraying the potato plants once or twice during the growing season with a fungicide.

Potato disease shows itself by certain well-defined symptoms. First, a few of the leaves begin to go yellow. Then these leaves curl up and turn brown. Ultimately the whole plant dies down. The tubers, when dug, show soft patches, the skin sliding off from these when rubbed. Diseased potatoes will not keep.

Either Bordeaux or Burgundy mixture may be used for spraying. See under their own headings for recipes.

For preventive spraying the following are the best times : South-west of England, spray by end of June; South, early in July; Midlands and East Coast, second week in July ; all other districts, end of July. Only when disease has a firm hold is it necessary to spray a second time, in the ordinary way. But should heavy rain fall immediately after spraying, a second application will be advisable.

Choose the evening of a calm day for the spraying and see that every scrap of the foliage—the under as well as the top surface—is thoroughly wetted.

A gallon of either fluid will be sufficient to do a rod ($30\frac{1}{4}$ sq. yds.) of potatoes.

Killing Potato Haulm at Harvesting. In recent years it has been the custom to kill potato haulm a week or two before

POTATO DISEASES AND THEIR TREATMENT

Disease	Symptoms	Critical Period	Treatment and Remarks
Leaf Curl .. Leaf Roll ..	Foliage curls at edges—haulm stunted—tubers small.	All stages of growth	No Cure. Virus-infected plants are always late starters—hence boxing and sprouting are chief methods of control. Dig crop cleanly and burn. Do not save "seed" from diseased plants, or from any that have been attacked by aphides which disseminate the virus.
Mosaic ..	Plants stunted. Leaflets crinkled, with yellow or grey spots or stripes.		
Common Scab ..	Surface patches of scab on tubers—no deep-seated injury. No injury to haulm.	As tubers begin to form	Disinfect sets by plunging for 2 hours in solution of formalin. Give a lining of leaves or grass mowings in planting drills.
Corky Scab ..	No damage to haulm. Tubers with pustular scabs, which eat inwards. Tubers are distorted, and unfit for food.	From first formation of tubers.	No Cure. Disease worst in wet seasons. Treatment as for Common Scab.
Black Scab or Wart Disease	Spongy, greenish masses on nodes of haulm near base. Tubers with spongy, grey outgrowths round eyes. These turn black and tubers rot.	All stages of growth	No Cure. Disease may be carried over with "seed" tubers. Grow only immune varieties. Dig and burn all parts of infected plants.
Black-leg or Black Stem-rot.	Stunted, pale green or yellow haulm. Base of stem black and rotten. Tubers with black spots at "heel" end.	From half-way stage of growth	No Cure. Disease brought over with "seed." Fork cleanly and burn. Do not save seed from infected plants.
The Potato Disease, or late Blight	Grey-brown blotches and streaks on leaves and haulm. Later, leaves blacken and die, and haulm rots. Tubers with brown streaks, rotting later. Definite faintly acid smell to plantation.	July	Spray 2 or even 3 times with Bordeaux or Burgundy mixture, first spraying to be given when plants are 10 in. high. Do not spray in industrial districts, but dig cleanly and burn all diseased haulm.

harvesting the crop, by spraying with a 10 to 15 per cent solution of commercial sulphuric acid. This destroys the haulm, and annual weeds, so making harvesting easier ; while it goes far to prevent damage by outbreaks of late blight. It is, of course, less of an economic proposition in gardens than on areas over an acre.

Harvesting Potatoes. Potatoes are ready to be harvested when the haulm ceases to grow and dies down into a yellowish tangle—that is, unless disease is present. Disease will cause the haulm to die down prematurely. In that case the haulm should be cut off close and the tubers dug up immediately they are of usable size—as proved by lifting an experimental root here and there.

If in doubt as to whether a crop is ready to lift, dig up a root or two and see if the skin of the tubers has "; set." A " set " skin is one that is tough enough not to flake off when it is rubbed between the tips of the fingers—*not* scratched up with the thumb-nail. It is essential that the skin be " set " when the potatoes are to be kept.

To dig potatoes, use a fork. Drive it in beside, and then right beneath, a root and lever upwards. Turn over the soil and pick out all tubers that have become detached from the plant.

After digging leave the potatoes lying on the ground, in the sun, for a few hours. Then gather them up and take them to a dark place. They must not be exposed to light too long or they will be "greened" and largely spoiled.

Storing Potatoes. In a normal garden, where only a small area is grown, the simplest way to store is in a box, or sack, placed in a cellar, shed, or loft, or spread out in a single layer. The shed, or other place used for storage, should be more or less dark. Where one must store in a light place, the tubers should be covered with a good layer of straw. The potatoes, moreover, should not rest actually on the floor, but on a layer of dry straw.

When the potato crop is reckoned in hundredweights or more, then the best way to store the tubers is in a clamp in an open, dry and unshaded corner. See Clamping.

Potato Diseases and Pests. For details of and remedies against diseases, see chart opposite.

In regard to pests, *Wireworms* may cause holing of the tubers (see Wireworms). The caterpillars of the *Rosy Rustic Moth* sometimes attack growing plants from June to August, causing sudden flagging. Such growths (which contain a fat grub) should be cut off and burnt.

POTATO ONIONS. See Onions.

POULTRY MANURE. See Manure.

PUMPKIN. Vegetable that is becoming more popular in this country. The fruits are usually globe- or pear-shaped and, with the Mammoth variety, weigh anything up to 80 or 100 lb. The culture of pumpkins is as for marrows. See Marrows.

PURSLANE. Annual herb, young shoots and leaves of which are used in salads, soups, and for pickling. Sow seeds in May, in drills 12 in. to 15 in. apart. Thin plants to 6 in. apart. The leaves and shoots will be ready for gathering in about two months from the date of sowing. Also see Herbs.

PYRETHRUM. Good insecticide for use on vegetables. May be used in powder form or as a solution. This is prepared by dissolving one teaspoonful of pyrethrum powder in a little boiling water and adding to a gallon of water in which a tablespoonful of soft soap has been dissolved.

QUASSIA SOLUTION. Useful as a preventive of aphis attacks. Is made up from : quassia chips, 1 lb. ; soft soap, 1 lb. ; water, to 8 gallons.

The quassia chips are soaked in water for a couple of days and simmered over a slow fire for two or three hours after the soaking—to extract the bitter principle. The strained liquid and the dissolved soap are then mixed and churned up together thoroughly. The quantities given will make 8 gallons. Alternatively, quassia in liquid form can be bought, and this simplifies the preparation of the insecticide.

RABBIT MANURE. See Manures.

RADISHES. Indispensable crop for every food garden and also a saleable crop for market-growers if crisp, small and mild.

The secret of success in obtaining good radishes is to grow them quickly. This means sowing in a well-prepared bed, deeply dug, manured early in the winter, and given a sprinkling of fertilizer when the surface is being finally raked.

Another requirement for good summer radishes is a cool, moist soil—which you will probably find on the north border. Failing this, choose a place that is in semi-shade.

Make fortnightly sowings from early March to late autumn. Either scatter the seeds broadcast or sow them in drills 1 in. deep and 6 in. to 8 in. apart. Sow thinly to obviate any thinning and, at least until the roots penetrate to a good depth, water the plot if it dries up badly before the plants get big. Also keep the dutch hoe going between the rows all the time. Water during very hot and dry spells or the radishes will bolt.

The final point in growing good radishes is to pull them at the right time. All too quickly, after they come to maturity, they pass beyond the crisp state.

The long radish varieties are best for summer work, the round for early work. Reliable varieties are Scarlet Round and Scarlet Globe, French Breakfast and Long Red. For autumn sowing select Black Spanish.

Forcing Radishes. Radishes can be ensured all the year round by making the winter sowings in boxes or beds in the greenhouse, and, later, in a frame either heated or unheated. An invariable custom is to sow radishes between the rows of vegetables forcing in a frame— between potato rows, for instance, or carrot rows.

RAMPION. Useful salad plant, the root being cut up and used together with the leaves. A hardy biennial, the seed should be sown at the end of May in very fine soil in a shady corner. Sow in rows 1 ft. apart and thin to 4 in. apart in the row.

RED CABBAGE. See Cabbage, Red.

RHUBARB. One of the most useful of all crops. Has a good market demand if the sticks are of reasonable size.

Planting Rhubarb. Rhubarb grows best in deeply-dug, well-manured soil in an open sunny position. It is best planted in autumn but can also be planted in spring. Plant deep enough so that the tips of the main crowns are just keeping above the surface. In a single row let the " sets " be 2 ft. apart for small varieties like Cherry and 3 ft. asunder for larger sorts such as Royal Albert, Victoria and Linnæus. If there are several rows they should be 3 ft. apart in all cases.

Top-dress the crowns with rotted manure every autumn.

If any flower stems appear during the summer, cut them off short.

Do not pull any " sticks " the first year after planting.

Forcing Rhubarb. Most people prefer the forced rhubarb that is available in the early months of the year to the naturally grown outdoor produce which comes along from about the end of March onwards. Not only is the forced rhubarb richer in colouring, but it is better, more delicate, in flavour.

For earliest supplies lift some clumps from the outdoor beds, in succession, in November, December and January. Choose the strongest clumps with good, fat crown buds on them. Weaklings or

roots less than two years old from the original planting set are not of much use. After lifting leave the clumps lying on the ground for a few days to be "frosted". They are then ready for forcing.

In the Greenhouse. One good place to force rhubarb is under the stage in the heated greenhouse.

Place a board about 6 in. to 8 in. away from the heating pipes and pack the space between board and pipes with fallen leaves, which must be kept moist throughout. Stand the roots on the floor as closely together as possible and cover the exposed outer roots with soil. Then hang mats or sacking round so as to keep the rhubarb quite dark. Water with tepid water before you close in.

The only other attention needed is to repeat the sprinkling of tepid water whenever the roots look dry.

In Frames. For succession supplies forcing may be done in a frame.

Stand the frame in a sunny corner and pack the roots together on the floor of it. There must be at least 18 in. of head room —2 ft. is better—and this can easily be obtained by excavating some of the soil. Glass lights are not needed ; the plants can be covered with a hurdle or boards. Also the sides of the frame may be improvised out of boards. Having fixed the frame, put in the rhubarb roots and watered with tepid water as before, heap fallen leaves and manure generously over the frame and pack it up round the sides. The more of this put on, the more heat will be obtained and the quicker supplies will be available.

In the Open. For later supplies still rhubarb may be forced in the bed in which it is growing, without any disturbance of the roots.

Pick out as many as required of the best clumps and cover them with leaves and strawy manure. Over the manure place half tubs or boxes.

If you are able to pack more leaves and manure around the outside of the tubs or boxes, the sticks will be ready to pull still earlier.

Rhubarb forwarded in this way needs no watering and no after-attention of any kind.

Propagating Rhubarb. Rhubarb can be grown from seeds sown outdoors in spring. The variety sold under the name of perpetual will provide sticks in about six months from sowing. The usual method, however, is by root division. Lift the stools in November or February and divide each up into pieces, each piece having a section of the solid root stock, a few of the thick fleshy roots and at least one good bud, or crown, with, round it, several smaller subsidiary buds. Replant the sections as advised above. Usually the centre section is worn out and not worth replanting.

No rhubarb should be pulled from divided crowns in the first season.

RIDGING. Method of cultivating soil. See Digging and Manuring.

ROCAMBOLE, or SPANISH GARLIC. Uncommon vegetable related to the onion, with bulbs, or cloves borne in clusters, like garlic. Requires the same general treatment as shallots. See Shallots.

ROTATION OF CROPS. Term denoting the changing about of the different vegetable beds for different crops in order that the utmost use may be made of whatever manure has been put into the ground and plant-diseases may be given every opportunity to die out of the soil. It is a secret of successful gardening not to let the same crops occupy the same site two years in succession.

The following table shows the crops which best follow one another :

Beans may follow borecole, broccoli, cabbages, parsnips, carrots or potatoes. They may be succeeded by celery, leeks, lettuce, turnips and any of the cabbage tribe.

Beet may follow the cabbage tribe and any other crop except spinach, turnips, parsnips, carrots, salsafy and scorzonera. It may be succeeded by peas, beans, cabbages, cauliflowers, lettuce or any other spring-sown crop except spinach, turnips, parsnips and carrots.

Borecole may follow peas, beans, lettuce and potatoes. It may be succeeded by peas, beans, beet, carrots, parsnips, onions, potatoes, celery, kidney beans.

Broccoli may follow peas and beans and may be succeeded by any crop requiring to be sown or planted when it is cleared off, except the cabbage tribe.

Brussels Sprouts. As borecole.

Cabbages may follow or be followed by peas, beans, potatoes, lettuce, onions and any other crop not belonging to the cabbage family.

Carrots may follow or be followed by any but root crops, celery and parsley.

Cauliflowers. As cabbage.

Celery may follow any crop which is cleared off the ground in time to allow the trenches to be prepared early in the season. It may be succeeded by peas, beans, onions, potatoes, turnips or any of the cabbage tribe.

Endive may follow potatoes, peas, beans and any member of the cabbage tribe and may be succeeded by any crop except lettuce.

Leeks may follow any crop but onions, garlic, shallots, rocambole or chives.

Lettuce may follow peas, beans, potatoes,

the cabbage tribe and any other crop with the exception of endive, chicory, salsafy, scorzonera, artichoke and cardoons.

Onions may follow the cabbage tribe, celery, potatoes, peas and beans. They may be succeeded by cabbages, coleworts, peas or beans.

Parsnips. As carrots.

Peas. As beans.

Potatoes may follow any crop except carrots, parsnips, beet, salsafy or scorzonera and may be succeeded by any crop.

Seakale may follow potatoes or almost any other crop and may be succeeded by potatoes, peas, or beans.

Shallots may follow peas, beans, potatoes, cabbages, lettuce, endive and spinach and may be succeeded by any crop except onions.

Spinach may follow peas, beans, cabbage, cauliflower, lettuce or any other crop, beet excepted. Winter spinach may be succeeded by any spring crop.

Turnips. As carrots.

RUE. Perennial herb, the leaves of which are used for flavouring and for herb " tea." It is an evergreen plant. Propagated by seeds sown in spring, by division and by cuttings. Grow 1 ft. apart in rows a foot asunder, preferably in chalky soil. Also see Herbs.

RUNNER BEANS. See Beans.

RUSSIAN KALE. See Chou de Russie.

SAGE. Perennial herb, the leaves of which are much used for seasoning. Is of a shrubby nature and evergreen. Will grow in any soil which does not hold too much moisture in winter, though preferring chalky ground. Propagated by seed sown in spring (in rows 2 ft. apart, plants to stand 1 ft. apart), by cuttings in May, or by layering in August. The shoots are available for use all the year round. Also see Herbs.

SALSAFY, or VEGETABLE OYSTER. Uncommon vegetable of parsnip-like appearance. Cooked in the same way as parsnips, it has a distinctive flavour. Seeds may be sown in the open in April, but the better way is to sow in a greenhouse in February and plant out in April. Give the lightest position possible, also rich ground deeply manur , with the top spit improved by the addition of wood ashes, weathered soot and decayed leaf-mould.

The crop is ready for lifting when the foliage is quite dry. To test the foliage, twine a main leaf round the finger and note whether any moisture exudes.

The roots may be stored in layers of sand or soil in a dry shed.

SALT. The chief use for agricultural salt on the vegetable plot is as a stimulant for asparagus and seakale. For method of application, see Asparagus and Seakale.

SANDY SOIL. For methods of treatment to make sandy soil fertile. See Digging and Manuring.

SAVORY. Useful kitchen herb. There are both annual and perennial kinds. Sow annuals (summer savory) outdoors in spring, thinning seedlings to 6 in. apart. Sow perennials (winter savory) as annuals. Also propagate by cuttings in spring or root division in spring. Also see Herbs.

SAVOY. See Cabbage, Savoy.

SCORZONERA. An uncommon vegetable with long narrow roots which are used like carrots. Sow in March in shallow drills drawn 15 in. apart in ground which has not recently been manured. Thin out to 10 in. apart. The roots mature about November, but they can be left in the ground through winter and drawn as required for use, like parsnips. Alternatively the roots can be lifted and stored in dry soil or sand, like carrots.

SEAKALE. Considered a luxury vegetable, seakale can yet be quite easily grown in any food garden. As a market crop it is supreme. Prices are invariably good and supplies rarely exceed demand.

The crop will flourish in any piece of ground which has been well dug and to which plenty of old leaves or other vegetable rubbish has been added.

The best way to start is to buy roots in March and plant these—in a bed in an open sunny position, prepared during the winter—upright, with 2 in. of soil over them, and 18 in. apart, the rows being 2½ ft. apart. The rows should be dressed with rotted manure in April and the soil should be given a good sprinkling of agricultural salt in June.

Blanching. The shoots have to be blanched for use. For early supplies put special seakale pots, or large ordinary pots, or boxes, over the plants in November and cover these with a mixture of half fresh stable manure and half leaves. For ordinary supplies merely cover with leaves in January or February.

Forcing Seakale. Particularly early supplies can be obtained by lifting roots in November-January, planting them close together in boxes or pots, covering with other boxes or pots to keep dark and bringing them into a warm greenhouse. Can also be forced in a heated frame kept dark, or under the staging of a warm greenhouse, the position being kept dark with sacking. Roots forced in this way are of no further use so it is customary to grow a supply of plants from cuttings specially for forcing.

Propagating Seakale. By Cuttings. The simplest method is by taking cuttings —6 in. long pieces cut from the side-roots of the main " thongs "—in spring. To ensure that the cuttings will be planted right way up, the bottom end should be cut sloping and the top end straight across. Plant as follows :

Make a narrow trench about 1 ft. deep, in rich soil. In this space put the roots upright, 2 ft. apart. Let the tips of the roots come 2 in. below the surface. Fill in the soil and tread it down firmly. Thin the resulting growths to one per plant.

The plants so obtained can be dug up and forced the following winter.

By Seed. Alternatively seed can be sown at any time between August and March. Sow thinly, in 1-in.-deep drill (or drills 1 ft. apart) in soil sunnily situated, dug deeply and manured, and which has been lightened (if inclined to be heavy) by mixing in sand or sharp grit. If the seed-drills have to be drawn in rather lumpy soil, pack the bottoms with dry woodash and place the seeds on that.

In June thin out the young plants to 6 in. apart, and when they are one year old transplant them 15 in. apart in rows 2 ft. apart, in really rich soil, if for outdoor forcing. For indoor forcing the strongest of the crowns can be taken under cover direct from where they were thinned out.

SEAKALE BEET. See Beet, Seakale.

SEAWEED. For use as manure, see Manures.

SEEDS AND SEED SOWING. The methods of sowing the various kinds of seeds, together with special hints to secure the success of the sowing are given under each vegetable concerned. Here are general suggestions applicable to the sowing of all vegetables.

To Make Seed Drills. Most vegetable seeds are sown in V-shaped drills. These drills are conveniently drawn with any one of the following tools :

(1) With the corner of the ordinary draw hoe ; (2) with the point of a triangular hoe ; (3) with the back of the rake laid flat and used as a " plough " with the taut line as guide ; (4) a notched end to the hoe handle, the notch being fitted over the taut line and used plough fashion.

No. (3) method is recommended for carrots, kohl rabi and turnips, and No. (4) for onions and shallots (seeds). For drawing the flat drill in which peas and beans are sown use the draw hoe. Be certain that the drills are of the same depth throughout, to ensure even germination.

Always measure up the distances and fix the positions of the respective rows with stakes at each end before you do any sowing. Also to ensure correct spacing always make the drills on the same side—the left or the right—of the garden line.

Wherever possible let the rows run from north to south rather than from east to west. The former direction gives the plants maximum exposure to sunshine.

For onions, leeks, turnips and other dark-coloured seeds it helps one to see how the seed falls if a little finely crushed chalk—enough to whiten the ground—is first dusted in the drills. Against this the dark-hued seeds show up plainly. For light-coloured small seeds, such as carrots, parsley and radishes, the same purpose is achieved by giving the soil a dusting with *weathered* soot.

The Right Way to Sow. If there is a breeze blowing when seed-sowing get the hand well down close to the soil surface. A mere puff of wind will blow small light seeds far away from where they are wanted to fall.

The usual method of sowing with small seeds is to tip the seeds into the palm of the left hand and then, taking up a pinch at a time with the thumb and first finger of the right hand, sprinkle them thinly by a movement of the finger and thumb.

With larger seeds the seeds may be tipped into the right hand and worked out over the first finger by movement of the thumb.

Thin and even sowing is most desirable ; sowing too thickly means overcrowding, additional work in thinning and waste of seed.

Testing Old Seeds. Before using seeds left over from the previous year it is well to test them for germination. Old seed sometimes "goes wrong" and if sown yields but a disappointing crop. The testing is done by germinating a sample of the seed. Distribute a sprinkling of the seed on a piece of damp flannel on a plate and put the plate in a fairly warm place.

Results can be judged in three or four days. If the germination is under 50 per cent the seed is not worth using. Nor is it of use, whatever the germination pecentage, if the "sprouts" are obviously weak.

SEWAGE MANURE. See Manure.

SHALLOTS. Shallots are a crop that every garden should grow and they are also a fair crop for market though a better channel for the disposal of produce would be the bottling or canning factory.

February-March is the usual planting time, though some gardeners like to make a planting in November. Two pounds of shallots will plant a 50-ft. row.

Preparing Ground for Shallots. Shallots *will* grow in any soil, but if the latter is poor, most of the cloves will probably only be of pickling size.

The best place for shallots is a piece of ground which has been deeply dug and enriched with a fair dressing of farmyard or hop manure.

Planting Shallots. Before planting, rake the surface of the soil into a fine level tilth. Select your bulbs before you plant. If they are not uniform, sort them out in different sizes and plant all of one size together ; if you sort them out from bulk, let your selection be of uniform bulbs, medium in size, firm, well shaped and nicely coloured.

Shallots should be put in drills from 9 to 12 inches apart and 8 or 9 inches apart in the drills. The best way to plant is to scoop out a little hole with the trowel, place the bulb in that and then firm the soil round it. The common plan, however, is to press the bulb in with a screwing motion and simply leave it ; if there is a change of moisture conditions before the bulb sends out roots to anchor itself, however, the bulbs are then very apt to jump out of the ground and to require resetting.

Shallots from Seed. Shallots can be raised from seed sown early in the year. The bulbs so raised are of no use for planting, but they are excellent eating. The seed is sown in well prepared soil, in drills $\frac{1}{2}$ in. deep and 8 in. apart. Thin sowing is desirable ; 1 oz. of seed will be sufficient for 100 ft. run of drill.

As soon as the seedlings show dust them with soot. Afterwards it is necessary only to keep the soil well stirred between the rows and to keep down weeds.

Protection from Birds and Frost. The bulbs begin to grow almost at once, and in a few days green tips will show. Sometimes birds peck at the tips, and they may loosen the plants. If there is trouble of that kind, dust a little soot over the bulbs after you plant them. If hard frost comes along, there is no danger, for the bulbs are exceptionally hardy ; it may, nevertheless, loosen the bulbs, and in that case, they must be firmed again when the thaw comes.

Feeding Shallots. There are two kinds of shallots commonly grown, the brown and the red. These need different feeding. Give the brown a weekly dose of soot-water until the outer leaves begin to turn yellow, then finish with two weekly dressings of sulphate of potash—$\frac{1}{4}$ oz. per plant. Red shallots also require weekly feeding, but in this case consisting of an application of sulphate of potash for two weeks running, soot-water being given the third week, the programme then being repeated and continued until the leaves begin to turn yellow.

To Prevent Thick Necks. See that the mother bulbs never become more than half buried. As the little side bulbs form, carefully scrape the soil away from them with the fingers, so that they, too, are only half buried.

Ripening, Harvesting and Storing Shallots. Shallots should be ready for harvesting—as shown by the withering of the foliage—some time in July. When the foliage falls, scrape as much soil away from around the cluster as possible, so leaving the bulbs exposed to the drying influence of the sun.

Lift the bulbs when ready, and take them just as they are in the cluster to a dry shed or greenhouse, there spreading them out in a single layer. Break up the clusters when drying is complete and store in single layers in trays in a dry, frost-proof shed.

SILVER BEET. See Beet, Seakale.

SLUGS AND SNAILS. These pests are often very troublesome indeed, and may do severe damage to crops. They are best combated by attending to soil conditions, to ensure good drainage—slugs especially favour damp situations; the use of lime and soot will help ; a mixture of 14 lb. lime and 2 lb. naphthalene will destroy many ; and, if it can be obtained, Meta may be admixed with bran or similar meal that will attract the pests, which will be poisoned. Hand collecting during the late evening is sometimes helpful.

SOAKAWAY PITS. For methods of making, see Drainage.

SOIL. For methods of improving the fertility of various types of soils, see Digging and Manuring.

SOIL FUMIGATION. Fumigation consists of applying to the soil a substance which, giving off fumes, gases to death the insects living in the soil. It is an excellent method of cleansing plots which are infested with wireworms, cutworms, millipedes and other insect pests.

There is a choice of several good fumigants. One of the cheapest and best is naphthalene. Scatter it over the soil surface at the rate of about 2-3 oz. per square yard and lightly fork it in. The fumes penetrate deep down into the soil and kill all insects encountered.

When properly applied, crude carbolic acid is also effective as a fumigant. Dilute ¼ pint in 2½ gallons of water and pour this solution on the ground after it has been dug. Give a good soaking ; the better

the treatment, the more satisfactory is the result.

Carbon disulphide is yet another soil-fumigating substance which, however, is rather expensive. To make it go further it may be mixed with an equal bulk of paraffin oil. To apply this liquid, dibble holes at 1 yd. intervals and 9 in. deep ; pour a table-spoonful of the chemical in each and at once plug the hole with a piece of clay. No lights must be brought near this liquid—it is highly inflammable. Nothing can be grown in the ground for two months. Also see Soil Sterilising.

SOIL STERILISING. In recent years nurserymen in particular and a number of other growers have been making extensive use of sterilised soil. They have found that sterilised soil ensures stronger plants, healthier growth, often an entire absence of disease. Soil sterilising might also be said to be the salvation of the tomato industry, enabling tomatoes to be grown in glass houses where hitherto the crops were always destroyed by disease. It is recommended now that sterilised soil be used for all greenhouse work—for all pots and boxes, for all seeds, seedlings and plants.

There are various methods of sterilising soil, as shown below.

The Formalin Method. This is the method that is almost exclusively resorted to by gardeners nowadays. It definitely results in the soil being cleansed of all disease spores.

Formalin, which is a 40 per cent. solution of formaldehyde, is obtainable at chemists for the purpose.

Having prepared the compost, spread it out in a 4-in. thick layer on the potting bench, or if mixing a large quantity, on the floor. An open shed is all right provided there is no drip from above. Measure out the compost in barrowfuls. It is necessary to know exactly the quantity which is to be treated.

To sterilise about 1 cwt. compost of ⅛th pint of formalin is needed. Either sprinkle it on the compost in a crude state or mix it well with half-gallon of water. The latter method is preferable because it gives a better distribution. After watering, turn over the soil into a dome-shaped heap and cover with newspaper or sacking. Leave thus for a fortnight without disturbance, then remove the covering, wait another week and the compost will be ready for use.

To Sterilise Tomato, etc. Borders. It is possible to sterilise the soil of a tomato or other border whilst it is *in situ.* The sterilising agent may be either formalin or cresylic acid. The sterilising is done at the same time as the border is being dug and manured in readiness for cropping.

Take out at one end a trench 18 in. wide and 1 ft. deep and move the excavated soil to the other end for filling in the last trench.

Now work the subsoil a foot deep, and mix with each square yard ¾ pailful of littery manure. Turn over the top spit and mix with each square yard ¾ pailful of well-rotted stable manure. When turning over the top sprinkle the soil thoroughly with the sterilising solution, the quantity needed being 4 gallons per 10 to 12 square yards.

The solution is made by dissolving 1 quart of formalin in 49 quarts of water.

Sterilisation with cresylic acid is done on quite different lines. The top is thoroughly soaked, and to do this approximately 4 gallons of solution per square yard are needed. The strength of the solution is 1 gallon of cresylic acid to 40 gallons of water.

First pour the acid into a tub, make up to 40 gallons, and stir thoroughly. Trench the soil as advised above, and as the work proceeds treat with solution at the rate advised.

It is not safe to plant in sterilised soil for at least a month. For the first week after using either formalin or cresylic acid keep the ventilators of the glasshouse almost closed, but afterwards open out widely to clear the air of fumes.

There is no need to cover the borders with sacking or tarpaulin.

There should be no plants of any kind in the house during the sterilising process.

Fire and Water Methods. The fire method is convenient for treating a small quantity of soil.

Make a fire in an old bucket or a grid, and place over it a sheet of iron. On the latter spread out your soil, and leave it, with a good fire burning beneath, for three or four hours. By this time the soil at the bottom of the heap will be thoroughly baked. Turn it over, and leave the top part heating for another three or four hours.

The soil will be ready for use after cooling though, naturally, it will need to be moistened.

The boiling method is to fill the soil into cans or pails and let these stand in a copper of boiling water until they are heated through and through. On the big scale special methods of sterilisation depend upon various appliances for treating the soil with live steam.

Also see Soil Fumigation.

SOOT. After it has been weathered, by being exposed to the air for a few weeks to take out poisonous elements, soot is a valuable fertiliser for practically all vegetables. Sprinkled over the soil and lightly forked in around growing plants it gives them a healthy vigour and deepens the green of their leaves. It can also be applied in liquid form, a bag of it being suspended in a tub of water, the solution being of the right strength when the colour of weak tea.

Soot is also a good deterrent to insect pests. Dusted over seedlings it will safeguard them from slugs and other enemies.

SORREL. Herb whose acid leaves are used for flavouring sauces and soups. Raised from seeds sown in spring in drills 1 ft. apart. Thin out to 6 ins. apart in the row. Also see Herbs.

SPAWN. Name given to the "bricks" from which mushrooms are grown. The "bricks" consist of cow manure in which the spores have been partially started into growth. See Mushrooms.

SPINACH. Spinach is a health-giving crop, and should be grown in every garden. It is also a good commercial crop.

Several kinds of spinach are commonly grown—the Round, Prickly, New Zealand and Perpetual varieties. Every grower should grow all, making successional sowings of the Round from January to June, sowing the Prickly in August to provide winter crops. New Zealand spinach is sown in a heated greenhouse in March and planted out in the garden in May-June.

Summer Spinach. A good place for sowing for the first supplies of spinach (the Round) is on a sheltered south-facing border where the soil has been well dug and manured. For the later sowings a shaded or semi-shaded position is essential. The soil, too, must be light and porous ; otherwise there is wholesale bolting—running to seed. Half an ounce of seed will sow 60 ft. of drill.

A fortnight after sowing thin the seedlings to 9 in. apart. A tablespoonful of dried poultry manure per yard of row is an excellent stimulant when the plants are 4 in. high.

Winter Spinach. Prickly spinach will grow readily in any reasonable garden soil. What it dislikes is a soil which lies soaked in water for weeks on end, a cold, water-logged, sour soil. In such a site it gradually fades away so that in spring there is no spinach at all.

The correct position is one out in the open where the soil is well drained. Dig the soil well and deeply, add gritty matter to it to open it up if it is not already porous, add leaf-mould to it to enrich it and raise it 6 in. above the general level of the garden. Thus is assured the drainage so necessary to keep the plants fresh.

Spinach should be gathered by taking a few of the larger leaves from each plant as soon as they are sufficiently developed.

New Zealand Spinach. This is a

variety specially suitable for hot and dry summers. Produces succulent leaves in great abundance, in spite of drought. Sow under glass in March-April, in small pots or boxes of fine soil. Transplant seedlings three or four weeks later to open, sunny positions, placing them 1 ft. apart.

Perpetual Spinach, or *Spinach Beet.* Variety of spinach which produces a generous supply of edible leaves not only through the autumn, but also from the following spring onwards. Sow seeds outdoors at any time, in the spring and summer, from March. Gather the leaves as they reach useful size. Pick frequently; if leaves are left on to grow coarse young growth is checked.

Feeding Spinach. It is not desirable to feed winter spinach. If the soil is well prepared that should provide all the nourishment necessary.

Summer spinach thrives best with fortnightly waterings with weak liquid manure. Occasionally a little dried poultry manure may be sprinkled alongside the plants and hoed in.

Spinach is very seldom troubled by either pests of diseases.

STORING VEGETABLES. See under separate headings; also Clampings.

SUCCESSIONAL CROPPING. Term applied to the sowing of crops in immediate succession to others so that the ground remains constantly occupied. See Rotation of Crops.

SUGAR PEA. See Peas.

SULPHATE OF AMMONIA. Chemical manure of value for most crops. See Manures.

SULPHATE OF POTASH. Chemical manure that fosters the development of most crops. During the war, shortage of sulphate of potash made necessary the use of substitutes. The value of wood ashes (which see) lies in the potash content. See Manures.

SUPERPHOSPHATE OF LIME. or Superphosphate, or Super. Chemical manure that encourages the growth of plants and their roots. See Manures.

SWEDES, GARDEN. Swedes differ from turnips in that the flesh is often sweeter ; they rarely get tough or woody ; are free from a " biting " taste ; and the " tops " are even more tasty than turnip tops.

Seedsmen now offer selected strains particularly suitable for cultivation in the ordinary garden. There is little to choose between the three popular sorts —yellow, white, and purple-topped—

except that the last-named is best for poor soil or an exposed district.

Swedes should be sown thinly in May in well-moistened soil, the rows being 2 ft. apart, as these plants require plenty of room. Half an ounce of seed will sow 50 ft. of drill. Thin seedlings to 15 in. apart.

Use as required but store late crops in a shed as recommended for turnips.

SWEET BASIL. Useful kitchen herb. See Basil and Herbs.

SWEET CICELY. One of the lesser grown herbs, the leaves of which are occasionally saved for their fragrance. Raised from seed sown in March, in rows 10 in. apart. Thin out to 6 in. apart in the row. The aromatic leaves can be gathered in late summer and autumn. Also see Herbs.

SWEET CORN. See Maize.

SWISS CHARD. See Beet, Seakale.

TANSY. Perennial herb, the leaves of which are used for seasoning and flavouring. Propagated from seed sown in March or April—in rows 1 ft. apart, the plants being thinned to 1 ft. Also propagated by division in spring or autumn. Also see Herbs.

TARRAGON. Perennial herb, the leaves of which are used in seasoning and the young shoots in salads. Grow 2 ft. high. Seed may be sown in March, but for quicker results young plants may be bought then and set out in rows 15 in. apart, with 8 in. between the plants. Increase can be effected by taking root or stem cuttings or by separating offsets from the parent plant. Also see Herbs.

THINNING. Term denoting the removal of some of the seedlings in a row in order to give the remainder adequate growing room. It should be a progressive operation, the rows being thinned gradually until the plants stand the appropriate distances apart.

Thinning should always be done when the soil is moist, so that the surplus seedlings can be drawn up with their roots intact. In general, if desired, the lifted seedlings can be used to fill gaps or form other rows. After thinning press back the soil around the seedlings which are left. For the correct distances at which to thin out vegetable seedlings see under separate headings.

THYME. Evergreen herb, the leaves of which are used for flavouring. Seed is sown in March or April and the seedlings planted out in June in rows 1 ft. apart with 1 ft. between plants. Or the plants may be set 6 in. apart to form an edging for one of the vegetable plots. Thyme can be increased by means of cuttings or root division in spring. Also see Herbs.

TOMATOES. Tomatoes should be grown by every cottage gardener, allotment holder, kitchen-garden owner, even by those who possess no more than a flower garden. In normal times they are also a fine market crop though naturally supplies coming in early and late in the season are the real profit earners, midsummer crops not showing so good a return unless of exceptional quality.

As is well-known, tomatoes can be grown equally well in heated and unheated greenhouses and frames and in the open. Naturally the heated greenhouse supplies are earliest—or latest (in the case of autumn crops).

When to Sow Tomatoes. For the earliest fruiting sow at the end of October. For normally early supplies (for fruit in May, that is), sow early in January. For outdoor plants sow mid-March. For late summer supplies sow the middle of May. For frame crops, sow mid-March. For indoor cultivation Potentate is a reliable variety, Pride of the Market for outdoors. Sunrise does well indoors or out.

Raising Tomatoes from Seed. Cleanliness is all-important. Not only must the seed vessels be scrubbed and disinfected but it is almost essential to use sterilised soil (see Soil Sterilising). Using unsterilised soil, there is grave risk of disease decimating the crop.

The best compost to use is one consisting of well-rolled turf rubbed through a sieve with about one-third its bulk of silver sand. Another good compost is half sifted loam and half leaf-mould with a pint of silver sand to each peck of the other two ingredients. For winter sowing the compost should be warmed before use.

Use 6-in. pots for preference, draining them with 2 in. of crocks covered with 1 in. of rotted leaves. Fill up with compost to within an inch of the top of the pot and press firmly.

Sow the seeds separately, pressing them into the surface soil. Space them out an inch apart over the whole surface. Then sprinkle fine soil over the seeds until they are covered to a depth of ¼ in. Cover each pan or box with a sheet of glass and some brown paper. Germination takes from six to ten days.

A minimum temperature of 55 degs. F. by night is necessary for the seed ; 58 degs. to 60 degs. F. is better.

Keep the seedlings on the dry side ; only give enough water to prevent them from flagging. Provide the necessary moisture by holding the base of the seed vessels in water until moisture works through to the surface of the compost.

As soon as the seedlings appear water them with Cheshunt Compound (which see) as a safeguard against damping-off. Repeat the application on three occasions at fortnightly intervals, using 1 oz. of the compound in 2 gallons of water.

Pot off the seedlings when they have reached the third-leaf stage—not before—again using a 50-50 compost of sifted loam and leaf-mould with ½ pint of sand per peck. Use small 60 (3 in.) pots—one plant per pot. Plant the seedlings so that the seed leaves are just level with the surface soil.

Grow on in a temperature of 55 degs. to 60 degs. F. on a shelf near the glass. Shade for a week after potting, but after that give all the sun possible.

Stop overhead watering as soon as possible, for tomatoes do not like to have their foliage wetted, although young plants have to put up with it.

As soon as growth is going ahead strongly the plants should be moved on to 5-in. pots filled with a well-mixed compost of 1 bushel of rich, fibrous loam, 1 quart of bone-meal, 1 gallon of ashes from the bonfire (wood ashes, etc.), ⅛ of a bushel of thoroughly decayed manure, and a final dusting of lime.

Water rather sparingly after potting. Give each plant a supporting stick.

The plants are ready to go into their fruiting quarters (in greenhouse, frame or outdoor bed) when the first truss of flower buds is seen.

Planting Tomatoes in the Greenhouse. For the production of fruits for home use, tomatoes may be grown in boxes measuring 15 in. long, 10 in. wide and 8 in. deep or in 9-in. pots. For market work they are either grown in ridges made upon the staging or in beds on the floor.

In regard to soil for the pots, boxes or beds, three parts of fresh, chopped loam and one part of nearly rotted manure makes a good tomato compost. If the bed is on the ground 15 in. depth of good soil is needed. If the beds are made up on raised platforms or stages start with 9 in. depth of soil placed in the first case as a ridge for each row. The sides of the ridges may be kept in place by rows of bricks which can be moved outwards to allow room for the top-dressings of soil that will be wanted for these raised beds.

For pots or boxes, sufficient soil to bring the level up to within 3 in. of the top is required, this to allow for top-dressings, several of which will be needed (not only in the case of pot plants but all plants), one when the first truss of fruit has set, another when the second truss has set;

others as and when roots show on the surface.

Keep the plants 15 in. apart. If there are two or more rows keep these 2 ft. apart.

From the time that the young plants are well under way right to the end, they should never be allowed to get root-dry. Root-dryness is the most fruitful cause of flower-dropping.

Tomatoes must not be coddled. They want all the sun they can get, but airy conditions as well.

The temperature of the houses after the middle of May, when fire-heat can be dispensed with, will vary in accordance with outdoor conditions. Up to then the following minimum temperatures should rule : February 55 degs., March 58 degs., April 60 degs., May 60 degs.

Tomatoes must be supported—even when they are young. Give these supports at the 8-in. high stage. Use bamboos or hazel or ground ash stakes or strings. In the latter case soft string is looped round the plants 3 in. from the ground, carried up and tied just *straight* —*not strained*—to the roof wires. String and plant are twisted round each other.

Planting Tomatoes in a Frame. Use the same soil mixture as recommended for greenhouse tomatoes. Make up a ridge along the front of the frame and plant the plants in this at 2 ft. apart. Fix canes along 6 in. under the light, running from front to back of the frame. Train the plants along these canes, looping them to the canes with soft twine. Pinch off the top of each plant when it reaches the back of the frame.

Planting Outdoor Tomatoes. For preference choose a position along the foot of a sunny fence or wall or building. Alternatively plant in the open but back the plants with screens of some sort— thatched hurdles, for instance.

Work the soil two spits deep. With each square yard of lower spit incorporate 6 lb. of clean straw chopped into 6-in. long pieces. Set the straw almost upright in the soil. This is a new and very successful method of tomato ground preparation ; the straw makes the best possible humus for tomatoes. In addition it aerates and draws the subsoil, thus encouraging strong root growth. Also add to each square yard of subsoil 2 oz. of bone meal.

A still later development in the use of straw s particularly applicable to greenhouse tomatoes, though it can also be applied to outdoor plants. The method is to make 1 ft. deep, 4 in. wide trenches near each row of plants. The trenches slope towards the plants. These trenches are filled with straw cut' to a length of 15 in. The straw is placed upright, so that when soil is returned to pack the straw fairly tightly. 3 in. of straw stands above the surface. When water is given it is poured on to the straws and easily finds its way to the roots.

With each square yard of surface soil incorporate ¾ pailful of stable manure and 2 oz. of bone-meal. As an alternative to stable manure prepared hop manure or shoddy can be used at the rate of 8 oz. to the square yard.

Plant out about mid-May.

For walls and fences the plants should be 12 in to 14 in apart in single row and there should be 2 in of clearance between plant stem and wall (or fence). Put the stakes in first ; then plant firmly—in all cases an inch deeper than the plants were in the pot. Leave a saucer-shaped depression around each plant.

Tie each plant up securely. It is important not to let tomato plants flop over and lie on the ground—even for an hour.

If any spare frame-lights are available rear these up against wall or fence to cover the young tomato plants for a few weeks.

Training Tomatoes. The training of tomatoes, which is vitally important, consists of limiting each plant to a single stem. This means nipping out all side-shoots as they are produced. The side-shoots will not be confused with flower trusses as they are quite distinct. They appear at all the points where a leaf-spray joins the main stem. They should be taken out in infancy, *not* allowed to grow some inches long and then cut off.

Another item in the training is to nip off the tips of the plants when they have set as many trusses of fruit as required. Five or six trusses can be allowed with greenhouse plants, usually only three or four with outdoor plants. In the case of frame plants the tops are taken off when the plants have grown the length of the frame.

If leaves form at the end of fruit trusses pinch these off at once. If leaves shade any fruits shorten them to half length or towards the end of the season remove them entirely.

Training Fancy Tomatoes. The ordinary yellow tomatoes are trained as above. The small " fancy " tomatoes resembling bunches of currants or other fruits require different treatment. They are naturally of dwarf and bushy habit. They should be given a single stout stake, 3 ft. high. The main shoot should be. stopped at the 3-ft. height and all side-shoots allowed to develop. These side-shoots should also be stopped below the 3-ft. height. If the bushy heads become crowded, a shoot here and there may be cut clean out.

Pollinating Tomatoes. Greenhouse and frame tomatoes need pollinating as each truss of flowers opens. Pollinating can be done in two ways : the stems can be tapped vigorously enough to set the plants trembling or a rabbit's tail tied to a cane can be passed from flower to flower. A sunny day when the blooms are wide open should be chosen for the operation.

Feeding Tomatoes. Give outdoor tomatoes each a pinch of sulphate of potash a month after planting; at the same time give a teaspoonful each of super-phosphate, repeating the dose in August. From the time the first fruits form until the end of August give weak liquid manure every ten days.

In the case of greenhouse and frame tomatos, these should not be fed till they are fruiting; then they çan be given liquid manure once a week. Occasionally give instead a mixture of 1 part nitrate of soda, 1 part nitrate of potash and 2 parts super-phosphate, dissolving 1 oz. of the mixture in 1 gallon of water and sharing this among three plants.

Tomato Troubles. *Brown leaf-mould.* This disease, which is also spoken of as *Mildew, Rust* and *Leaf-mould,* ultimately covers the leaves with a brown mould after starting as dirty grey spots. The spots spread and, if the leaf is badly attacked, it browns and dies.

Something can be done in the early stages of the attack by cutting off and burning those portions of leaves which show the grey spots, and later on the dried-up leaves must be served so. Also spray with liver of sulphur solution (which see), preferably on a hot bright day, with the house closed and kept so all day.

Blight. The fungus which causes "The" potato disease is a common trouble of outdoor tomatoes, but rarely troubles plants under glass. The first signs are black, water-soaked spots on the stem and leaves. Later on the fruit is attacked and the undersides of the leaves become covered with a white, downy mildew.

Spraying with Bordeaux mixture (which see) at the first sign of attack is the remedy. Where tomatoes have been sprayed with Bordeaux it is necessary to wash and wipe the fruits before use in order to get rid of the blue-green film left by the spray.

Black Rot. This trouble is also known as *Water Rot* and "*Buckeye*". The symptoms are discoloured patches—varying from grey to red-brown in colour—on the fruit. The disease usually appears on the first truss, but it may spread to higher trusses. A soft rot is the last stage. There is no spray for controlling this disease, and removal and burning of the attacked fruits before they produce their crop of spores is the only plan to follow.

Blossom-end Rot. A black spot appears at the end of the fruit remote from the stalk. Glossy green, sunken areas appear on other parts of the fruit. Trouble is caused by low temperatures, lack of ven-

tilation and overwatering. There is no remedy save picking off and burning the attacked fruits.

Canker. The leaves droop, the plants collapse and there is blackening of the stem joints. The remedy is to spray at ten-day intervals with liver of sulpher solution (which see).

Stripe Disease. The symptoms are brown longitudinal stripes on the stem and leaves and, later on, shrivelling of the leaves, and the development of brown sunken pits on the fruits. Such fruits should be picked and burnt, but it is not necessary to destroy the plants unless they are seriously crippled.

If the tomatoes are planted out apply sulphate of potash at the rate of 2 oz. per square yard of bed, and repeat the dose in a month's time. If the plants are in pots or boxes give a teaspoonful of sulphate of potash per plant, and repeat the dose in a month's time. This will induce an attacked plant to mature its crop.

Sleepy Disease. In this disease plants that have looked healthy over-night are found to have flagged in the morning and completely wilted by the end of the day. There is no cure. Affected plants must be immediately pulled up and burnt, to prevent the disease spreading.

Sleepy Disease usually attacks plants *via* the soil in which they are growing. For this reason it is recommended to use fresh soil for sowing and potting tom-atoes (never the same soil two years in succession), and, as a further precaution, to sterilise the soil.

TOOLS, GARDEN. Certain tools are essential to the work of a vegetable garden. They are a good spade and fork, a steel-toothed rake, draw and Dutch hoes and a trowel.

It is false economy to by cheap tools; work cannot be done efficiently with them and they soon break or wear out, and must be replaced. Good tools are an economy in the long run.

When choosing these tools select a spade with a spring-temper, all-steel blade of a suitable length and weight. The fork should have four prongs, square in shape. The ten-toothed type of rake is best. With the hoes a 6-in. wide blade is handi-est, as also is a short-handled trowel.

Other essentials are a garden line for marking drills, a dibber (made from an old spade handle or piece of broom-stick), and a good sharp knife.

There are other tools which, while not as essential as the above, facilitate garden work. They include such items as a triangular hoe for drawing seed drills; a potato fork, with four flat prongs; a hand fork; and a watering-can, preferably with detachable roses of varying degrees of coarseness, and not too short a spout.

A wheelbarrow is a very useful addition to the equipment, as also is a spraying syringe and a garden sieve.

A measuring rod is often required. A batten or the rake handle can be adapted by marking on it a succession of nicks to indicate 1 ft., 6 in. and 3 in. spaces.

All tools should be kept clean. They then do their work better and more easily, and last longer. Soil should be scraped off spades and other tools after use, the steel parts then being rubbed over with an oily rag, which should also be used occasionally on the wooden handles.

TREE ONION.. See Onions.

TRENCHING. See Digging and Manuring.

TURNIPS. Useful crop for both home use and, if good quality and early, for market. Reliable varieties for summer use are Early Milan and Early Snowball ; for winter storage Green Top Stone and Orange Jelly (yellow fleshed). Half an ounce of seed will sow 60 ft. of drill.

Turnips grow best in well-dug, fairly rich soil. The presence of lime in the soil, also, is an important factor in ensuring success. Lime is best forked into the ground in winter.

If poor soil has to be used, a little rotted manure may·be forked into the ground preliminary to cropping. Alternatively open a shallow trench where each drill is to be and bury in it the manure. Failing manure, rake 3 or 4 oz. of a general garden fertiliser into each square yard or set the garden line and rake in 1 or 2 oz. per yard run.

Sow an early variety in early April in drills ¾ in. deep and 15 in. apart. Make successional sowings at intervals of about three weeks until early July, pulling the roots for use when about the size of a tennis ball.

The late varieties for storage are sown from late July to mid-August. Lift in autumn.

Thin the early turnips 6 in. to 8 in. apart, the later sorts to 12 in., starting when the first pair of true leaves form. If the soil is not damp at the operation, give a good soaking immediately afterward, lest sunshine and drying winds get to the remaining plants and check them. After thinning also give a stimulant in the form of 3 oz. superphosphate per yard of row.

The seedlings are liable to attack by flea-beetles or turnip fly. Watering every evening after sunset lessens the risk of attack. Dusting with derris powder is also a useful deterrent.

Turnips may be stored in a shed in dry sand or soil—a layer of the sand or soil being placed between each layer of roots —or in an outdoor clamp.

Turnip Tops. The young tops of turnips make a very tasty green vegetable. One of the best varieties to grow purely·for the tops is Hardy Green Round, which yields big bunches of succulent foliage. Sow thinly and thin only sufficiently to prevent overcrowding.

As this turnip is very quick growing it can be sown in between rows of broad beans, or between early peas and broad beans.

WATERCRESS. Useful and healthful salad. Is best grown in a running stream of deep-spring water, but can be grown quite satisfactorily in any ordinary shallow stream, as well as on ordinary ground. The special value of spring water is its evenness of temperature, variation between summer and winter being at a minimum.

Given a shallow stream (2 in. depth of water), dibble in rooted pieces (the ordinary bunch of watercress, obtained from the greengrocer, invariably contains some pieces having roots) into the mud at the bottom of the stream, setting them 6 in. apart in May. Crop will provide pickings from August onwards. Fresh beds should be planted in September and thereafter replanting should be done twice a year, in May and September.

Growing Watercress Without Water. Failing a running stream, take out a trench 18 in. deep in a sunny portion of the garden free from overhanging trees. break up the sub-soil and fill in about 9 in. with well-decayed manure. This should be trodden down and a covering of about 3 in. of good soil placed on top. On this sow watercress seeds thinly and finish with a sprinkling of earth placed on top.

The remainder of the soil should be formed into two ridges on either side. If plenty of water is given, a crop will be obtained in less than two months.

WATERING. As a general rule, little watering need be done with vegetables grown in the open. The only crops needing water in long dry spells are lettuce, radishes, and other saladings, marrows, cucumbers and tomatoes. Of course, seedlings and newly-planted-out plants may need several waterings in dry weather to get them established. Further during dry spells it is a good plan, to follow each application of fertilizer with a watering.

WEEDS. Where troublesome perennial weeds like bindweed establish themselves in vegetable ground they can only be eradicated by combing out every scrap of root during the winter digging. They can, however, be kept under during the

summer by hoeing off their tops as they
appear.

The regular use of the hoe will keep down
the annual weeds, such as groundsel and
chickweed.

Weed killers cannot be used among
crops; they would kill the cultivated
plants as well as the weeds.

Certain weeds are classed as noxious,
and it is an offence to permit them to
grow to such an extent that they damage
neighbouring land. These weeds are
spear thistle, creeping or field thistle,
curled dock, broad-leaved dock and ragwort.

The owner of land infested with these
weeds may be required to cut down or
destroy the weeds within a stated time,
under penalty if he fail to do so.

WIREWORM. The most practicable
means of treating vegetable beds against
this pest (the larvæ of "click" beetles)
is to dress them with a mixture of 14 lb.
lime and 2 lb. of naphthalene.

Those who have small areas to deal
with may skewer potatoes, carrots, turnips,
etc., on wires and bury these here and
there, with the wire protruding from the
soil. Pull up these "traps" once a week
and kill the wireworms. Or get an old
tin, pierce it with holes and fill up with
potato peelings. Bury this so that it is
just covered with soil. Pull up weekly
and drop the contents into very hot water.

WOOD-ASHES. Name given to the
material left after a garden bonfire of old
soft vegetable rubbish, weeds, etc., has
burnt right through. It is a valuable
fertilizer, being rich in potash (see
Manures). The ashes should always be
stored under cover until wanted. If
soaked by rain, much of the manurial
value is washed out.

WOODLICE. These pests can do
serious damage to seedlings, and often
invade frames and mushroom beds.
They breed mainly in damp wood and
rubbish. Burn all such waste material,
and, when cleaning frames, pour plenty
of boiling water into crevices. Good
traps are flower pots half-filled with moss
and horse manure and placed on their
sides where woodlice are seen.

WORMWOOD. Perennial herb, with
very bitter foliage, which is used for
seasoning. The whole plant also has a
medicinal value. Propagated from seeds
sown in March or April, and by division
of the roots in spring. Also see Herbs.

FRUIT CULTURE

AGE OF FRUIT TREES. The oldest tree to buy (except in special circumstances) is one of 7 years of age. After that age fruit trees do not transplant well.

Taking everything—including the price factor—into consideration, the best ages at which to buy fruit trees are:

Apples, Pears, Plums, etc.—As standard trees; 4 or 5 years old; half-standards, at 4 years old; bushes, 3 or 4 years old.

Fan-trained trees of any fruits. At 4 or even 5 years old; espalier, at 4 or 5 years old; cordons, from 2 to 3 years old.

Gooseberries and Red and White Currants. As bushes, from 2 to 4 years old; as cordons, at 2 years old.

Black Currants. As bushes at 1 or 2 years old.

Loganberries and Blackberries. As young plants.

Strawberries. As layers rooted in pots during the current summer.

AMERICAN BLIGHT, or WOOLLY APHIS. Insect that attacks fruit trees, particularly apples, forming patches of white wool-like substance on the branches and trunks of the trees.

In spring and summer, when the insects have festooned the branches with wool, the most effective treatment is to paint the patches with methylated spirit.

In winter, spray the trees with a tar-distillate wash (see Spraying). Another effective step, also to be taken in winter, is to gas the insects in the soil, in which they spend the winter, coming down from the branches in autumn.

The gassing is done by applying carbon bisulphide to the soil round the tree. Make four holes, about 6 in. deep and about 2 ft. or so out from the tree stem and equidistant round it. Then pour into each hole about 1 oz. (not more) of the carbon-bisulphide fluid. Immediately stop the hole with a clod of turf, so that the gas cannot escape. The fluid is highly inflammable, so let there be no lighted cigarettes or matches about.

Another method of treating the soil is to apply naphthalene—about ¼ lb. to each square yard of ground occupied by the affected trees' roots, just pricking this into the top 2 in. or so of soil.

APPLES. The apple is a fruit for every garden, besides being the " standard " tree in commercial orchards. In normal times the quality of the fruit largely decides the price received for it—hence the importance of good culture.

Recommended Apple Varieties. There are some hundreds of apple varieties. Following is a brief selection of good growers and croppers. Some varieties of apples are Self Sterile, others are Self Fertile (for further details see these entries). With the self sterile varieties it is necessary to plant a "mate" to ensure good cropping; suitable companions are noted below.

Dessert Apples: Self Fertile: Laxton's Exquisite (season of use August); Ellison's Orange (Sept.–Oct.); Worcester Pearmain (Sept.–Oct.); Egremont Russett (Oct.–Dec.); Lord Lambourne (Oct.–Dec.).

Self Sterile: Beauty of Bath (August); suggested "mate" D'Arcy Spice (Mar.–May); Cox's Orange Pippin (Nov. Feb.) and James Grieve (Sept.–Oct.); Allington Pippin (Nov.–Feb.) and Chas. Ross (Nov.).

Cooking Apples: Self Fertile: Early Victoria (July–Aug.); Stirling Castle (Sept.–Oct.); Lord Derby (Oct.–Dec.); Bramley's Seedling (Nov.–April).

Self Sterile: Grenadier (Sept.–Oct.); suggested "mate" Lord Grosvenor (Aug.–Oct.); Royal Jubilee (Oct.–Jan.), and Gascoyne's Scarlet (Nov.–Jan.); Newton Wonder (Jan.–Mar.) and Crimson Bramley (Dec.–Apr.).

Preparing Soil for Apple Trees. Any good soil will grow apples, always providing it is deeply dug before planting, and properly manured *after the trees are in.* Do not dig in a lot of manure, but work in plenty of rotted or burnt garden refuse, leaf-soil and such like, especially if the ground is light and sandy.

Lime the ground if it is stagnant and sour.

In heavy ground, plenty of old mortar rubble and brick rubble can be mixed in with the bottom soil, to help drainage.

Planting Apple Trees. Late October and November are the best planting times.

The first stage in planting is to prune the roots. Cut off broken and bruised pieces and shorten the very long roots to 18 in. or less.

Next take out the planting holes, making them deep enough to allow each tree to be at the same depth as it was in the nursery. If there are 3 in. or 4 in. of good soil over the topmost roots, when planting is finished, that is ample.

Spread the roots evenly, work good soil among them and tread firmly.

When planting is finished, put two or three good forkfuls of farm or stable manure on the ground all round the tree, covering a 3-ft. circle with a thick, even layer.

The trees will need no special manuring, other than the surface mulch each spring, until they are starting to crop heavily. Then they will need artificial manures as well as the mulch (see Manuring). Also see Planting Fruit Trees.

Pruning Newly-Planted Apple Trees.
Trees planted in autumn need some pruning the following spring. If you have planted maiden, or one-year-old trees, these have to receive their first shaping—into bushes or standards or cordons, and so on. If the single stem maiden tree is to be grown in bush form, cut it off at about 2 ft. from the ground ; if it is to be trained as a half-standard, cut it off at about 4 ft. or 4 ft. 6 in. from the ground ; if as a full standard, do not shorten it at all unless it exceeds 6 ft. in height—then cutting it back to 5 ft. 6 in. or 6 ft.

Most people, however, plant three- or four-year-old trees—trees that have received their first shaping in the nursery and are well furnished with branches. With these trees the leading shoot at the end of each branch is shortened by about half, and all side-shoots along the branches are cut back to within two or three buds of their base. That method applies to all forms of trees.

Summer Pruning Apple Trees.
The grower of fruit who is really out to get the very best from his apples, should undertake a certain amount of pruning early in July in addition to that which he tackles in the winter. Only by so doing can he prevent his trees from making a lot of superfluous growth at the expense of the fruit to be harvested. Only thus also can he let the correct amount of light and air into the fruit to ensure that it ripens properly.

With bush, cordon and espalier apples shorten all long side-shoots to within four or five leaves of their base. Open out the middle of each bush and expose fruits to sunlight.

With standards and half-standards it is only necessary to remove unhealthy growths.

Winter Pruning Established Trees.
Nearly all apple trees bear fruit on short spurs that form along the branches. The purpose of the pruning, then, is to put more spurs on the tree by cutting away the unproductive laterals or side-shoots along the branches.

Pruning Cordon Trees. Cut back each side-shoot or lateral to within two or three buds of the base. Do not cut the leading shoot hard back, this being the shoot right at the end of the cordon stem. Just " tip " the leader. Later, when the tree is old and has filled its allotted space, the leading growth can be cut clean out and

no further extension allowed, but for the first half-dozen years or so just " tip " the leader.

Pruning Espalier Trees. Cut back all side-shoots to two or three buds and just " tip " the leader-shoot right at the end of each branch. Keep the spurs regulated —spaced about 5 in. apart along the branch and when they get very long, shorten them to three good " fruit " buds.

Pruning Bush-shaped Trees. Cut hard back any very strong-growing laterals and small branches that crowd the middle and shut out sunlight. Then, taking each branch in turn, prune the laterals. Any side-shoots less than 6 in. long can be left unpruned, but all others should be shortened to within about three or four buds of their base.

The leading shoot at the end of each branch must not be cut hard back. As a general rule, if the branch is making over-strong growth, just " tip " the leader ; if growth is moderately vigorous, shorten the leader by about one-third ; if growth is weak, shorten the leader by about half its length, or even a little more than that. If growth is *much* too strong, leave the leaders alone for a season.

Keep the branches spaced out evenly and the tree nicely balanced—and cut off any " suckers " from the stem or ground well below the head of branches.

Pruning Standard Trees. It is impossible to shorten every lateral in these big trees—nor is there any need to. When pruning a standard apple tree the idea is simply to keep the head of branches well spaced out and regulated—by pruning back awkwardly placed shoots and small branches each winter—so that there is no overcrowding in any part of the tree.

The leader-shoots need no shortening, unless very long, but they must be cut as required to keep the branches well spaced out, to stop the branch growing too much toward the middle of the tree, or too much to one side or other.

Very strong laterals can be shortened, otherwise they will grow into small branches and soon choke up the tree.

Pruning for Quick Fruiting.
The war-time need to produce all fruit possible led to the adoption by many growers of a different system of pruning. The object of this system is to encourage the production of maximum crops as quickly as possible. Under it, pruning is reduced to a minimum—broadly, only dead and diseased wood, and the weaker of crossing branches are removed. There is no general shortening of side-shoots ; the shorter laterals are left untouched ; the longer are so curled back that the tip of the shoot touches the point where the shoot emerges from the branch, and there it is tied. This treatment results in the formation of fruit buds along the curled shoot.

Propagating Apple Trees. The only way to raise vigorous heavy-bearing trees is by budding (see Budding). Even if cuttings could be induced to root, the resulting trees would be almost useless.

Pruning by Lorette System. See Pruning—Lorette System.

Thinning Apples. A mature apple tree normally sets more fruitlets than it can mature. These fruitlets must be thinned out, the time to commence the thinning being early June. Clusters of five or six apples should be reduced to two fruits—later to one for dessert apples, two for cookers. Before taking out good fruitlets, remove all under-sized, malformed and pest-scarred specimens. Be specially careful to take out all "holed" apples, as these carry pests.

Stimulants for Apples. For Free-blooming but Weak-growing Apples : The best mixture, where the ingredients are available, is one consisting of 2 parts nitrate of soda and 3 parts sulphate of potash. Apply ¼ lb. to each square yard of ground occupied, in spring.

For Strong-growing, but Unfruitful Apples : Mix 6 parts superphosphate and 2 parts sulphate of potash, and give ¼ lb. to each square yard of ground, in spring.

Harvesting Apples. See Harvesting Fruits.

Storing Apples. See Storing Fruits.

Diseases and Pests of Apples. Apple Scab. This disease causes round scabs and gaping cracks in the skin of the fruit. Fallen leaves must be collected and burned. Dead snags and diseased-looking growths must be cut off the trees and all prunings burned. Pay particular attention to old spurs. The trees affected must be sprayed with Bordeaux mixture (which see) in spring, just before the flowers unfold.

Brown Rot. Attacked fruits become quite rotted. The disease starts as a brown spot—a sort of soft rot—on the apple skin, but later forms circular patches with pustules of white or grey " mould " here and there. Eventually the entire apple rots away. The disease spreads from apple to apple at an alarming rate and whole clusters of fruit may be spoiled if any diseased apples are allowed to remain on the trees. Keep a close watch during August, and pick off and burn all the spotty brown-patched specimens, leaving only the clear-skinned, sound fruits to ripen.

Maggots in Apples. In some instances pretty well the whole of the apple crop every year is spoilt by the presence of grubs in the fruits. They are the larvæ of the Codlin Moth and the Apple Sawfly (see also Caterpillars and Codlin Moth). Once the trouble has become chronic only

a regular programme of anti-maggot work will ensure a clean crop.

The programme to follow is this : (a) Every apple that falls to the ground in summer and every holed and grub-eaten fruit to be seen on the tree, *must* be collected and burned *before* the grubs escape —or they may be chopped up and given to pigs or poultry.

(b) Dress the ground all round about the trees in early autumn with a good soil fumigant.

(c) Just as the last flowers are fading in spring spray the trees with a strong insecticide, well drenching all the flower clusters.

(d) Fasten a band of hay or straw or a piece of roughly-wound sacking round the tree trunk in May. Leave it in position until August, then carefully remove it and soak it in strong disinfectant. Scores of the pests will hide in the folds of this " summer band " and be destroyed.

(e) Spray the trees in late December or early January with a tar-oil winter wash.

Other Pests. Apples are also attacked by blight, caterpillars, capsid bugs and American blight. See under separate headings for remedies.

APRICOT. This choice stone fruit requires firm soil containing plenty of lime rubble. It should be grown either under glass or in a sheltered position. In the open, even with shelter, the blossom will need protection from frost ; hence the advisability of growing it against a wall, when netting can be hung over it during frost.

Plant apricots in November, February, or March. See Planting Fruit Trees.

Recommended Varieties. New Large Early; Peach; Royal. These are all regular croppers, and hardy.

Protecting Wall Apricots. As soon as the first buds begin to open on wall apricots, protection from cold winds and frost is urgent. Herring netting in double or treble thicknesses, hanging from the top of the wall and supported by stays to prevent it dragging on the tree, is an effective protection. The protective material must be raised during the day time except when severe frosts are imminent and there are cold winds—this to give the pollinating insects and bees an opportunity to reach the flowers.

Pollinating Apricots. Apricots growing in the greenhouse must be pollinated by hand. The best procedure is to touch each flower in turn with a rabbit's tail tied to a cane. Although insects and bees can be relied on in some measure to pollinate the out-door trees, it is safer to hand-pollinate these also with the rabbit's tail.

Thinning Apricots. Apricots need heavy thinning, sufficient fruits being removed to start with to leave the remainder standing at 4 in. apart. The surplus should be taken off when they are the size of cherries. A further thinning is necessary when the fruits are the size of a small plum, only one fruit then being left to each square foot of space.

Pruning Apricots. Apricots bear fruit on sturdy young shoots *and* on young spurs along the main branches.

Shorten the lateral or side-shoots that are at all crowded or badly-placed, each to within two or three buds of their base, and leave in short, sturdy, uncrowded young shoots wherever there is ample room.

As the tree ages, cut out a certain amount of two- and three-year-old wood every year ; that will keep the tree furnished with young, fruitful branches. But be especially careful never to prune *too* drastically. If big boughs and most of the young growths are removed, the fatal Die-back disease is certain to start in the tree and it will be ruined.

Whatever the tree, be sure to leave no dead wood behind or the Silver Leaf trouble will surely come and ruin the tree. Also cover over every large pruning cut, immediately, with a dressing of good lead paint.

Troubles of Apricots. The oyster-shell scale—resembling minute oyster-shells on the bark—sometimes troubles apricots. The remedy is to spray, either with lime-sulphur or a strong paraffin and soft soap wash.

There is also a somewhat mysterious malady which causes whole branches suddenly to wilt and die back. Lack of lime, too free use of the pruning knife, sour soil or unlimed borders are all contributory causes and attention to these points should keep down the trouble.

Apricots are also subject to the silver leaf disease that attacks plums and other trees. See Silver Leaf.

ARSENATE OF LEAD. Poisonous wash much used for the purpose of destroying caterpillars on fruit tree leaves and fruits.

It should be bought in ready-prepared paste form. Use 1 lb. of the paste to 20 gallons of water.

B **ARK-RINGING.** The cutting of a strip of bark from the trunk as a means of making hitherto unfruitful trees more productive. The operation is only performed on very strong-growing and unfruitful apple and pear trees. It is *not* recommended for plums and cherries.

The time to operate is flowering time and the following is the method :

Take out two half-rings of bark on the fruit tree stem—choosing a point just below the lowest branches. The half-rings are made *on opposite sides of the stem,* one about 4 in. above the other. They are made so that the half-ring of bark removed encircles just half the circumference of the stem, so that the top half-ring will begin and end immediately above the one below—with 4 in. between them.

Use for the ringing either a penknife, a small ½-in. chisel, or the tool sold specially for the purpose.

The strip of bark cut out for each half-ring should be about ½ in. or ¾ in. wide. First cut round the edges with a knife point and then just peel off the strip of bark right down to the hard white wood of the stem. Immediately cover over the areas with a generous coating of white lead paint—to keep out the germs of such troublesome and dangerous diseases as canker and silver leaf.

Every spring the rings should be cleaned out again, this being continued until the tree has settled down to good fruiting and restriction is no longer necessary.

BIG BUD. Common trouble with black currants, due to mites infesting the buds. See Currants.

BIRD DAMAGE. Bud-pecking by wood pigeons and other birds is one of the fruit trees' worst troubles in spring. Usually the birds go for the currant and gooseberry buds and the plums and greengages, in preference to apples and pears, but once they commence experimenting, no fruit tree bud seems to come amiss.

The best of all methods of checking the birds is to cover the boughs with small-mesh netting. This is easily done with well-trained trees and gooseberries and currants.

Another way is to weave strands of black thread along the branches—a method that answers very well if the network of thread is fairly complete around the bush. It is not the tiresome business it might at first seem if the reel is fastened to the bottom of a long cane and the thread payed out through an eyelet screw at the top end, in fishing-rod style. Or a cottoning "spinner" that works on much the same idea, can be bought quite cheaply.

Spraying Against Birds. If preferred fruit trees can be sprayed with a wash that will make the buds distasteful to the birds. A quassia wash, a lime-sulphur solution or a mixture of lime and water sprayed on the trees will have the desired effect.

BLACKBERRY. Cultivated blackberries are a profitable market crop and are also worth growing for home use.

They can be grown on strong wire or wooden trellis in the open or against a fence.

Plant in October or November.

The soil preparation, method of planting, general cultivation, pruning, etc., are the same as for loganberries. See Loganberry.

The standard varieties are the Parsley-leaved for quality berries, the Himalaya Giant for strong growth and good berries and The Pollards for moderate growth and good-flavoured fruit in quantity. Two newer varieties are Bedford Giant, very large and early—it ripens in July—and John Innes, which ripens in August and continues to fruit until frosts come. The latter variety was raised by crossing from the thornless Rubus *Inermis*.

BLEEDING. Exudation of sap which sometimes follows pruning of grape vines. See Grape Vines.

BLIGHT. Spraying is the only effective method of overcoming the " blight " that infests most fruit trees in early summer. The trees should be sprayed as soon as any of the insects are seen, with a non-poisonous wash such as paraffin wash or quassia solution, the spraying being repeated ten days later. Well wet the foliage, particularly the undersides of the leaves, getting good pressure behind the jet of spray.

A tar-distillate wash applied in winter will often secure immunity against blight during a whole season, as it destroys the eggs on the trees and so prevents the hatching of broods. See Spraying.

BLOSSOM WILT. The symptoms of this crippling disease are as follows :

About a fortnight after the fruit blossom opens the leaves surrounding the clusters begin to flag or wilt and a few days later turn brown. Then the blossoms wilt and die, looking for all the world as though suddenly killed by frost.

After killing the blossoms the fungus passes down the spurs. As a result, many spurs die and the fungus then travels into the wood of the branch, there forming small, cankerous-looking wounds at the base of the spurs ; which wounds may or may not entirely ring the branch at that point after a few months.

The cutting out and *burning* of every infected spur and canker is the most successful line of treatment.

When possible, the wilting and dead spurs should be cut off in summer, when they are much easier to spot; but the trees must be gone over again in the winter—two or three times.

Sprayings are also essential.

1. Spray with caustic soda wash (which see) *immediately before* the swelling of the buds.

2. Spray with Bordeaux mixture (which see) immediately before flowers open wide and again as soon as flowering is over.

BORDEAUX MIXTURE. For recipe see under this heading in Vegetable Section.

BOTTLING FRUIT. It is an excellent plan to bottle some of the summer fruits for use during winter, especially when the summer fruits come in at a glut rate. Apart from the value of the fruit in the home, some growers have found bottling a profitable sideline.

Elaborate equipment is unnecessary, especially for fruit to be bottled for home use.

Old jam-jars can be used for containers, and a saucepan, small bath, bread pan, or similar utensil will serve quite well as a steriliser. The bottles, should, of course, be thoroughly tested and cleaned before use and any defective containers should be rejected. If no sugar is available fruit will keep quite well bottled in water.

The fruits should be firmly ripe. To make certain, each individual fruit should be examined for perfection, those, or the affected parts of them, which are damaged, decayed, diseased or " squashy " being discarded.

It is always wise to cleanse large fruits by wiping them with a cloth ; small fruits by allowing water to run through them in a colander.

To free raspberries and similar soft fruits of maggots, soak them for one hour in salted water (1 teaspoonful salt to the pint) and then rinse in clear water.

Apples and pears are prevented from " browning " by placing the prepared fruit, immediately it is cut, into a bowl of water, to which has been added 1 teaspoonful of salt, or ½ saltspoonful of citric acid, to each pint.

If it is specially desired to retain the colour of fruits when bottled, then sulphur-dioxide treatment is required, and this needs some time and trouble.

Space will be saved in the bottles if fruits such as plums are stoned and halved and apples cored and sliced. Pack layer by layer, from the bottom, pressing the pieces together one against another with a long-handled spoon or stick.

Fill each bottle level to the top. To fill the neck, slice the fruits.

Small fruits, such as raspberries, are best shaken down into place by jarring the packed bottle on the palm of the hand and then filling up again.

When the jars have been packed full, they should be filled till they overflow with cold water. Then the caps or lids are put on.

" Snap " lids are slipped on and left

thus. Screw tops are screwed firmly down and then loosened half a turn to allow the steam to escape during sterilisation.

The packed and capped bottles are now sterilised. The vessel used for this operation must be deep enough to allow the bottles to be almost covered with water.

On the bottom of the steriliser, a " false bottom " of double-thickness wire or slatted wood or cloth must be placed, to prevent the jars cracking.

For the sterilising process, the flame under the steriliser is set low, so that the maximum temperature, whatever it may be for the particular fruit you are bottling, is *not* reached in less than 1½ hours. The bottles are then kept at that temperature for the required period, and sterilising is done.

The temperatures and " holding " period for various fruits are as follows :

	deg. F.	min.
Apples, Apricots, Black-berries, Damsons, Goose-berries, Loganberries, Plums (ripe, whole), Raspberries, Rhubarb, Strawberries ..	165	10
Plums (halved or unripe) ..	165	20
Currants	180	15
Cherries	190	10
Pears and Quinces	190	20
Tomatoes (covered with brine, made from ½ oz. salt to 1 qt. water. About ¼ oz. sugar may be added if desired)..	190	30

It is advisable to use a thermometer for this operation, since under-heating leads to the fruit not keeping, over-heating to the produce being spoiled.

Immediately after sterilising the bottles are removed from the hot water.

Bottles with " snap " lids need no more immediate attention. Screw-tops, however, must be taken out one at a time and have the lids or bands tightly screwed down.

Test for Sealing. Twelve hours after sterilising all jars should be tested for sealing. The spring clips or screw tops are removed and each jar is lifted up by the metal or glass cap, grasping it firmly with the finger nails. The sealing should be good enough to bear the weight of the filled jar. If a lid does come off re-sterilising is necessary.

The best place for storage is a dry, cool, well-ventilated, dark shelf, where the fruits will keep in perfect condition for years. If a jar should " go wrong " it is always due to improper heating.

A new method of bottling fruit with the aid of sulphur-dioxide tablets was introduced during the war. The tablets are dissolved in water and the solution is poured into the fruit-filled bottles.

Under this method the fruits are not cooked and they lose colour, which is,

F

however, restored during the cooking when the fruit is used.

The sulphur-dioxide tablets are available, with instructions for use, at most chemists.

BOYSENBERRY. An American introduction, with dark crimson fruits, which are intermediate between a loganberry and a dewberry. The fruits are highly flavoured and very juicy. Culture is as for loganberry.

BROWN ROT. See Apple Diseases and Pests.

BROWN SCALE.. A pest like a dark-brown hard blister, which infests currants, gooseberries, peaches and nectarines, sucking the sap and causing unfruitfulness. The method of control is to rub over the scales with Gishurst Compound, which can be bought ready for use.

BUDDING. " Top " fruit trees such as apples and pears are propagated by budding or grafting. The stocks are planted in winter and the buds of the variety of apple, pear, plum, cherry or damson, etc., are grafted on the stocks in summer—about July.

The Stocks to Use. Apple Stocks. For cordons and pot trees, use Jaune de Metz Paradise (Malling No. 9). For espaliers, cordons and small garden bushes, use Doucin Paradise (Malling No. 2). For large bushes in the permanent orchard and for use when weakly varieties are grown as cordons, select Broad Leaf English Paradise (Malling No. 1). For large orchard standards and half-standards and for large bushes of weakly growing varieties on poor land, use Crab or Free Stock.

Pear Stocks. For cordons, espaliers and small bush or pyramid trees, use Angers Quince. For standard trees and large pyramids, use Wild Pear or Free stock.

Plum Stocks. For small bushes and most wall trees, use Common Mussel stock. For larger bushes and standards use Myrobalan, Brompton or Pershore.

Cherry Stocks. For bushes and trained trees, Mahaleb. For standard trees, Gean or Wild Sweet Cherry stock.

Peaches and Nectarines. These are budded on to plum stocks, the best usually proving to be Brompton, St. Julien or Mussel.

Apricots. These are usually budded on the Mussel or common plum.

Preparing the Stock. It is usually best to insert the bud in the stock at a point a few inches above ground level. Cut off all side-growths from the stock where the bud is to go and at the chosen spot, where the bark is clean and fresh, make a T-shaped incision in the bark.

Make the downward-cut of the T about 1 in. long, with a shorter cross-piece at the top—just cutting through the bark to reach the white wood below *and no deeper*. With the thin handle of the budding knife slightly raise the bark on either side of the down-cut, so, later, the bud can be slipped down between.

Preparing the Buds. Each bud must be a well-formed wood bud. It must be healthily plump, and must come from a sturdy young shoot of the current summer's growth. Cut it off the shoot with a shield-shaped piece of bark attached. Slip out the tiny piece of wood attached to the back of the bud—without pulling out the " body " or the bud with it ; cut off the leaf attached, leaving just a short stalk. (For closer details of preparing the bud, see Budding, in Flower Section.)

When prepared, slip bud down between the raised bark of the T-shaped cut on the stock.

If there is any overlapping of the " shield " of wood attached to the bud, trim it off neatly so the bud fits snugly in place and is held close against the white wood. Press the raised edges of the stock bark close around the bud, and then bind round with raffia or soft string.

After-care. Keep the budded stock well watered, and if the weather is very hot and dry, shade the buds with pieces of evergreen stuck in the ground on the sunny side.

In about three or four weeks' time loosen the raffia ties or the swelling buds will be strangled. If a bud has not by then made a good union with the stock, the chances are that it never will.

The bud will not break into growth until the following spring. Just before then the stock should be cut off to within 6 in. or so of the " bud-spot," and all growth made from the stock below that point kept pulled off.

BULLACE. A fruit similar to the damson except that it is round instead of oval and white in colour. In all respects the culture of the bullace is the same as for the damson. See Damson.

CANKER. Dangerous disease that attacks most fruit trees, usually as a result of the wounds left when pruning and lopping, or as a result of attack of American blight. To prevent an outbreak, all cuts should be pared smooth with a sharp knife, and the surface of the wound painted over with lead paint or smeared with clay, to exclude air and assist quick healing.

When trees are already cankered, affected branches should be cut away.

When the trunk is affected, cut out the diseased part and paint with Stockholm tar.

CAPSID BUG. An insect that sucks the sap from fruit trees and causes wart-like swellings on the fruit. The method of control is to spray immediately blossoms open, and again directly flowering is over, with nicotine wash.

CATERPILLARS. When the young leaves are unfolding watch must be kept for the hordes of caterpillars that may then be hatching out from eggs laid in autumn and winter. The pests eat the leaves of young shoots ; they ruin bloom trusses and bite and bore into the small fruits which do " set " and swell.

The best way to tackle the invaders is to spray all the trees and bushes with a good insecticide directly the flowers have all fallen. Give the branches, and especially the fruit clusters, a good drenching. If the trees are apt to be " grubby " in spring, or if they had a bad time the previous season, repeat the spraying a fortnight later.

Many gardeners use arsenate of lead wash (which see) and this is very effective for the purpose. It is very poisonous stuff, however, and must be handled and applied very carefully. Apply it within a week of the blossoms fading away. Be sure that the wash is not allowed to drift on to maturing vegetables or soon-to-be-eaten salad plants.

Apple Sawfly Caterpillars. The caterpillars of the apple sawfly need special treatment. The caterpillars burrow within and partly consume the fruits, which are ruined. The affected trees should be sprayed within three to eight days after the last petals fall from the blooms, with nicotine wash (which see).

Dusting the trees with insecticide powder is preferred by some to spraying with liquid. It answers well if the trees are a handy size and a proper appliance—a small hand-dusting machine—is used to apply the nicotine or other dust employed.

CAUSTIC SODA WASH. Much used for cleansing fruit trees of fungoid and other diseases. The recipe is : 12 oz. caustic soda ; 9 oz. pearlash ; 8 gallons water ; 8 oz. soft soap. Dissolve and add the soap last. Rubber gloves should be worn when working with this fluid, which has a burning effect.

CHERRIES. Cherries are an ever-popular market crop, and when well-grown they always sell readily at good prices.

Recommended Varieties. Not only are cherries self-sterile, but many are sterile with the pollen of each other.

Hence it is important to plant "compatible" varieties. Suggested pairs of sweet cherries are: Bigarreau Schrecken and Biggareau Napoleon; Early Rivers and Elton; Black Tartarian and Emperor Francis

Morello and Kentish Red are the cooking varieties to plant.

Best Type of Trees. As a rule, the sweet cherry trees are best grown as standards in grass land. Unlike most other fruits, the cherry is at its best with turf over the roots, largely because it does badly under root disturbance. Deep cultivation over cherry tree roots may be ruinous.

In the garden, grow sweet cherry trees in the "fan" form of training, against a wall or fence. The sour cherry can be grown as a small, round-headed bush, or better still, as a fan tree against a north-facing wall.

Distance Apart for Trees. Allow a space of at least 12 ft. between bush trees, give each fan-shaped tree about 12 ft. or as much as 15 ft. of wall space, and if you are planting several standard trees in a grass orchard, put them quite 18 ft. apart.

Preparing Soil for Planting Cherries. To plant a cherry tree in waterlogged, acid ground is courting failure.

It is best to prepare separate stations for each tree as follows :

Dig out a wide, deep hole and remove the old soil. Put a 6 in. layer of brickbats and rubble in the bottom of the hole, fill up with turfy loam, leaf soil and best top-spit soil from another part of the garden, and mix with this plenty of crushed mortar rubble. Do not put in any manure at all.

Avoid planting too deeply ; spread out the roots to fullest extent ; work fine topsoil round and among them ; ram it firm and finish off with a mulch of manure laid on the ground all round each tree. Also see Planting Fruit Trees.

Pruning Newly-Planted Cherries. Both sweet and sour cherries should be pruned to some extent in the spring following the autumn of planting. The shaping of young trees and the cutting back of older trees is as recommended for apples.

Summer Pruning Cherries. The summer pruning of sweet cherries is the same as for apples (see Apples). With sour cherries the pruning consists of cutting away only crowded and awkwardly placed shoots, shortening each to three or four leaves.

Winter Pruning Cherries. When pruning a sweet cherry the first duty is to remove every dead and dying shoot. Then go over each branch in turn and shorten the very long, crowded side-shoots to within two or three buds of their base.

Big trees make very few long young shoots, so probably but little cutting back is required. But search every branch and be sure to miss none that have diseased or dead growths on them.

A wall-trained tree may need rather more cutting than a standard tree, but the principle is the same. Cut back to two or three buds the laterals or side-shoots that are more than, say, 5 in. long, and then shorten the leader shoot at the end of each branch by about one-third its length, or by rather less.

The sour-cherry tree merely needs thinning out every year, the object being to keep the tree filled with sturdy young shoots. The shoots of the previous year's growth are left practically untouched, the two and three-year-old pieces are pruned back, every year, to make room for the younger growths.

All well-placed young shoots are left intact, or just tipped if very long, and the crowded and unhealthy shoots are cut off close to their base.

Harvesting Cherries. See Harvesting Fruits.

Cherry Troubles. Cherries are remarkably free from troubles, their chief enemies being Blight and Birds. For remedies against these, see under separate headings.

CHESTNUTS. See Nuts.

COB NUTS. See Nuts.

CODLING MOTH. Destructive pest of fruit-trees, particularly apples. Caterpillars cause what are popularly known as " maggoty " apples, the caterpillars boring through to the core of the fruit and out through the side again. Caterpillars may be killed by spraying the trees with lead arsenate (see Arsenate of Lead) soon after the blossom falls. They may also be trapped as they leave the fruits by bands of sacking or hay tied round the tree trunks in June, and removed and burnt in autumn.

CORAL SPOT. Disease that attacks red currants and gooseberries. Recognised by pink wart-like eruptions on the bark of affected branches. All attacked parts must be cut away and burnt.

CRAB APPLE, or WILD APPLE (*Pyrus malus*). Grown mainly as an ornamental tree for its bright blossoms in May, and also as a pollinator. Often attains 15–20 ft. high. The fruits, however, are useful for jelly-making. Needs the same general treatment as the apple, except that the pruning should be very light.

CRANBERRY. Fruiting plant of creeping habit, with wiry stems and small leaves. Has small red flowers which are followed by deep red berries ripening in August. The berries are commonly used for jelly and wine making, as well as stewed and in pies. Cranberries grow best in peaty, wet soil. Propagated by layers in autumn in the same way as loganberries are layered. See Loganberry.

CURRANTS. Currants, especially the black varieties, are one of the most profitable of all market crops. Not only are the blacks in demand for culinary purposes but for jam-making and in the manufacture of dyes. The bottling and canning factories also require black currants—and red currants, too—in quantities.

Recommended Varieties. Black Currants: Boskoop Giant, Seabrook's Black; Baldwin, Daniel's September. Red Currants: Earliest of Fourlands, Laxton's No. 1, Laxton's Perfection, River's Late Red. White Currants: White Dutch, White Versailles, White Transparent.

Planting Currants. The planting season for currants extends from October to March. No special soil preparation is necessary; they will do well in any average well dug ground. (See Planting Fruit Trees).

Pruning Newly-Planted Currants. Red and White Currants. The first pruning is due just before buds begin to break into leaf about late March. The aim is to build up a stout frame of branches, making a goblet or basin-shaped bush, with, eventually, six or eight branches ranging from the top of the short stem or "leg." The centre should be wide open and free from growth.

To this end, at the first pruning, the leading or topmost shoot on each branch is shortened by about half its length and all side or lateral shoots not required to form branches, and all pieces growing inwards, are pruned hard back to two or three bottom buds.

Black Currants. The first pruning of the new bushes is due just before the buds break into leaf in March. Then, if one- or two-year-old bushes have been planted, cut back every growth to within two or three buds of ground level. If the bushes are older cut hard back three out of every five of the branches—taking these back to near the ground.

Summer Pruning Currants. The summer pruning of red and white currants consists of shortening all side-shoots along the main branches to within three or four leaves of their base. With black currants, all the old branches should be cut out as soon as the berries are picked, young shoots being left entirely alone.

Winter Pruning Currants. Red and White Currants. These bear their bunches of berries on fruiting spurs which form along the main branches. The chief aim of the pruner is to shorten all the side-shoots that grow out along the branches so that the bottom buds are strengthened and develop into these clusters of short fruiting spurs.

In the established bush, furnished with good branches, the pruner has simply to cut back each young lateral or side-shoot to two or three bottom buds. The "leader" shoot should be merely "tipped" if of normal length, or cut back by half if unduly long. Always be most careful to prune the leader back to a sound, plump bud.

In a very old bush there may be an odd branch that is weakly and unfruitful. Cut such a branch clean out and train up a strong young shoot from its base to take its place in the framework of the bush. Do this every year or so and you can gradually replace all old branches with young and give the bushes an entirely new lease of fruitful life.

Red and White Currants trained as Cordons. Every year, from the very first season onwards, cut back every side-shoot to two buds and just "tip" the leading shoot right at the end of the cordon.

Black Currants. Black currants fruit on the lengths of young shoots. The aim of the pruner must therefore be to encourage plenty of strong young growths from low down in the bush, by sacrificing older branches.

All healthy young shoots on a black currant bush, left full length, will bear fruit in the next summer—but the type of growth to encourage is that which grows up from the ground or from the lower half of the bigger branches.

The pruning of established bushes therefore consists of cutting out branches that have fruited the previous summer and leaving long, young growths in replacement. In very old bushes, which have very little young growth anywhere, prune out three or four big branches and leave the others intact. From the cut-back stubs will come strong growth the next summer, and, after fruiting, you can cut out more old branches and leave in young growth.

Propagating Currants. Currants of all kinds can be propagated by means of cuttings. See Cuttings.

Fertilizers for Currants. For Red and White Currants. Where the ingredients are available mix 3 parts sulphate of potash and 4 parts superphosphate, and give ¼ lb. to each square yard of ground in early spring.

For Black Currants. In early spring, apply 2 oz. of sulphate of ammonia per square yard and, if bushes are old, give ¼ lb. superphosphate per square yard in addition.

Harvesting Currants. See Harvesting Fruits.

Diseases and Pests. Reversion Disease of Black Currants. Reversion disease is a very common complaint. It is most easily detected in June and July. " Reverted " bushes bear leaves which, compared with leaves on healthy, fruitful bushes, are generally very narrow and elongated, sometimes pale in colour and *scantily veined.* Look at a sound leaf from a bush that is known to be healthy and fruitful and it will be seen that there are more than five sub-veins running from each side of the mid-rib to the central lobe of the leaf. Count only those sub-veins that run to the central lobe, taking no notice at all of those running to the lateral lobes at the lower corners of the leaf. There must be five, or more than five, of these sub-veins running from each side of the mid-rib to the central lobe if the leaf is healthy. When there are less than five, then the branch is to be regarded with grave suspicion. Where several leaves show short of veins are found the branch is " reverted " and the bush doomed.

There is no cure for Reversion Disease. The only course is to pull up and burn affected bushes.

Big Bud. A common trouble with black currants ; due to a mite which infests the buds Causes buds to swell to abnormal size and fail to break into leaf in spring. Every swollen bud should be picked off and burned in spring. Spraying with lime-sulphur wash (which see), or dusting the bushes with a mixture of equal quantities of slaked lime and sulphur, at intervals of a fortnight during April and May, will destroy mites before they can enter unaffected buds.

Currant Aphis. When the leaves of red, white or black currants are " pimpled " or blistered it is a sign that the currant aphis is at work.

The berries may be entirely spoiled—certainly they will be shrivelled and stunted if the pest is at all widespread—and the premature falling of the leaves, which follows as a matter of course, so seriously weakens the bushes that the quantity and quality of the following season's crop will also be undermined.

The blisters or pimples on the leaves are usually red or reddish brown.

Get the sprayer to work and make a really clean sweep of the aphides on the branches. Use any good insecticide.

CUTTINGS. Red, white and black currants, and gooseberries, can be grown successfully from cuttings. The time to root the cuttings is November.

In all these sturdy, healthy shoots of the current season's making should be taken.

Take growths 12 in. long whenever possible but never go less than 8 in. Either cut them off fairly close to the main branch or pull them away with a " heel " of old wood attached.

Snip off the top inch or so of unripened wood, and then, with gooseberries and red and white currants, rub off all buds but the top five or six. Do not damage these top buds, for they are to produce the branches of the new bushes. *Do not rub any buds from the black currant cuttings.*

Rooting the cuttings in a sunny spot in the garden, where the soil is light and sandy, is best. If the soil is heavy, dig it deeply and work in a quantity of road grit and sharp sand, also leaf-mould, to lighten it. Tread the ground firm before putting in the cuttings.

Take out a narrow trench and lay the cuttings upright in this, about 12 in. apart, the depth to plant varying with the class of cutting. Thus, black currant cuttings should go in to leave only the top three or four buds above ground level, but the red and white currants and the gooseberries should be put in so that the top buds and about 4 in. of bud-less stem are above ground. This means the cuttings will be some 6 in. out of the ground.

Put a little sharp sand or grit around the base of the cuttings before finally filling in the trench and trampling very firm.

Leave the cuttings undisturbed until the autumn of the following year, by which time they will be sturdy little plants ready for setting out in the fruit plot.

DAMSON. Good varieties of damsons are Merryweather, which produce the largest fruits, Bradley's King and Frogmore Prolific. The trees will grow in any average soil and their culture generally is the same as for plums. The winter pruning of damsons consists of thinning out crowded growths to keep the trees open and shapely ; anything in the way of drastic branch cutting should be avoided. Though the pruning is light, it *must* be done every year to keep the trees fruitful.

D.N.C. Common name for a wash containing di-nitro-ortho-cresol ; a very effective and complete spraying fluid. See Spraying.

DISEASES. For methods of treating diseases which attack fruit trees generally, see under separate headings. For diseases which affect only specific fruits (such as Gooseberry Mildew) see under that fruit.

ESPALIER. Term applied to that type of fruit trees so trained that two, three or four branches extend horizontally and parallel with one another on each side of, and at right angles to a vertical stem. The branches are called arms.

Occupying little space and casting practically no shade, this type of tree is specially suitable for planting along the sides of vegetable plots, being trained on wires stretched from one post to another at a suitable height.

Apples, pears and plums are amongst the fruits most commonly grown in espalier form.

FAN-SHAPED TREES. Term denoting fruit trees so trained that their branches radiate fanwise from a short stem.

FERTILIZATION. Term used to denote the union of the male and female organs in flowers, as when the pollen from the male organ has definitely made contact with the female organ and reached the immature seeds. Pollination (which see) is a necessary preliminary to fertilization.

FERTILIZERS. The fertilizers best suited for specific fruit trees are given under separate headings. For general feeding, see Manures.

FIGS. In some areas figs can be grown as bushes and even as standards in the open, and will crop regularly and heavily. In the average garden, however, they are best as fan-trained trees, growing against a sheltering wall facing south or south-west.

Three-year-old trees are the usual choice for planting, and these would fruit within a season or two of planting.

Planting Outdoor Figs. For outdoor planting, Brown Turkey is the outstanding variety. If others are to go in as well, choose White Marseilles for early fruit, and Negro Largo for later fruits.

It is best to prepare a special border in which to plant a fig tree. Drainage must be perfect, and tile drains covered with 6 in. or so of brick and mortar rubble must be laid if the ground is in the slightest degree heavy or cold. The border should then be filled with fresh, good loam and plenty of mortar rubble, to a depth of 2½ ft. Rich soil and rank manure are unnecessary, and indeed, check fruiting. Ordinary good garden soil, with mortar rubble added to it, is quite sufficient, with a manure mulch and copious waterings each summer.

The best times to plant are November and February.

Greenhouse Figs. Figs may be grown in a cold or warm greenhouse if desired, but unless for commercial work they are hardly worth the space they occupy. When greenhouse space *can* be spared the trees should be planted in a border against a wall, prepared as advised above. White Marseilles does well under glass.

Summer Pruning Figs. Summer pruning encourages better fruiting. It consists of pinching off the tips of all specially vigorous shoots in July-August. The little second-crop figs which appear in late summer can never ripen ; they should be picked off as they form.

Winter Pruning Figs. The aim when pruning figs is to save the sturdy young shoots and take out older growths and rank, coarse shoots to make room for these fruitful shoots.

Thus fig tree pruning, in brief, consists of cutting the very old worn out branches, back to strong young branches which are spaced out evenly over the wall. The spaces between the branches are filled by training in sturdy young shoots of the previous summer's making, wherever there is room.

Strong, suckerous growths from the base of the tree should be cut away ; they will only choke and smother the fruiting sprays.

The longest of the young shoots can be shortened a little, but the dwarfer, short-jointed shoots can be left practically intact.

Having a big, overgrown tree, do not cut back every branch there is. Just thin out the branches, cutting back the very big, coarse-growing ones and leaving in those of moderate growth. Space out the retained growth so that every shoot and branch in the tree is exposed to full light and all the sun there is.

Propagating Figs. Figs can be propagated by 6-in. long heeled cuttings planted in the greenhouse in autumn ; by layering in September ; and by potting up suckers formed in the autumn.

FILBERTS. See Nuts.

FROST, PROTECTING FRUITS AGAINST. Severe frost is often experienced in April, a time when many fruits are blossoming. It is essential, as far as is possible, to protect the blossom, otherwise it may be ruined.

Little can be done to protect big garden trees, though commercial growers find it worth their while—in normal times—to invest in a few orchard heaters. These are just tin cans holding about two gallons or so of crude oil, and they are placed on the ground here and there among the fruit trees.

Lighted last thing at night, when severe frost threatens, the heaters burn for eight hours or so and the heat generated is sufficient to prevent the blossoms freezing. Methods of " blacking-out " the heaters were devised for use during war-time, but generally they have been temporarily discontinued.

With pyramid and bush trees, or neat growing cordon and espalier trees, trouble can be prevented by fixing a canopy-like covering of small mesh netting (a double thickness is needed) or a sacking or canvas screen over the *top* of the trees—the frost strikes downwards, so the tops of the trees need most cover.

Fruit trees that are growing against a wall are easily protected with a light screening over the top and front of the branches.

Fix the sheltering " blinds " so that they can be drawn up on fine mornings and lowered at night. It does not do to enclose the flowering trees every day and all day ; they need sun and air and, of course, pollinating bees and insects must have free access to the flowers whenever conditions are at all favourable.

If Fruit Blossom is Frosted. If any fruit blossom should be frosted it can be saved if the frozen clusters are syringed hard and generously with cold water before the sun reaches them.

GISHURST COMPOUND. Insecticide used for the cleansing of grape vines and other fruits. It is bought ready prepared for use from the horticultural shop ; it cannot be made up at home

GOOSEBERRIES. One of the chief stand-bys of the commercial fruit grower and a crop that should be obtainable from every garden for home use.

The usual plan is to grow gooseberries as small bushes, each bush having a short, clean stem of 4 in. or so before branching. These are easy to manage and remarkably fruitful. In the garden, however, to save space, to simplify cultivation, and when the finest berries are wanted for exhibition, it is a good scheme to grow them cordon fashion. The cordons make an excellent edging for pathways and beds, or they can be planted against any fence or wall, or light trellis built up of three strands of galvanised wires.

As a rule it is best to buy three-year-old bushes for planting ; they will begin to crop almost at once.

Recommended Varieties. Lancashire Lad, Whinham's Industry, Warrington (red berries); Golden Drop, Cousin's Seedling, Leveller (yellow berries); Keepsake, Lancer, Profit (green berries); Langley Gage, Whitesmith, Careless (white berries).

Preparing Soil for Gooseberries. Gooseberries can be grown in any good garden soil, providing the ground is deeply dug and manured generously. What must be avoided, however, is deep shade and a wet, badly-drained spot, for in conditions of this nature the dreaded American Gooseberry Mildew will play havoc with the bushes.

Where well-rotted manure can be worked in so much the better. Where manure cannot be spared, fork in wood-ashes and ½ lb. of basic slag to each square yard of the ground to be planted. Also see Planting Fruit Trees.

Pruning Newly-Planted Gooseberries. Prune new gooseberries just before they start to break into leaf in early spring, by shortening the young shoot at the end of each branch by little more than half and cutting back all lateral or side-shoots, not wanted to form branches, to within two or three buds of their base. Wrongly-placed growths should be cut clean away.

With cordons, the first pruning consists of cutting back every side-shoot to two or three buds and just " topping " the leading, extension shoot at the end of the cordon.

Summer Pruning Gooseberries. Shorten to three or four leaves the side-shoots crowding the middle of the bush and any basal growths drooping near the ground. Just tip well-placed young shoots. On cordons shorten all side-shoots.

Winter Pruning Established Gooseberries. The bushes bear their berries on short spurs along the branches *and* on well-placed young shoots. Start by cutting back to two or three buds every young side-shoot that is at all crowded or badly placed or unhealthy looking. Be specially careful to open out the middle of the bush by this means. Sturdy young shoots for which there is adequate room without overcrowding should be left practically full length ; just snip off the top few inches from each.

On a weakly bush, cut each leader shoot to about half its length ; on a strong bush, shorten each leader only by about a third its length ; and on a very rank growing bush only just " tip " the leaders.

Some sorts of gooseberries make very drooping growths. Keep these up off the ground, and the bush shapely, by cutting each arched growth to a bud on the top side.

Other sorts make upright, dense-headed bushes. With these remove all the in-growing shoots and cut the leaders to a bud pointing outwards, away from the centre of the bush.

When rank, suckerous growths sprout right from the base of the bush, cut them clean out so that they cannot grow again.

Only occasionally will one of these basal growths come in useful—as when a worn-out or dead branch has to be cut out—and then a strong growth from the base can be trained up in replacement.

With cordon gooseberries cut back every side-shoot to within two or three buds of the base and shorten each leader by about one-third its length.

The best time to prune gooseberries is January, but where birds are troublesome pruning may be delayed until early spring.

Thinning Gooseberries. It is advisable to thin out the fruits on gooseberry bushes as soon as the berries reach a size at which they will be of use in the kitchen or for sale. Take out the largest green berries from the crowded branches, leaving the remainder of the crop evenly spaced out along the branches to ripen.

Fertilizers for Gooseberries. In early spring mix 3 parts sulphate of potash and 4 parts superphosphate, and apply at the rate of ¼ lb. per square yard of ground.

Harvesting Gooseberries. See Harvesting Fruits.

Gooseberry Diseases and Pests· *European and American Mildew.* The chief disease of gooseberries is mildew, which appears in two forms, one harmful, the other comparatively harmless. The harmful kind is American Gooseberry Mildew ; the other, European Gooseberry Mildew. The American form, which is highly contagious, starts as a thick white powdery mildew on the tips of the young shoots in spring and appears only on the *undersides* of the leaves. Late in summer it changes to a brown felt-like coating on the shoots, the tips of which are distorted and often killed. It seriously affects the berries, too, first as a white powdery coating and later as brown felt-like blotches. The European Mildew appears in late summer and attacks only the upper sides of the leaves, on which it shows in the form of a very thin white powder. The berries are not affected. American Mildew must be met at the onset by spraying with lime-sulphur wash, by liming the soil beneath the bushes in spring, and by cutting back all affected shoots to one-third their length, the prunings being immediately burnt. Lime-sulphur spraying in May is recommended for bushes which have been attacked by European Mildew.

Brown Scale and Caterpillars are the chief insect pests of gooseberries. For methods of control, see separate entries. The grubs of the sawfly will completely defoliate gooseberry bushes if given the chance. Prompt measures are therefore imperative (see Caterpillars).

GRAFTING. Grafting is the most usual method of propagating apples, pears and other fruits which cannot profitably be propagated by cuttings or layers. Unfruitful or aged trees can also be made productive, or rejuvenated, by grafting.

Young fruit trees are usually formed by grafting a " scion," or shoot, from any desired fruit on to a " stock," or rooted stem, of another variety of a similar kind of fruit.

Stocks for Grafting. There is not only a " best " stock for each fruit but also a " best " stock for the purpose in view. Thus there is one kind of stock particularly suitable for a standard apple, another particularly suitable for an espalier apple and so on. For the stocks to choose, see Budding.

For the production of new fruit trees a supply of stocks should be obtained and planted in autumn and these will be ready for grafting the following spring twelve-month.

Securing the Scions. The scions required for the production of new fruit trees should be cut from trees of the required varieties in the January preceding the spring when grafting is to be done. Select for the purpose healthy shoots off the previous year's growth. Cut them off to full length (they must contain a minimum of four good sound buds) and then bed them into a sheltered border to await the spring.

Preparing the Stock and Scions. The exact time to commence grafting is when the buds on the stock are just beginning to move—probably about the latter end of March.

When the time comes the stock must be prepared to receive the scions. This is done by trimming off all side-shoots and then cutting it down to within 4 in. to 5 in. of the ground.

Now, on one side of the cut-down stock make a sloping, upward cut with the knife, about 1½ in. long. Across the middle of this sloping cut make a thin " tongue," by a clean downward slit with the knife. That leaves the stock all ready.

Next prepare the scion.

Taking one of the shoots bedded in the previous January (see above), cut it to a length that contains just four good, sound buds—making the top cut close above a bud and the lower cut about ¾ in. below the bottom bud of the four.

Then, at the end of this scion and on the side opposite the bottom bud, make a long sloping cut and a corresponding slit or " tongue " to that made on the stock.

To get a good fit lay the scion against the cut made on the stock and mark with a nick of the knife just how long the sloping cut has to be and just where and how deep the " tongue " has to be ; for the two must be " faced up " accurately ; else they will not fit nicely together.

Fitting the Scion to the Stock. With the stock ready and the scion ready, fit the two together, pushing the " tongue " on the scion into (and behind) the " tongue " on the stock so they hold tightly together.

The rind or inner bark of the stock and the scion *must come in close contact at least along one side;* wherever possible—and with careful work this is usually possible—the two sides should exactly coincide. But if the scion is narrower than the stock, then make the inner barks fit exactly at least all along one side.

Then bind the two together with raffia and finally cover over the union with grafting wax or clay.

Work quickly ; do not give the cut surfaces time to dry out before fitting the two together. A damp sack is useful ; keep the scions in it and if necessary lay it over the cut-back stock while getting the scion ready. Dried surfaces will never properly knit together.

Top-grafting Old and Unfruitful Trees. Top-grafting is only advised for apples and pears, not for plums and cherries. The first operation is to cut back the branches of the tree to be operated upon to within 2 ft. of their base. Do this in January.

The next operation is to graft scions from selected fruits on to each of the cut back branches. Do this in early April.

One graft is put on to a stump only 2 or 3 in. across, two grafts on to each branch stump from 4 to 6 in. across. Boughs much bigger than 6 in. across do not take the grafts very well, so on a tree with very thick, old limbs, cut back less drastically, so as to leave younger, smaller branches to take the grafts.

Commence by freshening up the cut surfaces of the stumps by sawing off a further couple of inches or so from each. Pare the surface smooth and level and then make the slit in the bark that is to take the scion. Make this with the knife point.

Start at the surface of the branch cut and run the knife down to make a slit about 2 in. long, just penetrating the bark. Then the bark on either side of the slit should be slightly raised so that a scion can be slipped down between, to lie flat against the white, sappy wood.

It is best to put the scion on the outside edge of the branch stump. If there are to be two grafts on one branch, put them on opposite sides. Cuts should be made accordingly.

To prepare a scion for top-grafting, cut off the top of the shoot (shoots for the purpose as already mentioned, are secured in January and bedded into the ground to await grafting time) immediately above a sound bud and make the bottom cut about 1 in. below the fourth bud down the shoot.

F*

Then, with a downward stroke of the knife, make a long, sloping or oblique cut, about 1½ in. long, across the bottom end of the graft and on the side opposite the lowest bud.

To fit the prepared scion on to the branch, push it into position between the raised edges of the 2 in. slit in the bark of the branch, pushing it down until the top of the 1½-in. cut is level with the surface of the branch. The sloping cut on the bottom of the graft should fit close against the white wood of the branch.

Bind the scion in place with raffia and then cover over all with a moulding of grafting wax or clay. Where two grafts are put on one branch stump, fit both in place before doing any tying with the raffia.

Within a few weeks the grafts will be growing nicely—and then the raffia and wax should be removed to give the grafts room to swell.

If they make a quick, secure union they will produce several feet of growth before autumn and within three seasons the new branches will be bearing fruit.

GRAPE VINES. During the war grapes have been selling at "luxury" prices. They are, however, a worth-while proposition in normal times, when good bunches are sure to sell at a satisfactory figure.

Excellent grapes can be grown in any greenhouse that is weatherproof and well ventilated, and which gets plenty of sun and light. Artificial heating is not absolutely necessary (except when the grapes are wanted early) though it is a big help in a cold, wet spring and summer. Grapes can also be grown outdoors—see below.

Planting a Grape Vine. Planting is best done in January or February. It is best to buy a fruiting cane, which will give a few bunches the first summer.

The vine will probably arrive from the nurseryman in a pot. The roots must be uncurled, and swilled in tepid water to clear them, and the vine can then be planted in the prepared bed.

Deep planting is a mistake ; the top roots should be only 2 in. below the surface soil, and none ought to be deeper down than 6 in. If there are to be more than one vine there should be 6 ft. between them.

If the vine is planted in an outside border, all the buds on the cane outside the house should be rubbed off, and that part protected against frost with a hayband. The hole in the wall through which the cane enters should be stuffed with hay to keep out frost, too.

After planting an outside cane, it should be shortened to the two lowest buds inside the house ; the vine planted inside should be shortened to two big buds low down near the ground.

Good varieties—whether the greenhouse is heated or not—are Black Hamburg, Gros Colmar (the popular market black grape) and Foster's Seedling (one of the finest white grapes).

Pruning Grape Vines. Newly-planted vines are cut back after planting, as just mentioned. Only one main stem is allowed to grow that year and this is cut back by one-third the following December or January. Thereafter pruning follows on the lines adopted with established vines.

The idea with established vines is to prune back each lateral (or young side-shoot) making the cut just above the second bud counting from the base.

The leader shoot, at the end of each stem, must not be cut so hard. Shorten it to within about 18 in. of where it began growth last spring.

"Bleeding" of the pruning cuts is always a danger even when pruning is timely (in January). To be on the safe side, seal over the surface of each pruning cut with painters' knotting, or shellac, at once.

Pruning finished, collect and burn all the prunings, along with all dead leaves and rubbish littering the house and then get on with the cleaning.

Cleaning Grape Vines and Vineries. First wash down the greenhouse. Scrub the glass, the woodwork, ironwork, brickwork and staging. Put a wineglassful of paraffin and plenty of soft soap in a bucketful of warm water and beat it up well. With this wash and scour out every crack and crevice anywhere inside the greenhouse.

Then make up a thick creamy mixture of lime and water, adding a handful of flowers of sulphur and rather more than a pint of skim milk to every bucketful. Put this on with a brush or with the sprayer, covering every inch of brickwork.

While this cleaning is going on, cover the soil with sacks or papers.

To clean the vine itself, start by removing all the loose dead bark that is peeling from the stems. Rub it off with the hands, or use a blunt table knife.

After scraping off as much of the dead bark as possible, get a tin of Gishurst Compound (obtainable from any horticultural shop) and, with this, paint over every inch of the vine stems and spurs, missing only the actual buds.

To put final touches to the vinery, take the rake, scratch off the top 2 in. or 3 in. of old soil from the border (if an interior one) and replace it with fresh, fibrous loam. Mix some crushed mortar rubble and plenty of dry wood-ashes with the loam.

When this is down, give a dressing of basic slag—about ½ lb. to each square yard of border—and then on top of that put a 2-in. layer of decayed farm or stable manure, leaving it there to rot in.

Starting Grape Vines into Growth. Given a heated greenhouse, vines start into growth in January or February, to produce fruit in June or July. In an unheated house, growth starts in March, for the fruits to ripen in July-August.

Desirable temperatures are: until the buds are bursting into leaf, about 35 to 40 deg. F.; after bud breaking and until flower bunches show, 40 to 45 degrees; during the flowering period, 50 to 55 degrees; after flowering, about 60 to 65 degrees. The house should be some 10 to 15 degrees warmer with sunheat during the daytime. The great thing is to avoid any *sudden* rise or fall in the temperature.

Ventilate slightly directly the sun shines on the house in early morning, more freely as the sun gains power towards mid-day, and close up when the sun ceases to shine, or in late afternoon to prevent the temperature falling too low.

Syringing and Watering. As soon as the buds are swelling and unfolding, syringe the vine morning and early afternoon—using tepid water—and continue to do so until flower bunches open. Then cease syringing for a time, but damp down the floor and staging on warm days.

As regards watering, give the border a good soaking—about six gallons of water per square foot of border—when the vine first begins to grow; when young shoots are 5 or 6 in. long; just before flowers open; and when berries have set.

Training Grape Vines. To get the buds to break into leaf evenly all along the vine stems, loosen the rods from the wire trellis, and bend them down with their ends pointing to the floor, for a week or so. As soon as the buds are leafing out evenly, right along the stems, carefully return the rods to their normal position on the trellis.

When the new young shoots are 2 or 3 in. long, start limiting the number of new shoots on each "spur." Leave only one new fruiting lateral—or at very most, two—per spur. Choose the strongest looking shoot on each spur, preferably one showing a small bunch of embryo flowers, and pinch off the other shoots.

Disbudding finished, start training the lengthening shoots, tying them down well away from the glass, to prevent the leaves scorching. Put a piece of matting or raffia round them and the nearest wire and draw them down in easy stages—pulling each shoot down a little further every other day or so until it lies flat along the wires and can be finally secured in place.

Stopping Grape Vine Shoots. As the shoots lengthen the growing point of each must be pinched out. Go over the vine every few days and pinch back the more forward shoots on each occasion. Nip off the shoots just past the second leaf made beyond the flower bunch. On shoots that are not bearing bunches, nip off just beyond the sixth or seventh

leaf along from the base of the shoots. *Grape Vine Troubles.* The following chart gives the chief troubles to be encountered when growing grapes, with the symptoms and treatment of each :

GRAPE VINE TROUBLES AND THEIR TREATMENT

Trouble.	Symptoms.	Treatment.
Powdery Mildew	Patches of white Mildew on leaves, berries and shoots—as though powdered with flour. Leaves turn brown, shoots die and berries are cracked and spoiled.	Dust vine with flowers of sulphur after flowering. Avoid draughts and too-close atmosphere in spring. Renovate border and attend to drainage in winter.
Shanking	Black rings round berry stalks. Berries shrivel and turn brown; often whole bunches shrivel up.	Renovate border and drainage in winter. Avoid over-cropping. Thin berries and bunches *early*. Apply stimulating manures. Perfect drainage is essential.
Scalding	Berries shrivel and appear "scorched," eventually rotting.	Avoid irregularities in temperature, especially after "stoning," by careful ventilation and damping-down. Regulate shoots very carefully and uniformly.
Berries Failing to Set	Healthy-looking flower bunches which fail to mature berries.	Keep greenhouse cool and rather dry during flowering period. Pollinate flower bunches with rabbit's tail or by shaking rods daily to distribute pollen.
Mealy Bug Insects	Clusters of grey or whitish insects on shoots and among berries. Shoots and foliage stunted and berries disfigured.	Daub each cluster of insects with methylated spirit. Clean greenhouse and scrape vine in winter, and scrub with soap and nicotine wash. Then spray with 2 or 3 per cent. tar distillate wash. On the large scale fumigation with hydrocyanic acid gas may be practised.
Red Spider	Yellow, semi-transparent patches on leaves, with minute red mites clustered on undersurface. Premature dropping of leaves and crippled growth.	Spray vine in summer with Volck. Keep atmosphere moist and well ventilate house. Clean vine and greenhouse thoroughly in winter.
Vine Weevil	Bitten leaves and shoots. Large grey-black bettles on foliage at night and in soil or rubbish below during day.	Shake vines over sheet at night and collect and destroy beetles dislodged. Lay sacking on border in evening and next morning destroy beetles found hiding beneath. Dress soil round roots with sifted ashes containing 1 pint carbolic acid per bushel.
Bleeding..	Blobs of sap oozing from pruning cuts; general weakening of vine.	Prune early in winter. Seal over pruning cuts with painters' knotting, shellac or French polish. To check bleeding, sear pruning cut with hot iron before redressing with knotting.

Pollinating the Flower Bunches. When the flowers in the bunches open, pollinating must be done. Use a rabbit's tail, lightly brushing over the open flowers in the bunches, about noon, every few days. Alternatively, give the vine stem or the trellis wires a shake or a sharp rap every day while the flowers are wide open.

Limiting the Grape Bunches. It is never advisable to leave more than one bunch to each lateral shoot. If there are two bunches formed on one shoot (often there *are* two, sometimes three) clip off the least promising-looking and leave only one to develop.

Thinning Grape Bunches. Thinning out surplus berries from the bunches is most important. A start should be made when the berries are no bigger than sweet pea seeds. Use for the purpose a pair of long-pointed scissors and a short, forked stick, the latter to steady the bunch when working upon it. On no account must the berries be fingered or knocked or even bruised with the coat sleeve ; the slightest touch may produce an unsightly blemish on the skin.

The order in which the berries are removed is as follows. First, unfertilized berries are cut out. Then the berries tucked away in the heart of the bunch are carefully snipped out—remembering that just one berry jammed in the centre may rot and start a decay which will ruin the entire bunch. Next thin the " shoulders "—that is, the two top side branches which most bunches have, and after that begin at the bottom of the bunch and work upwards—spacing out berries uniformly over the whole bunch. The berries should all stand about ½ in. apart. It should be possible to push a pencil among them anywhere without touching them.

Feeding Grape Vines. All vines should be given a top dressing of 2 in. of rotted manure in early spring, when growth commences (see Cleaning Grape Vines and Vineries). When the berries form give a good soaking with weak liquid manure and repeat when the berries show colour. If the vine is not making very strong growth give 1 oz. per square yard once a week, from time berries are size of pea until they show colour, of following mixture : 1 part nitrate of soda, 2 parts superphosphate, 3 parts kainit.

Propagating Vines. The best way to propagate a vine is by buds, or " eyes " in January. From a stout shoot cut off at pruning time, cut out a piece of wood about 2 in. long including a fat bud in the centre. Cut away a strip of bark on the side underneath the bud, about ½ in. deep, this to encourage free rooting. The bud is then ready to pot-up. Fill a 3 in. pot with loam and leaf-mould and press the prepared eye into this, cut side down, so that the top of the bud just peeps through the surface. Firm the soil round it, then place the pot on a shelf in a warm greenhouse. Alternatively, plant the eyes in boxes, placing these in a propagator and standing this over the hot-water pipes. Do not over-water, but syringe the bud several times during the day with tepid water, and shade it during bright sunny spells. It will quickly root and make leaf, when it needs a 6-in. pot.

Storing Grapes. Ripe grapes will not keep long if left hanging on the vine. They may be kept for some weeks, however by cutting the bunches when fully ripe, with a piece of shoot attached, the shoot being long enough to support the bunch when inserted in a bottle of water. A narrow-necked bottle—or special grape-storing bottle—should be filled with clear rainwater and the lower end of the shoot placed in this. The bottle is then placed on a shelf in a cool and rather dry room, being inclined so that the bunch hangs clear.

Outdoor Grape Vines. Vines that yield excellent crops of good grapes can be grown outdoors. Varieties for outdoor culture are Black Cluster and Buckland Sweetwater. The vines should be planted against a wall or fence, preferably facing south, west, south-east or south-west (*not* north). November is the best planting month, with spring as the alternative.

GREASE-BANDING. Grease-banding is now adopted almost universally as a means of preventing damage by the caterpillars of species of winter moths. Its slight cost is saved many times over by the increased value of the fruit gathered.

It is carried out by applying paper bands coated with sticky grease around fruit-tree trunks and stakes so as to provide a barrier against winter moths (which are wingless) and other pests whose habit it is to climb up the trees into the branches in autumn, for the purpose of spending the winter there or of laying countless eggs to hatch into fruit- and shoot-spoiling caterpillars. The pests are unable to pass over the sticky bands, are trapped in them and killed.

Grease-banding is advised for apples, pears, plums, damsons and cherries. The bands *must* be in position by the end of September and remain until spring.

It does not pay to use home-made materials for grease-banding. Not only should one of the proprietary brands of grease be purchased but also the special grease paper.

To fit grease-bands first scrape off any loose or rough bark on the tree stem, where the band is to go. This may be anything from 2 ft. to 4 ft. from the ground ; certainly it should be a foot, at least, below the lowest branch. The band is then tied in place, top and bottom, with two strings, and the grease is smeared on between the strings. Smear it on with a flat-ended piece of wood, not too thickly, but just a thin, even, *complete* coating. If a bush-tree has branches radiating out from a stem so short (less than 18 in.) that mud would splash up on to any band placed around it, band the lower extremities of all the branches. When a tree is provided with a stake or stakes to support it band these also.

The above is the normal procedure. During the war, however, it became common practice to place the grease direct on to the bark, the paper not being used. By this method the grease is worked well into the bark in a band about 3 in. wide.

GREENFLY. See Blight and Spraying.

GREENGAGE. The greengage is very closely related to the plum and in every respect the culture of the two fruits is similar. See Plums.

GUMMING. A trouble to which all " stone " fruits (plums, cherries, etc.) are liable. The first sign is a thick, resin-like gum which exudes from the bark. This may be followed by the death of that particular branch.

Often " gumming " is directly attributable to a too-free use of the pruning knife in winter, though unhealthy conditions at the root, a lack of lime in the soil, over-vigorous growth and inattention to wounds may also be responsible.

Where too vigorous growth is the trouble this may be checked by pruning the roots of the tree in late October. Where soil conditions are bad, the remedy lies in proper drainage, in providing a sufficiency of lime and so on. Timely spraying will keep down pests and prevent *them* causing the trouble.

HAILSHAMBERRY. This is just an autumn-fruiting raspberry—the best of them—and is quite distinct in habit and fruiting from the hybrid berries. The fruit, which is borne on the tops of the strong, young canes, ripens in October. Treatment is as for raspberries, which see.

HARVESTING FRUITS. *To Test Fruits for Ripeness.* Apples are usually ready for picking when sample fruits part from the spur or branch when given the slightest lift-and-twist movement.

Another test for ripeness is to take a sound sample apple or pear and cut it open.

If the pips have turned brown or dark-coloured, then the fruit is practically ripe.

Hints on Picking Fruits. Even if fruits are to be used in the household, or for bottling, the old-fashioned method of snatching them off is to be condemned. These are the commonsense lines to work on :

Pick all fruit when it is dry. Gathered wet, it quickly deteriorates and loses its attractive appearance.

Pick every fruit carefully and separately and handle it as little as possible.

Never gather at one time more than can be used, marketed or stored immediately.

Overhaul the bushes and trees systematically, and take only those fruits that are " just right " at each visit.

Whenever possible, pick straight into a basket or a market package.

Strawberries and raspberries for dessert should be gathered with a short piece of stalk attached. The " small man " will do well to use a pair of scissors and clip them off. This is better than to risk squashing them in the fingers.

Gooseberries pay for very careful harvesting. The bushes should be gone over several times, only the largest and best berries being taken at each visit. Never on any account strip any bush at one go. At the first two or three pickings confine attention to the lower and centre branches, thinning rather than stripping them, and leave the outside branches until last.

Currants are a somewhat different proposition. As a rule the whole crop on the bush is ripe at the same time, and for this reason the market man finds that it pays to strip it at one go. It is bad work, however, to start stripping before every single berry in the bunch is ripe.

Cherries are best picked as soon as they are well coloured. It pays to go over the trees twice in gardens, but the market man usually favours the " all-at-one-picking " plan. However, in any case, wait until the fruits are fully coloured and developed before starting on them.

Cherries should be picked with the stalk attached and always when dry.

Apples and pears should be picked when they are in season. They can be tested for ripeness in the manner mentioned above.

Every little bruise and knock the fruits receive at picking-time will show up and later start a " rot," so pick them carefully into felt, canvas or paper-lined baskets or boxes, or into a pocketed apron. As you pick proceed to grade the fruits into " blemish-free," " slightly-damaged " and " bad "—using three picking baskets. That will save endless sorting over and unnecessary handling later.

Harvesting Nuts. See Nuts.

HAY-BANDING. Many troublesome pests, including the deadly apple blossom weevil, sawfly and codling moth caterpillars, can be trapped by the simple device of placing bands of hay round the trunks of apple trees in spring, after flowering. The pests use the bands as summer quarters, and all that has to be done to dispose of them is to burn the bands and their " residents " at the end of summer.

Alternatives to the bands of hay are pieces of folded sacking or folds of corrugated paper tied to the trunk just below the branches. Special traps are also made and can be purchased cheaply.

HIMALAYAN GIANT BERRY. This is an American blackberry. It is a very rank grower, making many very stout, thorny canes 10 ft. and more long. It is a tremendous cropper, and although the berries are not nearly so well flavoured as the English blackberries, they are produced so plentifully that for culinary purposes it is a most valuable sort to grow. Treat as for blackberry, which see.

I NSECT PESTS. See Pests.

INSECTICIDES. See under separate headings : Lime-sulphur wash, caustic soda wash, nicotine wash, etc.

L ACKEY MOTH. Destructive pest of fruit trees. Apples, pears, plums, and cherries all fall a prey to the voracious caterpillars, and if a colony becomes established the trees may be stripped almost bare of foliage and the developing fruits ruined. The blue-grey, hairy, and gaudily striped caterpillars appear in May and live in large silken webs spun on the branches, their presence being quite conspicuous. Immediately the webs are observed, prompt measures must be taken. Spraying with arsenate of lead (which see) is effective if carried out early enough, but in small fruit gardens it is often simpler and more effective to destroy the nests and caterpillars by hand, care being taken that caterpillars that fall to the ground when disturbed do not escape.

LAXTONBERRY. A useful fruit introduced a few years ago. Forms canes 7 ft. or 8 ft. long and bears medium-sized round, red berries rather like big raspberries. The berries are sweet but are rather soft and apt to crumple up when picked. The fruit does not " set " well unless the plants grow near raspberries or loganberries.

LAYERING. Method of propagating loganberries, blackberries and similar fruits, and strawberries. The layering of the canes is done in late summer or early autumn, by taking a long young cane and arching it over so that its top just reaches the ground, digging a small hole in the soil and burying the top 3 in. or so of cane in the ground. The cane is held in place with a peg.

The buried top soon throws out roots and some time during winter one can dig it up, sever the rooted end from the long cane and transplant it elsewhere. The cane is tied back in place to the wires.

The layered cane-tops take root readily enough in most soils, but with very heavy ground, or very light sand, it is better to sink pots filled with light loamy soil in the ground around the parent clump and to layer a cane-top into each of these.

Strawberries are layered in July, as follows : Peg down the best of the runners into the soil (or into pots of loamy soil) where they will root quickly and freely. Within a few weeks they will be entirely self-supporting, the connecting stalks can be severed and the rooted plantlets lifted and transferred to fruiting quarters.

Up to four or five runners from each healthy plant may be used, the rest being cut off at once.

When two or three plantlets have formed on any runners, layer only one plantlet from each runner, snipping off the runner stalk immediately behind the first plantlet on it.

LIME-SULPHUR WASH. Spraying fluid much used against woolly aphis, scale, mildew, etc., and for rendering fruit buds distasteful to birds. Most fruit-growers prefer to buy the wash, ready prepared ; it is somewhat troublesome to prepare at home. For those who prefer to mix their own wash, however, it may be prepared by mixing 5 lb. slaked lime and 5 lb. flowers of sulphur, adding water, boiling and stirring well. Bring up to 25 gallons with water.

LOGANBERRIES. The loganberry is a useful and profitable fruit demanding the minimum of attention and producing big crops of berries which command a ready sale. They are also invaluable for preserving and bottling for home use.

Loganberries can be grown against wire or wooden trellis, either in the open or against fences or walls. They need a sunny position ; the more sun they get, the greater and better the crop.

October and November are the best planting months.

Preparing for and Planting Loganberries. First, the ground must be deeply dug. If the soil is at all poor, take out a hole 3 ft. wide and 2 ft. deep where each plant is to go, cart away the soil removed and replace it with fresh top-spit soil from another part of the garden, along with a supply of garden refuse, leaf-mould and rotted manure.

The planting site should be reasonably well drained.

Do not plant too deeply, or there is a danger of smothering and killing the root-buds from which the new canes are to grow up in spring.

The question of how far apart to set the plants depends rather upon how the canes are to be trained—whether cane growth will be spread low and almost horizontally along the background or trellis, or whether it can run to some good height before being trained horizontally. On a 4-ft. or 5-ft. fence plant at least 12 ft. apart; with a tall fence or trellis plant 8 ft. or 9 ft. apart.

One-year-old canes are best. If planting old clumps cut back all the cane growth to within a foot or two of the ground before growth starts in spring, leaving no long canes to fruit the first summer.

Pruning Loganberries. The logan-berry bears its fruits on long young canes of the previous years' growth. These may grow right from the base of the clumps, or branch out from old stubs a foot or less above ground.

The pruning of the established clumps consists of cutting clean out all old canes that have fruited, and tying in the long, new canes over the trellis in replacement. Do this directly the ripe berries have been picked.

Feeding Loganberries. In spring cover the soil with a good mulch of manure laid in a thick circle around each plant. When the plants get to the heavy fruiting stage give a dressing of sulphate of potash, ¼ lb. per square yard each spring, followed by basic slag, ½ lb. per square yard in autumn.

Layering Loganberries. See Layering.

Loganberry Pests. The cause of maggoty berries is the Raspberry Beetle. It lays eggs on the flowers and the grubs eat into the berries. The remedies are to shake the flowering canes over an open umbrella or sheet and collect and kill the beetles caught. Also spray the flowering canes with insecticide.

LORETTE SYSTEM OF PRUN-ING. See Pruning—Lorette System.

LOWBERRY. Often known as the Mammoth Berry. Bears very large, long, jet-black berries, which are tender and sweet. Is a very strong grower, making canes anything up to 15 ft. long. Does not stand winter well unless in a sheltered spot. Its general culture is similar to that of the loganberry, which see.

MAGGOTS. For methods of pre-venting maggot trouble in fruits. See Apple Diseases and Pests, and Raspberries.

MAIDEN. Term applied to a fruit tree in its first year. Maiden trees are much cheaper to buy than three- or four-year-old trees, but most people prefer to buy the latter to avoid the somewhat " tricky " work of shaping the young tree into the required form. Also, after planting maidens there is a wait of some six or seven years, in most cases, before fruit is produced.

MANURES FOR FRUITS. Mulching. A mulch of manure put down over the fruit tree roots in the early part of the year (see Mulching) is the mainstay of fruit tree feeding. Where animal manure is scarce, chemical manures will give every satisfaction. For the special chemical manures to apply to the different fruits in normal times, when supplies of artificials are readily available, see under the fruit heading—i.e., under Apples, Pears, etc.

A General Fruit Tonic. Where a general manure for use over the whole of an orchard or fruit garden is required, measure out, by weight, and mix together : Superphosphate, 4 parts ; sulphate of potash, 3 parts ; and sulphate of ammonia, 1 part. When artificials are not readily available any compound fertilizer that can be obtained will serve as a substitute.

Apply the mixture to the soil at the rate of ¼ lb.—no more—to each square yard of ground covered by the trees or bushes. Spread it evenly, and then, with the hoe or fork, scuffle it into the top soil.

For Trees and Bushes Cropping Heavily but Making Little Growth. At any time during the early summer water with weak liquid manure, applied once or twice a week, after first soaking the ground with clear water. The manure water can be made from fresh liquid from the farmyard watered down to the colour of weak tea ; from 1 oz. of guano dissolved in each gallon of water ; or from 1 oz. of nitrate of soda or sulphate of ammonia in 3 gallons of water.

For Trees and Bushes Growing Strongly but Fruiting Sparingly. During the summer give sulphate of pot-ash, using 3 oz. to each square yard of ground occupied by the trees, followed by bone-meal, at the rate of 2 oz. per square yard, or superphosphate at the rate of 6 oz. per square yard. Hoe in.

For Trees and Bushes with Un-healthy Brown and Scorched-looking Foliage. Apply sulphate of potash in summer at the rate of 3 oz. per square yard, followed by sulphate of ammonia, 1 oz. per square yard. A fortnight later put a thick mulch of manure round each tree.

To Help Fruit Blossom to Set. In early March apply nitrate of soda or sulphate of ammonia all round the tree at the rate of 2 oz. per square yard. Hoe it in. Old plum, cherry and pear trees need this more than most other fruit trees—and often it is the making of the crop.

MARKETING FRUIT. At the outbreak of war a very comprehensive National Mark Scheme was in operation for fruit. Growers who undertook to pick, grade and pack their fruits in the manner laid down were permitted to label their packages with a National Mark, which in turn was recognised by the buying public as a certain guarantee of the contents.

The scheme had to be suspended for the period of the war, when the high standards in all directions—in picking, grading and packing—achieved before the war could not possibly be maintained under war-time conditions.

It will probably be some time before the old standard can be re-introduced after the war. In the meantime growers should do everything they can in prevailing circumstances—as a duty to the public and to maintain good will—to ensure that the fruit they market is of the best quality possible.

MEALY BUG. Injurious insect freely infesting grape vines, etc. Insects are pinkish brown and covered with a whitish mealy coating. They cause much loss of sap, followed by canker and other fungoid diseases of the bark of attacked plants.

One method of control is to hunt for the bugs along the vine shoots, before they get to the bunches and touch each with a brush dipped in methylated spirit. Also see Grape Vines.

MEDLAR. The two most widely grown varieties of medlar are the Nottingham and the Dutch. The former bears small, well-flavoured fruit and is a very upright grower ; the latter is a very large fruited sort which makes a rambling tree. As a rule, the fruit is not ready for eating raw until two or three weeks after gathering in late October, though it can be used straightway for jellies and preserves.

Medlars succeed best in somewhat moist soil. Autumn is the time to plant.

Pruning Medlars. Medlars require very little pruning. The branches of a young tree should be pruned and trained in the way they should go, in just the same manner that a young apple tree is built up (see Apples).

Once a tree is furnished with well-spaced-out branches—be it a standard, pyramid or wall tree—all the pruning

required each winter is to cut away dead and diseased pieces and any small branches that are choking up the tree.

MELONS. Melons are a most profitable crop to grow for market and a highly appreciated crop to grow for home use.

For hot-bed cultivation the best varieties are Hero of Lockinge and Monro's Little Heath. For the greenhouse these two are also quite suitable, or others are Royal Sovereign, British Queen and Scarlet Gem.

Melons will do well in any warmed greenhouse, while it is not difficult to grow them in a frame on a hot-bed, providing the district is a reasonably warm and sheltered one.

Sowing Melons. Wherever intending to grow the plants, sow the seed in April. Use 3-in. pots, filling them with a mixture of 3 parts loam and 1 part sharp sand, and cover each seed with about 1 in. of fine soil. The pots can be sunk to their rims in the hot-bed frame, or placed in the greenhouse on a shelf near the glass and covered with brown paper until the seedlings appear. This will be within a week if the temperature is kept steady at round about 65 or 70 degrees.

Within three weeks of sowing the seed, the young plants will be large enough to transplant to their fruiting quarters.

Planting Out Melons. Where it is the intention to grow melons in the greenhouse, some fresh turfy loam and a little decayed manure will be needed. To every three barrowloads of the loam, add half a barrowload of old mortar rubble, crushed small, and ½ lb. of basic slag.

When making the bed or border, first put down a layer of turves, grass-side down over the slates or tiles covering the hot-water pipes and on to this heap the compost in the form of a ridge about 1 ft. high and 18 in. or so wide. On to these ridges the small plants should be set, at about 18 in. apart, planted firmly, to within an inch or so of the seed leaves. Later, the plants must be trained up over the trellis wires—the bed then being made on a raised staging over the hot-water pipes and within 18 in. or so of the bottom wires of the trellis.

Where a hot-bed is to be used, this can consist of fresh strawy stable manure and leaves in equal parts or, better still, of manure alone. The material must be heaped and turned once or twice and allowed to ferment for a week before the bed of soil is put on.

The frame is then placed on top, facing due south so as to catch all the sun, and a 2 in. deep layer of dry, loamy soil is spread on the surface of the manure bed. Half a barrowload of the prepared compost is placed on the bed in the centre of each frame-light, in the form of a mound,

on which the young plants will be set—
two plants to each mound.

After a few days the soil will be warmed
through and, if large enough, the plants
can then be transplanted from the seed
pots to the mounds.

If the weather is cold at the time of
planting, the frames must be covered with
mats or straw each night.

Training Melons. When each plant
has developed its second rough leaf, the
growing point should be nipped out to
encourage laterals to form, two or three
side branches being allowed to grow out
on each. When each of these lateral
growths has made some six or seven leaves,
again pinch out all growing points and on
the further new shoots made the flowers
and fruit will form.

Management of Growing Plants.
Always use tepid water when watering
and take great care to avoid wetting the
stems of the plants ; otherwise " canker "
may set in and ruin them. Syringe the
foliage with tepid water twice a day until
the flowers are open, then only once a day.

Top-dress the plants with rich soil when
flowers are forming and feed the plants
freely with liquid manure.

Fertilizing Melon Flowers. Melon
flowers must be hand-pollinated. Do
this, preferably about midday, when four
or five female flowers are wide open, and
pollinate all on the same day. That is
important. The easiest way is to pick
off a male flower, remove the petals, and
thrust the pollen-laden stamens into the
centre, or cup, of the open female flower—
the female being distinguished by a bulge
or tiny melon-shaped swelling immediately behind the yellow petals.

Ripening the Fruits. As soon as the
melons attain any size, raise them slightly
off the ground by placing tiles or inverted
flower-pots underneath them. Cut off
backward fruits as useless and leave only
two or three evenly sized specimens to
mature on each plant.

Outdoor Melons. A variety known
as the Hardy Open-air Melon can be
grown in the open garden. It is pink
fleshed, and is a fruit with a very good
flavour. It is cultivated in exactly the
same way as the vegetable marrow.

Seeds can be sown outdoors in late May,
but the better plan is to sow in a greenhouse or hot-bed frame in early April.
The seeds should be sown separately in
small pots. From this sowing sturdy
young plants will be ready for planting
out at the end of May.

MILDEW. This disease attacks all
kinds of fruit trees and bushes, coating
the leaves with a white or grey powdery
mould. It usually reaches its height in
late summer. Badly affected growths

should be picked off and the remainder of
the tree sprayed with liver of sulphur
solution—1 oz. of liver of sulphur (potassium sulphide) and 4 oz. soft soap to
3 gallons of water.

A very troublesome form of mildew
attacks gooseberries. For methods of
control, see Gooseberries. Grapes are also
attacked by mildew. See Grape Vines.

MULBERRY. The black or common
mulberry is the best to plant. It does
quite well as a standard tree in the open in
the south, but in the north it needs the
protection of a south wall, where it should
be trained flat.

Mulberries are very slow in growth
and seven-year-old trees should be planted.

Mulberry fruits are at the dessert stage
of ripeness when they are nearly black.
At this stage they drop from the tree very
readily—which is the reason why mulberries are so often planted in grass land,
the grass saving the fallen fruit from harm.
In harvesting the crop, the usual plan is
to place clean cloths under the tree and to
shake the branches over these.

Mulberries require but light pruning.
It is sufficient each winter simply to cut
away dead and diseased wood and any
growths which are causing overcrowding.

MULCHING. All fruit trees and
bushes benefit from a mulch of manure
over their roots some time in February.
This is particularly important where the
trees are growing in light gravelly soil
which dries out quickly.

Farmyard or stable manure is suitable
for the mulching. Strawy horse manure
is best; cow manure is quite good ; pig
manure less valuable, though better than
nothing.

Before putting on the mulch loosen the
soil well over the roots of the trees and
remove weeds.

MUSSEL SCALE. Pest which infests
apple and pear trees. So named because
it resembles a minute mussel shell fixed
to the bark. Destroyed by spraying the
trees with lime-sulphur or paraffin emulsion, which see, and by winter spraying
with tar oil washes.

NECTARINE. In all respects, nectarines are given the same treatment as peaches. See Peaches and
Nectarines.

NEWBERRY. Recently-introduced,
and bears fruits much like loganberries,
though the berries are sweeter and better
flavoured. It does not " do " too well in
light, sandy soil, unless generously manured every year. Culture is the same as
for loganberries, which see.

NICOTINE WASH. Excellent insecticide for use on fruit trees. A good recipe is ¼ oz. nicotine (95 per cent. purity), 1 lb. soft soap and 10 gallons water. Dissolve the soap first in hot water, allow to cool, then add the nicotine and dilute as instructed.

NUTS. Nuts are a remarkably profitable crop, requiring comparatively little attention and mostly finding a good market at high prices.

Cobs and Filberts. Some of the best nut varieties for commercial use or private garden are : Prolific Filbert, an early variety with curiously frizzled husks; Cosford, a large and very thin-shelled nut ; Kentish Cob; the best for all-round use ; Merveille de Bollwyller, probably the largest nut and of fine flavour. The distinction between Cobs and Filberts is this : In the Filbert the nut is entirely covered with the husk, whereas in the Cob the nut is only partly covered.

Planting Cob and Filbert Bushes. Nuts will thrive in almost any soil, though in moist or very rich ground they are apt to make wood too freely to be prolific in crop. If planted in rows the bushes should be not less than 10 ft. apart each way. When established and cropping, they will respond to liberal manuring with farmyard manure or shoddy. Planting may be carried out during autumn or spring. Well-formed bushes that will commence to bear almost at once should be chosen whenever possible.

Chestnut. The eating chestnut is Castanea. It is produced on tall flowering trees which need little attention once planted.

Walnuts. These are produced on tall trees which, planted in autumn or spring in well-dug soil in a sunny situation, need little further care.

Pruning Nuts. Cobs and Filberts should be pruned in January or February —while they are in blossom. The trees produce male and female flowers separately, on the same tree. The catkins are, of course, the male flowers which produce the yellow pollen; the female flowers, which produce the actual nuts, are the tiny, deep crimson tufts or plumes which protrude from the end of fat buds along the twiggy side growths.

When pruning keep close watch for the red-tipped buds and avoid cutting off shoots bearing them ; otherwise the crop will suffer.

Starting with a young tree, train in six or eight good branches and on these shorten the leading shoot at the end of each by about half its length. Also cut back all strong lateral or side-shoots to three or four buds. Any growths not wanted as part of the permanent framework of the young tree should be shortened to within a few buds of the base.

With an established bush, first remove the strong, suckerous growths growing up from the bottom of the tree. Next, the strong side-shoots along the branches should be shortened, back to within three or four buds of its base. The shorter, twiggy side-shoots should usually be left untouched but may be shortened when they get too long and whippy. Then cut back to good buds showing the red flowers at their tips.

The leading shoot right at the end of a branch should be shortened by about one-third.

Chestnut and Walnut trees require very little pruning at any time. Simply see that all dead and unhealthy pieces are removed and that small branches threatening to choke the middle of the tree or spoil its shape are cut back. Do this in autumn. Both take some years to reach fruiting stage.

Harvesting and Storing Nuts. Cobs and Filberts. Gather when they are ripe and brown, which is usually in September. Separate the nuts from the husks—by shaking them to and fro in a sack—and then spread out the nuts in some airy, dry place for a few days until the shells are quite dry.

If the nuts are to be stored, pack them into stone jars—or new flower pots—sprinkling a little salt among them as they are packed in. If the jars cannot be tightly stoppered, finish off with a good thick layer of salt—for you must keep them free from damp and mustiness.

Store the jars in any cool, dry, frost-proof shed and the nuts will turn out at Christmas as sound and even crisper than when they went in.

Another way is to pack the nuts in tin boxes—biscuit boxes are excellent—making each box airtight and damp-proof by pasting a strip of paper around the edge of the lid.

Walnuts. These are gathered for dessert, when a few of them have fallen naturally, by shaking or beating the branches. Nuts required for pickling are handpicked when the shell has formed but has not hardened.

Dessert nuts are placed in a shed in a single layer and left thus till the husks have dried. Then they are placed, a few score at a time, in dry sacks and are shaken vigorously to separate husk and nut.

Chestnuts can be stored in sand in a dry shed.

Propagating Nuts. Cobs and Filberts can be propagated by suckers in autumn, these suckers being dug up with roots and the lower buds being rubbed off from the stems. Also propagated by layering in autumn. See Layering.

Nut Bush Pests. Among the few

pests that attack nut bushes is the bud-mite, which is " cousin " to the mite that causes the big-bud disease of black currants. Similar damage is done to the nut bushes as in the case of black currants, infested buds becoming abnormally swollen and failing to make leaf in spring, most buds being killed outright. It is unusual for the nut bud-mite to cause serious harm to the bushes, but should an attack be observed, handpicking and burning infested buds will be advisable.

The nut weevil is sometimes very injurious. Eggs are laid on young cob, filbert and other nuts, and the grubs grow within the shell, consuming the kernel and finally escaping through a circular hole they make in the shell. Observation may indicate the presence of the brownish weevils (having a curved and rather long snout) at the time of early nut formation, and they may be shaken off on to tarred paper. Spraying with quassia in paraffin and soft soap emulsion may deter the weevils from egg laying.

ORANGE TREES. These will not fruit outdoors in this country, but are sometimes grown as pot plants in a greenhouse. They are raised from orange pips sown in pots in the spring.

OYSTER SHELL SCALE. Growths on bark of trees somewhat resembling minute oyster shells. See Apricots, Troubles of.

PARAFFIN EMULSION. Good spray for use against aphis, etc. For recipe see Vegetable Section.

PEACHES AND NECTARINES. These are best grown as fan-shaped trees. In warm districts they do well in the open against a south wall, but in bleak localities they are best grown only in the greenhouse.

The same treatment applies to both peaches and nectarines, and the following remarks refer to both fruits.

Planting Outdoor Peaches. A sunny, sheltered position is essential. Dig the ground deeply, incorporate rotted manure with the lower spit and lime with the upper spit. Plant in October, November, February, or March. Also see Planting Fruit Trees.

Recommended Varieties. Peaches for greenhouse: Duke of York, Peregrine, Royal George; Outdoors: Hales Early, Peregrine, Dymond. Nectarines for greenhouse: Early Rivers, Humboldt, Lord Napier. Outdoors: John Rivers, Early Rivers, Lord Napier.

Protecting Outdoor Peaches from Frost. Peach flowers open early, often when there is frost about. It is essential to protect the flowers during such frosty spells. Either hang fine fish-netting over the trees or else sheets or tiffany. In each case raise these every morning or when frost has ceased.

Planting Greenhouse Peaches. Peaches to be grown in a warm or un-heated greenhouse are best planted in a border against the wall. Prepare the soil of the border as advised for outdoor peaches. Choose fan-shaped trees.

Managing Greenhouse Peaches. The peaches will flower soon after the turn of the year and will be making leaf-growth in April. With the changeable weather and uncertain temperatures of the early spring days the ventilating of the house needs the most careful attention. Regulate the ventilators many times in the day, especially when fitful spells of bright sunlight alternate with heavy cloud. A steady temperature should be maintained, and the harmful effects of a sudden rise and fall avoided. For some time after growth has started the ventilators should be closed early in the afternoon each day, before the sun loses power, so that the temperature of the house rises to 75 or even 80 deg. F. for a short time. The top ventilators should be opened a little as the morning sun touches the house, and on warm days more air should be gradually given towards noon. Even when the sun is at its highest the thermometer ought not to register higher than 70 degrees, except just after closing in the afternoon.

The young foliage, the pathways, and walls, must be syringed with tepid water two or three times during the day, especially in bright, sunny weather. The first syringing should be done as the temperature rises in the morning, and the last just before closing in the afternoon. Syringing should cease during the short time that the blossoms are open.

Feeding Greenhouse Peaches. Every spring mulch the soil with rotted manure. Also, in early spring, apply 2 oz. of bone-meal to each yard of border. This is a good stimulant for these trees.

Pollinating Peaches. The flowers of both outdoor and greenhouse peaches and nectarines need pollinating. For methods see Pollinating.

Thinning Peach Shoots and Fruits. A certain amount of shoot-thinning is necessary on both outdoor and indoor peach-trees. All shoots growing from the front of the branches, that is, outwards and away from the wall, should be rubbed off. Sufficient of the new side-shoots on the branches of the previous year's growth should be removed to leave the remainder

6 in. apart. Do this by a gradual process during May and June.

Fruit-thinning. is equally necessary. Give the first thinning when shoots are the size of hazel-nuts, leaving only one fruit every 4 in. When fruits are as big as a golf ball, thin again, leaving them 10 in. apart.

Pruning Peaches and Nectarines. Only a light pruning is needed in the first season. All that is necessary is to remove any shoots that are obviously awkwardly placed and to shorten the young leader shoot right at the end of each branch by about one-third its length.

With established trees the less the pruning knife or saw is used, within reason, the better. The disbudding of surplus young shoots in spring and early summer, when properly carried out, should leave very little winter pruning to be done.

The two- and three-year-old growths should be pruned back to strong young shoots near their base, leaving the young shoots practically full length.

Training Peaches. Before pruning, unfasten all the branches from the wire trellis and give the brickwork or wood-work a really good scouring—to get rid of pests and diseases. Then when all the old unwanted wood has been pruned out, re-fasten the retained branches in place.

Fasten the two lowest side branches to the bottom wire, then arrange the remainder to form an orderly, spaced-out framework in the form of an evenly-balanced fan.

The spaces between the main branches can be filled with the sturdy young growths that are to fruit in summer, these being tied in at 3 in. or 4 in. apart all over the tree, and the surplus and weakly pieces pruned back to one or two buds.

Peach Diseases and Pests. Peach Leaf-curl is the most serious of all peach troubles. It attacks both outdoor and indoor trees in spring and early summer, the symptoms being crumpled, curled, pinkish-red leaves. Affected leaves should all be picked off and burned, even if whole branches have to be stripped. A new crop of healthy leaves will quickly be produced to replace them. Spraying with Bordeaux Mixture (which see) in spring is the method of prevention.

Outdoor peaches may be attacked by the usual run of pests—Blight, etc. Give the treatment recommended under separate headings.

PEARS. Well-grown pears of popular varieties are an excellent market proposition. Every garden, too, should grow pears for home use.

Recommended Pear Varieties. Following is a brief selection of free-growing and good-cropping varieties. Some varieties are Self Fertile, others are Self

Sterile (for further details see these entries). With the Self Sterile varieties it is necessary to plant a "mate" to secure good cropping. Suitable companions are noted below.

Self Fertile Varieties: Dr. Jules Guyot (season of use Sept.); Williams Bon Chretien (Sept.); Louise Bonne of Jersey (Oct.); Conférence (Oct.–Nov.); Pitmaston Duchess (Oct.–Nov.).

Self Sterile Varieties: Doyenné du Comice (Nov.–Dec.), and suggested "mate" Glou Morceau (Nov.–Jan.); Beurré Bosc (Oct.–Nov.) and Doyenné du Comice (Nov.–Dec.); Souvenir du Congrés (Aug.) and Beurré d'Amontes (Sept.); Beurré d'Amantis (Sept.) and Beurré Diel (Oct.–Nov.).

Soil Preparation and Planting. When preparing the sites for pears, do not dig in a lot of manure or there will be too much growth and too little fruit. Where each tree is to go, dig out the old soil to a depth of about 2½ ft. and, in a circle quite 5 ft. wide, put in a 6 in. layer of brickbats, coarse rubble and such like at the bottom, and then fill in the hole with good top-spit soil. Mix in plenty of wood ashes, crushed mortar rubble and, if available, some turfy loam.

Never plant a pear tree very deeply. If the topmost roots are 3 in. underground that is quite deep enough. Also see Planting Fruit Trees.

Summer Pruning Pears. The summer pruning of pears is exactly as for apple trees. See Apples.

Winter Pruning Pears. *Young Trees.* Whether the branches are trained in the form of an espalier, bush or pyramid, the early pruning consists of cutting back all side-shoots not wanted to form new branches, each to within two or three buds of its base, and in shortening the leader by about half its length.

Cordon and Espalier Trees. Cut back to two or three buds every young shoot grown out along the side of the stem and shorten the leaders by about one-third the length. Spurs form very freely on cordon trees and they must be kept spaced out quite 5 in. apart.

Bush and Pyramid Trees. Shorten all side-shoots along the branch to within two or three buds of their base and reduce the leader at the end of each branch by half or one-third its length. If growth is strong, prune the leader lightly; if growth is only moderate or weakly, shorten each leader by as much as half.

Keep spurs thinned and shortened as with the cordons, and never allow branches to become crowded.

Standard Pear Trees. The pruning here is the same as for bushes and pyramids.

Pruning Fruit Spurs. On established trees the knobbly, woody spurs along the branches often get much too crowded and much too long. The result is dense masses of blossom in spring, and no fruit. Spurs must be spaced out along the branches so that they are at least 5 in. apart : very weak spurs should be pruned clean off. When a spur is much branched and very long, shorten it so that four or, at most, five plump buds remain.

"Tip-Fruiting" Pears. Two varieties of pears—Jargonelle and Josephine de Malines—do not fruit freely on spurs, but instead bear blossom and fruit on short side-shoots. When pruning these, don't cut back all the laterals. Shorten only the very long and crowded ones and leave all short, sturdy side-growths untouched.

The special method of pruning for quick and heavy fruiting that was recently introduced for apples was also applied to pears. For a brief description of the method see under Apples. Pruning by Lorette System. See Pruning—Lorette System.

Feeding Pear Trees. Most young trees do well with a mulch of rotted farm manure laid over the roots each spring, but with fruiting trees more intensive feeding is necessary.

In autumn apply a dressing of basic slag at the rate of ¼ lb. to each square yard of ground. In late winter give sulphate of potash at the rate of 2 oz. to the square yard, and in late spring give bonemeal at the rate of 2 oz. to the square yard.

Thinning Pears. The fruitlets on established pear trees must be thinned out in early June. At the first thinning it is sufficient to reduce the clusters of five or six pears to the three best fruits ; a further reduction to one or two is desirable later. Take out in particular all under-sized, malformed fruitlets and those which show signs of pest attack, particularly those blotched with black, sooty patches.

Havesting Pears. See Harvesting Fruit.

Storing Pears. See Storing Fruits.

Diseases and Pests of Pears. One of the commonest troubles with pears is Scab —a fungus disease which causes round scabs and gaping cracks on the fruits. The disease is the same as that affecting apples and the treatment is the same (see Apples).

The Pear Midge pest causes wholesale dropping of half-grown pears. The best remedy is to pick up and burn all fallen fruitlets, doing this daily ; otherwise the maggots escape and are a plague the following year. Poultry allowed to run beneath pear trees will rid the ground of numbers of the midge pests.

PESTS OF FRUIT TREES. For the methods of combating general pests see under separate headings. For methods of dealing with the pests attacking specific trees see under separate headings.

PHENOMENAL BERRY. Uncommon fruit. The berries usually are larger, longer, darker and a little sweeter than loganberries. It is a strong grower and a good bearer, the fruit ripening at about the same time as that of the loganberry. Treatment is as for loganberry, which see.

PICKING FRUITS. See Harvesting.

PLANTING FRUIT TREES. Brief directions on preparing the soil and planting fruit trees are given under the separate headings. Below are general instructions which apply to all fruits.

How to Plant. The fruit-planting season extends from the moment the leaves have fallen (late October or early November) until early April, but always it is a good rule to plant as early as possible and to choose fine weather for the work. Should the trees arrive during a break of bad weather, dig a trench of sufficient depth to accommodate the roots of the trees, these being, if necessary, laid in aslant, so that all roots are covered.

For the actual planting, wide, shallow holes, seldom deeper than 12 in. and at least a foot broader than the root-span of the tree to be planted should be taken out. When digging them, throw the very few inches of best soil in a heap on one side and the coarser sub-soil on another ; then you can return the top soil close among the roots and the less fertile sub-soil on the top. The bottom of each hole must be well broken up, to the depth of the fork-prongs, and then lightly trodden, before the tree is put in. Make the bottom of the hole somewhat cone-shaped, that is, slightly deeper round the edges than in the centre.

Trees which were injured in lifting and handling, or on the journey from the nursery, should have all broken roots or branches cut back several inches behind the torn, jagged end. Long, fibreless roots and down-striking tap-roots should also be shortened by half.

When it comes to putting the tree into the hole, place it exactly upright, in correct alignment and with all the roots evenly spread out to their limit.

As a general rule, 3 in. or 4 in. of good soil over the roots is ample, even on light land.

Before you commence any filling-in drive in the tree's supporting stake but do not tie the stake to the tree until planting is finished. To fill in, take some of the fine top soil and shake it among the roots, arranging them in natural tiers as the filling-in proceeds. Lift the tree up and down—only a slight motion—now and again, so that the fine soil works close in among all the fibrous roots. If the ground is poor import some good loam, leaf-mould, old potting soil, burnt refuse and soil, etc., to work among the roots, to ensure a favourable start in spring.

When all the roots are just covered out of sight tread that layer of soil fairly firm. Fill in the rest of the soil to within an inch or so of the top, then tread or ram it *quite firm*. Afterwards put enough soil round the tree, so that there is a suspicion of a mound, this deterring water from " standing " round the tree.

Planting Fruit Trees in Turf. Press in a cane where the tree is to go and then mark out a circle around it with a diameter of about 3 ft. 6 in. Cut through the turf from centre cone to edge with a spade to form V-shaped wedges. Lift these turf wedges and set them on one side.

Remove the exposed soil to a depth of 4 in. and fork the subsoil to a depth of a further 8 in. to loosen it. Plant and stake the tree as outlined above.

Return half of the removed soil over the roots, and then replace the turf, grass-side downwards. The turf will rot down and provide valuable food for the roots. Fill in the balance of the soil.

Distances Apart at which to Plant. *Apples, Pears, Plums, etc.* As standards, allow 20 ft. from tree to tree ; as half-standards, allow 15 ft. ; as bushes, allow at least 10 ft. ; as espaliers, allow 12 ft. to 15 ft. space for each tree ; for fan-trained trees, allow 15 ft. lateral space ; and for cordons, allow from 2 ft. to 5 ft., according to the number of stems—whether they be single-, double- or treble-stemmed cordons.

Red and White Currants and Gooseberries. As bushes, allow 5 ft. for each ; as cordons, plant at 15 in. to 18 in. apart.

Black Currants. Allow 5 ft. of space for each bush.

Raspberries. Plant canes 18 in. to 24 in. apart, 5 ft. between rows.

Loganberries, Blackberries, etc. Allow 8 ft. space for each plant—or 12 ft. each if the wall or fence is a low one.

Strawberries. Allow 18 in. between plants, 20 in. between rows.

Also see separate entries.

PLUMS AND GREENGAGES.

Plums are one of the fruits that can be relied upon to command a ready sale at satisfactory prices.

Recommended Plum and Gage Varieties. Following is a brief selection of reliable plum and gage varieties. Some varieties are Self Fertile, others are Self Sterile (for further details see these entries). With the Self Sterile varieties it is necessary to plant a "mate" to ensure good cropping. Suitable companions are noted below.

Self Fertile Varieties: Denniston's Superb (season of use August); Early Transparent (Aug.); Czar (Aug.); Prince of Wales (Aug.); Monarch (Sept.).

Self Fertile Varieties: Jefferson (Sept.), and suggested "mate" Early Rivers (July); Kirke's Blue (Sept.) and Victoria (Aug.-Sept.); Pond's Seedling (Sept.) and Transparent (Sept.); Coe's Golden Drop (Sept.) and Early Transparent (Aug.); President (Oct.) and Early Rivers (July).

Soil Preparation and Planting. For plums drainage must be particularly free and open, and lime is also imperative. Where each tree is to go, mix in a liberal quantity of old mortar rubble or plasterer's refuse. Failing the mortar rubble, dress the ground with slaked lime or ground chalk, and fork it in. (See also Planting Fruit Trees.)

Pruning Plum Trees. Plum trees will not tolerate hard pruning. While obviously badly-placed, crowded or diseased wood must not be left on the trees, only the lightest pruning is needed. It is safer to amputate big branches in spring or summer, rather than during mid-winter, and it is essential to seal over the saw-cut —and all other large pruning cuts—with white lead paint.

Starting on any plum tree, the first thing is to cut out any dead and diseased wood there is. Next ply the pruning knife among the crowded, spindly shoots choking the middle of the tree and the fruiting branches—shortening these to within three or four buds of their base. There is no need to cut back all the side-shoots along the branches. Nor is there need to cut back the leader shoots hard. Very long leader shoots need just a light topping, and the stocky, twiggy side-shoots can be left alone.

Pruning Young Trees. With a young tree there is need to prune a little harder. Here all but the shortest side-shoots should be shortened to within two or three buds of their base and the leaders shortened by nearly half their length.

Pruning Wall Plums. With a fan-shaped plum or gage tree against a wall, all the young growths not convenient for training in over the space should be shortened to two or three buds, and sturdy young laterals trained in wherever there is room. If they are too long to train in neatly, shorten them by a little.

When building up a wall tree, shorten the leaders by about one-third their length each year. When the tree is well-established, leave the leaders practically full length.

Summer Pruning Plums. Leave at full length well-placed shoots for which there is ample room, but shorten to five or six leaves the crowded and badly placed side-shoots. On wall trees similarly shorten the fore-right shoots—those which grow out at right angles to the fence or trellis.

Thinning Plums. Plums need especially heavy thinning—sufficient being removed to start with to leave the remainder standing at 4 in. apart. The surplus should be taken off when they are the size of cherries. A further thinning should be given when the fruits are nearing mature size, the aim being to leave each fruit full room for development.

Fertilizer for Plums. Where artificials are available mix 2 parts superphosphate, 1 part sulphate of potash and 1 part nitrate of soda and apply in early spring, at the rate of ½ lb. to each square yard of ground occupied by the trees.

Harvesting Plums. See Harvesting Fruits.

Diseases and Pests of Plums. The commonest disease affecting plums is the deadly Silver Leaf (which see.) Plums are also attacked by Blight and other pests (see separate headings).

POLLINATION. Term used to denote the transference of pollen from one flower to another, which is essential if there is to be fertilization, or a good " set " of fruit. In most cases the transference is effected by the wind and by insects, but some fruits (grapes and peaches are examples) are best pollinated by hand, even when growing outdoors. Greenhouse fruits must always be pollinated by hand.

The best method of pollination is to fasten a rabbit's tail to a cane or stick and with this very lightly brush over the wide-open flowers. About noon is the best time.

To Ensure Pollination of Single Trees. When there is only one tree of some particular fruit growing in a garden, it may well be that faulty pollination will cause it to bear no fruit. One way of overcoming this difficulty is to get a few in-flower shoots or a small branch from a fruit tree of the same kind (but not of the same variety) and put these into a jar of water suspended in the tree.

The bees will then carry pollen from the imported shoots to the flowers on the tree and thus the necessary mating will be accomplished, and a good set of fruit should follow.

Also see entries Self Fertile, Self Sterile.

PROPAGATING FRUITS. See Budding, Cuttings, Grafting, Layering.

PRUNING. For details of winter and summer pruning all fruits see under respective headings.

When to Prune. The season for winter pruning extends from leaf-fall to mid-March.

How to Make the Cuts. When removing a growth the cut must be made cleanly, always just above and sloping slightly away from a healthy bud pointing in the direction you require the branch to extend. Sharp secateurs are the best tool for quick and efficient pruning.

Leaders and Side-shoots. A " leader " is the young growth which terminates or " leads " each main branch, while " side-shoots " are the shoots that grow out along the sides of the branches.

PRUNING—LORETTE SYSTEM. The Lorette system is a system of pruning apple and pear trees named after M. Lorette, the famous Frenchman who introduced it.

If properly carried out season after season, excellent results follow, but it is not the slightest use to follow the system one year and miss the next ; it is the accumulated effect of the " Loretting " over several seasons that brings its reward in perfectly shaped trees and remarkable fertility. Only well-shaped trees with well spaced-out branches pay for treatment.

The essence of the system is that pruning is carried out *only in spring and summer.*

Towards the end of April, when the trees are just sprouting new growth along the branches, begin the pruning of the leaders made in the previous year. If the buds on these are well developed and strong, shorten each leader by about one-third its length ; if buds are small and growth weakly, shorten each leader by about half its length.

About the middle or end of June the more forward of the lateral growths along the branches are shortened. All new shoots that are about 8 in. or 10 in. long, half-woody and *about the thickness of an ordinary lead pencil*, are cut right back to within ½ in. or so of their base—back to the bottom cluster of leaves. Some shoots will not be long enough for shortening by then, and these you leave until they *have* acquired the right thickness and length—in July, or later, perhaps.

From each of the shoots cut back in June one or two new growths called " darts," will grow out, and as these in due course attain the thickness of a pencil these, too, are cut back to near their base. This will probably be in August.

During this first shortening of laterals, if there is a gap in the branchwork to be filled, you leave in a strong, well-placed shoot to fill up, treating this as a " leader " the next April.

In August and September the trees are gone over again and the woody, pencil-thick shoots each cut back to the cluster of leaves near their base. Weakly " darts " are then shortened to about three leaves, so that by the end of September there is no growth on the trees but short " darts " and short spurs, furnished with big bud-building leaves with fruit. buds forming in the centre of each cluster.

Practise the Lorette pruning only on varieties of apples and pears that are naturally spur-fruiters ; the few " tip-fruiting " varieties are best if only lightly thinned out at the normal pruning-time each autumn.

PRUNING, ROOT. See Root Pruning.

QUASSIA WASH. See under Quassia in Vegetable Section.

QUINCE. Fruit that finds favour with gardeners primarily for its beauty when in flower, but also for the fruits, which are valuable for cooking and preserving purposes.

The common quince includes both the apple-shaped and the pear-shaped quince. Of the two, the former is the better flavoured. Portugal is a very strong-growing variety, although it does not fruit so freely as the above. The fine Serbian quince, Bereczki, is a good grower, a prolific cropper, and the fruits are tender and well flavoured.

The quince will do well in almost any soil, but likes best of all a rather damp spot in the garden.

Quinces ripen in October, but should be allowed to hang on the tree as long as there is no danger from frost. Gathered dry and stored in a cool, frost-proof room, they will keep in condition for two or three months. They should not be stored with apples and pears, or the flavour of the latter fruits will be disagreeably tainted.

Quince trees should be pruned as soon as the leaves begin to fall in autumn. All that is required is to thin out crowded, crossing and misplaced growths.

RABBITS. PREVENTING DAMAGE BY. In some districts young fruit trees—especially apples—are gnawed each year by rabbits. It is during the hard spells in winter that most of the danger is done—when the natural green-food of the rabbits is too frozen to eat.

The best way to safeguard the fruit trees is to ring them round with small mesh wire netting, some 4 ft. high.

Failing netting, use one or other of the proprietary pastes specially prepared to keep away rabbits. Such preparations are smeared over the stems of the trees.

To Repair Damaged Trees. If any trees have already been " barked " they should be doctored immediately—before disease gets in. Cover the gnawed stem with a coating of lead paint or Stockholm tar.

In very bad instances, where all the bark has been stripped off all round, grafting wax is even better than paint or tar as a covering, for beneath the wax the injured tissues heal over more quickly and completely.

RASPBERRIES. Raspberries are a crop all should grow. Commercially, they can show an excellent profit when well grown and marketed quickly. They are in large demand by jam-making and canning factories.

Recommended Varieties. Summer fruiting : Lloyd George,Pyne's Royal, Red-Cross. Autumn fruiting : Hailsham, November Abundance, October Red.

Good yellow varieties are : Yellow Antwerp, (summer fruiting) and October Yellow (Autumn fruiting).

Planting Summer Fruiting Raspberries. November is the best time to plant raspberries, but they can go in any time in the winter.

Choose a sunny, open spot. Raspberries do no good in deep shade.

If the soil is very heavy, lime it well and dig in long, strawy manure ; if it is light and sandy, work in plenty of bulky, rotted manure, spent hops, old leaf-mould and the like to add humus and hold moisture. With a shallow soil, or in chalky ground, work in plenty of farmyard manure, all available burnt garden refuse, woodashes, old compost and such like.

Fertilizers are usually best applied after the canes are established, but in ground well dug and manured for a previous crop all that is required before planting is a dressing of basic slag. Apply ½ lb. to each yard run of row.

Usually raspberries are grown in long, straight rows and that is the best way where there is room. Let the row run as nearly as can be due north and south. In a crowded garden it may be more convenient to grow them in clumps.

When planting in straight rows set the canes 18 in. to 24 in. apart, and have the rows 5 ft. to 6 ft. apart.

For clumps, set the canes in a group or triangle of three, with 9 in. between the canes. Leave 5 ft. between clumps.

Avoid planting too deeply. On the roots will be seen a few spear-shaped buds which make the new canes in spring. These must not be smothered deeply or they will die. Two or 3 in. of soil over the top-most roots is ample.

The easiest way to plant a long row is to dig out a wide, shallow trench and place the canes, evenly spaced out, against the side, ensuring correct alignment with the garden line. Plant firmly, putting good top soil over and among the roots and treading it down. Finish with a thin mulch of manure laid on the ground, covering a strip a foot wide along the row, and leave this to rot in.

After planting, shorten each cane to within about 24 in. or 30 in. of the ground. Leave them at this height until buds are growing out in spring, then shorten each cane to a live, sprouting bud within about 8 in. or 12 in. of the ground. ·

Planting Autumn-Fruiting Raspberries. Autumn-fruiters must be given a sunny, open spot—preferably a warm site—or the late berries will rot on the canes instead of ripening properly.

The canes are planted in exactly the same way as the summer fruiters and the first pruning follows the same lines—each cane being shortened to within 24 in. or 30 in. of the ground for a start.

Supporting Raspberries. The best method is to secure the canes to galvanised wires strained between stout end posts. Two wires may be sufficient. But in a long row smaller, intermediate posts will be needed every 12 ft. or so, to keep the wires taut and the support strong.

Many growers get good results without providing permanent support ; they simply stretch strands of twine along either side of the row, in June and July, to keep the canes upright and trim.

Renewing Raspberry Plantations. To maintain supplies of good quality fruit a fresh planting should be made every sixth, seventh or eighth year, all the old canes being scrapped when the newly-planted ones come into bearing.

The best plan is to buy in new canes but if desired supplies can be secured from the number of new canes seen coming up among and around the old canes in summer. These can be dug up in early autumn—with their roots intact—and planted where required to make a new row.

Pruning Raspberries. The first pruning after planting consists of cutting back the canes to within 24 in. to 30 in. of the ground, as mentioned above. Subsequently each year after fruiting, the old canes of the summer fruiting raspberries are cut down to ground level. The thin, young canes (which will bear the following summer's fruit) are left alone unless they are crowded, in which case they are thinned out by digging up the surplus canes.

The young canes are " tipped " in early April, from 6 in. to 12 in., being taken from the top of each, so that the rows are of an even height of 3 ft. 6 in. or a little more according to variety.

Autumn-fruiting raspberries are pruned in February, when each cane is cut down to ground level.

Feeding Raspberries. Give a dressing of sulphate of potash, 3 oz. to each yard run of row, each spring, and on top of that put a mulch of farm manure—to ensure adequate nourishment.

To Ensure Maggot-free Fruit. It is important to take preventive measures or at picking time the fruit may be found full of maggots. The canes should be sprayed or dusted with a derris wash or powder when they are nearly in full flower and again when flowering is practically over.

Harvesting Raspberries. See Harvesting Fruits.

RED SPIDER. Pest that attacks all kinds of fruit trees. Numerous red " dots" which are really the tiny mites, appear on the under-side of the leaves. The foliage becomes sickly and rusty in appearance and the young fruits drop. The outbreak of the pest is favoured by hot, dry conditions and, where practicable, it can be kept away by syringing with clear water—particularly wall trees which usually grow in hot, dry positions.

Where an attack matures, spraying with nicotine wash (which see) or weak lime-sulphur (1 in 100) after blossoming is the remedy. Also see Grape Vines.

There are also special proprietary washes for use against Red Spider, these being applied in winter.

REVERSION DISEASE, or Nettlehead Disease. Disease which affects black currants. See Currants.

ROOT PRUNING. When fruit trees make very luxuriant growth but bear no fruit, pruning the roots often proves effective treatment. Apples and pears are the fruits mostly root-pruned. Currants and gooseberries never need it, however strongly they grow.

The time for root pruning is when the foliage begins to take on its autumn tint and before leaves begin to fall.

With a tree up to, say, five years old, the simplest plan is to lift it clean out of the ground, cut back the coarse fibreless roots by half their length, and then replant the tree in the same spot.

· With older trees proceed as follows : Dig out a trench right round the tree about 2½ ft. to 3 ft. out from the stem, and, as big roots are encountered, fork away some of the soil and cut back each—using a sharp knife, not the spade—to within about 2 ft. of the bole of the tree. Do not damage or shorten, or even disturb the bunches of fibrous roots you will come across ; only the very thick, fibreless roots need shortening.

Work all round the tree and then, with the spade tunnel under the tree and find the deep-diving tap roots, if any, and chop them through with a very sharp spade.

Pull out all the pieces of root pruned off, and see that the end of every spade-chopped root is trimmed off neatly and smoothly with the knife.

Pruning completed, return the soil to the trench, treading it quite firm.

Robbed of its main " anchor " roots, the tree will perhaps be a little shaky for a time, so at once secure the stem to a stout stake driven in firmly.

Only rarely do big, oldish trees call for root pruning treatment. When they *do*, the safest plan is to take out a trench only halfway round the tree, pruning the roots on one side only one year and leaving the other half for treatment the following year.

SCAB. Disease that causes scabs and gaping cracks in skin of apples. See Apple Diseases and Pests.

SCALDING. Trouble that causes grapes to shrivel. See Grape Vines.

SCALE INSECT. Another name for the Mussel Scale pest. See Mussel Scale.

SELF-FERTILE. Term applied to fruits whose flowers are able to " set " without the aid of pollen from other similar fruit trees. Self-fertile varieties must be chosen when single specimens of apples, pears, etc., are to be planted. Among the varieties that are self-fertile are :

Apples : Laxton's Epicure, Lord Lambourne, Laxton's Superb, Rev. Wilks, Lord Derby, Bramley's Seedling.

Pears : Conférénce, Louise Bonne of Jersey, Laxton's Superb.

Plums : Czar, Victoria, Laxton's Bountiful.

SELF-STERILE. Term applied to fruits which are not self-fertile and cannot set fruit unless they have a " mate." The apple Cox's Orange Pippin, the pear Doyenné du Comice, and the plum Coe's Golden Drop are examples of self-sterile varieties. Also see Pollination.

SHANKING. Serious trouble of grapes. See Grape Vines.

SHOT-HOLE BORER. Injurious insect pest. The beetles burrow deeply into the wood of fruit trees, tunnelling up and down, and in and out among the delicate tissue, with the result that they quickly upset the food and water channels, so that the branch or tree is eventually killed. In these tunnels the beetles lay eggs from which grubs hatch, these eventually becoming beetles which continue tunnelling. The remedy is to cut out the attacked branches, this being done as soon as wilting is noticed, or what look like shot-holes are found in the bark. The parts removed should be burned so that any beetles in them may be destroyed.

SILVER LEAF DISEASE. A very deadly and highly-contagious disease causing the rapid death of attacked trees.

When the disease first broke out, some years ago, so terrible were the ravages that it was causing that the Government made an Order that all wood, be it never so lightly affected, should be removed from diseased trees before July 15th of every year. That Order is still in force, but has been so effective in practice that outbreaks are now far less frequent.

The disease mostly occurs in plums but may also attack apple, pear, cherry and damson trees. It usually starts on one or two odd branches or twigs, but very quickly spreads until the whole tree is killed.

The First Symptoms. The first intimation of the presence of the disease is the foliage on a branch or two losing its natural deep healthy green colour and taking on a silvery tone. During the winter, when the branches are bare of leaves, fungus growths are to be seen on the bark of diseased branches.

Treatment of Silver Leaf Disease. There is only one way to deal with Silver Leaf Disease—to cut out and burn every " silvered " branch and shoot *and* every scrap of dead wood in the tree. Each branch to come out must be cut well behind the diseased part until, in fact, no brown stain can be seen in the centre of the wood—the brown stain denoting the presence of the fungus.

If the entire tree is attacked it should be cut down and burned, root and branch.

The fungus responsible for the Silver Leaf trouble can only spread from *dead* wood, and it can enter a healthy tree only through *an abrasion in the bark.* So, by cutting out silvered parts and all dead wood, and taking the precaution of immediately painting the cut surfaces with lead paint, the disease can be arrested and the tree saved.

It is a good plan to dress the ground about an attacked tree with a mixture of equal parts sulphate of ammonia and sulphate of potash, applying this at the rate of about 2 oz. to each square yard of ground occupied by the tree. Spread the manure evenly and just hoe it in lightly.

SPRAYING. Spraying is now recognised as the most important item in the fruit-grower's routine. Only by regular spraying can fruit free from the blemishes of pests and diseases be obtained.

Tar Distillate Spraying. The spraying campaign opens in mid-winter with the use of one or other of the Tar Distillate washes. There is nothing to equal these for ensuring freedom from spring-time plagues of blight and caterpillars and such-like crop despoilers. Actually they *kill the eggs of insects, kill all insects hiber-*

nating in crevices on the boughs and, incidentally, clear off moss, lichen, etc.

The only proviso with Tar Distillate sprays is that spraying be completed well before spring growth commences.

To estimate the amount of wash to buy, remember that you require, of the properly diluted wash, about : ¼ gallon for two medium-size currant or gooseberry bushes ; ½ gallon for a tree about 4 ft. in diameter ; 1 gallon for a tree 7 ft. or 8 ft. in diameter ; 1½ gallons for a tree about 12 ft. in diameter ; 2 gallons for a tree about 14 ft. in diameter, and so on in proportion.

A fair average strength to use a tar-oil winter wash is 6 per cent, that is, 6 gallons of neat liquid make 100 gallons of wash ready for use. Using a 6 per cent strength wash and having to make up 16 gallons of spray, 1 gallon tin of the wash must be bought. If the trees are going to take 50 gallons of spray then a 3 gallon tin will be required, and so on.

Winter Spraying. Apart from Tar Distillate washes there are other washes customarily applied in winter. Caustic soda wash is a winter wash much used against caterpillars and fungus diseases. Lead Arsenate is another recognised winter wash, being particularly effective against Codlin moth and other moths.

Proprietary washes against Red Spider and Capsid Bugs, again, are applied in winter.

A new winter wash containing di-nitro-ortho-cresol, and commonly called D.N.C., has the merit that it can be used even when the buds have started to develop. It is a very effective and complete pest controller; it smothers the eggs of aphides, apple sucker, kills the caterpillars of the winter moth and tortrix moth, and destroys the eggs of the capsid bug and red spider.

Spring Spraying. Most growers find it necessary to supplement winter spraying, the sprays being applied just before blossoming and immediately after the petals have fallen.

No spraying of any kind may be done whilst the blossom is out ; otherwise the bees and other insects working on the flowers to pollinate them would be killed.

Here is the spring spraying routine usually found best for the different fruit trees :

Plums. Spray with a good insecticide, such as nicotine, against leaf-curling blight. Or, if there is red spider on the trees, spray with lime-sulphur at 1 in 100 strength, adding ¾ oz. of liquid nicotine to every 10 gallons of the diluted lime-sulphur.

Cherries, Peaches and Nectarines. Spray with any good insecticide to check blight and small caterpillars.

Gooseberries and Currants. Spray with a proprietary insecticide or lead arsenate to prevent gooseberry sawfly caterpillars eating the foliage. Or dust the bushes all over with insecticide dust (see Dry-spraying, below).

Black Currants. Spray with lime-sulphur when the unfolding leaves are the size of a sixpence—not later—this as a remedy against Big Bud.

Pears. Spray with Bordeaux mixture to control the pear scab disease.

Apples. Spray with lime-sulphur at 1 in 100 strength to kill red spider and control scab disease, mildew, brown rot, etc. If blight threatens add ¾ oz. of nicotine to each 10 gallons of the diluted lime-sulphur wash. Or spray a second time three days later with insecticide.

Summer Spraying. Any non-poisonous insecticide may be used at any time during the summer (except immediately before harvesting) as a measure of control against blight, caterpillars, mildew, etc.

Spraying Appliances. For the garden the ordinary type of sprayer will serve. For a small orchard the pneumatic knapsack type of sprayer is recommended. For the large orchard the wheeled type of sprayer, operated by a pump is most serviceable. Before choice is made of a sprayer, catalogues issued by the appliance makers should be consulted.

Dry-Spraying. There has recently come into popularity a new form of spraying against insects and diseases—powder, or dry-spraying or dusting. Dusting is claimed to be easier and quicker than normal spraying. It has this limitation, of course ; it is no use for winter work ; it can only be done when there are leaves on the trees to hold the dust. Dusting is chiefly employed as a remedy in early summer against blight, caterpillars and mildew and other diseases.

There are *insecticide* dusts, like nicotine, derris dusts, etc., and there are *fungicides* like sulphur and copper in finely-ground powder form, prepared by specialist firms —something for each and every purpose. To apply them you must have a proper dusting appliance or " blower," and of these there are plenty from which to select.

SPURS. Term used to denote a small cluster of fruiting buds or a fruit tree branch.

STOCK. The stem or trunk upon which shoots are budded or grafted to make new " top " fruit trees such as apples and pears. See Budding.

STORING FRUITS. Apples and pears that ripen late in the season can be stored for use during winter, but those which ripen earlier should not be stored for more than a few weeks, as they will not keep.

It is usual to grade apples and pears before they go into the store. Small and medium-sized fruits will usually keep longer than will very large specimens and it is best that they be stored separately, the first grade going into one box or on one shelf, and the " seconds " into another, and so on.

The Storage Room. The ideal store is a well-ventilated, thatched house with double walls and earthen floor and fitted with central and side staging at convenient heights. The next best storage place is a basement cellar that is cool and dry, well ventilated and airy. An attic or loft is not particularly suitable, being rather too much exposed to extremes of heat and cold. However, fruit often *is* successfully stored here. An iron-roofed shed with a wooden floor should only be used as a last resort, but almost any other form of outhouse, provided it has a good roof and an earthen, brick, or concrete floor, may well be utilised.

Whatever is used, the place must be clean and wholesome, frost-proof, dark, or capable of being darkened, well ventilated and not naturally wet or dank. An earthen floor is always better than tiles, but a tiled floor is better than one of wood, which tends to cause excessive dryness.

After fruits have been placed in the store, plenty of air will be required for the first two or three weeks. The doors and ventilators should be left wide open for a short time, the reason for this being that the fruits, for the first week or so, " sweat " profusely, throwing off a considerable amount of moisture. Only when the " sweating " has ceased and the fruit is again dry should the store be closed up and just ventilated as required to maintain an even, cool atmosphere.

How to Sterilize a Fruit Store. All the storeroom woodwork ought to be sterilized before any fresh fruit is placed in the store. Get some soft soap and sulphur powder, mix the two together in hot water and then scrub every corner of the shelves and boxes with this; scrub the doors and trays and anything else likely to be contaminated with the fruit-rot fungus.

The Paper-wrapping Method. With very choice keeping varieties, such as Cox's Orange Pippin, Lane's Prince Albert, Bramley's Seedling, etc., an excellent method of storage is to wrap each apple in clean white tissue paper, or the specially prepared oiled or sulphite wraps to be bought for the purpose, and then pack the apples in wooden boxes—preferably boxes made of new wood.

Wrapped apples not only keep better but retain a better appearance and far better flavour than fruits stored in the ordinary way.

The Clamping Method. When ordinary storage space is limited " keeping " apples may be " clamped " just like potatoes, and will keep perfectly well thus for as long as required.

On a level, well-drained piece of ground put down a layer of clean straw and heap the apples on this, leaving them to " sweat " and dry for ten days or so. Then, after picking out any that are " going off," cover the heap with clean straw and over this put a 5-in. or 6-in. layer of fine, dry soil, smoothing off the sides so that the rain will run off.

Put wisps of straw or small drain-pipes along the top of the ridge, to act as ventilators, the bottom of the straw wisp or pipe touching the apples and projecting through the soil coating as airholes, so that moisture and heat will be liberated.

Finally, dig a small trench all round the heap to drain off surplus water.

Gas Storage of Apples. Apart from the normal storing of apples under dry and frost-proof conditions, or in oiled-paper wraps, gas storage may be practised by those who can fulfil the necessary conditions.

For successful gas storage over the longest possible time the apples should be gathered over a period of about two weeks just before they reach the stage of vital change that takes place when growth ceases. The storage is essentially a system of restricted ventilation, this being due to storage in a (variable) concentration of carbon dioxide gas.

The best results can only be obtained by using equipment providing rather precise conditions, and these are not often applicable or economical for small producers. Large producers who wish to take up gas storage on an economic basis should consult the Director of the Low Temperature Research Station, Cambridge.

Storing Apples by Refrigeration. Apples may be also kept in cold storage, where the atmosphere is moist, the temperature does not rise above 42° F. nor fall below 35° F., and ventilation can be suitably adjusted. This form of storage is also regarded as on the whole suitable only for large-scale use. The outlay for cold storage plant is considerable.

Storing Nuts. See Nuts.

STRAWBERRIES. Strawberry culture is an excellent proposition for those who grow for home use. It is also in some years a good proposition for market growers. In districts serving a jam factory or a canning factory, strawberries should always be considered.

Recommended Varieties. The best varieties for general use are Royal Sovereign (early); Western Queen (second early); Sir Joseph Paxton (mid-season); Tardive de Leopold (late).

Planting Strawberries. It is best to

buy plants which have been layered and rooted into small pots during the summer. Plants layered into the open ground are quite suitable, however—and cheaper—if they have been well looked after and come from a reliable grower.

There are two seasons for planting, August and March. The former is preferable, for the plants will then give a crop the following summer. If planted in March the plants should not be allowed to fruit the first summer.

The best position for strawberry growing is a sunny position where the soil has been well cultivated. Ground which the previous year grew a crop of potatoes suits well.

Preparatory to planting, the bed must be deeply dug and, if it has not been manured recently, a liberal dressing of manure should be worked into the top spit of soil during the digging. Failing good manure, dig in plenty of burnt or well-rotted garden refuse. After the digging, fork in some basic slag, spreading it over the soil at the rate of $\frac{1}{4}$ lb. to each square yard.

For planting, mark out the rows at 24 in. or, on heavy ground, 30 in. apart, and set out the plants at 18 in. apart in the rows. Plant with a trowel. Spread out the roots and set each plant firmly, and at such a depth as to allow the "crown" or "heart" of each to be just at soil level.

Certification Scheme. With the object of reducing disease and ensuring stock true to type an official scheme of certification of stocks was introduced some years ago. Under it, Officers of the Ministry of Agriculture inspect stocks while growing, and if satisfied that they are true to type and reasonably healthy, grant a certificate accordingly.

By The Horticultural (Cropping) Amendment and Consolidation Order, 1942, it is laid down that, except with the written consent of the County Committee, no occupier of agricultural land exceeding one quarter of an acre shall plant on his holding any strawberry plants other than such as are certified by the Ministry.

Feeding Strawberries. Mix 5 parts superphosphate and 3 parts sulphate of potash and apply in early spring at the rate of 4-6 oz. to each square yard of bed, spreading in between the plants. Alternatively, dig in rotted manure between the plants. Generally speaking, strawberries need plenty of organic matter, adequate potash and phosphate, and only moderate nitrogen.

Strawing Strawberries. In May spread clean straw among and around the plants, tucking it well beneath the plants. This not only serves to protect the flowers from frost damage, but later keeps the fruit off the soil, and so prevents it becoming gritty.

Remove the straw when fruiting is finished.

Straw "mats" specially made for the purpose are obtainable.

When Strawberries Have Finished Fruiting. First pull up or hoe out all weeds; also pick off dead or decaying leaves. Then cut off any runners not required to make new plants (see Layering). Finally, loosen the soil surface between the rows and also between the plants.

Strawberry Troubles. Birds are the principal enemies of the strawberry grower. The only possible ways to safeguard the fruit are either to place string netting over the whole bed, propping it up on sticks to keep it clear of the plants but pegging down the outer edges close to the ground (otherwise birds will creep under); or to have a permanent cage of wire-netting over the bed.

Small beetles sometimes attack ripening strawberries. In this event, prop the fruit trusses clear of the ground on forked twigs.

Mildew frequently attacks strawberries. When it is noticed, spray the plants with liver of sulphur solution. (For recipe see Vegetable Section).

Strawberry Leaf Spot is the most serious strawberry disease. In an attack the leaves are speckled with reddish black-edged spots. The treatment—to be undertaken directly the fruit is all picked—is to cut off all the old foliage, so that no browned or spotted leaves remain, and burn it. Dust the plants all over with flowers of sulphur, and give a second sulphur dusting just before the plants come into flower the next spring.

Growing Strawberries in the Greenhouse. It is easy to obtain strawberries in spring. The procedure is to root some runners in small pots in summer (see Layering) or buy rooted runners for the purpose, grow on the plants outdoors until October, then place them in a cold frame and finally take them into a heated greenhouse in January placing them on a shelf near the glass.

The chief trouble to guard against will be greenfly, and the plants must be dipped into weak insecticide the moment the first few pests are seen.

The plants should be fed once a fortnight with liquid manure when the flower buds form. When the flowers open pass a camel-hair brush from bloom to bloom.

The plants should be planted into the outdoor strawberry bed when fruiting is finished. They cannot be grown two years in succession in pots.

Harvesting Strawberries. See Harvesting Fruits.

Layering Strawberries. See Layering.

SUCKERS. Term applied to the soft sappy growths which spring up around fruit trees during the late summer and autumn. These suckers are harmful; they rob the tree of nourishment and weaken the cropping ability of the tree. They must be removed—right from their point of origin on the roots. If merely chopped off at soil level they will come up stronger than ever. The soil should be scraped away until the root from which the suckers spring is exposed. Then the suckers should be cut off flush with the root.

Suckerous growths springing from the stems of fruit trees should also be removed.

TAR DISTILLATE WASHES. Spraying fluids which, applied to fruit trees in winter kill all insect eggs on the trees. See Spraying.

THINNING FRUITS. Established apple, pear and plum trees, etc., in a good season, set far more fruitlets than they are capable of maturing. The fruitlets must therefore be thinned out.

For methods of thinning see under respective fruit headings.

TOP-GRAFTING. A method of bringing old or unfruitful apple and pear trees back to youth and fruitfulness. See Grafting.

VEITCHBERRY. Uncommon fruit bearing large, mulberry-coloured berries, combining the flavour of the raspberry and blackberry. The berries ripen after raspberries are over and before blackberries come in. They are fairly well flavoured, are useful for stewing and make good jam. It is a strong, sturdy grower.

VINE WEEVIL. Large grey-black weevils which eat vine leaves and shoots. See Grape Vines.

WALNUTS. See Nuts.

WASPS, PROTECTING FRUIT AGAINST. Wasps in some seasons do a great deal of damage to fruits, especially plums, and when they are plentiful measures must be taken against them.

To Destroy Wasps' Nests. In the commercial orchard the only practical measure is to search for and destroy the nests, with their inmates. The nests will usually be found in a hedge-bank, probably in an angle formed by a large tree root. The nests can be destroyed without the slightest risk of being stung if you do it late at night.

There are various methods. Some people use cyanide of potassium, putting a small piece the size of a cherry inside the entrance and sealing it up with a piece of wet clay, when the nest can be removed and burnt in the morning. Cyanide is, however, a very dangerous poison, a small grain being fatal to human life, and less risky materials are recommended.

An earth nest, for instance, can easily be disposed of some time after dark— say at 11 p.m.—by smashing up ½ oz. of rock sulphur into small fragments, dropping these into the entrance, pouring a gill of petrol on and setting it alight, when the nest can be safely dug out in the morning.

Other methods of destruction are to pour into the entrance hole at night two or three gallons of boiling water; or a can of hot water containing half a pint of creosote.

Protecting Choice Fruits. Fruits for show or other specially choice specimens are usually enclosed in light muslin bags as a safeguard against wasp damage.

Trapping Wasps. In small fruit gardens some good may be done by setting traps for the wasps.

Glass jars half-filled with a mixture of beer, sugar and water and hung in the branches of the trees form an irresistible attraction to wasps. Use ordinary jam-jars with a small flower-pot fitted into the neck. The wasps crawl in through the drainage hole at the bottom of the pot but cannot find their way out again.

WATER GROWTHS. Name given to weak sappy growths that often appear in numbers on old, neglected trees. They are useless for fruiting and should be cut out at their point of origin during winter pruning.

WAX, GRAFTING. TO MAKE. A suitable wax for grafting purposes can be made at home from the following recipe; Melt together tallow, 1 part; beeswax, 2 parts; resin, 4 parts. Stir the mixture thoroughly. This should be used warm. A cold wax recipe is: Burgundy pitch, 1 part; paraffin wax, 1 part; tallow, 1 part. Melt them together, stir well, and use when cold.

WINEBERRY. Useful fruiting bramble. Has rich, orange-scarlet berries ripening in August. The habit of growth is the same as that of loganberries, and the plants should be treated the same as these. See Loganberries.

WOOLLY APHIS. See American Blight.

WORCESTER BERRY. Uncommon fruit bearing berries in trusses just like black currants—the berries being black and a little larger than currants. This berry should be grown just as are Gooseberries, which see.

YOUNGBERRY. Cross between loganberry and dewberry, with very sweet, juicy fruits. Culture is as for loganberry.

FLOWER GROWING

A CHIMENES. See Greenhouse Plants.

AMARYLLIS. See Greenhouse Plants.

ANCHUSA. See Herbaceous Border Plants.

ANEMONE. The anemones, among the most useful and brilliantly coloured flowers of the garden, enjoy a very ready sale, especially when they come in early. The several different kinds of anemones can all be grown from seed, though the usual way to start with them is to buy corms or so-called " bulbs ".

Anemones, Coronaria. The Coronaria, or Poppy-flowered anemones, are the most beautiful. Included among them is the St. Brigid anemone, with semi-double and beautifully shaped blooms ; St. Bavo anemones—single, but as brilliantly colourful as the St. Brigid ; Caen anemones, single and mostly in carmine and scarlet shades ; and the double French form, called Blue-Gown, mauve-blue in colour. His Excellency is a giant-flowered variety, bright scarlet and specially fine for cut flowers.

The position to choose for these anemones is one where they will be sheltered from cold winds at the time their early flowers are opening—a non-shaded place but a sheltered one. The soil must be rich and not too heavy.

The best time to plant is in September. Set the corms or " bulbs " in the soil, 3 in. deep and 6 in. apart, on a thin layer of dry sand. The corms should be arranged with the " buds " upwards and the " claws " downwards.

This outdoor planting will give flowers in spring and summer. For winter blooms, plant the corms in a cold frame, prepared by putting in a drainage layer of broken bricks and pots, covered with a layer of rough turf and, finally, a soil-mixture of loam, leaf-mould, sand, and rotted manure.

Do not put the " lights " on until October. Even then plenty of air must be given in all but really frosty or foggy periods.

Coronaria anemones can also be grown in pots. For compost, use a mixture of turfy loam, 2 parts, and 1 part of leaf-mould or old hot-bed material, with a sprinkling of silver sand. Plant the tubers just clear of each other and about 1 in. to 1½ in. deep. Stand the pots in the cold frame, moving them into the greenhouse when required.

The corms are lifted after flowering.

Anemone Japonica. See Herbaceous Border Plants.

ANNUALS. Hardy Annuals. Being so bright flowered and easily grown, hardy annuals make a big appeal to amateur gardeners. Interest in hardy annuals so far as market growers are concerned is limited to those kinds which are popular as cut flowers. They include some of the best-selling flowers on the market—calendulas, coreopsis, cosmea, godetia, clarkia, nigella, cornflower, gypsophila, scabious and sweet sultan in particular. There is also some demand for poppies—especially the Shirley varieties.

To ensure the production of good quality blooms the seed must be sown in well prepared soil. Dig 1 ft. deep and stir the subsoil as digging proceeds. Work in a good dressing of rotted manure. Finish by making the ground firm.

Before sowing rake the surface fine, at the same time working in a dressing of bone-meal at the rate of 1 lb. per 16 sq. yards of ground. There is in this fertiliser a wonderful property that stiffens growth, hastens flowering and improves the quality of the flowers. When raking, work in the bone-meal 4 in. deep.

To speed up the germination of godetias and sweet sultans soak the seed in cold water for two hours before sowing. Minister to gypsophila's love of lime by whitening the seed drills with freshly-slaked lime immediately before sowing.

Thin out the plants—from 4 in. to 1 ft. —according to the height to which they grow. Start to feed in July, with liquid manure and a good general fertiliser in turn.

For specially fine flowers there are certain cases in which disbudding pays. For instance, in calendulas take away two out of every three buds, leaving the strongest. Cornflowers and scabious may have their buds reduced by half, while annuals producing branched spikes, like clarkias, should have the shoulders or side-spikes taken off.

Although early spring is the normal sowing time for hardy annuals, many growers make a practice of sowing in September, finding that the plants stand the winter well and bloom early.

Half-hardy Annuals. Boxes of half-hardy annuals find a ready sale to amateur gardeners for summer bedding work. Certain half-hardy annuals—Zinnias are a notable example—are also widely grown for cutting. For details of the most popular kinds see Bedding Plants.

ANTIRRHINUMS. See Bedding Plants.

AQUILEGIA. See Herbaceous Border Plants.

ARUM LILY. See Greenhouse Plants.

ASPARAGUS, ORNAMENTAL. See Foliage Plants.

ASTERS. See Bedding Plants. For Perennial Asters (Michaelmas Daisies) see Herbaceous Border Plants.

AUBRIETIA. See Rock Plants.

AZALEAS. See Greenhouse Plants.

BEDDING PLANTS. *Half-Hardy Annuals.* In normal times boxes of half-hardy annual bedding plants, and of other plants usually treated as half-hardy annuals, are in big demand by amateur gardeners who have no facilities for raising their own bedding plants at home. Good profit can be made by the commercial grower who can cater for this market.

Following is a list of the half-hardy annuals most popular among amateurs for summer bedding work.

Raising Half-Hardy Annuals. Seed is sown in the heated greenhouse in January or February. When large enough to handle the seedlings are pricked off into other boxes. The plants are kept in the house until April, when hardening-off starts in readiness for sale from the beginning of May.

Biennials. The biennials and other

HALF-HARDY ANNUALS FOR SUMMER BEDDING

Name of Flower.			When to Sow.	Height.	Flowering Season.
Ageratum (Floss-flower)	..	..	Early March	6 in. ..	July–Sept.
Alyssum (Sweet Alyssum)	..	..	,,	4 in. ..	June–Oct.
Antirrhinum (Snapdragon)	..	..	January ..	Tall, 2–2½ ft. Med. 1½–2 ft. Dwarf, 6 in.	,,
Arctotis *grandis*	..	..	Mid-Feb. ..	1½–2 ft. ..	June–Sept.
Aster, Ostrich Plume	..	..	Early March	18 in. ..	July–Oct.
,, Southcote Beauty	..	..	,,	,,	,,
,, Sinensis ..	..	..	,,	,,	,,
,, Giant Comet	...	..	,,	,,	,,
,, Victoria ..	..	..	,,	,,	,,
,, Dwarf ..	...	..	,,	12 in. ..	,,
Cosmea ..	..	..	,,	2½ ft. ..	,,
Golden Feather ..	..	..	Mid-Feb. ..	3–5 in. ..	Ornamental foliage plant.
Heliophila (Cape Stock)	..	..	Early March	15 in. ..*	June–Sept.
Ipomœa (Morning Glory)	..	..	,,	Climber ..	July–Aug.
Kochia (Summer Cypress)	..	..	Early March	2–3 ft. ..	Ornamental foliage plant.
Lobelia ..	..	..	Early Feb.	4–6 in. ..	June–Oct.
Marigold, African	..	..	Early March	2–3 ft. ..	July–Sept.
,, French	..	..	,,	6–9 in. ..	,,
Mignonette	..	..	Early March	1 ft. ..	July–Sept.
Nemesia ..	..	..	End Feb. ..	9 in.–1 ft. ..	June–Sept.
Nicotiana (Tobacco Plant)	..	..	Mid-Feb. ..	3–4 ft. ..	July–Sept.
Perilla *nankinensis*	..	..	Early March	18 in. ..	Ornamental foliage plant.
Petunia ..	..	..	End Jan. ..	1–2 ft. ..	June–Sept.
Phlox *Drummondii*	..	..	Early March	9 in. ..	,,
Salpiglossis	..	..	,,	2–3 ft. ..	June–Oct.
Salvia ..	..	..	Late Jan. ..	15–18 in. ..	June–Sept.
Stocks, Brompton	..	..	July ..	2 ft. ..	May–Aug.
,, East Lothian	..	..	January ..	15 in. ..	June–Sept.
,, Ten Week	..	..	Mid-Feb. ..	18 in. ..	June–Sept.
Tagetes ..	..	..	Early March	9 in. ..	June–Oct.
Verbena ..	..	..	Early Feb.	6–12 in. ..	June–Sept.
Zinnia ..	..	..	Early March	2 ft. ..	July–Sept.

plants treated as such include most of the flowers popular for spring bedding and a number of other plants which enjoy a ready sale to amateurs. Following is a list of the best-known biennials :

August or September, and winter these. The latter plan is the better, for the cuttings can be rooted outdoors, form nice strong plants by winter, and usually come through the cold period well. Further,

Kind.	Colours.	Flowering Season.	Height.
Canterbury Bell ..	Blue, rose, white	Summer	2 ft.
Cheiranthus ..	Orange, mauve ..	Spring and Summer ..	2 ft.
Double Daisies ..	Pink, white, etc. ..	Spring	6 in.
Forget-me-not ..	Blue, pink ..	Spring	8 in.
Foxglove ..	Pink, white, yellow	Summer	4–8 ft.
Honesty	White, purple ..	June and July ..	2 ft.
Iceland Poppy ..	Various	Summer	2 ft.
Indian Pink	Various	Summer	1 ft.
Oenothera (Evening Primrose)	Yellow, purple ..	July and August ..	3 ft.
Polyanthus	Various	Spring	1 ft.
Scabious	Mauve, pink ..	Summer	3 ft.
Stocks (Brompton and flowering)	Various	Spring and early Summer ..	2 ft.
Sweet William ..	Red, pink, white	Summer	2 ft.
Wallflower	Yellow, red, purple	Spring	2 ft.

Raising Biennials. Seeds of biennials are sown in May-July, in the open, either in very shallow drills, or broadcast on seed beds. When the young plants are large enough to handle they are transplanted to a nursery bed, being planted in rows 1 ft. apart, with 6 in. between the plants. From this bed they can be lifted for sale in the autumn or early spring.

Other Bedding Plants. Begonias. Tuberous begonias are very popular for use in bedding displays. The tubers are started into growth in boxes in late February and potted up separately when they show pink shoots. They will be ready for sale in May.

Calceolarias. These are half-hardy perennials. The usual way to increase them is by means of cuttings. Take these in September and either plant in boxes for the greenhouse or direct in sandy soil in a cold frame. Here they will survive until planted the following spring, only needing the "light" to be matted over during frosty nights and days.

Dahlias, Bedding. See Dahlias.

Geraniums. Bedding " geraniums " (they are really pelargoniums) are required by the hundred thousand in spring.

Supplies of young plants are very easily obtained from cuttings, obtainable by two methods.

The first method is to lift and store a stock of plants at the end of summer, winter them in a frost-proof greenhouse and take off the young shoots as they are produced in spring.

The second method is to take the cuttings direct from the bedding plants, in such plants often themselves each provide a cutting in spring, the tips being clipped off and rooted.

In regard to saving stock plants, foliage varieties are more tender than the flowering kinds. Lift plants before they have been seriously seared by frost, shake off all the soil, shorten the straggling roots to half length and cut back the top growths to 2 in. stumps. Then pot up singly into small pots in dry soil, and stand them on a dry, sunny shelf in the heated greenhouse.

Flowering varieties will take no harm if they are left out until frost has had a *little* nip at them, for all the top hamper has to be cut away, roots trimmed and so on. Then they should be potted or boxed up in dry soil. The heated greenhouse is also the best place for these. The soil used in pots or boxes must be poor, and it must be quite dry.

After January, pot up the plants properly and resume regular watering.

In regard to the cuttings, choose for the purpose short, firm, well-ripened shoots. Trim them by shortening back the stem to immediately below a joint and removing the lower leaves. Expose them to the sun for a few hours, then plant them in pots filled with very sandy soil. Pick off any leaves that fade whilst the cuttings are rooting.

BEGONIAS. See Bedding Plants, Greenhouse Plants.

BIRD DAMAGE. Sparrows in particular, are a nuisance in the flower garden,

G

picking out the tips of carnations, pulling polyanthuses to pieces, nibbling sweet pea seedlings, wrecking yellow crocuses, eating lawn seed and so on.

For protecting flowers and plants, strands of black cotton stretched above the soil are effective. For protecting seed-beds use fish netting, raised a foot above the soil. For protecting sweet peas, use the wire guards sold for the purpose.

BLIGHT. See Greenfly.

BORDEAUX MIXTURE. One of the best fungicides available. For formula, see Bordeaux Mixture in Vegetable Section.

BUDDING. Method of propagation, mainly applied to roses, consisting of grafting a bud (not a flower bud but a shoot bud) from a cultivated rose oñ to a rooted stem (or stock) of a wild or other rose. Stocks of various kinds are available, but Rosa Rugosa and Wild Briar stocks are most often used, these giving the best rooting system for garden purposes. Stocks can be bought or, in the case of Wild Briars, they can be obtained from the hedgerows. They should be secured in autumn, rooted stems, some 6 ft. long, being obtained and planted in a sunny position. These will be ready for budding the following summer. Stocks can also be obtained by rooting cuttings of Briar or Rugosa in autumn.

The best time to bud roses is July. Suitable buds are obtained from half-ripened rose shoots of the current year's growth.

Cut off the shoots 6 in. or so long and remove about an inch of the tip, which will not be ripe. Remove only a few growths at a time and, to prevent the buds drying out, stand them in a jar containing $\frac{1}{2}$ in. depth of water—no more or, instead of just keeping fresh, the buds will become sodden.

Remove the leaves from one of the shoots, allowing the leaf-stalks to stay on. There is a dormant growth bud in the axil of each stalk where it joins the stem, and it is these buds that are required. Insert the knife blade about $\frac{1}{2}$ in. below one of the leaf-stalks and draw it upwards so that it comes out about $\frac{1}{2}$ in. above the leaf-stalk. In cutting out the bud penetrate the shoot sufficiently to avoid cutting the base of the bud.

This operation provides a small shield-shaped portion of shoot bearing a dormant bud and a leaf stalk—which is useful as a " handle ". Turning over the portion, a little strip of the wood of the shoot will be observed at the back. Holding the leaf stalk firmly with the left hand, raise the end of the strip of wood with the point of the blade and then pull it out cleanly. All being well, the inner surface will show clean juicy green tissue, with the base of the dormant bud showing as a little projection in the centre. If instead of a projection there is a little hole, that means the bud has been dragged out and that portion of shoot is useless.

The buds are now inserted in cuts made in the bark of the stocks. To prepare the stock to receive a bud, make a T-shaped incision in the bark. Penetrate only the soft bark—avoid cutting into the hard wood beneath.

Gently lift the edges of the upright cut of the T with the handle of the budding knife. Then slip in the " shield " bearing the bud so that its inner surface lies snugly against the inner tissues of the stock. If necessary, trim the top of the shield so that it fits close against the top cut of the T.

Bind the bud in position with soft raffia. See that the whole of the cut is covered—leaving just the bud and leaf-stalk exposed—and tie firmly without tying too tightly. Allowance must be made for the subsequent swelling of the stock.

For bush roses, insert three buds low down on the stem of the stock—below ground level, if possible, soil being removed from the base of the stock to permit this. For standards, the top branch of the stock, as close as possible to the main stem, is the place selected for budding, this and any other branches on the stock being shortened for the purpose by one-third to a half. Insert one bud on each branch.

When the buds have " taken," the raffia should be removed. When the buds grow and produce shoots—they should flower the following season—support them with little canes tied to the branches to prevent them being knocked or blown out.

BULB GROWING. One of the most profitable lines the commercial gardener can take up is that of growing bulbs in pots and bowls. Christmas and shortly after is the time when sales are biggest and prices at their maximum ; but the trade continues until well into April.

Recommended Varieties for Forcing. Certain varieties are established favourites with the bulb-buying public. These varieties are as follows :

For the Christmas Market. White Roman hyacinths ; Cynthella hyacinths ; Paper White narcissus ; Duc Van Thol tulips ; iris *tingitana* ; crocuses in variety ; the choicer snowdrops.

For January. Tulips : De Wet, orange and scarlet ; Prince of Austria, red ; Rose Grisdelin, pink ; Frederick Moore, terracotta. Daffodils : Golden Spur ; Princeps ; Cervantes ; the yellow narcissus Soleil d'Or.

For February. Tulips : Pink Beauty ; the white La Reine ; Yellow Prince ; Peach Blossom ; the pink Murillo ; the lavender William Copeland ; the pink Princess Elizabeth ; the red William Pitt. Three fine daffodils for this month are : Glory of Leiden ; Empress ; Tresserve.

For March. Hyacinths come in now, sorts which always sell well being : Yellow Hammer ; L'Innocence (white) ; Enchantress (blue) ; Lord Balfour (mauve). Among cottage tulips : Artus (red) ; Diana (white) ; Chrysolora (yellow), are good sellers, and among Darwins : Loveliness (pink) ; Bartigon (red) ; Rev. H. Ewbank (mauve). Of daffodils, choose King Alfred, the smaller Sir Watkin and the white Mrs. Langtry.

For April. Choose any good late variety of hyacinth, also such fine daffodils as Van Waveren's Giant and the usual run of Darwin tulips.

Recommended Varieties for Cut Flowers. Favoured Daffodil market varieties are : Yellow Trumpets—King Alfred, Van Waveren's Giant, Emperor and Golden Spur ; White Trumpet sorts —Alice Knight, Madame de Graaf and Mrs. George H. Barr. Miss Ellen Terry is a popular bicolour—deep yellow trumpet with white perianth. In the Barrii narcissus group, first place is given to the old but still very popular *conspicuus*. Pheasant's Eye, another old favourite, in the Poeticus group, is also esteemed as an ideal cut flower.

The single sweet-scented jonquil, with its dainty heads of small golden-yellow flowers, is another wanted variety.

Among tulips, recommended varieties are :

Early singles : White Swan, Chrysolora, yellow ; Primrose Queen, primrose ; Van de Neer, purple ; McKinley, deep rose ; Vermilion Brilliant ; Prince of Austria, orange-scarlet ; and General de Wet, orange.

Darwins : Clara Butt, salmon-rose ; Farncombe Sanders, geranium scarlet ; Rev. H. Ewbank, soft heliotrope ; Zwanenburg, pure white ; Bartigon, fiery red ; and La Tulipe Noire, maroon-black.

May-flowering : Inglescombe Yellow, Orange King, Pride of Inglescombe, white ; Gesneriana *lutea*, deep yellow ; Rosabella, carmine-rose ; Inglescombe Scarlet, and John Ruskin, a mixture of apricot-orange shaded rose edged with soft lemon-yellow.

Parrot tulips : Cramoisie Brilliant, Fantasy and Sundew.

The bulbous iris has come very much to the fore as a spring cut-flower. These bulbous irises contain varieties known as English, Spanish and Dutch, all very beautiful. They provide a welcome succession to the daffodils, tulips, hyacinths, and other spring flowers.

While all the sorts are alike in style and colours, the Dutch irises flower a fortnight earlier than the Spanish type ; the English a fortnight *later* than the Spanish. The flowers of the Spanish iris are the smallest of the three. The English iris is the tallest and has the largest blooms.

The bulbs are planted in August or September.

Growing Bulbs in Bowls of Fibre. A bushel of peat or fibre will fill sixteen bowls, 7 in. in diameter.

Begin by lining the bowl with a 1-in. layer of lump charcoal. Then fill the lower half of the bowl with lightly-pressed fibre or peat. Place the bulbs in position on the fibre, at such a depth that their tops are level with the bowl rim. Then pack in more fibre or peat around them. Just the tips of the bulbs should be left exposed.

Water thoroughly after planting, and in half an hour tilt the bowls on one side to drain away surplus moisture.

The bulbs must be started into growth in darkness. The best plan is to place them in a sheltered—but not covered— corner of the garden and heap weathered ashes or sand over them, so that there is a layer 6 in. deep above them. Where neither ashes nor sand are available sifted soil will serve as a covering. Here they must remain until the shoots are showing ¼ in. above the fibre. This will be in from five to eight weeks, according to the kind and variety of bulb.

If facilities for covering the bowls with ashes are not available, the bulbs can be started in a fairly dry, airy cellar, or dark cupboard.

When the time comes to bring the bowls out of their dark quarters the pale shoots should be greened gradually by exposure to a dim light for three or four days.

Growing Bulbs in Pots. An ideal potting soil mixture consists of 3 parts loam, 1 part leaf-mould, 1 part well-decayed manure and 1 part sand, with a 5-in. pot of bone-meal to the barrowful of soil.

The bulbs should be planted so that their tips are level with the surface. Cynthella hyacinths should be planted three in a 6-in. pot ; multifloras, two in a 6-in. ; large-flowered, three in a 6-in. ; cottage and Darwin tulips, four in a 5-in. ; narcissi, five in a 6-in. ; daffodils, four in a 6-in. ; and grape hyacinths six in a 5-in. pot.

The pots of bulbs should be plunged in a bed of ashes to a depth of about 6 in. as advised above for bowl bulbs.

Growing Bulbs in Boxes. For cut bloom it is usual to grow the bulbs in boxes. To allow for vigorous rooting the boxes should be quite 6 in. deep. Place an inch layer of manure in the bottom of each box. For soil use a mixture of loam 4 parts, granulated peat, well-rotted manure and sand 1 part each. After planting plunge the boxes as already advised.

Outdoor Bulb Culture. The best month for planting all spring-flowering bulbs is October, but the planting season may be extended, if need be, to mid-December.

Very stiff loams and clayey soils will want additions of sand and leaf-mould. If the soil is on the poor and sandy side, a few barrow-loads of the chopped-up stuff from old marrow, cucumber, general hotbeds and mushroom beds, can be added with advantage.

The correct planting depth for hyacinths and daffodils is 5 in. ; for tulips, 4 in. ; for snowdrops, 3½ in. ; for scillas, 3 in. ; for crocuses, 2½ in. ; for grape hyacinths and chionodoxas, 2 in.

The space to allow between hyacinths is 9 in. ; daffodils and tulips, 6 in. ; crocuses, scillas, and grape hyacinths, 4 in. ; snowdrops and chionodoxas, 2 in.

Bulbs which have been grown in bowls or pots will not be suitable for such culture again the following year. They can, however, be planted outdoors. They should be placed in the greenhouse or frame and watered regularly until the weather becomes warm, then planted just as they are, without removing them from the fibre.

Growing Bulbs for Sale. There is a steadily-growing demand for British-grown bulbs for British gardens, in place of the imported bulbs hitherto used almost exclusively. Experiments carried out in various parts of the country—in Cornwall, Devon, Sussex, Middlesex, Norfolk, Lincolnshire and elsewhere—have disposed of the old idea that our soil is not suitable for the production of " seed " bulbs.

Sandy loam is the best soil for bulbs, especially for tulips, though daffodils will succeed in heavier ground.

CALCEOLARIAS. See Greenhouse Plants and Bedding Plants.

CALENDULAS. See Annuals.

CAMPANULAS. See Herbaceous Border Plants.

CARNATIONS. There is always a good demand for young carnation plants, while carnation blooms—especially the perfect specimens grown in the greenhouse—command high prices at all times.

Outdoor Carnations. To grow choice blooms it is necessary to provide an open, sunny situation and a light and well-drained soil.

The way to prepare the bed is first to dig the soil deeply. Old manure can be

added in small quantities if thought desirable, but it is not essential. Old mortar rubble or lime, however, must be put down as a top dressing after the ground is dug—also bonfire ashes.

If the soil is particularly heavy it is a good plan to raise the carnation bed a few inches above the general level of the ground.

When planting, only the roots should be buried—none of the stem. The soil should be pressed very firmly around the roots. The correct planting distance is from 8 in. to 10 in. apart. (Also see Disbudding.)

Greenhouse Carnations for Winter Flowering. The classes chiefly grown for this purpose are the Perpetual-flowering and Malmaison carnations.

The plants are potted up in spring, singly in 3-in. pots to start with, afterwards being transplanted to 5-in. pots, a compost of loam, 4 parts ; rotted manure, 1 part ; and silver sand, 1 part, with a sprinkling of mortar rubble or slaked lime being suitable. The plants grow on in a frame during the summer, any flower-buds that appear being picked off and stakes being inserted when plants need support. They should be brought into the greenhouse in September and grown on into flower under cool (50-55 degrees) conditions. The plants should be stopped several times during growth. The first stopping should be given when the plant has started to go ahead after its first potting and subsequent stoppings as the branch growths appear. ·

To Feed Carnations. Ordinary liquid manure suits them very well but best of all fertilisers is Guano, distributed at the rate of ½ oz. per square yard every ten days during the flowering period and watered in.

Propagating Carnations. Carnations can be propagated equally well from seeds, cuttings and layers, layering, however, being the method usually adopted.

Layering. This is usually done in July or August.

First loosen the soil around the plant to be layered, and then put 3 in. of fine, sandy soil on the top. Next take the first outside growth intended for layering (a strong, healthy shoot which is not carrying a bloom or a bud) and strip off the leaves from the bottom up to the fourth pair from the top of the growth. Then cut a " tongue " in the stem of the growth. Make a clean cut only half-way through the joint, which is the lowest of those from which the leaves have been removed, and then, with a turn of the knife, continue in an upward direction as far as the next joint, thus forming the tongue.

An important thing to do now is either to cut off the slither of stem that has been separated from the main part of the stem, cutting it level just below the top joint to which it runs, or to keep the " tongue " open by inserting a match-stick. Both ways aid and quicken rooting.

The layer prepared, bend the growth and press it down until the cut surface of the stem, or the complete " tongue," is inserted about ½ in. deep into the fine soil. Pin it securely down into position with a small peg and also make the soil firm, so that the stem will not rise.

As many layers can be treated thus as there are suitable shoots or room for pegging them down.

Roots will form on the layers in from three to five weeks' time. Then sever the layers from the parent plant by cutting through the stem that joins them, taking care to cut well away from the point where the roots will be. The layers may be lifted for potting or planting elsewhere in the garden a week (no less) after they have been isolated from the parent.

Pot carnations can be similarly layered by pegging down their shoots into other pots, or, alternatively, by plunging the plant's pot to the rim in fine soil in a cold frame and pegging down the shoots all around it.

Raising Carnations from Seed. The trouble with this method is that named varieties cannot be obtained by it ; the plants that are produced will flower in all colours. None the less, seed-sowing is a simple and cheap means of obtaining a stock of plants. Sow in boxes in a heated greenhouse or hot-bed frame in April, pot the seedlings when large enough to handle, transfer them to a cold frame when they are growing nicely and in summer stand them out of doors or, in the case of plants destined for pot culture, in a cool, steady frame. Outdoor plants can be planted out in their flowering-quarters in September.

Hardy border carnations can also be sown outdoors in July.

Taking Carnation Cuttings. Carnations can be readily grown from cuttings. These should be taken in late June or early July. Three-inch long tips of young shoots make the best cuttings.

Prepare and root the cuttings in the ordinary way, preferably in a propagating-case.

CATERPILLARS. See under this heading in Vegetable Section.

CHESHUNT COMPOUND. Valuable remedy against damping off in seedlings. See Damping off.

CHRYSANTHEMUMS. Chrysanthemum blooms are always good sellers, especially the early outdoor varieties, the early greenhouse varieties and the late varieties. There are scores of excellent varieties suitable for commercial greenhouse work. The following are the pick of them. :

Nov.-Dec. Bacchus, scarlet; Coralie, lilac-white; Red Coralie, bronzy-red; Annie Currie, snowy white; Ondine, old ivory; Carisbrook, rich yellow.

Dec.-Jan. Gladys Payne, buff pink; Gracie Fields, chamois rose; Jane Ingamells, canary yellow; American Beauty, white; Friendly Rival, deep buttercup yellow; Colham Pink.

Outdoor Chrysanthemums. The best time to plant is May or June, according to whether the district is warm or cold. An open position is preferred and the soil should be in good condition following winter digging and manuring. Allow 2½ to 3 ft. of space per plant. The flowers will be much larger and better coloured if the plants are generously fed from the time they are nearing maturity ; then they should have weekly doses of weak liquid manure—one week ordinary liquid manure, another week soot water, then a mixture of ½ oz. guano in 1 gallon water and then back to ordinary liquid manure.

When the buds are beginning to show give weekly ½ oz. of the following mixture : superphosphate of lime, 6 parts ; sulphate of potash and sulphate of magnesia, 1 part each.

Buds are sometimes produced in hundreds and may well be reduced to from three to half a dozen per shoot.

Propagating Border Chrysanthemums. Plants may be divided in autumn or spring. Better flowers are obtained, however, when a fresh batch of plants is grown each year from cuttings. Clumps are lifted from the border after flowering is finished in autumn and are planted in moderately rich soil in a cold frame. In the early spring shoots appear which are taken off, potted up and rooted in the greenhouse in the same way as advised below with cuttings of greenhouse chrysanthemums.

Greenhouse Chrysanthemums. Taking Cuttings. Cuttings are best taken in December or January.

Stem cuttings do not make successful plants. Sucker shoots should therefore be chosen for cuttings, the best being those of medium strength—not too weak and not too fat and gross. These should be cut off at soil level so that they are from 3 in. to 4 in. long, and prepared in the ordinary way.

The soil for rooting the cuttings should consist of equal parts fibrous loam (chopped up and passed through a ¼ in. sieve), and beech or oak leaf-mould, rubbed through a ¼-in. sieve. To this should be added one-twelfth of the whole bulk of clean, sharp sand.

The cuttings should be inserted singly in " thumb " pots, being planted so that the lower leaves are just clear of the soil surface. Alternatively four cuttings may be planted in a 3½-in. pot, or cuttings may be spaced out 2 in. apart in 2½-in. deep seed-boxes. The cuttings should preferably be rooted in a propagating-case in the greenhouse.

As the cuttings grow and form a mass of roots they should be potted on into larger pots.

The first potting is required in March, a 3-in. pot being used ; the next will be due about mid-April, a 5-in. pot being used this time. The final potting is due some time in June, an 8-in. or 10-in. pot now being used. Here the plants remain for flowering.

Stopping Greenhouse Chrysanthemums. All varieties of chrysanthemums have two periods in which the main stems break into side-growths, each to give three or four side-shoots, instead of continuing the vertical growth of the original shoot. In both cases a flower-bud, called a *crown* bud, lies in the middle of the ring of growth shoots. If the growth shoots are allowed to remain, this crown bud either dies away completely later on or only produces a poor bloom. By stopping the growth of the main stem a few weeks before the break would naturally occur, an artificial break is caused, and this means the saving of time.

By saving time in this way it is possible to induce naturally late-flowering varieties to bloom earlier than they otherwise would do.

As to *when* to stop, if it is desired to have blooms towards the end of November —as is usual—stop from the middle to the end of April.

The stopping consists of pinching off not more than ½ in. of the soft top of the shoot.

Some varieties do best if allowed to grow naturally, without any pinching at all. The best-known instances of such varieties are Dawn of Day, Julia, Golden Glory, Mrs. Algernon Davis, Mrs. B. Carpenter, Edith Cavell, and Peace.

In no case should the stopping be done immediately after re-potting. At least a fortnight should elapse between one operation and another.

Greenhouse Chrysanthemums During Summer. Chrysanthemums should spend the period from mid-May until the autumn outdoors in a position which gets full sunshine. The pots should either stand on boards on a gravel path or on a bed of ashes, the latter being better. They should be in long lines. At the ends of each line six foot posts should be driven into the ground and a stout wire or cord run taut from one to the other.

When the plants are approaching 18 in. high each should be given a 4-ft. cane,

and the top of each cane should be tied securely to the wire above it.

As the plants grow they will be noticed to be producing side-shoots all the way up the stem ; these must be regularly nicked with the point of a sharp knife.

The big-bloom plants must be watched carefully for the " break " (branching of growths), due about the middle of July. Up to that time the plants have been carrying three main shoots each (after the April stopping). Each of these will produce several branches and the job is gradually to reduce these to one, so that the plant (after the July break) carries three shoots as before.

Taking the Bud. This is a highly important operation. It takes place from mid-August to mid-September, according to variety. (The time to take the bud of the different varieties is set out in most of the catalogues issued by chrysanthemum nursery-men.)

Examining an average plant in August it will be noted either that there is a bud at the top of each main shoot surrounded by little growth shoots, or there is a bud which has two smaller buds close beneath it. (There are usually two of these buds, but there *may* be more.)

In the first instance—where the bud is surrounded by little growth shoots—taking the bud consists of removing the small shoots so that the bud " stands alone " and nothing is left growing on the stem but the leaves of the bud. The small shoots must be taken off very carefully, to avoid damaging the bud or the stem, and the operation must be done gradually.

In the second case—where the main bud has smaller buds beneath it—taking the bud consists of removing these smaller buds.

On a two- or three-stemmed plant the buds may form at different times. Do not " take " them as they form, or the flowers will open on different dates. Equalise development by allowing three side-shoots to form near the top of the stem bearing the forward bud. This will slow down growth, and enable the backward bud or buds to catch up.

Feeding Greenhouse Chrysanthemums. The feeding programme commences in mid-August. One of the best-known foods to use consists of : 4 lb. of cow dung (enclosed in a canvas bag), 1 oz. of nitrate of soda, 1 oz. of nitrate of lime, and ½ oz. of sulphate of potash. Place the ingredients separately in 8 gallons of water. Allow the liquid to stand for three days before use, stirring well each day. Dilute with an equal quantity of clear water. Feed once a week for four weeks.

If it is inconvenient to prepare this food, give the following mixture of artificials weekly—½ teaspoonful per plant : 3 parts superphosphate of lime, 1 part each of sulphate of ammonia and sulphate of potash.

In the fifth week feed twice with diluted soot water. From then until colour shows in the bud, use special chrysanthemum manure once a week ($\frac{1}{2}$ teaspoonful per plant).

CINERARIAS. See Greenhouse Plants.

COCKCHAFER GRUB. Greyish-white caterpillar-like grub frequently turned up in the soil when digging. An enemy of the rose and many flowers. All grubs found should be destroyed.

COREOPSIS. See Herbaceous Border Plants.

CUCKOO SPIT. The cuckoo spit, or spittle-fly, is known to all by its habit of forming blobs of froth on various plants. In the blobs of froth lives a large green insect with a frog-like face. These insects can best be destroyed by vigorous syringing with a pyrethrum wash. Dissolve 3 oz. of pyrethrum powder in 1 gallon of water, and syringe the whole of the plant.

CUTTING FLOWERS. Blooms should preferably be cut early in the morning. If they must be cut at some other time of day it should be when the sun is not shining on them.

Never bunch flowers together if they are wet with dew or rain. Shake them carefully to remove as much moisture from the blooms as possible.

The only flowers which should be *pulled* when gathering are, generally speaking, bulbous ones, such as daffodils, lily-of-the-valley, cyclamen and so on. All others should be cut with a knife or sharp scissors. The cut should be made in a slanting direction, especially when the stems are woody, as with pæonies and chrysanthemums.

When it is necessary to cut long stalks for decorative purposes, immerse the long stalks deeply in water. Take care to remove any leaves that would be under water.

After gathering such flowers as poppies, geums, pentstemons and others inclined to droop the moment they are cut, dip the stalks in boiling water or hold the ends in a candle flame. This will seal the cuts and help them to live longer. After cutting woody-stemmed flowers slit up the bottom of the stalk for an inch or two.

Cut Shirley poppies and eschscholtzias before they have burst their green caps. Try to "catch" the poppy buds when they are whitish and pointing straight upwards; the eschscholtzias when they are showing colour through the "cap".

CUTTINGS, ROOTING. See Bedding Plants and separate entries; also see Hormones.

CUTTINGS, TYPES OF. There are several kinds of cuttings, the commonest being the stem cutting.

Stem Cuttings. These are taken when the plant is in full growth. Select ripe-looking, fairly-firm shoots. When preparing for insertion remove flowers and flower buds, and any bottom leaves that would rest on the soil. Finally, shave the stem across with a sharp knife immediately beneath the bottom joint. Geraniums, calceolarias, fuchsias, violas, pansies, ageratums, lobelias and pentstemons strike very easily from stem cuttings.

Leafless Stem Cuttings. Privet, golden elders, mock oranges, and other shrubby plants are propagated by leafless stem cuttings, which are best taken as soon as the leaves fall. Each cutting should be 12 in. long when prepared and must be taken from growth of the current year. After cutting off an inch of the tip, shaving beneath the bottom joint and rubbing off the leaf buds on the bottom two-thirds of stem, plant the cuttings to half their depth in a sheltered bed of sandy soil.

Heeled Cuttings. Roses, lilacs, buddleias, brooms, escallonias, lavender, cupressus, heliotrope and many other plants, strike best from heeled cuttings. In taking these you pull off a shoot with a " heel " or small piece of the parent stem attached. Prepare the cutting for insertion by just trimming off loose shreds of bark on the heel. Insert evergreens in April, those which lose their leaves, in autumn.

Root Cuttings. Many favourite border plants, like oriental poppies, pæonies, statice and anchusas increase best from little portions of roots, known as root cuttings. Autumn and early spring are the best times for inserting them. Roots about the thickness of the forefinger, and 3 in. long, form excellent material. Shave them cleanly across both top and bottom, plant in a sandy, sheltered border outdoors.

CYCLAMEN. See Greenhouse Plants.

***

DAFFODILS. See Bulb Growing.

DAHLIAS. There is a great variety of dahlias, and apart from the fact that plants are required for every garden from a decorative point of view, there is also a fair demand for them as cut flowers in the market. Young dahlia plants, grown from seeds or cuttings, are in good demand in spring.

The most suitable classes to grow for market are the Charm, the Pæony-flowered, the Pompon, the Collarette and the Single. •

Dahlias are grown from cuttings, seeds, or root divisions.

Taking Dahlia Cuttings. To obtain a supply of shoots suitable for use as cuttings the roots must be started into growth in a heated greenhouse in late January or early February. The roots are planted closely together in boxes containing a soil-mixture of equal parts loam, leaf-mould and decayed manure, all passed through a ¾-in. sieve. Shoots will appear in a few weeks and when these are 3 in. long they are suitable for cuttings. Each cutting should be detached with a sharp knife, being taken away with a tiny bit of the parent tuber attached.

The cuttings should be planted singly in thumb pots containing equal parts loam, leaf-mould and sand, passed through a ½-in. sieve. A ¼-in. layer of sand should be spread on the surface of the soil.

The cuttings should be kept in a warm position, preferably in a propagating-case, until they have rooted.

Propagating Dahlias by Root Division. If no facilities exist for producing and rooting cuttings, dahlia tubers can either be replanted as they are, in May, or divided up in March.

Every one of the tuberous roots will make a good plant. As, however, the buds are all clustered on and round the thin neck where it joins the stem, cut away a piece of the connecting stem with each.

It is better to pot up these divisions than to box them up. If necessary, cut them to half length, throwing the bottom half away.

Stand the pots, after planting, closely together in a hot-bed frame or greenhouse.

Raising Dahlias from Seed. Autumn-flowering dahlias can, if desired, be raised from seed sown in the greenhouse in January. One gets rather a mixed collection by this means, however, so that the other methods of propagation are recommended.

Bedding dahlias are on a different footing. It is quite in order to grow these from seed. Sow in a hot-bed frame or a heated greenhouse, in January. Use a soil-mixture of loam, 3 parts; leaf-mould, 1 part; sand, 1 part, all passed through a ½-in. sieve. The seedlings should be transferred singly to thumb pots. In April they should be transferred to a cold frame to be hardened off for planting out in mid-May.

Planting Dahlias. Dahlias like very rich soil. The sites where they are to be planted should be prepared in early spring,

some time before planting is to be done. The soil should be dug thoroughly, decayed manure being mixed in about 1 ft. below the surface. When the digging is done, over the soil should be scattered a mixture of equal parts soot, wood ashes and bone-meal, 2 oz. per square yard, raked in.

The plants must not be put in until late May or early June.

Ample space should be allowed the plants, taller kinds going 5 ft. apart, medium 3 ft. to 4 ft. and bedding kinds 18 in. apart. For the first week it is advisable to cover the plants with an inverted pot at night.

Care of Dahlias During Growth. Directly the plants are beginning to grow all but the bedding kinds should be staked. Most growers nowadays limit the shoots when a large number is produced. Three or four shoots are sufficient for any plant, the remainder being cut off. As the plants grow they will make a number of side-shoots ; these also should be limited to a reasonable number, say three or four per stem.

Each shoot may produce several buds. For good quality flowers limit the buds to one per shoot, picking off the others in infancy.

Feeding Dahlias. Feed the plants fortnightly with liquid manure during the summer, alternated with soot water. From the beginning of September give each plant ½ oz. of ichthemic guano every eight days, and soot water half-way between.

Storing Dahlia Roots in Winter. The best programme to follow is : (1) Plants to be cut down to within about 6 in. of the ground two or three days after the foliage has been seared by frost. (2) Lifting of the roots, intact, to be done five to seven days after the cutting down. All soil to be washed off and the roots left exposed to the open air for a few hours. (3) The roots to be placed, points or " fangs " downwards, on the floor of any airy shed for a week or ten days. (4) The dried roots to be packed into boxes, covered with sand or dry soil and stored in a cool, moderately moist and dark place for the winter.

DAMPING-OFF. Fungus disease which attacks seedlings and young plants in pots and boxes, causing them to wilt, rot, and die. A spraying with what is known as Cheshunt Compound will prevent attacks of the disease. For recipe, see Cheshunt Compound in Vegetable Section. Another useful preventive is to dust powdered charcoal around the plants.

DELPHINIUMS. See Herbaceous Border Plants.

DIGGING. For full details of the various ways of digging soil and the best methods of dealing with particular types of soils, see Digging and Manuring in Vegetable Section.

DISBUDDING. The following are among the popular flowers which can be improved by disbudding, with the method to follow :

Border Carnations and Picotees. Allow each growth to carry one flower. This means the removal of the two side buds at the apex of the stem, and any buds which are thrown out from the joints.

Sweet Peas. Assuming that the growths are strong, allow two flower stems to develop at the same time. Remove surplus buds when they are at the " pot-hook " stage.

Delphiniums. Take out the lateral flowering stems of all varieties that make branched spikes.

Begonias. Double-flowered tuberous begonias often produce female buds at the side of the double, or male flower. To enable the double flower to reach maximum size, these female buds should be pinched off as soon as they appear. They can easily be distinguished by the presence of the small triangular seed vessel immediately behind the embryo petals of the bloom.

Roses, Chrysanthemums, etc. The disbudding of these is dealt with under the respective headings.

Also remove the short, weedy spikes called shoulders, at the base of the main spikes of verbascums, lupins and aconitums ; the small buds that appear a few inches below the central buds of oriental poppies, pæonies, anemones, heleniums, rudbeckias and eryngiums ; the weak spikes that arise at the base of campanulas *pyramidalis* and *persicifolia.* Remove the unwanted buds (while quite young) with the finger and thumb or penknife point.

Most annuals, too, benefit from disbudding. See Annuals.

DISEASES OF PLANTS. See Fungicides.

EARTHWORMS. In the garden earthworms are the friends of the gardener; they drain and aerate the soil. When, however, they invade flower pots they are a nuisance and must be driven out. Simple measures are to water the soil in the pots with lime water or a weak mustard and water solution. Sometimes a sharp rap on the outside of the pot will bring the worms to the surface, when they can be picked off.

G*

EARWIGS. Trapping is the simplest method of dealing with these pests. A small roll of corrugated cardboard, capped with a small tin lid to keep out the wet makes the best trap. These rolls can be be tied to short stakes set out among the plants.

Half-open match boxes hung among the plants, and the familiar small pots half-filled with hay and placed on stakes are other useful traps. Less familiar, but very effective, is to lay pieces of damp flannel on the ground near the plants. The pests congregate under these to enjoy the cool conditions they provide in hot weather.

EELWORMS. Small pests that burrow into plant stems and tubers. For remedies, see Eelworms in Vegetable Section.

ERICA. See Greenhouse Plants.

FERNS. Ferns in variety are an excellent market line. They are always in demand.

Varieties of Ferns. Outdoor varieties. Athyrium *corybiferum* ; Blechnum *spicant cristatum* ; Adiantum *capillus veneris* (the hardy maidenhair) ; Scolopendrium *crispum grande* ; Scolopendrium *nanum cristatum* ; Woodsia *alpina* ; Lastrea *Felix-mas cristata*; Lastrea *dilatata lepidota*; Osmunda *regalis* (the Royal Fern) ; Polypodium *calcareum* ; Polypodium *phegopteris* ; Aspleniums *adiantum nigrum, ruta-muraria, viride* ; Scolopendrium *vulgare* (the Harts-tongue) ; Nephrodium *montanum* (Mountain Buckler) ; Polystichum *angulare* (the Soft Prickly Shield Fern) ; Polypodium *vulgare* ; Cystopteris *alpina* ; Hymenophyllum.

Greenhouse varieties. Of maidenhairs, the favoured varieties are Adiantum *elegans, elegantissimum, tinctum, Williamsi* and *Capillis Veneris imbricatum.* Of the Pteris ferns the favoured varieties are Pteris *Childsi,* which grows a foot high; P. *tremula* with 2 ft. long fronds, P. and *serrulata.*

Other very suitable kinds are the Aspleniums *Colensoi* and *Hilli* ; Davallias *canariensis* and *tenuifolia stricta* ; Osmunda *palustris,* and Polypodium *Mayi.*

Propagating Ferns. Division. Most kinds of ferns may be increased by root divisions in spring.

The most convenient way is to thrust two forks into the clump, back to back, and then lever them apart.

A rare and choice variety may well be divided into single crowns, for the purpose of working up a good stock. When the replanted ferns are expected to make a fine show the following summer, the clumps should each have three or four crowns.

Other Methods. Ferns can also be propagated by "seeds" or by "planting" the diminutive buds and plantlets produced on the fronds.

The "seeds," or spores, are found beneath the small, usually brown "scales" on the backs of the fronds. They are sown in October as though they were ordinary seed, in pots filled with good soil and surfaced with a trifle of finely powdered charcoal, a thorough watering being given *before* the spores are sown. Sowing completed, the pots are stood in saucers of water under a handlight or in a glass-topped box on the green-house stage.

Germination may be a matter of days or months. The first growth to come up has the appearance of a green scale lying flat on the soil. The green scales vary in size, ranging round about ⅜ in. across. They should be moved from the pots and dibbled in, ½ in. apart, in other pots filled with gritty and fairly moist soil. The transplanted green scales give rise in due course to typical fern fronds.

As the young plants crowd one another they must be transplanted into small pots, the most suitable compost for many ferns being loam, leaf-mould, rotted manure and sand.

Fern propagation by means of rooting the buds and plantlets that break out from the fronds is carried out during October. One or more tiny fronds may be produced by the small outgrowth before the young fern falls away from its parent frond. The buds or plantlets that fall may be dibbled into small pots of soil, further growth and rooting being encouraged to proceed apace by placing the pots in a propagating-case for three weeks or so.

Where the outgrowths remain attached, the parent frond, or that part which carries the offspring, can be pegged down into a pot of soil placed alongside, until the young fern is sufficiently rooted to allow of its severance. A variation on this method of producing offspring is noticeable in Nephrolepis *cordifolia tuberosa* and others, small tubers appearing on the roots. These may be detached in autumn and stored in damp sand and sphagnum moss, until signs of growth appear in early spring.

Cutting Back Ferns. Ferns whose foliage dies down for the winter should have all brown fronds clipped back to a level with the pot. Fronds that go off during the summer should also be promptly removed.

Feeding Ferns. Occasional doses of weak liquid manure are beneficial to ferns once they are growing strongly.

FLOWER POTS. The following are the sizes of the standard flower pots :

Name of Pot.	Diameter of Top.	Depth.
Thimbles	2 in.	2 in.
Thumbs	2½ ,,	2½ ,,
60's (without rim) ..	3 ,,	3½ ,,
Large 60's (with rim)	3¾ ,,	4 ,,
54's	4 ,,	4 ,,
48's	4½ ,,	5 ,,
32's	6 ,,	6 ,,
24's	8½ ,,	8 ,,
16's	9½ ,,	9 ,,
12's	11½ ,,	10 ,,
8's	12 ,,	11 ,,
6's	13 ,,	12 ,,
4's	15 ,,	13 ,,
2's	18 ,,	14 ,,

To Prepare New Pots for Use. Before new flower pots are used the "fire" should be taken out of them. This is done by soaking the pots in water until all air-bubbles cease to rise.

The shortage of clay pots during the war resulted in the increasing use of composition pots. Some of these pots are made of strawboard and similar material which rots down readily in the soil. Thus tomatoes and other plants can be planted just as they are in the pots and there is no root disturbance.

FOLIAGE PLANTS. The following are among the most useful and popular plants for providing foliage and for mixing with cut flowers:

Asparagus, Ornamental. The so-called asparagus fern is raised from seed sown during summer, the plants being kept under glass.

Grasses, Ornamental. These are receiving increasing attention from market growers for their use as foliage with cut flowers. Produced from seeds sown in May. Some of the best for decorative purposes are Apera *arundinacea*, which has arching plumes of a bronze-purple colour ; Avena *elatior* with long, slender stems and light panicles ; Agrostis *spica-venti*, or Bent Grass ; Briza *maxima*, with pretty drooping heads ; Briza *media*, the purple-plumed Quaking Grass ; Eragrostis *elegans*, or Love Grass, a very pretty type often used to mix with sweet peas ; and Hordeum *jubatum*, a handsome Barley Grass with a pronounced beard.

Gypsophila. There are annual and perennial kinds. The perennial class is gypsophila *paniculata*, which is raised from seed sown in heat early in the year. The plants are set out in June. The annuals are sown outdoors in April and transplanted in June.

Maidenhair Fern. See Ferns.

Smilax. Propagated by suckers from the roots, being struck in a little heat.

FREESIAS. See Greenhouse Plants.

FUMIGATION. Soil beds, etc. For methods of fumigating, see under Soil Fumigation in Vegetable Section.

Greenhouse Fumigation. There are several methods available: One is to burn liquid nicotine compound. For this a special apparatus—consisting of a methylated spirit lamp and metal saucer—is required. Another method is to burn fumigating cones made of nicotine compound which burn slowly and give off a dense smoke when lighted. A third method is to burn tobacco shreds—coarse shreds of strong tobacco. These latter are very simple to use. Placed in a flower-pot and lighted, they will burn away, without attention, to ash. Commercial calcium cyanide is widely used for greenhouse fumigation. On exposure to the air it gives off hydrocyanic acid gas. The calcium cyanide treatment must be carried out with great care. Briefly the method is to sprinkle the chemical along the paths of the greenhouse in the evening. The dosage varies between $\frac{1}{8}$ oz. and $\frac{1}{4}$ oz. per 1,000 cubic ft. according to the nature of the pests present and the stage of growth of the plants in the house.

Points to bear in mind are to fumigate in the evening, to close all ventilators, to see that any breakages in the glass are covered with sacking, to close the door tightly as soon as you leave the greenhouse and allow it to remain closed until the following morning.

It is important that the foliage of the plants be dry and that the house be not too much charged with moisture when fumigating. If there are ferns bearing new fronds it will be better to remove them or to put them on the floor, where they will not get the full strength of the fumes.

After fumigation open the ventilators early the following morning.

FUNGICIDES. In some instances the best treatment against diseased plants is to destroy the plants by fire—as when Antirrhinum rust is prevalent. In general, however, it is to be assumed that disease is noticed in its early stages, when curative treatment may be applied.

Fungicides commonly used are flowers of sulphur, liver of sulphur, Cheshunt compound, permanganate of potash, Bordeaux mixture, ammonium polysulphide, and a considerable number of reliable proprietary preparations.

Flowers of Sulphur. A mixture of 2 parts flowers of sulphur with 1 part freshly slaked lime is used against such troubles as chrysanthemum mildew, delphinium mildew, downy mildew of roses and sweet pea mildew.

Liver of Sulphur. At a strength of about 1 oz. in 3 gal. water is used

extensively against rusts of carnations, hollyhock rust, leaf-spot of mignonette, rose rust, sweet pea rust, viola and violet rust, and wilt of sweet williams.

Cheshunt Compound. This may be made of 2 parts best copper sulphate and 11 parts ammonium carbonate, mixed in a corked glass or stone jar for 24 hours. This is used at the rate of 1 oz. in 2 gals. of water. It is used to treat wilt of asters, foot-rot of antirrhinums, and damping-off diseases generally.

Permanganate of Potash. Sufficient crystals are used to make a deep rose-coloured solution in water. This may be used against leaf-spot of dahlias, and as a general disinfectant for greenhouses, pot washing and soil treatment.

Bordeaux Mixture. This is best bought ready prepared and used as directed; but it may be made in a *wooden* vessel by dissolving 1 lb. of copper sulphate in a little water, and 1 lb. quicklime in more water—then pouring the latter slowly into the former, making up to 10 gallons. It is used against Botrytis disease of lilies, rust of pæony, black-spot of rose, sweet pea spot, violet smut and other diseases.

Ammonium Polysulphide. This material at a strength of 1 part in 50 of water, is employed against leaf scorch of chrysanthemums and roses, leaf spot of violet and gladiolas, sweet pea blight, white and black moulds of violets and mildew of wallflowers. Also see Insecticide Recipes.

GAILLARDIA. See Herbaceous Border Plants.

GERANIUMS. See Bedding Plants.

GEUM. See Herbaceous Border Plants.

GLADIOLUS. The gladiolus family is now one of the most popular of all flower families. Gardeners need gladioli to provide colour during summer and autumn, and gladioli, too, are gaining enormously in popularity as cut flowers. Great numbers of bunches are sent to market every season.

There are three groups of gladioli, the large-flowered (growing 2 ft. to 2½ ft. tall), the Primulinus (growing 1½ ft. to 2 ft. tall) with flowers only half the size of the former but rather daintier, and the Intermediate coming between the two others in size and height. The Primulinus and the Intermediate are best for cut-flowers.

Planting Gladioli. Gladioli do best in a sunny position in well-drained, fairly rich soil.

For the general summer display, planting is done from mid-March to April. For early flowering outdoors planting is done in November or February.

Large-flowered varieties must be spaced 9 in. apart, Primulinus hybrids 6 in. apart, and Intermediates 6 in. apart.

In heavy soil, plant large-flowered sorts 3½ in. deep ; in light soil, 4 in. The other plants 3½ in. in light soil and 3 in. in heavy.

Make the hole for each bulb 4 in. in diameter, and 1 in. deeper than the given planting depths. Put 1 in. bed of sand in the hole, set the bulb firmly on it and cover it completely with sand. If the ground is very heavy, fill up with following mixture of prepared soil : loam 2 parts, sand 2 parts, well-rotted manure 1 part, and leaf-mould 1 part. Should the ground be very light, fill up the hole to within 1 in. of the top with riddled leaf-mould. Fill up the top inch with soil.

Growing Gladioli in the Greenhouse. There is a section of gladioli known as the Colvillei section. Varieties in this section are suitable for potting up for early flowering in the greenhouse.

For potting mixture use : loam, 3 parts ; leaf-mould, 1 part ; well-rotted manure, 1 part ; and silver sand, 1 part, passed through a ½-in. sieve. Pot three to five bulbs equidistant in a 5-in. pot.

Stand the pots in a cold frame to start with, but when the shoots are an inch long, take them into a cool greenhouse where the plants will flower. Weekly waterings with dilute soot-water after the flower-sheaths show will immensely improve colour, size and quality.

Propagating Gladioli. Gladioli can be grown from seeds sown on the surface of pans of rich soil in a warm greenhouse in February. The plants form little bulbs which are allowed to grow on for three years, by which time they will have reached flowering size and may be treated as advised above.

An alternative method is to watch for and collect the small " bulbils " which are often found attached to the base of the big bulbs—or in the soil—when harvesting them. Store these little " bulbils " in sand as found. At the end of October " sow " them in small trenches 3 in. deep. Leave them till autumn and afterwards treat as ordinary corms. They will reach flowering size in two years.

GLOXINIA. See Greenhouse Plants.

GRASSES, ORNAMENTAL. See Foliage Plants.

GREENFLY. The best of all remedies for this troublesome sap-sucking pest is to spray plants likely to be attacked (notably roses) with soapy water—say 2 oz.

soft soap to the gallon of water—regularly every fortnight from the time growth starts until greenfly risks have ceased.

When greenfly have actually attacked plants in numbers an insecticide must be used. The proprietary brands are preferable, but for those who desire to use home-made mixtures one to be thoroughly recommended is nicotine wash.

Quassia and soft soap is another fine preparation for use against greenfly. The recipes for these sprays are given under separate headings in the Vegetable Section.

Whatever is used, repeat the spraying a few days after the first application. This second spraying is essential if a thorough clearance of the pests is to be made—to deal with the next generation of greenflies hatched out after the first spraying. Also see Insecticides.

GRASSES, ORNAMENTAL. See Foliage Plants.

GREENHOUSE MANAGEMENT. A greenhouse is a valuable possession for the ordinary gardener, indispensable to the man who grows plants and flowers for sale.

An unheated greenhouse enables the gardener to grow many choice plants and do much plant propagation and plant forcing which would otherwise be denied him. Heated, it enables him to grow practically any and every plant known to horticulture. The foregoing, of course, refers to normal times. War time restrictions prohibit the use of commercial greenhouses for flower-growing. Most amateurs in war time similarly concentrate on food-growing in greenhouses. Fuel restrictions would not justify the use of coal, etc., for flower-growing. The following information will be of value to those starting or renewing greenhouse flower-growing in peace time.

Types of Greenhouses. The span-roof type is the better as it is lighter and offers more shelf and bench space, but the lean-to type is cheaper and, going against an existing wall, does not take up so much room in the garden.

It is better to buy either type of house in section form from one of the well-known makers rather than to buy the materials separately, and endeavour to make the complete building oneself.

Heating Greenhouses. Many types of greenhouse heater at prices to suit all purses are available to-day. The fire and boiler apparatus, burning coke or coal, is the one most extensively used.

Where the stoking necessary with a coke or coal stove might prove a difficulty, a gas boiler is an alternative. Gas apparatus is very clean and simple to work, and the boilers are small and easy to fix up. It is probable that the cost of gas would exceed that of coke fuel.

Even more modern, electric greenhouse heaters can be obtained. These are highly efficient and, in districts where electricity is cheap, economical to run.

Another means of heating is provided by the specially made portable oil heaters fitted with a hot-water system. They are very handy in small greenhouses, being perfectly free from smoke and smell, requiring no chimneys, flues or fixtures, and holding sufficient paraffin oil to burn from sixteen to twenty-four hours.

It is unwise to use an ordinary household oil heater. This gives too dry a heat for plants, is liable to produce fumes and, above all, it may smoke—and one hour of smoking will seriously harm, if not kill, probably all the plants in the greenhouse.

Managing the Greenhouse Fire. In frosty weather. Start the fire going at 2.30 to 3 p.m., and keep it steadily running until nightfall to get the pipes comfortably warm. If the threat of moderate frost continues keep the fire going until 9 p.m. Bank up—that is, regulate the dampers to give slow but clean combustion—at 9 p.m.

If very sharp frost threatens defer the banking up until 10 p.m. Next morning clean out and restart the fire to run for an hour or so at 7 to 7.30 a.m. for very frosty spells ; at 8 a.m. for moderate frosts.

In mild weather. Start the fire (after a clean out) at 3 p.m. Check it at nightfall and bank up at 8.30 or 9 p.m.

To Prevent Frost Damage in Unheated Greenhouses. Where only a little heat can be introduced, much can be done in other directions to prevent plants from being frost-bitten. One of these helps is to cover the plants with several thicknesses of newspaper, which may be removed when the frost has gone. To keep the frost from special plants put them on the floor where they will be farther from the glass.

Where it is possible to put mats on the roof of the greenhouse, this will be a great aid. Where blinds are used in summer for shading, these may also be utilised against frost.

Shutting up the greenhouse early in the afternoon bottles up a certain amount of sunheat.

Winter Temperature Control. October, November, December and January : day, 48-50 degs. ; night, 45-48 degs. February : day, 48-52 degs. ; night, 46-48 degs. Early March to mid-March : day, 52-54 degs. ; night, 48-50 degs. Mid-March to end March : day, 54-60 degs. ; night, 50-52 degs.

Winter Ventilation Control. If plants are denied fresh air, sickly, yellow, flimsy growth is the result. Ventilation

must be given by both top and bottom ventilators, but always so as not to cause cold draughts. Only on very cold nights, and when fog reigns, should a greenhouse be closely shut. At all other times at least a " chink " of air, preferably from the top ventilators, should be given. Take advantage of all warm, or sunny spells to increase the allowance of air.

Winter Plant Watering. All watering should be done in the late forenoon—from 11 a.m. to 12 noon from November to mid-February, from 11 a.m. to 2 p.m. from mid-February to mid-March, and from 11 a.m. to 4 p.m. from mid-March to the end of that month. No water must be slopped about on the floor of the greenhouse, although, in warm houses (temperature 56 degs. F. to 60 degs. F. by night), foliage plants may be syringed almost daily in the late forenoon.

To Prevent Fog Damage. One of the best ways to keep the greenhouse free of fog is to use ammonia-water—2 teaspoonfuls of ammonia to the pint of water. One jar of ammonia-water will do in the small house ; two or three may be stood about in the larger houses.

Keeping Cuttings Safe in an Unheated Greenhouse in Winter. A simple plan is to place the pots or boxes of cuttings in other larger pots and boxes and over the top lay a sheet of glass. The cuttings will come unscathed through a really severe frost.

Summer Greenhouse Ventilation. During the summer the upper ventilators should scarcely ever be entirely shut· It is the top ones which you start to use in the morning when the sun first catches the house. As the sun increases in power allow more air at the top and also increase the lower ventilation till about eleven o'clock, when you should have the maximum amount of ventilation. In the late afternoon lessen the ventilation as the sun loses power by reducing the opening of the lower vents, and by about seven o'clock reduce ventilation to the minimum.

On dull days less ventilation will, of course, be needed, while on exceptionally hot days it may be necessary to open all the vents to their utmost capacity, and occasionally in excessive sunshine you may need to prop open the door.

Damping-down the Greenhouse. Damping-down means to sprinkle the floor and the exposed surfaces of the staging with water. It is done (on hot days) to ensure a moist atmosphere and to counteract the drying heat of the sun. Merely moisten the floor ; there is no need to flood it with water. Wet also the space beneath the staging and near and beneath the hot-water pipes.

Summer Greenhouse Watering.
The main watering should be given in the
late afternoon or early evening of each
day. In a general way that will last the
plants until the next evening. There are
some exceptions, however—coleuses, pelargoniums and fuchsias are particularly
thirsty subjects and may will need a
second watering given just before midday
during hot weather.

In addition to watering, foliage plants
with leathery leaves should be syringed
twice daily for the summer months at
8 a.m. and 5 p.m. Ferns should never
be syringed, or their fronds will turn
brown.

Shading Greenhouses in Summer.
The plants which like full sunshine include
coleuses, geraniums, carnations, schizan-
thuses, fuchsias, crotons and genistas.
Those which need shade include ferns,
begonias, gloxinias, cyclamen, calceolarias,
primulas, cinerarias, streptocarpi and
orchids.

The easy way of dealing with the shading problem, and the way which must be
adopted by the ordinary amateur is to
cover the whole of the glass with a coat
of shading material. The sun-lovers
don't get quite as much sun as they would
like, but they do not suffer to the same extent as would shade-desiring plants denied
that need.

There are several proprietary washes for
applying to greenhouse glass to hamper
the rays of the sun. Alternatively a wash
may be made at home. The following are
good recipes : (1) Place 7 lb. of size in a
pail and dissolve it over a fire. When
dissolved, stir in a knob of whitening,
crushed fine, and 1 lb. of Brunswick green.
Apply the mixture while still warm to the
outside of the glass, using a painter's
brush. (2) Obtain some quicklime and
slake it by wetting it with water. When it
crumbles down and while still warm, add
water until a milky fluid results. Then
add ½ lb. of salt for every gallon of lime
wash, stirring it in well. The salt will
prevent the wash from flaking off.

Better than washes, however, and saving considerable trouble in regard to
shading, are roller blinds, permanently
fitted to the greenhouse roof. With these,
those parts of the greenhouse in which
sun-loving plants are placed can remain
unshaded, whilst other parts containing
shade-loving plants can be shaded.

Lath roller blinds are the best, but
lengths of tiffany stretched over that
portion of the greenhouse where shade is
required are quite satisfactory. Another
plan is to make some wooden frames of
suitable size and cover these with canvas,
so that they may be laid on the glass roof
and secured in position.

GREENHOUSE PLANTS. Follow-
ing are brief details of some of the most
popular plants for greenhouse work.

Achimenes. Grown from rhizomes
planted in pots or boxes in late winter.
May also be raised from seed sown in early
spring. Roots are dried off after flowering and started into growth again in
February or March.

Amaryllis, or Hippeastrum. The
bulbs of the greenhouse amaryllis are
started into growth in 6- or 7-in. pots in
December.

Arum Lily, or Richardia. The white
lily-shaped flowers are much in demand at
Easter. The plants are started into
growth in June or July. They are divided
each year after flowering.

Azaleas. There are two groups of indoor azaleas—the Indian azaleas, which
are dwarf *evergreen* shrubs and not fully
hardy ; and the Ghent azaleas which are
hardy *deciduous* shrubs. The Ghent
azaleas are very popular nowadays for
forcing to provide bloom in the early
months of the year.

Roots should be potted in autumn and
grown on in the greenhouse until ready
for sale.

Begonias. Tuberous begonias for
greenhouse work are started in February
in boxes; when the tubers show pink
shoots they are potted separately and, in
May, transferred to 6- or 7-in. pots. The
fibrous begonias include both summer and
winter flowering varieties, the handsome
Gloire de Lorraine being the best known.
The summer flowering kinds are propagated by seed sown in a heated greenhouse
in February ; the winter flowering kinds
by cuttings taken in early summer.

Calceolarias, Herbaceous. These
richly-coloured biennials are raised from
seed sown in the greenhouse in May-July.

Cinerarias. There are three distinct
kinds—the large-flowered type, the intermediate, and the star-like *stellata.* All are
raised from seed sown in heat during
April-June for flowering the following
spring.

Cyclamen. Grown from corms potted
in 5-in. pots in July or August. Sowings
made in a heated greenhouse in October
will provide corms that will flower in the
following season.

Erica, or Heath. Pot ericas are much
in demand in the early part of the year.
The plants should be potted up very
firmly in peaty soil in March for autumn
and winter flowering varieties, in September for spring and summer flowering
varieties. Plants should stand out of
doors from July to September. Soil needs
to be kept moist, particularly during
flowering. Propagated by inch-long cuttings in greenhouse in spring.

Freesias. Flowers can be obtained in

mid-winter by potting up bulbs in July or August. They can be raised from seed sown in heat in January.

Gloxinias. Tubers are potted up in spring. Can also be grown from seed sown in August for spring flowers, in January for autumn or winter flowers.

Hydrangeas. Pot hydrangeas have become one of the most popular spring-flowering plants and are sold in very considerable quantities when just coming into bloom.

The pink variety *hortensis* is the one usually selected for pot work, the other variety *paniculata* being the variety that makes big shrubs in the open garden.

All hydrangeas are very easily propagated by cuttings in pots in the greenhouse in spring.

Splendid little plants, each bearing a large head of bloom, can be grown by taking off and potting up shoots which have not flowered and which show no signs of doing so. Take them in the usual way by severing them just below a joint and trimming off the lower leaves. Root them in a propagator in small pots.

When the cuttings have nicely filled their pots with roots, move them on to a size larger pot.

In January, if the pots are nicely filled with roots, move the plants into 6-in. pots.

There are many plans for turning the normal pink flowers of hydrangea to a pretty blue. Special " blueing " powder is sold by nurserymen. Probably the best " home " method is with the aid of alum water.

Mix 1 oz. of ordinary alum with each gallon of water used. Water the plants with this solution twice a week, starting the treatment just as the buds are forming.

Pelargoniums. With the handsome show and regal types cuttings are taken in July or August, and rooted singly in pots in a cold frame. When rooted they are moved to 4-in. pots and taken into the greenhouse. They go into 5-in. pots, for flowering, in January.

Schizanthus. One of the most beautiful of the half-hardy annuals, excellent alike for heated and unheated greenhouses. It is usual to make two sowings—one in February or March for flowering in summer, the other in August for flowering the next spring.

GYPSOPHILA. See Foliage Plants.

HARDENING-OFF. Term used to denote the preparation for an outdoor life of plants that have been raised or sheltered in greenhouse or frame.

The first step in hardening-off plants is to transfer them to an ordinary cold frame and, after keeping them rather close in this for a while, gradually to increase the amount of air given until the frame lights are removed altogether and the plants are receiving full air.

The final stage in hardening-off consists of transporting the plants from the frame to some sheltered part of the garden where they can wait a few days before being planted out.

HARDY ANNUALS. See Annuals.

HELIANTHUS. See Herbaceous Border Plants.

HERBACEOUS BORDER PLANTS. The necessary eclipse of the flower garden during the war will be followed by a great revival of interest in flower gardening with the resumption of normal peace-time life. Vast numbers of border plants will be required to restock the countless gardens from which flowers have been partially or completely banished.

Border flowers are legion in number. Below is listed a brief selection of those whose popularity is perennial.

Anchusa, or Borage. 2-4 ft. high, hardy perennial with blue flowers. Raised from seed sown in June, and by division of old plants in autumn.

Anemone Japonica. This perennial anemone blooms in early autumn and is ideal for cutting. Propagated by root division in autumn or spring.

Aquilegia, or Columbine. Hardy perennial, 2-3 ft., available in many bright colours. Raised from seed sown outdoors in summer, pricked off into a nursery bed and planted in flowering quarters in autumn or spring.

Campanulas. Choice 2-5 ft. hardy perennials that can be grown easily from seed sown outdoors in June to August. Propagated also by division in autumn and spring.

Coreopsis. The perennial varieties, such as the golden-yellow *grandiflora*, are very popular for borders. Best propagated from seeds sown in a frame in April, and by root division in autumn.

Delphiniums. There is a good sale for young seedlings or divisions in the autumn. Seed should be sown in June-July—either outdoors in shallow drills or, preferably, in boxes.

When large enough the seedlings should be pricked out into deeper boxes. They and the seedlings raised outside should be planted in a well-dug nursery bed when they have become sturdy little plants.

Root division is done in autumn. Also, young shoots can be taken from the plants in March and treated as cuttings.

Gaillardia. 2-ft. tall perennial mainly with orange and crimson flowers. Raised from seed sown in greenhouse in February or outdoors in June, and by root division in autumn or spring.

Geum. Bears scarlet or orange flowers in sprays on 2-ft. stems. Propagated by seeds sown in June-July, and by root division.

Helianthus, or Sunflower. The perennial varieties grow 4-6. ft., and bear their bright yellow flowers very freely. Propagated by sowing outdoors in spring or by root division in autumn or spring.

Hollyhocks. These fine 6-10 ft. perennials are best treated as annuals, as plants in their first year rarely fall victim to hollyhock disease. Seed is sown in greenhouse or frame in January or February, preferably one seed per small pot.

Lupin. Lupins, especially the Russell hybrids, are indispensable border flowers, not only for their own colour but because they blend so perfectly with other flowers. Propagated from seed sown outdoors from June to August and by root division in autumn and early spring.

Michaelmas Daisies, or Perennial Asters. The many sections range from 1 ft. to 6 ft. in height. The free-flowering dwarf types were gaining great popularity before the war. Easily propagated by root division—in autumn for most types, but in spring for the dwarf varieties, which tend to lose their dwarf character if divided in autumn.

Pæony. 3-3 ft. 6 in. tall perennial bearing large white, pink, carmine and other richly-coloured blooms in June and July. Propagated by seeds sown in a cold frame in July and by root division in March.

Pansies. Pansies are propagated by seed sown in boxes in July, by cuttings struck in boxes in July, and by pulling off and planting pieces which have a few roots on them, also in July.

Phlox. The perennial varieties grow 3 ft. tall and bear white, pink, crimson, mauve, etc., flowers. Propagated by seed sown outdoors in June and by root division in autumn or spring.

Pinks. Perennial pinks of the Mrs. Sinkins variety are always in demand. They can be raised from seed—sown either in the greenhouse or frame in February or outdoors in April—but the more usual methods of propagation are by taking cuttings or by root division in early autumn.

Poppy. Best known among the perennial poppies are the brilliant Oriental Poppies (mainly bright scarlet) and the Iceland Poppies (in varied colours). Propagated by seed sown from May to August, either in the open or in boxes.

Scabious. The perennial scabious is highly esteemed as a flower for cutting. Plant may be propagated from seed and by division of the roots in autumn or spring.

Growing Perennials from Seed. Practically all the popular perennials can be grown from seeds sown during June and July. To prepare the seed-beds, dig the soil 1 ft. deep. Except where otherwise advised, mix leaf-mould with the seed-bed at the rate of half a bucketful per square yard.

Scatter the seed thinly over the surface of the bed and just cover it by raking lightly.

When the seedlings have made three or four rough, or normal leaves, transplant into a nursery bed, 6 in. apart each way. Dig a foot deep and mix in well-rotted manure at the rate of ½ bucketful per square yard.

In autumn transplant into a rich, deeply dug border where they are to flower.

HERBS. See entries in Vegetable Section.

HOLLYHOCKS. See Herbaceous Border Plants.

HORMONES. Hormone preparations are used to speed up and make more certain striking of cuttings. A fine root system is produced resulting in plants going ahead more rapidly than untreated cuttings. True hormones are found in living tissue—animal and vegetable—and induce or control growth. The hormone preparations, produced synthetically, serve the same purpose. There are various proprietary hormone preparations. Where a liquid preparation is used a solution is made and the cuttings are immersed in it for some hours before they are planted. With a powder preparation the cuttings are dipped and straightway planted.

HYACINTHS. See Bulb Growing.

HYDRANGEAS. See Greenhouse Plants.

INSECTICIDE RECIPES. The most generally useful home-made insecticides in the flower garden (there are, of course, many excellent proprietary brands) are quassia solution, paraffin emulsion, and nicotine emulsion. For recipes for these insecticides, see separate entries in the Vegetable Section.

Insecticides and Fungicides. Certain spraying fluids can be made up which, in addition to destroying insect pests, also kill the spores of such diseases as Mildew.

Rust and Black Spot. (For fungicides proper, see Fungicides.)

Sulpho-quassia mixture. Steep 1 lb. of quassia chips in hot water for twelve hours. Do not boil it but keep the water warm. Melt ½ lb. of soft soap in ½ gallon of water, strain off the quassia extract and mix with the soap solution. Then bottle. For use, dilute 1 pint in 1 gallon of rain-water, and dissolve ½ teaspoonful of liver of sulphur and stir this in just before use.

Sulpho-nicotine mixture. Make stock soap solution as above, without quassia. Just before use, dissolve ½ teaspoonful liver of sulphur in a little warm water, and add 1 teaspoonful of nicotine to this. Stir into 1 pint of the stock soap, and dilute to make 1 gallon of spray fluid. Also see Greenfly.

INSECTS, HARMFUL. See separate entries—Greenfly, Wireworm, etc.—in this section and in the Vegetable Section.

IRIS. See Bulb Growing.

L AVENDER. Lavender is well worth the attention of the commercial grower, as there is a ready market for it. In normal times large quantities of lavender were imported. During the war wholesale druggists had to turn to the home market for supplies, and it is probable that the market will be held by home growers provided they can produce quantities large enough to meet trade needs.

Only the flower heads are wanted ; the stems must be cut off before supplies are forwarded to the wholesale druggist.

Lavender should be planted in March or September in a sunny position in light soil, the plants being set 1 ft. apart. Propagation is simple ; small shoots pulled off in spring or autumn and planted in a shady position will root very readily. The plants should be kept trim by clipping them back every August—after the spikes have been gathered—with the shears.

Harvesting Lavender. The right stage at which to cut is immediately after the flowers lose their colour and turn brown, but before they fall. Cut each stem separately. Cut in dry weather. If there is moisture in the heads, these will contract grey mould and be ruined. On no account cut when the sun is actually shining on the bushes, for at that time their perfume or essential oil content is at its lowest ebb.

After cutting, seal the stem ends in a candle flame, holding them in the flame long enough to char them. This prevents them "running" or losing sap and, incidentally, value.

Drying Lavender. The lavender must be dried very carefully in a cool, shady, airy room. Spread out the lavender in a single layer on sheets of brown paper. Open the windows sufficiently to keep the air healthily sweet, but not so much as to hasten drying. The slower the process, so long as it is continuous, the better.

Turn over the stems every four days for a fortnight to three weeks. At the end of this time the sap will have gone, the petals fallen, and the good qualities of the lavender will be well preserved for at least three years.

Also see Herbs in Vegetable Section.

LAYERING. Term applied to a form of plant propagation in which suitable branches or shoots are pegged down to the soil and so induced to form roots, thus converting the branch or shoot into a separate self-supporting plant. There is also another form of layering in which a plant's leaves are placed flat on soil to produce new plants.

Stem Layering. This consists of cutting a slit, or tongue, in a stem of the plant, the slit passing through a joint, the slit part then being pegged to the ground, with soil heaped over it. Carnations are layered in this fashion.

Shoot Layering. The majority of flowering and evergreen shrubs are propagated by this method ; rambler roses can also be so propagated. The usual season for layering is between July and October. Select low-placed shoots of one- or two-year-old wood that can easily be brought down to ground level and cut a notch on that side of the branch which will come into contact with the ground. Make it ⅛ in. to ½ in. across and in the middle of the shoot. Strip off any of the leaves near-by that will interfere with the pegging-down.

The notching done, take out a hole 3 in. deep with the point of a trowel, put the notched part of the shoot in it, peg down, and pack good soil round it firmly.

Bulky shoots, such as those of rhododendron, should be staked and tied in addition to pegging down, to keep them upright and undisturbed. In the event of the weather being dry at the time, a little watering should be done.

With most kinds, the layers will have rooted well and be ready for cutting from the parent plant in the autumn planting season of the following year.

"Tip" Layering. This is a method practised with rambler roses. It consists of embedding the tip of a shoot in the soil.

For each layer, choose a long young cane of the current summer's growth and bend it over so that the tip just reaches the ground. At that point take out a spadeful of soil and replace it with a mixture of rotted leaf-mould and sandy loam. Press the top 3 in. or so of the cane tip into this, and peg it securely in place with a forked stick or piece of bent wire, after covering it, lightly, with fine soil.

Water the layer if the ground is very dry.

In a few months' time the buried tip will have taken root and can be lifted, severed from the parent plant and transplanted wherever you may want it.

Leaf Layering. This method is made use of with gloxinias, begonias and other plants having fleshy bulbous leaves. It must be carried out in a heated greenhouse. It consists of laying ripe, mature leaves on the soil, and securing them there, when they will form new bulbs.

Provide a separate pan or shallow box for each leaf to be layered. Fill the pans or boxes with sandy, leafy loam, after they have been properly drained, make firm and water.

Select fully-grown leaves from good plants, being certain that the chosen leaves are free of blemish. Turn each leaf over and, with a keen-bladed knife, make cuts across the main ribs or veins, about 1 in. apart. Lay each leaf flat on the moistened soil in its pan or box and ensure that it makes close contact with the soil by means of small stones placed between —not over—the cuts.

Place the pans or boxes in a glass-topped box, or propagating-case, which stands in the warmest part of the greenhouse.

Shade the box lightly with a single sheet of newspaper to prevent direct sunlight falling on the leaves, and see that, whilst the interior of the box is always moist, it is never so moist as to cause the leaves to rot.

In a few weeks' time bulblets will appear beneath each cut. The leaf should now be cut through here and there in such a way that each bulblet can be lifted carefully from the compost with a plant label. Then the odd pieces of leaf can be trimmed away with a pair of scissors.

Plant the bulblets separately in small pots containing a compost of sandy peat or leafy and sandy loam ; keep them in a temperature which never falls below 55 degs. F., shade from strong sunshine and syringe them each day when the weather is warm and bright.

Move into larger pots as the need arises, until the 4-in. or 5-in. size is reached. Begonias, gloxinias, etc., grown in this way, must not be dried off in winter, but must be kept growing on steadily to form sturdy plants for blooming the following season.

LEAF-MINER MAGGOTS. The chrysanthemum leaf-miner is typical of these pests. It tunnels in the leaves, causing white lines to appear on them.

The best preventive measures are to spray plants likely to be attacked, with soot water or quassia wash, which makes the foliage distasteful to the fly laying the eggs which hatch into the maggots.

If the pest has actually attacked plants, pick out the maggots, which show as lumps in the structure of the leaf, with the point of a penknife blade—which is the best plan—or crush them, where they are in the leaves, with finger and thumb.

LEAF-SPOT. The foliage and stems of a good many plants are liable to a fungus disease which shows itself in the form of pale brown or almost white roundish spots with tiny black dots on them. This trouble is apt to spread and seriously affect the plants unless action is taken. The remedy is to spray the plants with a solution of 2 oz. liver of sulphur, ½ lb. soft soap, and 8 gallons of water.

LEATHER - JACKET. Leather-jackets are a great pest, destroying the roots of various plants. Plants can be protected against them by sprinkling naphthalene around their roots. Another plan is to mix 1 lb. of bran with 2 oz. of Paris green and lay this poisoned bait in little heaps between the plants that are being attacked. It must not be forgotten, of course, that Paris green is poisonous.

When leather-jackets are a nuisance on the lawn lay down strips of old linoleum, tarpaulin, or similar material on the infested spots, after watering the turf there thoroughly. The pests will come to the surface out of the wet ground and will congregate in large numbers beneath the lino, or whatever else is used, and so can be gathered up and destroyed by burning.

LILIES. Beautiful as they are in the garden, lilies offer only limited possibilities for the commercial grower. The best demand is for forced lilies grown in pots under glass.

The potting period is from September to March.

With the golden rayed lily of Japan (L. *auratum*) and its varieties, the bulbs should be potted singly in 7-in. pots. L. *speciosum* can be planted in threes in 8-in. pots, or singly in 6-in. or 7-in. pots.

Other lilies which can be grown in pots include *longifolium, Brownie, regale,* and the ordinary outdoor kinds.

The best compost for all lilies is one made up of equal parts of chopped turfy loam, peat and sifted leaf-mould, with a dash of crushed charcoal and a half pint of sand to each peck of soil. Plant 1 in. deep.

After potting, the lilies can go right away into the cold frame. Even if the frost comes all you need do is to sprinkle a little dry bracken or straw over them *inside* the frame. Then they will be quite safe and will gradually be making roots. They are taken into the greenhouse when they have made too much growth to remain longer in the frame.

Propagating Lilies. Lilies are usually propagated by offset bulbs, which are produced freely when a clump is left undisturbed.

Certain lilies, notably *sulphureum, tigrinum,* and *bulbiferum* produce small black bulbules at the axils of the leaves in autumn. These bulbules can easily be grown to flowering size bulbs.

Remove them from the stems and " sow " them about 2 in. apart each way in boxes of good fibrous compost, containing some leaf-mould and sharp sand. Cover the bulblets with a full inch of soil. The boxes should be placed in a cold frame for winter.

Each bulblet will bear a single leaf the following spring, but they should not be disturbed until the same time next year, when they may be transplanted to a sheltered bed, where they can be protected during winter. These varieties will flower when two years old, but others take three years to reach flowering size.

LILY - OF - THE - VALLEY. When produced early, these are a useful market crop.

Choose for the plants a slightly shaded spot, dig the ground 18 in. deep and work in a mixture of equal parts manure, leafmould and sharp sand, at the rate of 7 lb. per square yard.

Secure a good class of crowns. Soak them thoroughly before planting. Make trenches 6 in. apart and leave 4 in. between the crowns, which must be buried just below the surface. Do not plant the thick, long mat of roots straight down, but lay it horizontally in a 5-in. deep trench.

Planting finished, cover the bed with an inch layer of leaf-mould. Renew this top-dressing every autumn, or, better still, top-dress with equal parts rotted manure and leaf-mould.

Propagating Lilies-of-the-Valley. The usual method is to divide up the crowns when they become overcrowded. Lift the crowns in autumn, sort them into fat, plump crowns (which will flower the following season), medium-size and small crowns and replant them in a freshly-prepared bed, keeping the three grades separate.

Usually division is necessary every five years.

Forcing Lilies-of-the-Valley. Crowns can be lifted from the garden in autumn, potted up, half-a-dozen together, and brought into the greenhouse, there to flower. Moss packed over the soil surface helps blooming.

Alternatively, retarded crowns may be bought specially for the purpose. They should be planted in boxes for the production of cut bloom. Pots will accommodate six or eight crowns. Crock the pots or boxes well and then half-fill with good, light soil. Hold the crowns in position and fill in more soil. When potting is finished the buds at the top of the crowns should be just above the level of the soil. Stand the pots or boxes in a frame for three weeks, after which take them to the greenhouse.

The retarded crowns are not of much use after flowering; others can be planted in the garden when their display is over.

LIME. For details of the different types of lime useful in the garden and methods of application, see Lime in Vegetable Section.

LIME-WATER. Serves two useful purposes. First, it helps to improve sour soil and gets rid of slugs, wireworm, etc. For this purpose freshly-slaked lime is added to water at the rate of 1 oz. to the gallon. Secondly, it is a useful tonic for carnations and other members of the Dianthus family, for which purpose 2 oz. of lime are added to each gallon of water. It is supplied to plants at the rate of 1 quart per plant, the application being made monthly throughout the summer.

LIQUID MANURE. For methods of preparing this valuable plant food, see Liquid Manure in Vegetable Section.

LUPIN. See Herbaceous Border Plants.

M **ANURES.** For general advice on manuring soils, see under Digging and Manuring in Vegetable Section.

MARIGOLDS. See Annuals.

MARKETING FLOWERS. The return of peace will bring with it renewed interest in flowers and flower growing. Market growers who have necessarily devoted all their land to food growing in war-time will find a ready source of profit in the production of cut bloom and plants.

The best border plants to grow for market work are Michaelmas daisies (Little Boy Blue, Little Pink Lady, Powder Puff, etc.) ; delphiniums (Betty, Dorothy, Richardson, Dusky Monarch, Mrs. H. J. Jones, Queen Mary, etc.) ; gaillardias ; moon daisies (Beauty of Bath) ; pyrethrums (Jas. Kelway, Eileen Mary, Mrs. B. Brown, May Robinson, etc.) ; scabiosa *caucasia* and one or more of the perennial gypso-philas.

Gladioli are always in demand and a few of the good market varieties are : The Bride (pure white), Achermanni (sal-mon-red), Blushing Bride (delicate pink, with white), Nymph (white, carmine-blotched), Peach Blossom (light salmon-pink), Spitfire (scarlet), etc.

Sweet peas sell readily, provided they are well grown—that is, with large blooms and long stems. Wallflowers are another good market crop. With roses only the choicest blooms produced out of season under glass stand a good chance for market work.

Chrysanthemums are another extremely profitable line in the case of the early outdoor varieties, the early greenhouse varieties and, given sufficient glasshouse accommodation, the late varieties too.

Certain hardy annuals are also in de-mand as cut flowers. See under Annuals.

Early lilies-of-the-valley and violets are two more first-rate market lines.

In addition to cut flowers there is the bulb market to cater for. From Christ-mas onwards there is a huge demand for pots and bowls of well-grown bulbs just coming into flower. The florists and larger-scale growers who cater for this market cannot nearly meet the demand.

Also, given the space, it is a most profit-able line to grow bulbs to cut for the London or other flower markets. The flowers must be of good size, and long stalks and substantial foliage are essential. Many fruit growers now take a crop of bulb flowers in the spring, the bulbs being planted beneath the fruit trees.

Still another good line for the market grower who has the necessary facilities is that of producing bedding plants for sale by the boxful at bedding out time. So many amateurs lack the conveniences for growing their own bedding plants that a good local sale for boxes of well-grown stocks, asters and the rest should easily be possible. The same market can also be catered for with rooted cuttings of geraniums, dahlias, chrysanthemums, etc.

There is illimitable scope, too, in the pot plant trade. Thousands of cinerarias, cyclamen, primulas of various kinds, solanums, pelargoniums, and other plants pass through the markets in normal years.

MICHAELMAS DAISY. See Her-baceous Border Plants.

MILDEW. The term " Mildew " covers a multitude of fungi that, in one form and another, attack the whole run of cultivated plants. The most familiar mildew in the flower garden is that which covers the leaves and shoot-tips of roses with a whitish or grey powder. The next is that which performs similar disservice for sweet peas. The mildew of violets,

violas and pansies and chrysanthemums probably comes next, less common being that disfiguring the foliage of shrubs, such as berberis and hawthorn. Under glass the flowering plant which is the most frequent sufferer from mildew is the cineraria.

Mildew may be checked at once if the first leaves and shoot tips to show the downy patches are nipped off and straight-way burned. Neglect to do this will re-sult in rapid spreading of the fungi, with consequent increased trouble to eradicate it. Spraying thoroughly and at intervals with one or other of the advertised prepara-tions, or with liver of sulphur solution is admirable, especially when the spraying is carried out as a preventive measure.

MULCHING. Term denoting the placing of a layer of manure, etc., over the soil to prevent the evaporation of moisture. Mulching also keeps the soil temperature even ; prevents the soil from being sun-scorched in very hot weather and the heat from leaving it in a cold, sunless time. Finally, mulching feeds the plants, for every shower washes down to the roots some of the food it contains.

Well-decayed manure makes the best mulch for the mixed flower border and roses on light soils. Break it up well, and spread a 3-in. layer round the plants.

Phloxes are helped to resist disease by a 3-in. mulch of leaf-mould. Mulch sweet peas with a 2-in. layer of broken up cow manure, which does not, like other natural manures, predispose them to the dreaded streak disease. Spread a 3-in. layer a foot wide on each side of the plants, and 3 in. away from the main stems.

Mulch lilies in the border with a 4-in. layer of equal parts riddled leaf-mould or old bulb fibre and spent hot-bed manure.

Gladioli must be mulched with crude manure, or they will have " the yellows ". Equal parts mellow manure, leaf-mould, and sand (2 in. thick) is excellent for them.

Violas like a 2-in. mulch of leaf-mould, mixed with a little steamed bone-flour. It enables them to resist eel-worm and ensures a succession of fine, large blooms.

Never spread a mulch on hard ground. Always hoe and, if necessary, fork before-hand. Leave a clear ring of soil 2-in. wide round the plants, thus enabling watering and feeding to be done without difficulty. Rake the mulch occasionally during sum-mer, or it will cake, turn sour, and rain water will be thrown off it.

NAPHTHALENE. Valuable soil fumigant. For method of applica-cation, see Naphthalene in Vege-table Section.

NARCISSUS. See Bulb Growing.

NICOTINE EMULSION. Useful insecticide. For recipe, see Nicotine Emulsion in Vegetable Section.

PÆONY. See Herbaceous Border Plants.

PANSIES. See Herbaceous Border Plants.

PARAFFIN EMULSION. Useful insecticide. For recipe, see under Paraffin Emulsion in Vegetable Section.

PELARGONIUMS. See Greenhouse Plants.

PESTS, INSECT. See under separate headings.

PHLOX. See Herbaceous Border Plants.

PINKS, BORDER. See Herbaceous Border Plants.

POPPIES. See Annuals and Herbaceous Border Plants.

PROPAGATION. See separate entries—Cuttings, Layering, Root Cuttings, Seed Sowing, etc.

RED SPIDER. Minute pest attacking many flowers, notably carnations and violets. It will not make an appearance on plants that receive plenty of moisture as the pests like hot, dry conditions.

Little white spots appear on the foliage of attacked plants, and there is a kind of web and, with it, a minute spider, which is red in colour, on the under surface.

To get clear of Red Spider keep the syringe going, using a forcible jet of water, to dislodge the pest.

In a bad attack the worst leaves may be cut off and burned and the plants sprayed with a solution of nicotine extract.

ROCK PLANTS. Rockeries and rock gardens are one of to-day's most popular garden features. The growing of rock plants for sale is a line that was formerly neglected and might be developed by many.

Most kinds of rock plants can be grown from seed sown in February in pans filled with fine soil, in a warm greenhouse. Plants of a tufted bushy habit are also easily propagated by pulling off small rooted pieces after flowering, in summer,

and planting them in rows in a shady place in the garden, transferring them to the rock garden in the autumn or following spring. Plants of a trailing habit can be layered. The trails can be pressed down to the soil, more soil being heaped over them. This will induce roots to form and the stems can then be severed and the new plants transferred to the rock garden. Many rock plants, particularly those of a shrubby nature, can be propagated by cuttings.

ROOT DIVISION. The method of propagating plants by division of the roots varies in accordance with the nature of the plant's roots, as follows :

Plants which increase in size by the production of a number of separate growths around the original central root clump. In this group are such plants as chrysanthemum *maximum*, doronicums, erigerons, Michaelmas daisies, rudbeckias, phloxes, red-hot pokers, the mauve or white flowered border geraniums, sunflowers, heleniums.

After lifting and shaking free of as much soil as possible, thrust hand-forks into the mass, between the crowns of buds, and ease them gently apart. The outside parts will then separate easily.

Plants having a more solid mass of roots with closely knit crowns at the top. Among these are achilleas, artemesias, coreopsis, funkias, Japanese anemones, lythrums, pyrethrums, tradescantias, veronicas, geums and potentillas.

The general way to break up a clump of one of these into portions is to drive two garden forks back to back into the centre, and to lever the handles together.

Plants having a short, solid rootstock, such as galega, spiræ and astilbe. Cut down between the clusters of white leafy crowns with a sharp knife. A plant 9 in. across will make two divisions ; one 12-in. across, three divisions.

With spiræa and astilbe it may be necessary to lay a clump on its side and chop through the centre with a small axe or chopper. Rub the cut part of the severed stock in fine ashes.

Plants in which there is a main rootstock with new growths or buds actually growing from it, not merely connected by interlacing roots. This growth is most often found at the base of a stem. A sharp knife is needed to effect division here.

Plants of this type include aconitum, anchusa, campanula, delphinium, eryngium, heuchera, hollyhock, lupin, Oriental poppy, pæony, sedum, and verbascum.

ROSES. Generally speaking the only roses that show a good return to the grower are those produced in glasshouses, and specialist growers aim to produce blooms for sale all the year round. It is only where bunch flowers can be sold locally that outdoor-grown blooms have anything of a commercial value. In normal times there is, of course, a huge demand for rose trees and bushes for planting in gardens.

Every gardener, commercial and otherwise, must be fully familiar with all plans of the culture of our favourite flower, and following, in brief, is the cultural programme :

Rose Planting. The soil where roses are to be planted must be mock-trenched to a depth of at least 2 ft. Farmyard or stable manure must be mixed with the lower foot of soil. If the top soil is light mix with it rotting leaves and decaying vegetable refuse.

If the soil is very sticky and heavy, mix with the lower soil sand or road scrapings as well as rotted manure.

Plant bush roses 18 in. apart if weak-growing ; 2 ft. to 2½ ft. apart if strong-growing.

Standard roses planted in a single row should not be any nearer to each other than 3 ft.

Wall roses should be kept from 6 in. to 12 in. away from the base of the wall ; otherwise they will receive no moisture.

Before planting, all broken and bruised roots must be cut off to where they are quite sound. Also any long roots that may be present must be shortened back until they are only about 8 in. long.

As to the actual planting, let the planting hole be wide enough for the roots to be spread out to their fullest extent, and in the case of bush roses, deep enough for the budded part to be an inch below the surface. Standard and other roses should have 4 in. of soil over the roots.

Pruning Roses. When to prune. *September—November.* All ramblers and weepers.

Mid-March: Roses on wall and fence, the final tipping of ramblers and weepers and the H.P.'s in bush and standard. Also sweet-briers, moss roses and rugosas.

Middle to end of March: The hardier H.T.'s in bush and standard.

First days of April: The tenderer H.T.'s.

First week-end in April: The teas and tender roses generally.

How to Prune Ramblers. Cut out all shoots that have flowered during the summer, making the cuts a few inches from the ground.

How to Prune Weepers. Cut out in autumn such flowered growths as can be spared. In spring, cut off a few inches of the unripe tips of main growths and spur side-shoots to two buds.

How to Prune Climbers on Walls. Start by cutting out dead and worn-out wood. Lay in last year's wood, shortening strong shoots by one-third their present length, moderate growers by half and weak

growers by two-thirds. Spur side-shoots to two eyes.

How to Prune Moss Roses. Cut out dead and worn-out wood. Cut cleanly out internal and crossing shoots. Shorten last year's wood by half their length for strong-growers, moderate shoots by two thirds and cutting weak shoots back to three-four eyes or buds.

How to Prune Rugosas. Dead and worn-out wood must be cleanly removed, likewise internal and crossing branches. Then treat last year's wood as for moss roses, save that it may be left a little longer to permit of the formation of 5 ft. high bushes.

How to Prune Sweetbriars. Very little pruning needed, save for the cutting out of all old and worn-out wood and the thinning of crowded heads. Strong shoots of last year's growth to be shortened to half-length.

How to Prune Standards. Cut out dead and worn-out wood and take cleanly out internal and crossing shoots. Rub out or cut out cleanly any brier suckers which may be round the point of union of rose bud with stock. Spur side-shoots to two eyes. Cut back main shoots, to finish to three-four eyes or buds of the base.

How to Prune Weak-growing Bushes— H.P.'s and H.T.'s. Take out cleanly all suckers coming from under or near ground level. Then proceed to shape head as for standards. Strongish shoots may be left to six eyes, moderate ones to four and weak ones cut back to two.

How to Prune Strong-growing Bushes— H.P.'s and H.T.'s. Preliminary clearance of suckers, old wood and internal spray as for weak-growing bushes. Then cut back—moderate shoots to four eyes, strong ones to eight or ten eyes.

How to Prune Weak Bushes—The Teas. These want generally lighter pruning than any of the H.P.'s and H.T.'s. The programme for them is to cut out very weak and worn-out wood, thin the heads a little and spur back to four-six eyes.

Summer Pruning Roses. As soon as each flowering shoot has borne its bloom, cut it back to a good bud about half-way down, one pointing outwards-away from the centre of the bud. The result will be that you will get three or four good shoots in August and September from each main shoot you shorten in July ; and each shoot will produce its truss of flowers in turn.

Disbudding Roses. If you want choice blooms all the buds except the central one must be pinched off. If you are content with smaller blooms provided you have more of them, leave two buds per shoot, but no more.

Removing Rose Suckers. Bush and standard roses have a habit, during summer, of producing brier suckers. Suckers are readily distinguishable from rose shoots proper. To begin with, they always arise below the point at which the rose was worked on to the stock. With many suckers (those from a Rosa *rugosa* stock in particular) most of the leaves have nine leaflets as compared with the three or five leaflets of most proper roses.

Suckers which spring from a Rosa *Manetti* stock are different from those which come from a Rosa *rugosa* stock, but they can still be distinguished from the rose proper by the fact that the growth is thin, and the spines small and dark.

Suckers on bush roses should be traced back to their source of origin—perhaps to the root, perhaps to the very base of the stem—and be cut out.

Sucker growths on standards should be cut clean away, the very core of the growth where it joins the stem being taken out.

Feeding Roses. In February all rosebeds, as well as the soil around ramblers, climbers, etc., whether newly-planted or old-established, should be given a mulch of rotted manure. The manure should be forked into the surface at the end of April.

In early June give each tree 1 oz. of the following mixture : superphosphate, 3 parts ; sulphate of iron, 1 part.

The mixture will keep the roses going for a fortnight. After that liquid manure, used alternately with soot water, is recommended once a fortnight.

Protecting Roses against Blight and Caterpillars. Spray the trees once a week from mid-June onwards with weak soapy water ($\frac{1}{2}$ oz. soft soap in 1 gallon water). This will frequently keep greenfly entirely at bay. If trees *should* be infested with the pests, spray with a good insecticide.

To keep down caterpillars, go over the bushes at intervals during the summer and search for the pests, dropping all that are found into a tin of paraffin.

Rose Diseases. Rose Mildew. Most troublesome in the early summer and towards the end of August. The first sign of attack is the appearance on the leaves of isolated spots of white powdery mould. Dust with flowers of sulphur every ten days.

Black Spot. Generally noticeable about the middle of summer and persists until autumn. An outbreak is heralded by the appearance of large purplish black spots on the leaves.

Pick all affected leaves from the trees and gather up fallen leaves from the ground and burn them. Also cut away infected wood. Spray the plants with Bordeaux mixture. Also water the ground around the trees with permanganate of potash solution.

Rose Rust. Shows in the form of yellowish, rusty-looking spots on the leaves.

These are followed by brown and black spots and attacked foliage falls early.

Fallen leaves should be gathered up and burnt. Spray the trees with permanganate of potash—1 oz. to 5 pints water—adding 2 oz. of soft soap to the solution.

Leaf Scorch. Attacks only the leaves, on which small yellowish-green patches occur.

Gather all affected leaves remaining on the tree and on the ground and burn them. Spray the trees with liver of sulphur.

Die-Back. Long shoots—5 ft. or 6 ft. in length in the case of ramblers—may be completely lost, the shoots dying and turning black.

All diseased growths must be cut back to sound wood. Spray any trees which have suffered with Bordeaux mixture in autumn and with ammoniacal copper carbonate the following early spring. A stock solution of the latter may be made from copper carbonate, 8 oz. ; strong ammonia, $1\frac{1}{2}$ pints ; and water, 1 pint. For use, take 1 pint of the stock solution and mix it in 3 gallons of water.

Growing Roses from Cuttings. There is no plant easier to grow at home, from cuttings, than the rose. Taken and inserted during September, the cuttings will have formed roots before the winter sets in. The following year they will develop into sturdy plants.

Rose cuttings should consist of sturdy, ripe shoots at least 9 in. long after they have been cut off, straight through, just below a joint.

Select a sheltered position in which to plant the cuttings.

Insert the cuttings to two-thirds their length. If you have more than one row of cuttings, make the rows 12 in. apart.

Layering Roses. Many kinds of roses, and particularly ramblers, are easily propagated by layering. From June to August is the best time for this work, which is done as indicated under Layering.

Budding Roses. See Budding.

RUST DISEASES. See Fungicides.

SCHIZANTHUS. See Greenhouse Plants.

SEED SOWING. *Outdoors.* The methods of sowing seeds outdoors described under Seeds and Seed Sowing in the Vegetable Section are generally applicable to the sowing of flower seeds. Following are sowing points which have special reference to flowers.

In many instances results are much better when the seeds are treated before sowing.

Godetias, larkspurs, cacalias, sweet sultans, forget-me-nots, cannas, cosmeas, delphiniums, aconitum, geum, bocconia, and oriental poppy come up more quickly if soaked in cold water for two hours before sowing. The germination of lupin seeds can be expedited by filing very lightly through the skin near the plainly visible scar.

Dark-coloured sweet pea seeds should have a little file mark made in the skin on the side opposite to the eye.

Sunflowers, lavateras, annual delphiniums and annual chrysanthemums will gain in vigour and beauty if the ground before sowing is dressed with bone-meal, 2 oz. per sq. yd.

Sweet sultans and nigellas appreciate riddled peat sprinkled thickly on the ground and raked in.

Ursinias grow much more vigorously when the seed-drill is lined with an inch layer of equal parts peat and sand.

Sowing depth is important:

As a general rule, the smaller the seed, the nearer it should be to the surface. If a uniform depth is adopted for all seeds, some of the tenderer seedlings will rot before reaching the surface.

The following seeds should be sown 1 in. deep: Clarkias, helichrysums, godetias, cornflowers, annual chrysanthemums, bartonias, candytuft, cosmeas, larkspurs, calendulas and mignonette.

Half-an-inch is the best depth for gypsophilas, dirmorphothecas, love-lies-bleeding, Californian poppies, leptosynes, nemophilas, alonsoas, collinsias, love-in-a-mist, ursinias, coreopsis, jacobæs, sweet sultans, cacalias, Shirley poppies, erysimums, and those lovely climbers tropæolum canariense and ipomœa Morning Glory.

All perennials and biennals should be sown in drills ½ in. deep.

Sow both climbing and dwarf nasturtiums 1 in. deep, the former 15 in. apart, the latter 9 in.

Box Sowings. It is not necessary to make up a variety of composts for different sowings. One good compost will serve for all. A very fine compost is that prepared from the John Innes formula:

Medium loam	2 parts	
Peat	1 part	
Coarse sand	1 part	
Superphosphate	1½ oz.	per bushel of compost
Ground lime-stone or chalk	¾ oz.	

An alternative mixture is:

Loam	4 parts
Leafmould or peat	1 part
Sand	1 part

with a small flowerpotful of equal parts wood ashes and soot to the pailful. Pass the ingredients through a ½-in.-mesh sieve.

All sowing vessels—boxes, pans or pots —must be well drained, and the compost filled in evenly and firmly, especially at the edges. Leave ¼-½ in. space for watering.

Sow very thinly and cover in with finely sifted soil. A good general rule is to cover the seeds with twice their own depth of soil.

Germination is quicker and stronger when the boxes, etc., are covered with glass and paper until the seedlings show through.

SHRUBS. There is always a steady demand in normal times for young shrubs, and where space permits a " Shrub nursery " can be very profitable.

The most useful method of propagation is by cuttings.

The season for taking the cuttings ranges from July to October. Shrubs easily propagated by cuttings include weigela, syringa, veronica, tamarix, flowering currant (ribes), forsythia, berberis, broom, box, cistus, cornus, cydonia, euonymus, deutzia, escallonia, hypericum, kerria, laurel, laurustinus, olearia, philadelphus, buddleia and privet.

There are two sorts of cuttings—nodal cuttings and heel cuttings. The former, which are taken from soft-stemmed subjects, should be made from the strongest, *young*, current year's growths. They are cut just below a leafy stalk with a very sharp knife, should be about 8 in. long, and end in a joint.

With some shrubs it is easy to get the short cuttings away with a " slither " of the parent bark, the " slither " being trimmed neatly to form a good heel. These are the " heel " cuttings spoken of, and generally they root more quickly than those cut through the stem in the ordinary way.

Hard-wooded shrubs are always best propagated by heeled cuttings.

Plant both types of cuttings 6 in. apart each way in the open border, shaded from the sun, the soil first being dug 1 ft. deep, sand added to it, and then firmed with the feet. The planting holes should be deep enough to take half the full length of the cutting, and a good pinch of sand or clean grit should be dropped into the hole.

Many of the cuttings will have rooted in a few months, when they may be transplanted to their appointed positions.

Golden privet cuttings always root more certainly when afforded the close conditions of a frame, as also do such conifers as cupressus, juniper, thuja and retinospora.

Certain flowering and evergreen shrubs can also be propagated by layering. See Layering.

SLUGS AND SNAILS. To cleanse a slug-infested bed mix together 1 lb. of builders' lime and 10 gallons of water, allow it to stand for a couple of hours, draw off about 5 gallons of the liquid, minus any sediment and mix this with 1 lb. of commercial aluminium sulphate. The resulting liquid can be sprayed over any bed or border, plants and all, without the slightest fear of damage to plants.

Another plan is to get 14 lb. to 21 lb. of kainit, a potash fertilizer, and mix with it 1 lb. bluestone, a cheap commercial sulphate of copper. Keep in a tin and dust lightly over the ground from time to time. Lime water (which see) is also effective against these pests. There are some excellent proprietory killers available.

SOOT WATER. This is a useful stimulant for roses, sweet peas and many other flowers, both outdoors and in the greenhouse. To make soot water, take an old two or three gallon watering can, or other available container. Enclose a quart of weathered soot in a piece of sacking and plunge it for a few days in the can of water. Pour off the *clear* liquid and add it to the water used for watering so as to turn the latter a *light brown*—no more. At that strength soot water will not harm the roots of even the most delicate plants. Larger quantities are made in a tub in exactly the same way, a sack of soot being needed, of course, for the larger volume of water. The value of soot water is improved if 1 oz. of sulphate of iron is added to the bag of soot before it is immersed in water.

STOCKS. See Bedding Plants.

SWEET PEAS. Sweet Peas are among the indispensable flowers for commercial growers. Provided they are well grown they always command a ready sale at satisfactory prices.

Preparing the Ground for Sweet Peas. To attain real success with sweet peas, prepare rich trenches in winter. Take out the soil to a depth of 2 ft. and a width of 1 ft. at least. Place at the bottom of the trench (after stirring it with the fork) a layer of broken bricks for drainage. Break up the sub-soil and mix with it littery manure at the rate of 3 parts soil to 1 of manure. Return the mixture to the trench and whiten with bone-meal before filling in the top soil. Well-rotted manure (cow manure for preference) must be mixed with the top soil, at the rate of 1 part manure to each 2 parts of soil. Fill to within 4 in. of the ground level.

This top 4 in. requires special treatment. Break it up as finely as possible, and mix with each bucketful a 7-in. potful of wood ashes and 2 oz. of superphosphate.

A month after completing the trenches blacken the ground with weathered soot, and fork in 4 in. deep.

Sowing Sweet Peas Under Glass. Sweet Peas can be sown direct in the prepared beds in March, but better results are obtained by sowing in autumn or January-February in a greenhouse or frame.

Seeds can be sown in 3 in. to 3½ in. deep boxes or five or six seeds around the edge of a 5 in. pot, or singly in thumb pots.

For soil, use a mixture of loam, 3 parts ; leaf-mould, 1 part ; sand, 1 part ; and a sprinkling of finely-broken charcoal, the whole being passed through a ½-in. sieve. Sow the seeds 1 in. deep.

When about an inch of growth has been made, the seedlings in a greenhouse should be prepared for a move to the cold frame by giving them still more air. They should soon be ready then for the shift. In the frame, the " light " should be left off except in very severe weather. At about 3 in. high they will require stout twigs to support them.

Put out the plants in the prepared trenches in March or April.

Space the plants 9 in. apart alternately thus . · . · . · Wash the roots free of soil before planting and make holes wide and deep enough to allow the roots to go straight down into the ground. Stake the plants with little twiggy sticks immediately after planting.

Sowing Sweet Peas Outdoors. Good results can be obtained by sowing the seeds direct in the prepared beds as soon as the weather is warm enough—in March or April. Sow the seeds 9 in. apart, thus . · . · . · and 2 in. deep.

If mice or birds are troublesome in the garden damp the seeds with linseed oil and then dust them with red lead powder before sowing.

Give the seedlings twiggy sticks when they start to grow.

Stopping Sweet Pea Seedlings. Stopping—the nipping off of the tip of the plant—is not necessary with sweet peas sown direct outdoors but it is definitely advantageous with plants raised under glass.

Stop when the plants are 5 in. tall. Nip out the growing point cleanly with the finger and thumb.

Staking Sweet Peas. With plants trained to a single stem use strong bamboo canes 8 ft. long, thrust 1 ft. into the ground, 3 in. from the main stem. The canes should be linked up with stout wire.

For a row of ordinary sweet peas stake with pea sticks or brush-wood—not too thickly placed. A good distance apart is 1 ft. if the sticks are bushy, with a few twiggy branches at the bottom to keep the growth from falling out.

Training Sweet Peas. After sweet peas have been stopped; or, in the case of unstopped plants, when active growth commences, the plants branch out into

four or five growths. Taller, stronger
plants and finer blooms are obtained of
the growths are reduced to one in the case
of plants supported by bamboos, or two
or three if the plants are supported by pea
boughs. The bottom shoot is the strongest and this should always be retained
unless damaged.

A further part of the training of sweet
peas consists of removing the side-shoots
that spring out from the stem, at the
leaf-joints, throughout the summer. In
the case of plants supported by canes remove all the side-shoots when they are
quite young. With plants trained to
pea sticks, leave a few side-shoots to
develop.

It is also a good plan to cut off all the
tendrils by means of which sweet peas cling
naturally to their support, the plants
being tied, as they grow, to their supports
with raffia. Tie them by the leaves, not
by the stem.

The Care of Sweet Peas in Summer.
The only regular attentions sweet peas
will require (apart from training and tying) are watering twice a week during dry
spells, spraying overhead with rain water
or sun-warmed water on the evenings of
fine days, weeding and regular hoeing.

In dry summers mulch alongside the
row with rotted manure or grass mowings.
(Also see Disbudding.)

Feeding Sweet Peas. Give weekly
waterings, from budding time onwards
through the flowering season, of weak
animal manure water. On three occasions during the season—one in May, the
next in June, and the third in mid-July—
give superphosphate, a teaspoonful per
square foot of ground.

If the plants go a sickly green and stand
still during the growing season sprinkle
nitrate of soda on each side of the row,
½ oz. to the 3 ft. run, keeping it off the
leaves.

SWEET SULTAN. See Annuals.

SWEET WILLIAM. See Bedding
Plants.

THRIPS. These are very active,
thin and wiry-looking insects. While
usually black in colour, they may
also be either brown, grey or, in a young
state, yellow. Thrips not only bite the
foliage and cause whitish spots to appear
in it, but they also directly attack the
flowers and spot them. The thrips will
not easily be seen on the leaves, but a
shake of a bloom will set them running.
The pests can be killed by spraying with
nicotine insecticide. This spraying should
be done before the bloom buds begin to
open.

TULIPS. See Bulb Growing.

VIOLAS. A good market plant,
boxes of young violas being in
enormous demand in spring.

Plants for flowering in the summer can
be obtained by sowing seeds in February,
or spring-flowering plants can be obtained
by sowing the previous July. In each
case raise the plants in a greenhouse.

Sow in pots or boxes. The compost
should be half loam, half leaf-mould (both
sifted), with silver sand mixed in.

Scatter the seed very thinly and sprinkle
over it just a little sifted compost—no
more than is sufficient to hide the seed.

The young plants should be removed
separately—for germination is not usually
absolutely uniform—as soon as they can
be fingered. Place them 3 in. apart in
properly drained boxes about 4 in. deep,
filled with the same leafy compost as
before. They will be magnificently developed plants when the time comes—in
April or September according to whether
sown in February or July—to plant them
out, after a hardening-off period in the
cold frame. Set them 9 in. apart.

Violas can be increased by means of
cuttings or by division, as for pansies.

VIOLETS. Violets are among the
most certain-to-sell of all cut flowers, a
most profitable line for the market grower.

Good varieties are Princess of Wales
and White Czar among singles, and
Parma, Marie Louise and the Czar among
doubles.

Put out young plants in April in a shady
bed or border where the soil has been
deeply dug, manured, and, if clayey,
lightened by adding plenty of sand.
Plant 9 in. apart, with 1 ft. between the
rows.

Propagating Violets. During the
early summer a number of strong young
growths are borne on short runners.
These will make splendid new stock.

Lift the healthier clumps and shake off
all the old soil. Then with a knife detach
these outside growths.

Choosing a north or east border, give
the bed a dressing 2 in. to 3 in. deep of
leaf-mould or rotted farmyard manure,
dig it a good spit deep and mix the leaf-
mould or manure with the top soil. The
day before planting give ½ lb. of crushed
chalk per square yard of bed and a final
forking over.

Plant 12 in. apart in rows, the young
plants 6 in. to 7 in. apart.

Layering Violets. An alternative
method of obtaining new violet plants is
to peg down the runners around the old
plants. This should be done as soon as
possible after the runners have formed.
The runners will root, when they may be
detached from the old plants and planted
where required.

Still another plan where there are not many runners, is to divide the old plants. These divisions should be planted in a bed prepared as for the runners.

To Flower in Winter. Either new plants grown from runners or divisions may be used for the purpose, or old plants may be lifted from an established bed. Only the shelter of a cold frame is required. The time for planting in the frame is August or September.

The frame should be placed with its back against a sheltering wall or hedge, and so that it faces south. Put a bottoming of rough soil into the frame and make it firm. Mix some well-rotted manure with this rough soil. Over this foot or so of rough stuff place a mixture of loam, 2 parts and two-year-old leaf-mould, 1 part. See that the surface of the bed, when firmed down, is about 1 ft. from the glass of the frame light and follows the slope of the frame.

Set out the plants in the frame from 9 in. to 1 ft. apart.

When the plants have ceased to flag, remove the lights altogether, except during fog or frost.

The care of the violets during winter consists of keeping them clean, weed-free, pest-free and disease-free. Dead and dying leaves should be removed and the soil always kept nicely moist and loose.

WALLFLOWERS. See Bedding Plants.

WHITE FLY. One of the most troublesome greenhouse pests. The fly sucks the sap from the plant foliage, which becomes mottled; in severe infections the leaves wither and die. During the process of feeding the insects exude large quantities of honey dew, over which a sooty fungus develops. The usual method of treating an infested greenhouse is by fumigation, one of the special white fly fumigants or h drocyanic acid gas evolved from cal_ium cyanide. See Fumigation.

WIREWORM. For methods of dealing with this troublesome soil pest, see Wireworm in Vegetable Section.

WOODLICE. See entry in Vegetable Section.

ZINNIAS. See Annuals.

SMALLHOLDING

A **BORTION.** See Cattle, Diseases of.

ALSIKE. Species of clover succeeding well on all soils but particularly adapted to heavy land. Particularly free from Clover Sickness.

ARTIFICIAL MANURES, MIXING. With so-called "chemical" manures, it is necessary to know definitely what artificial manures may be mixed and also those that should not be mixed. If unsuitable manures are mixed, chemical action will take place and valuable fertilising ingredients may be lost. Other manures, when mixed, consolidate into hard blocks that have to be broken up finely again before being used.

Safe Mixtures. (1) Superphosphate with sulphate of ammonia ; (2) nitrate of soda with basic slag ; (3) nitrate of soda or sulphate of ammonia with bones, except nitrate of soda with dissolved bones ; (4) basic slag with bones ; (5) fish guano with mineral manures ; (6) phosphate guanos with nitrate of soda or sulphate of ammonia ; (7) organic manure with any mineral manure.

Unsafe Mixtures : (1) Superphosphate with nitrate of soda ; (2) sulphate of ammonia with basic slag ; (3) dissolved bones or dissolved guano with superphosphate ; (4) dissolved bones or superphosphate with basic slag ; (5) potash salts with superphosphate (except for an hour or two before using).

B **ARE FALLOW.** A "rest period" for the soil, during which no crops are grown.

The disadvantage of bare fallowing is the fact that, in wet seasons particularly, much of the soil-nitrogen may be washed away. To combat this "catch-cropping" is practised, certain crops being raised during the fallow period to hold the nitrogen in the soil and thus save the necessity of replacement by expensive manures (see Catch Crops).

BARLEY. A cereal valuable as a stock food. Good varieties are Spratt Archer, Plumage Archer and Victory. For cultivation and manuring details, see chart, page 220.

BASTARD FALLOW. A short fallow (see Bare Fallow) at the end of summer, as after a crop of clover or other green-meat and before the autumn-sowing of a cereal.

BEANS. Two kinds of beans are commonly grown on arable land—the familiar horse beans for feeding to stock and the many garden varieties. For cultivation and manuring details, see Vegetable Section, also chart on page 220.

BINDER. A modern and somewhat complex machine, drawn by horses or tractor, to cut cereal crops and tie sheaves automatically with string.

BOT FLIES. Insect pest that deposits eggs on parts of a horse. See Horses.

BRACKEN. Often named "fern." One of the most plentiful of our wild plants. It does great injury by invading what might otherwise be useful grazings, and should be reduced and eradicated whenever possible by means of repeated mowing for three years, by chain harrowing to break it down, by bracken cutter and bracken breaker.

This plant contains a fair proportion of potash, and is therefore at all times valuable as a fertilizer. If it can be cut and carried without undue expense or labour, it makes an admirable litter for all classes of stock, including poultry, and will add materially to the value of the manure.

BRAN. The outer husk of wheat grain obtained during milling. Very useful for compounding into rations for all kinds of stock. Often available in two grades, bran and broad bran.

Bran Mash: How to Make. Take a given quantity of bran (6 lb. or 7 lb. in the case of horses or cattle ; 3 lb. or 4 lb. for sheep or goats) and soak in boiling water. Serve to the animal at about blood heat, when it is a relished laxative. The mash loses its efficacy if it is too moist or too dry. Its consistency should be such that, when a quantity is squeezed in the hand, it will form into a ball, but no moisture will exude.

BROADCASTING. Method of sowing seeds without the aid of a drill. May be done by hand, or by brush machine as is usual in sowing grass and clover seeds.

BRONCHITIS. Affects all kinds of farm stock. See under diseases of various animals.

BUCKWHEAT. A crop grown for its seed; suitable for poor, light soils ; the grain useful for poultry, and relished by some people for buckwheat cakes. See chart, page 220.

BUTTER & BUTTER-MAKING.

Ripening the Cream. The best butter is made from ripened cream—that is, cream matured to the correct degree of acidity.

Cream ripens best at from 60 degs. to 65 degs. F., and the butter from such cream has the best flavour and texture.

In summer it is advisable to churn at least three times a week and in winter twice ; there will then be little fear of under- or over-ripened cream.

A point of great importance, often ignored, is the thorough aeration of the cream before churning. Many a sample of butter has been tainted with a nasty flavour solely because the cream was left unstirred. Stirring not only admits sweet, purifying air to the cream, but ensures an evenness of ripening in the bulk and brings down the loss of fat in the butter-milk to the lowest possible point.

How to Churn. Before attempting to churn cream, thoroughly clean the churn with boiling water, cooling afterwards, if necessary, with cold water.

The temperature at which the cream is churned should be 55 to 56 degs. F. or so in summer, and 59 or 60 degs. F. in winter.

One important detail to bear in mind when churning is the thorough ventilation of the churn and the admittance of plenty of pure, fresh air, until after such time as all the dangerous gases and ferments generated in the churn have been expelled.

The churn is turned until the cream just breaks into butter, then the butter-milk is drained off and a douch of cold water given to harden the grains and separate the small particles of curdy matter from the fat globules. A few more rotations of the churn are then given until the grains are observed to be about the size of wheat or small shot.

One washing with cold water is usually sufficient to rid the butter grains of curdy matter, which, if allowed to remain, would ruin the flavour of the produce, spoil its texture, and impair its keeping ; but sometimes it may be necessary to use two, and even three washing waters before the butter grains are sufficiently clean to be transferred to the worker.

The best method of testing this is to keep on washing the butter until the water comes out of the churn as clean as it went in.

Salting Butter. The salting of butter is best done with a brine made of 2 lb. of dairy salt to 1 gal. water. Put the brine into the churn after the final washing of the butter, close the lid, give half-a-dozen turns and leave for ¼ hour.

Dry salting is cheaper, but not so satisfactory. The salt is worked into the butter whilst on the worker.

Mottled and streaky butter is commonly the result of improper salting or the use of impure or adulterated salt. Only that quality of salt specially prepared for dairy use should be employed.

Working the Butter. Butter is worked in order to expel most of the water. The operation is carried out with a *worker*. This must be scrubbed with scalding and cold water, and a final rubbing with salt given both to the roller and the table before use.

Great care must be taken not to over or under work the butter. Over-working spoils the grain and causes the butter to become too hard, whilst under-working does not press out enough moisture and the butter will not retain its shape when made up.

Making-up the Butter. When the working is complete, the butter should be cut and patted into shape with a pair of " Scotch Hands," these also being scalded and salted before use.

The butter should then be wrapped in clean, grease-proof paper. If for private sale, it is a good plan also to use neat cardboard containers suitably printed.

Preserving Butter. If butter is to keep it must be made from ripened cream, churned into fine grains, and thoroughly well washed, salt being added at the rate of 1 oz. to each 1 lb. of butter. The salt must be well worked in, and the butter left for 12 to 24 hours. Again work out all surplus moisture, and pack into thoroughly clean glazed earthenware crocks, taking care that the butter is well consolidated.

The crocks or pitchers should be well filled, leaving just sufficient room for a layer of 1 in. or 2 in. of salt on the top, after which tie down with grease-proof paper and store in a cool dry place.

Causes of Bad Flavour in Butter. An excessive quantity of roots, cabbage or oil cake in the cow's rations will cause bad flavours, as also will certain wild plants.

Dirty milk and dirty or improperly ripened cream will also deteriorate the butter. Again, the finished product is apt to assimilate bad odours if it is stored in proximity to such strong-smelling things as onions, paraffin, cheese, etc.

BUTTER-MILK. The residue from cream after churning has taken place and the butter has been removed.

CABBAGES. On the farm, a very valuable fodder crop. For details of cultivation and manuring, see Vegetable Section and also chart on page 220.

CALVES AND CALF-REARING.

Methods of Rearing. There are three methods of rearing a calf : (1) on its own mother, (2) on a foster-mother, (3) artificially by means of milk substitutes.

The first of these methods may be ruled out as entirely uneconomical except in few instances such as the rearing of exhibition bulls ; the second and third methods each have their adherents among farmers, and there is no doubt that both are profitable and practical if properly managed.

Rearing by Foster-Mother. Under this system one cow is made to rear up to 6 or even 8 calves. Choose a suitable cow—one whose yield is at least 600 or 700 gallons a year—and immediately after she has calved allow her calf (No. 1) and another calf of the same age (No. 2) to suckle.

The cow is brought into the calf pen three times a day ; at 7 o'clock in the morning, at noon and again at about 5 p.m.

Calf No. 3 is introduced at the end of the fourth week after calving. At the end of the tenth week Nos. 1 and 2 are taken off the cow. As soon as they have gone, calves Nos. 4 and 5 are put on. Four weeks later No. 3 comes off. At the end of the twenty-first week Nos. 4 and 5 are taken off, and Nos. 6 and 7 may be put on. These may be allowed to suckle until the cow goes dry, or they may be removed at the end of the thirty-third week, and an eighth calf put on if the cow is still giving sufficient milk to support a calf.

It will be noticed that each calf is allowed to suckle for from six to ten weeks, at which age it should be weaned, but since all changes in calf-feeding and rearing must be introduced very gradually if stomach disorders are to be avoided, it is as well to introduce a dry meal mixture to the calf-pen at least a fortnight before weaning. The best mixture to use is one of 4 parts broken linseed cake, 5 parts crushed oats, and 1 part white fish meal.

At a month old the calves will start to nibble solid food. Once they have taken to the meal mixture the process of weaning from the cow may begin.

Each visit of the cow into the pen is shortened until one visit (the mid-day one first) is cut out. This shortening of the suckling must be spread over the last fortnight, and not in any way hurried.

Besides the meal mixture, a small amount of succulent hay should be fed to the weaned calves. The amount of both meal and hay which will be eaten at first will be small, but gradually, as the calves take to it, the amounts may be increased until, at the end of the tenth week, 2 lb. to 2½ lb. of meal are being fed.

Clean water should always be available in each pen after weaning.

Artificial Rearing. The most econ-omical way of rearing calves artificially is by means of suckling, or hand-feeding with milk for a month, and introducing in the third week gruel made from calf meals, often with cod-liver oil when the calves are taken off the cow.

The calf meal should be used in accordance with the directions of the manufacturer.

When butter is made the skim or separated milk may be used for calves, with the addition of cod-liver oil ; gruels may be introduced gradually after three or four weeks.

National Calf Meal. A meal specially designed for replacing milk for calf rearing, commonly resulting in the saving of 15 to 20 gallons of milk for human consumption. It is used by making a gruel, which gradually replaces the milk of the dam, and is itself in turn so replaced as the calf reaches the weaning age.

Whey for Calf Rearing. Whey is not extensively used as a milk substitute in calf rearing, but it can be fed successfully to established calves that are growing well. It should be introduced gradually up to 1 gallon a day. When the calves are used to it, up to 1½ or 1¾ gallons per day may be given in two feeds.

To neutralise the acidity of whey, ½ oz. of precipitated chalk should be added to each gallon.

Diseases of Calves. Ringworm. Common and very contagious ailment. The contagion remains in woodwork and fittings of places where the calves have been kept ; therefore the first step in stamping out the disease is thorough disinfection.

Affected calves should be taken out into the sunshine as much as possible. The places should be dressed with one of the many excellent proprietary ringworm ointments.

Scours, or persistent diarrhœa. Very infectious. Castor oil is the best remedy. Dose : From 2 oz. to 4 oz., according to the size of the animal.

Hoven. See Cattle Diseases, under Cattle.

CARROTS. Grown either for stock feeding or as a vegetable for market. Cultivations will be carried out with plough and tillage implements.

On the field scale carrots are best suited to light, deep soils. The seed is drilled in a fine tilth about April, at the rate of 8 or 10 lb. per acre, about 18 in. between the rows. The seedlings are thinned to 3-4 in. apart, hoeing must suffice to keep the crop clean, and harvesting takes place from August to October. They follow a cereal in the rotation and do not receive farmyard manure. The yield may be anything from 9 to 16 tons per acre. Good field varieties are James's Intermediate and Long Red Surrey.

CASTRATION. Term applied to the removal of the reproductive organs, or testicles, from male animals, the advantages being that, so treated, the animals are made sterile, are quieter and more tractable, and are more suited to the uses to which they are put, i.e. the production of beef, mutton, etc.

It is very necessary that this class of operation should be carried out in a humanitarian way, with the least possible suffering to the animals.

Lambs and boar piglings are quite often castrated by skilled shepherd or pig man, but the operation for all stock is best done by a veterinary surgeon.

CATCH CROPS. Term applied to crops that, needing only a short period of growth, may be sown and harvested between the gathering of one main crop and the setting of the next.

A good example of catch-cropping is the taking of trifolium or crimson clover after cereals, by broadcasting the seed after merely harrowing the surface of a cereal stubble. The crop will be fit for cutting or feeding to sheep the following May or June, after which the ground may be broken up and sown with roots.

Again, if a winter barley or oats crop has been harvested in early August, the ground may be quickly ploughed and a crop of six-week turnips taken. These will be off the ground in time to allow winter wheat to be sown.

CATTLE. The best breeds of cattle for milk-production are Dairy Shorthorns, British Friesians, Ayrshires, South Devons, Red Polls, Lincoln Reds, Kerrys, Dexters, Jerseys and Guernseys.

The best breeds for beef-production are Sussex, Herefords, Galloways, Aberdeen Angus, West Highland and Devons.

The Kerry, Dexter and Jersey breeds are small cattle, but all very suitable for smallholdings or for family use. For the rest it is often very satisfactory to keep the kind of cattle most common in the district.

Cattle Ailments. *Contagious Abortion.* Disease responsible for more losses on farms than any other. It is highly contagious and only the most rigid disinfection can arrest its spread.

Whenever a cow aborts, she should be isolated immediately and blood tests taken of every other member of the herd. All cows that respond as positive to the test should be inoculated with a special serum and very closely watched.

Veterinary aid should be sought; nothing can be done by the owner beyond giving special care and attention.

Actinomycosis. Popularly known as " Wooden Tongue," the chief symptom of which is the appearance of hard, woody growths on the tongue, the growths often spreading to the roof of the mouth, the lungs and other parts of the body. The tongue of affected animals gradually increases in size until it may become so large that the animal is unable to keep the tongue inside the mouth.

Sometimes the disease begins as a swelling on the side of the jaw, forming the so-called " wen " or lumpy jaw.

Treatment should be left to the veterinarian.

Anthrax. Extremely deadly and infectious disease. An affected animal may die in a very few hours.

In regard to symptoms, the animal may be off its food, slightly " blown " and suffering from colicky pains. The temperature will be several degrees above normal, probably 106 or 107 degs. F.

No cure is possible and immediately an animal dies from an unknown cause or an outbreak of anthrax is suspected, notification must be made to the local police (see Notifiable Diseases).

Since every particle of an animal suffering from anthrax or that has died of the disease, as well as the dung, urine, blood, hide, etc., is a possible seat of infection, rigid isolation and disinfection are necessary.

Blood issues from the several natural apertures after death. All natural orifices should be plugged with tow steeped in strong carbolic. Every drop of blood constitutes a medium of infection, and the slightest cut upon the hand or any other portion of the skin of a human being may lead to infection, and anthrax is as deadly to man as to cattle. Extreme care is therefore necessary to avoid contamination.

Apoplexy. Cattle are occasionally struck down suddenly through the rupture of a small vessel upon the brain.

The animal becomes unconscious, and falls to the ground, where it remains, either in a comatose condition or else becomes delirious.

When cattle are in this condition they usually die, but a small percentage recover; therefore, as there is a certain amount of hope of recovery, professional assistance should be sought.

Black Quarter. This disease is very common amongst calves and young stock from a few months up to two years of age. It is possible to inoculate animals against attack. This is a wise proceeding on farms where the disease is known to be prevalent.

There is no difficulty in the diagnosis of this trouble. Either a fore- or hindquarter is the usual seat of attack, especially where the skin is loose. If the part is examined and the skin pressed, there is a crackling sensation beneath the latter, as though there is air or gas beneath the skin. The swelling is small at first, but soon the whole of the quarter is affected. The beast usually succumbs in a few hours.

If a farmer is troubled with this complaint, he should top-dress his pastures with salt and allow them to lie ungrazed for a time.

Black Water or *Red Water*. Common among cattle kept on moorlands. The most noticeable symptom is the discharge of dark, reddish-coloured urine. It sometimes occurs in cows that have calved.

A good saline laxative, to which a stimulant has been added, is always beneficial, and may be followed up by iron and vegetable tonics.

Unless the after-calving type of black water is promptly treated, the animal will probably die, hence the advisability of securing professional aid.

If this trouble occurs in young stock, take them off the land, house them, and give them plenty of good food. A dose of linseed oil to each beast will do good, and it should be given at once, followed by stimulants and tonics.

Catarrh. A mild disease which, under careful management, usually yields to treatment within a week. The symptoms are shivering, loss of appetite, discharge from the nose and eyes and general indisposition.

The animal must be comfortably housed and warmly clothed, and given warm food; 2 oz. of Epsom salts should be added to the food or drinking water once a day until four or five doses have been given.

Cow Pox. Comparatively mild disease but contagious. The udder is the seat of trouble, which commences with soreness, increased heat, and swelling of the teats, followed by hardness of the skin at the bases of the teats. Small blisters form on the udder and teats, burst and leave sores which heal very slowly and cause a lot of pain.

The milker may contract the disease unless he adopts precautionary measures, in the form of disinfecting the hands both before and after milking.

Isolation and disinfection constitute the broad principles of management. The hands of attendants may readily convey the disease from one cow to another, so that disinfection is essential after each is milked.

Foot-and-Mouth Disease. At the first hint of an outbreak of this dangerous and very contagious disease, the local police must be notified (see Notifiable Diseases) and the strictest isolation and disinfection brought into force. The rest may be left to general direction of the Ministry of Agriculture, Veterinary Division.

The first symptoms of foot-and-mouth disease are that the affected animal will be dull, off its food, with a high temperature. Very soon there will develop a

certain lameness and the animal will start slavering at the mouth. Not long after the characteristic lesions or sores of the disease will commence to develop. They consist of blistering growths on the pad of the mouth, on the insides of the lips and on the tongue. About the feet they are usually found round the coronet, at the junction of the skin with the hoof, at the base of the supernumerary digits and in the soft tissues between the claws. They are also commonly found on the teats in females.

When the local authorities are informed of the existence of foot-and-mouth disease, certain very rigid rules will be enforced on the farm (see Notifiable Diseases) and, once the outbreak has been proved, the slaughter of all cattle, sheep, pigs and goats, whether healthy or diseased, will be ordered. No curative measures are of any avail.

Foul in the Foot. This trouble causes suppurating sores in the cleft of the foot, from which it gradually extends until, in extreme cases, the whole hoof sloughs and falls off. The horn becomes overgrown, softens and disintegrates.

The trouble is most prevalent on marshy lands, and when it occurs the cattle should be housed, the foot or feet put into an antiseptic bath, pared, and dressed with antiseptic ointment, tow and a bandage. The great thing is to keep the feet dry until they are cured.

Garget or *Mammitis*. A common trouble, especially in cows shortly after calving. It consists of inflammation of the udder and usually affects one quarter of it.

Loss of the quarter or sometimes the whole gland may result, involving the destruction of the animal.

The inexpert should never attempt treatment, as it is very variable and calls for the best skill obtainable.

Hoven or " *Blown* ". Fermenting within the stomach, the resulting gases causing distention and great discomfort. Caused by feeding on frosted roots or cabbage, or excessive consumption of succulent wet pasturage or green crops.

When a beast is blown the belly becomes as tight as a drum and the breathing distressed.

Veterinary aid should be sought at once, but, where this is not possible, the farmer must undertake the cure himself. For this purpose, a double instrument, known as a Trocar and Canula, is inserted into the patient's side at a point mid-way between the point of the haunch-bone and the rear-most rib. When right home, the Trocar is withdrawn, leaving the Canula (a hollow metal tube) in place to permit the escape of the accumulated gases. The Canula may be left in position for several hours if necessary.

A dose of 3 oz. of turpentine mixed with 1½ pints of linseed oil will be found very useful, and may be administered as a trial before attempting the above operation or pending the arrival of the vet.

Husk or *Hoose.* Particularly prevalent during the autumn. Quite common among calves and young cattle at any time of the year. Due to thread worms in the bronchial tubes.

Bronchitis is often present, and the coughing fits frequently lead to the expulsion of mucus and parasites through the nose and mouth.

Young cattle should be kept off land over which affected stock have run.

The inducement of coughing by means of the fumes of burning sulphur will often result in the expulsion of the causal worm.

Indigestion. Caused by wrong feeding or poor foods, or it may be a symptom of some other disease in connection with the heart, liver, kidneys, stomach or intestines.

Vegetable tonics and saline laxatives sometimes do good, but one of the most important matters is to give a change of food and see that this is neither deficient in quantity, nor lacking in quality.

Milk Fever. At one time one of the most deadly diseases of cattle, but nowadays, thanks to the discovery of a new treatment, mortality does not amount to more than 5 per cent.

Milk fever usually comes on within forty-eight hours after calving and is denoted by the animal becoming uneasy, shifting the feet, and a want of control of the body and the limbs. The animal staggers about and then finally falls to the ground and eventually becomes unconscious.

A vet. should be sent for at once, but the farmer should himself attempt a cure if professional aid is not at once obtainable.

The cure consists of inflating the udder with air through the teats. This *may* be done with an ordinary bicycle pump but an apparatus that sterilises the air before injection is naturally preferable.

Having inflated the udder, the teats should be tied with clean tape to prevent the escape of the air.

With skilful handling, a cow with milk fever will usually be on her feet again within a few hours, or, at any rate, within two or three days.

Ophthalmia. Inflammation of the mucous membrane lining of the eye. Usually due to a piece of chaff, hayseed, etc., lodging on the ball of the eye or in the corner of the eyelid.

Whenever an accident of this kind happens, the beast will be found with its eye closed and tears flowing over the face.

Remove the obstruction with a camel's-hair brush dipped in strong gum brushed over the eye.

H

The eye may then be bathed with a little warm water two or three times a day.

Pleurisy. Nearly always an accompaniment to Pneumonia (see below). Treatment is similar to that required for pneumonia.

Pneumonia. A fairly common complaint. The symptoms are quick breathing, refusal of food and a temperature of about 104 or 105 degs. F. The muzzle will be dry and the beast will cease cudding. In the early stages rigors or shivering fits are common. The patient will usually be found lying down.

The animal should be comfortably housed and the body clothed, but plenty of fresh air is indispensable.

Treatment—which varies in accordance with individual cases—should be left to a vet.

Tuberculosis. Common complaint.

As to outward symptoms, cattle *may* be tuberculous without showing any obvious signs of it. Mostly, however, there is a cough, especially marked at feeding time ; a capricious appetite and a gradual wasting of flesh. As the disease advances the cough becomes more and more pronounced, and digestive disturbance will be observed.

" In-calf " cows often " slip their calves," and after doing so, fall away more rapidly than ever.

The presence of one or more affected animals in a herd is extremely detrimental.

For many years it has been customary, amongst dairy farmers, to have their cattle periodically tested with Tuberculin, which is, when properly employed, a very reliable agent for detecting the existence of tuberculosis, even in its most incipient form.

Dairy cows should be tested periodically, but all fresh animals introduced into the herd should be isolated until, or unless, they have been tested with Tuberculin.

The test should be carried out by a veterinary surgeon two or three days after purchase.

If an animal is tuberculous, it should be destroyed. If the disease only exists to a slight extent, the carcase, or parts of it, may be suitable for food.

The best preventives of tuberculosis are plenty of ventilation and light in the buildings.

Tymphany. Another name for Hoven or Hoose, which see.

Warble-fly. The eggs of the warble-fly are laid on the legs of the cattle early in the summer, hatch and penetrate the skin, the grubs working their way upwards and forwards for about nine months. From March to June they congregate beneath the skin on the backs of the cattle, where they pierce the hide from the inside to breathe, thus forming the well-known warble bumps.

Dressing the " warbles " with a Derris powder solution is effective.

Dissolve ½ lb. soft soap in a gallon of hot water, add 1 gallon of cold water and stir in 1 lb. of Derris powder. In dressing the cattle, " agitate " the wash so as to disperse the sediment, and rub the material vigorously into any evident lump on the back, removing any protective scabs, if they are present.

As a preventive of warble-fly attack in spring and early summer the legs of the cattle may be rubbed over with a cloth on which some waste oil has been poured.

Worms. Ground worms, tape worms and many other species are apt to attack cattle. The chemicals used in treating them are very dangerous if used without proper knowledge, and veterinary guidance should be sought.

Worms are usually picked up in the pastures, and some fields may be much more infected than others. To keep the cattle off a very bad field for a time and apply several dressings of lime may effect a " cure," but in very bad cases ploughing up and growing crops on the field is the best course.

CAVINGS. Term applied to the broken straw and empty ears which come from the threshing machine. Sometimes used for feeding (especially oat cavings), but more frequently as bedding.

CHAFF. Term applied to the husks of corn that are blown off the grain in the winnowing process of threshing, and also to cut-up straw. The latter may be anything from ½-2 in. in length. Oat-chaff in particular is used for compounding rations for sheep, goats, cattle and horses.

It is deserving of mention in this connection that very large quantities of threshing "waste," consisting of chaff, weed seeds, small grains of corn, broken clover and weed leaves, and the like, may be of real value for stock, especially perhaps fattening cattle. They may be best ground to a meal and used in small proportion in admixture with other feeding stuffs.

CHARLOCK. A pernicious, yellow-flowered weed common among potato and corn crops. Destroyed by spraying with a 4 per cent solution of copper sulphate or 15 per cent solution of iron sulphate, used at the rate of not more than 40 gallons to the acre ; or by a 7 to 10 per cent solution of commercial sulphuric acid (7 gals. to 93 gals. up to 10 to 90 gals. of water per acre).

CHEESE-MAKING. There are three different kinds of cheese—hard cheese, such as Cheddar, Cheshire, Stilton, etc. ;

soft cheese, such as Camembert, Pont L'Eveque and Brie ; and cream cheese.

Hard Cheese. The process of hard cheese-making is too complicated and lengthy to be described here. Readers desiring information are referred to Ministry of Agriculture Bulletin No. 43, obtainable from H.M. Stationery Office, Adastral House, Kingsway, London. The Bulletin describes the making of all kinds of cheese in detail, including the Smallholder cheese.

Soft Cheese. Soft cheeses require very much less elaborate equipment than the hard varieties, and are therefore more suitable for making on smallholdings. They ripen very quickly and are made from perfectly sweet milk, the acidity being developed in the milk after the rennet has been added.

Rennet of good quality only should be used and just sufficient should be added to the milk to bring a tender curd in about an hour or an hour and a half.

The moulds used in making the cheeses given below are circular in form, about 4 in. high and 3 in. across.

To make the soft cheeses :

Add 1 teaspoonful of rennet in 2 teaspoonfuls of water to each gallon of milk raised to a temperature of 84 degs. F.

One method of raising the temperature is to put the milk in an enamel pail and stand the pail in a large bowl of hot water. After adding the rennet stir the milk occasionally during the first ten minutes.

Cover the pail with a piece of clean muslin and at the end of about one and a half hours the curd should be ready to place in the moulds. Put the moulds in twos or threes on straw mats resting on draining boards that are slightly larger than the straw mats.

A tablespoon can be used as a ladle.

The curd at ladling should be distinctly firm ; if ladled out too soft, the cheese will not drain properly. Place about four ladlefuls of curd into each mould, cover and leave ten minutes to drain ; ladle more curd into the moulds and leave to drain as before. Continue in this manner until all the curd has been ladled into the moulds. Leave until the next day, when the moulds will be about half full and the cheeses firm enough to admit of turning.

The turning of the cheeses is rather a delicate operation but with a little practice it can be skilfully performed. Place a clean straw mat on the top of the moulds, also a draining board. Put your left hand under the bottom board, invert the whole —steadying the other board with your right hand—so turning the cheeses face downwards upon a fresh straw mat.

The upturned surface of the cheese should present an unbroken grooved appearance due to the straw upon which it has been resting.

A little salt should be sprinkled over the surface every day until the cheeses are ripe.

Cream Cheese. Both double-cream and ordinary cream cheeses are very popular, the former being made from extra-rich cream such as is obtained by the Devonshire way of cream-raising.

Sometimes rennet is used in making these cheese, but not usually. If rennet *is* used the cheese is generally sweeter, and more weight per given quantity of cream can be obtained.

The cream is drained twelve hours after separation in a fine linen cloth spread over a wooden form, a weighted loose board being used, if necessary, to press out surplus moisture. A form 18 in. long by 14 in. wide by 4 in. deep will accommodate 1 or 2 gallons at a time, and with this should be provided a hard-wood pressing board 1½ in. thick.

The cloth should be large enough to fold over the whole surface of the cream in the form, and must be opened out once or twice during the first hour and the sides scraped down, the cream being afterwards reweighted with a 14 lb. weight. Thick, well-cooled cream will usually be ready to mould in from three to four hours after it has been placed in the frame to drain.

Ordinary cream, containing from 25 to 30 per cent of fat, may be used to make single cream cheeses, and thirty of these cheeses can be made from a gallon of such cream, as a rule each cheese weighing about 4 oz. to 4½ oz. One cubic centimetre, or 3 or 4 drops of rennet is added to each gallon of cream, and the latter left to coagulate for eight to twelve hours. Salt may also be added to the cream, if necessary, at the same time as the rennet. After renneting, the cream is hung up in cloths, and scraped down at frequent intervals.

The action of the rennet will facilitate the escape of the whey, and drainage will consequently be hastened.

If rennet is not used it is advisable to strain about ½ pint of starter into each gallon of cream as soon as the temperature of the latter has reached 60 degs. F.

To prepare the moulds (which must be fitted with a movable bottom) line them with greaseproof paper or butter muslin, and stand them on straw mats placed on drainage boards.

After adding a little salt to the cream, it is ladled into the moulds and a little pressure exerted by means of weighted followers in order to preserve a firm and good shape, a round piece of paper or butter muslin being fitted on the top, and the cheese finally slipped out.

Note.—Milk producers who wish to make cheese for sale under war conditions should first consult the Milk Marketing Board, Thames Ditton, Surrey.

CLOVER. Important crop for feeding to livestock either green or as hay. The most common varieties are Red Clover, White or Dutch Clover, Alsike and Trefoil.

About 20 or 25 lb. of red clover seed are generally sown per acre, usually under a " cover-crop " of a cereal, the seed going in after the cereal is showing in row. It may either be broadcast or drilled, after which it should be harrowed and lightly rolled.

The different clovers form part, in various proportions, of mixtures of clover and grass seeds for permanent and temporary grass. These mixtures are usually purchased ready prepared to suit given conditions.

Clover Sickness. If clover is grown too frequently on any one piece of land, that land becomes " clover-sick " and the crop dies before it reaches maturity. The obvious remedy for this is so to plan the crop rotation that clover is not taken on the same piece of land more than once in, say, four years. Deficiency in lime and potash is also conducive to clover-sickness.

Clover Weevil. Common enemy of clover. Eggs are laid in spring close to roots. Larvae and adults feed on crop. To keep down the trouble all litter must be cleared in winter, together with vetches, etc.

COLT. A male foal. See Horses.

COMFREY. A valuable fodder that should be much more extensively grown, in particular the Russian type if available. May be grown from rooted portions of established plants, planting in rows 3 ft. apart and 1½ ft. apart in the rows. Crops first year, and at high-yielding stage in two to three years.

COMPOST. It is often of much value to rot down waste green stuff, weeds, straw and the like—often mixed with poultry manure and/or light soil—to form compost that will be almost, and often quite, equal to farmyard manure, for use in growing potatoes and roots.

CORN PESTS. *Midge.* Eggs are laid on wheat in spring. Larvae and pupæ remain on plant. In serious cases only remedy is to burn the straw to destroy the eggs.

Corn Aphis. Winter forms hide on broom and furze. Feed on plant juices. Remedy is rigorous cleansing of hiding-places in winter.

Corn Thrips. Feed on grain and breed all summer. Winter cleansing of harbourage is again remedy.

Corn Weevil. Eggs are laid on grain, larvæ feeding within. Adults breed in barns, etc. Scrupulous cleanliness in granary and barn is necessary to keep down pest.

CROPS FOR FARM AND HOLDING

Crop	Notes as to Seed and Treatment	Time to Sow	Most suitable soil	Preceding Crop	Manuring	When to Market and How Sold
Barley	3 bushels an acre.	Feb. to Mar.	Light loam.	Roots.	Superphosphate, 5 cwt. an acre.	Sold by the cwt.
Beans (Horse)	2 bushels an acre.	Oct. to April.	Heavy Soil.	Barley or Oats.	Heavy farm-yard manure. 10 cwt. of basic slag an acre.	By the cwt. Otherwise harvest in September and feed to cattle, pigs and horses.
Buckwheat	1-2 bushels per acre. 2½ bushels broadcast.	Early May.	Light.	Turnips or other cultivated crop.	Farm-yard manure. Does not demand rich soil.	About August. By the cwt.
Cabbage	2 lb. for plants for an acre. Transplant when ready.	July and spring.	Heavy soil.	Wheat or Oats.	Farm-yard manure. 5 cwt. of superphosphate and 1 cwt. of nitrate of soda, in two dressings.	All the year round. By the bag or per dozen. Feed green to cattle and pigs.
Clover, Trefoil or Alsike	Sow 24 lb. seed to the acre. On stubbles.	August or with winter-sown corn.	Any good sweet soil.	Any cereal.	5-6 cwt. per acre basic slag in autumn and 1 cwt. nitrate of soda in spring.	Cut for hay or feed green to stock.
Flax (Linseed)	2 pecks (28 lb.) for linseed 4-8 pecks for flax.	April-May.	Very clean, light loams.	Potatoes, lea or any cereal.		Grown for flax, or for linseed.
Kale	2 lb. of seed to acre, in seed bed.	Mar. and April. Transplant as large enough.	Any soil.	Peas.	Heavily farm-yard manured. Superphosphate, 5 cwt. Nitrate of soda, 1 cwt. an acre.	Autumn. ½ bag. Feed green to cattle and pigs.
Lucerne	Land deeply ploughed. 15 to 20 lb. an acre.	Sow in April - June.	Calcareous loams.	Mangolds, Turnips or fallow.	Heavy farm-yard manure. Basic slag 10 cwt. an acre.	Can be made into hay or fed green. Valuable for silage.
Maize	Sow 2½ bushels of seed per acre.	May.	Rich soil.	Roots or Potatoes.	Heavy farm-yard manuring in autumn, 3 cwt. superphosphate, and 2 cwt. sulphate ammonia per acre in spring.	Cut green and feed to stock. Valuable for silage.

			Prefers			
Mangolds	8 or 9 lb. to the acre.	Mid-April to early May.	heavy soil.	Oats or Wheat.	Heavy dressing farm-yard manure. 2 cwt. of nitrate of soda, top dressing.	Autumn, for use following spring. By the ton.
Oats	2–3 bushels an acre.	Spring or winter. Drill in Oct. or Feb.	Heavy lands and rich loams.	Beans.	Heavy farm-yard manure. Sulphate of ammonia. 2 cwt. an acre.	May and July. By the cwt.
Peas	3 bushels an acre.	Oct., Feb. and Mar.	Light land and light loams.	Fallow or root crop, or grass.	No manure. Use soot in early spring.	Harvest in September. By the cwt.
Potatoes	15 cwt. per acre.	March-April.	Open, Fen, loamy or sandy-loam.	Cereal.	Farm-yard and artificial.	By the ton.
Rye	Sow 3 bushels per acre.	August and September.	Light soils and loams.	After early fallowing and a catch-crop.	3 cwt. of superphosphate, 1½ cwt. sulphate of potash and 1¼ cwt. nitrate of soda per acre in spring.	Cut green and feed to stock. Grain by the cwt.
Sainfoin	4 bushels an acre (in pod) with Barley.	April.	Chalky soil.	Barley.		Used to make hay, or green feed.
Sugar Beet	10–15 lb. per acre.	April-May.	Medium, rich soil.	Cereal.	Heavy farmyard, 3-5 cwt. super, 2-3 cwt. potash, 2 cwt. sulphate of ammonia.	Autumn and early winter, to sugar factory, by the ton.
Tares or Vetches	3 bushels an acre.	Autumn, for spring food.	Any soil.	Barley.	Farm-yard manure. Superphosphates, 4 cwt. an acre.	Fed to cattle in spring. By the cwt.
Trifolium	On stubbles and harrowed in, 15-20 lb. per acre.	Aug.–Sept.	Calcareous loams.	Cereal.	5 cwt. basic slag.	For fodder early spring.
Turnips and Swedes	3–4 lb. an acre.	End of May and June.	Lighter soils.	Wheat or Oats.	Superphosphate. 5 cwt. an acre.	July onwards. By the bag.
Wheat	2–3 bushels an acre. Dress before drilling.	As early as possible after harvest.	Heavy soil, good loams.	Bean, Peas, Tares or fallow.	Heavy farm-yard manure. Nitrate of soda in spring. 1 cwt. an acre.	By the cwt.

For Details of Market Vegetables, see Vegetable Section.

COUCH GRASS. A very trouble-some weed on all classes of farm soil. Land that is over-ridden by this weed should be cultivated and harrowed several times after the autumn-ploughing in order to collect the dried weed into heaps to be burnt on the spot, the ashes being-returned to the soil.

COULTER. Part of a plough.

COWS. When buying a cow one should pay marked attention to the appearance of the animal. The eyes should be placid, to indicate a quiet dis-position. The ribs should be well sprung and the belly capacious. The udder, while its size is no guide to the milk yield, should not sag unduly, but extend well forward and up at the back.

It is a good sign if the milk veins travel prominently from the vessel right to the middle of the belly at least. Further, the teats should be a comfortable size. Plenty of breadth across the hips and a level back are also desirable features.

If a purchaser is undecided whether or not an animal is in a healthy condition, it is well to look for a nice, mellow, loose skin, a fine neck and shoulders, and warm ears and horns—the hallmarks of health.

Milking. For milking it is desirable to wear a clean, washable coat and wash the hands clean. It is also nowadays a common and useful practice to wash the cows udder before milking, so that dust and dirt do not fall into the pail.

Approach the cow quietly and place the milking stool close to the animal but leave room for the pail to be placed between the knees, close under the udder, the left knee being placed close against the right hock of the cow. On no account should a crouching attitude be assumed with the head rammed against the cow's side, and the stool and pail at some distance from the udder.

Now proceed to draw the milk quickly but not so as to cause discomfort to the animal. If the milk is drawn rapidly without producing discomfort, the flow is beneficially increased. It is considered the best modern practice to milk with dry hands, not with hands lubricated by dipping them in the milk.

Always hold the teat as high up as poss-ible, allowing it to lie across the centre of the palm of the hand. Close the tips of the fingers upon it so that the milk it contains is expelled into the pail. Then release the pressure and when the teat is full again is beneficially increased.

Times of Milking. Cows should be milked at regular hours. The more equally the times are divided, the more uniform will be the quantity and quality of the milk.

Heavy milkers may be all the better milked three times a day instead of twice, in order to relieve the pressure on the udder.

Feeding Dairy Cows. In making up a ration it is necessary to consider the foods available on the farm. For actual rations, see Feeding Stock.

Grooming Cows. In the main the grooming of cows follows along the same general lines as for horses (which see) except that hardly as much care in " fin-ishing off " is required. Cows should al-ways be groomed *after* milking, so that the disturbed dust does not settle on the milk.

Particular attention must be paid to the flanks since these animals have a habit of lying in their own dung. In normal circumstances, the dirt may be removed simply with the curry comb and brush, but occasionally it may be necessary to wash off the worst of the filth and finish with the brush.

CREAM. Cream may be obtained from new milk in two ways : skimming and separating.

In the former case, the milk is allowed to stand in wide, shallow " trays " for a few hours, the actual time largely de-pending upon the temperature—the cream rising more quickly as the temperature increases. The cream is then skimmed off by means of a special skimmer.

Clotted or Devonshire cream is thick cream obtained by keeping new milk in shallow pans for about twelve hours at 60° F., and then scalding on a stove at 180° F., until the cream thickens and wrinkles. It is normally extensively made in the south-western counties.

In separating, a special machine, many excellent models of which are on the market, is used. The milk is run through the separator whilst still warm from the cow.

For ripening cream and churning, see Butter.

CREAM CHEESE. See Cheese.

CROPS. In the table on pages 220–1 are notes relating to the growing of various crops. These, however, must not be followed too literally, as seeding varies considerably with district and time and method of sowing, while many crops will do well on a wide variety of soils if con-ditions are otherwise congenial. Further, some latitude is permissible in the order in which crops succeed one another ; while, during the war period, changes otherwise inacceptable may be essential, and the manuring indicated may well be imposs-ible owing to the shortage of fertilisers.

DRAINAGE. It is of the utmost importance that land shall be well drained if it is to grow satisfactory crops. Indeed, in reclaiming or improving farm land, it is the first matter that should be considered. Tile draining is nowadays almost out of the question owing to cost, but clay land can be vastly helped by mole draining, which may be expected to last in good condition for 10 years and more. Bush drains are also often made, trenches that will lead away water being about one-third filled with brushwood, on top of which the soil is returned. For assistance with Drainage the County Committee should be consulted.

DREDGE CORN. A mixture of oats and barley grown together, and used as ready-mixed grain after threshing. Beans or peas are now sometimes included. The mixture might be 2½ bus. oats and 1 bus. barley per acre; or 1½ bus. oats, 1 bus. barley and 1 bus. horse beans.

DRIED GRASS. It is now some years since investigations at Cambridge clearly showed the great food value of dried young grass, cut when only 4 to 6 in. high, as compared with hay made from tall grasses in the old way. So rich in protein is such dried young grass that it compares favourably with a good protein cake.

Wherever it is possible it is well worth while to take two or three cuts of young grass when it reaches a height of 6 in., dry it quickly and store it for winter use, when it is much more valuable, weight for weight, than ordinary hay. It is, moreover, rich in vitamins, those constituents that impart colour and flavour to milk and butter, and are so valuable for health. (See also Poultry Section.)

If conditions made it impossible to dry the young grass it could with almost equivalent advantage be converted into rich silage (see Silage).

Dried grass is used for stock in much the same way as hay, but it has to be remembered that it is much richer in protein, and can therefore be used in association with foods containing a much larger proportion of carbohydrates.

About 4¼ lb. of high quality dried young grass can replace 3¼ lb. balancer concentrates for milk production. For high-yielding cows the dried grass will help to keep down the quantity of meadow hay normally consumed and that might prove excessive; 4 lb. dried grass can replace 8 lb. meadow hay.

DUNG. See Farmyard Manure.

ELEVATOR. An implement for conveying straw or hay from threshing machine or wagon on to the stack or rick during building. A great economy in time and labour.

ENSILAGE. The method by which silage is made, in a container called a silo. See Silage.

EWE. Adult female sheep. See Sheep, Feeding Stock.

FARMYARD MANURE. Farmyard manure—often abbreviated f.y.m.— is usually formed of the dung and urine of the different classes of farm stock mixed with and absorbed by litter of various kinds, such as straw, cavings, bracken. When in the "fresh" state it helps to ameliorate very heavy soils, and when well rotted it helps to bind light soils. In addition, its content (somewhat variable) of nitrogen, phosphate and potash makes it valuable for a wide variety of crops, especially roots, potatoes, cabbages, kale and the like. For rate of application see table, pages 220 and 221.

FEEDING STOCK. The rations given to stock must depend largely on what is available on the farm or holding and what can be purchased economically. The following, however, may be taken as standard peace-time rations, a proportion being for "subsistence" and the rest for "production." In each instance roots may be substituted in whole or in part by silage.

Cattle, Fattening. Winter Ration (per head, per day):
88 lb. yellow turnips, or 66 lb. swedes; 14 lb. oat straw; 7 lb. meadow hay; 4 lb. crushed oats or maize meal; 2 lb. undecorticated cotton cake; 2 lb. linseed cake.
Summer Ration:
Fresh meadow grass which, if poor, may be supplemented by 4 lb. crushed oats and 1 lb. linseed cake per day.

Cattle, Store. Winter Ration (1½ to 2½ years old, per head, per day):
56 lb. mangolds or swedes; 14 lb. oat straw; 2 lb. to 4 lb. undecorticated cotton cake.
Summer Ration:
Good fresh grazing.

Cows, Milking. Winter Ration (per head, per day). Yielding 2 gall. milk per day:
30 lb. mangolds or swedes; 14 lb. chaffed oat straw; 7 lb. meadow hay; 2 lb. crushed oats; 3 lb. decorticated cotton cake.
Yielding 4 gall. milk per day and over:
45 lb. mangolds or swedes; 14 lb. oat straw; 7 lb. meadow hay; 4 lb. crushed oats; 4 lb. decorticated cotton cake.
Summer Ration:
Good, fresh grazing, supplemented, if necessary, by 2 lb. crushed oats and 2 lb. cotton cake per gallon of milk yielded.

Goats. *Winter Ration:*
2 lb. good hay per day ; $\frac{1}{2}$ kibbled maize and $\frac{1}{2}$ crushed oats at the rate of 6 oz. per day per gallon of milk yielded.
Summer Ration:
Good grazing and a handful of linseed cake in the morning and a double handful of crushed oats at night.

Horses. *Winter Rations* (At heavy work, per head, per day) :
8 lb. to 14 lb. oats ; 2 lb. to 3 lb. bran ; 15 lb. to 20 lb. straw chaff ; hay *ad lib.*
(At rest, per head, per day) :
5 lb. oats ; 8 lb. oat straw chaff ; 6 lb. meadow hay.
Summer Ration:
Good grazing with 4 lb. to 8 lb. per day good oats, with chaff, if working.

Pigs, Sows in Farrow. As much as they will clear up of a ration composed of 4 parts weatings ; 2 parts each barley meal and flaked maize ; 1 part each fish meal and bean kernel or soya bean meal.

Sows, Milking. As above.

Piglings Just Weaned, and Store Pigs. 5 parts weatings ; 4 parts barley meal and 1 part ffsh meal. This ration, fed at the rate of 1 lb. per day for every 20 lb. live-weight, will last the pigs till they are put on to fattening rations.

Porkers. 4 parts weatings ; 5 parts barley meal ; $\frac{1}{2}$ part fish meal, fed at the rate of 1 lb. for every 20 lb. live weight.

Baconers. 2 parts weatings ; 7 parts barley meal ; $\frac{1}{2}$ part fish or meat meal, fed at the rate of $\frac{3}{4}$ lb. for every 20 lb. live weight.

Sheep. (Fattening, *per head, per week*) :
120 lb. swedes ; 3 lb. hay ; 2 lb. chaffed oat straw ; 3 lb. maize meal ; 3 lb. decorticated cotton cake and linseed cake.

Sheep, Store.
100 lb. swedes ; 7 lb. hay ; 2 lb. crushed oats.

In-lamb Ewes.
120 lb. swedes ; 2 lb. hay ; 4 lb. chaffed oat straw ; 2 lb. oats ; 5 lb. linseed cake.

FEEDING STOCK IN WAR-TIME.

It would not be possible to give here a detailed statement as to the rations that can be used for farm stock during war-time. In the first place rations must depend upon the Government's coupon rationing scheme, which varies with the foods available for use ; and in the second place farmers must, to an extreme measure, give their stock foods that they have themselves produced from arable land, together with young dried grass, silage, pulp from sugar beet, etc. Pigs, rabbits and poultry may need to receive much " waste " to make up for the very reduced quantities of cereal offals that may be available. By the time these lines appear it may be necessary for farm livestock of all classes to be provided for independent of imported foods. Stock keepers will be wise to try and shape their farming accordingly.

FERTILIZERS. Apart from farmyard manure and various other organic materials, very many so-called " artificial" fertilizers are of great value for raising crops. For the use of lime see Liming. Fertilizers proper fall under the three heads Phosphates, Potash, Nitrogen.

Among the first are superphosphate, basic slag, bone manures ; among the second are sulphate of potash, muriate of potash, kainit ; and among the third are nitrate of soda, sulphate of ammonia, cyanamide, nitrate of lime.

The artificials are best used in association with farmyard manure or other organic manure, and are employed in various associations in accordance with the crops grown, the crops making heavier demands on one fertilizer than another.

During war-time some fertilizers are in extremely short supply or unobtainable, and it is necessary to follow the advice given from time to time in accordance with what is available.

FILLY. A young female horse, so called until she has her first foal, or, if not for breeding, is well past the normal age for producing a first foal.

FOALS. Young horses. For the first two or three months of its life a foal will obtain all the nourishment it needs from its mother's milk. At about four months it will begin to graze and nibble dry food, but should not be fully weaned until it has attained the age of six months.

Foals, like their mothers, should be housed at night if the weather is at all bad. At all other times, they are better when running on a good pasture.

As soon as foals show an inclination for dry food, they should be given a few handfuls of crushed oats daily, in addition to what grass they like to eat. During the winter they should be given the best of hay, crushed oats, and a weekly bran mash.

Joint-ill in Foals. Usually makes its appearance within a few days after birth. Infection starts at the navel. Preventive measures consist of tying two ligatures about $1\frac{1}{2}$ in. apart in the navel cord, which should then be severed between the ligatures.

It is further advisable to paint the end of the cord and round about the navel with iodine liniment. It is only necessary to do this once or twice.

Joint-ill comes on very suddenly and is indicated by lameness and swelling at or around a joint, such as for instance the shoulder, hock, or knee. These swellings are hot and painful, and the foal refuses to suck. Death may ensue in a few hours.

G **ESTATION.** The period during which pregnant animals carry their young. The approximate gestation periods of farm animals are as follows: mare 340 days, cow 285 days, ewe 150 days and sow 128 days.

GRASS, DRIED. See Dried Grass.

GRASS LAND. Sowing Grass. It is usual for the seeds mixture to be sown under cover of a nurse crop, barley being the most favoured.

As the covering crop may be 2 in. or 3 in. above the soil it will not be possible to undertake any vigorous cultural measures before sowing, but a fine tilth must be made for the nurse crop, and a light harrowing will probably produce a fine surface tilth without difficulty.

Under this surface tilth the soil should be consolidated, and to achieve this effect a rolling, with a Cambridge roller, will produce ideal conditions.

The selection of the seeds mixture will depend on the type of ley that is being laid down, temporary or permanent.

For a 1-year ley a good mixture is: Italian rye-grass 8 lb., perennial rye-grass 14 lb., broad red clover 2 lb., late-flowering red clover 2 lb., and Montgomery red clover, 2 lb. Total seed sown per acre 28 lb.

For a three-years' ley for hay and grazing (suggested by Stapledon): Italian rye-grass, 3 lb., perennial rye-grass, 4 lb., perennial rye-grass Aberystwyth S. 24 and S. 101, each 4 lb., cocksfoot, 6 lb., Timothy (commercial), 2 lb., Timothy (S. 51), 2 lb., rough-stalked meadow-grass and/or crested dogstail (on poor soils), 1 lb., late-flowering red clover, 2 lb., Montgomery red clover, 2 lb., white clover, 1 lb. Total seed sown, 31 lb.

For a permanent pasture: Perennial rye-grass 14 lb., cocksfoot 8 lb., timothy 4 lb., rough-stalked meadow grass 1 lb., late-flowering red clover 2 lb., and wild white clover 1 lb. Total seed sown per acre 30 lb.

The actual sowing of the seeds may be performed in many different ways. In some districts the " fiddle " is employed ; in others the seed-barrow ; while not a few prefer to broadcast the seed by hand. If a horse drill is used it must be set with the utmost caution, or the seed will be buried at a depth at which it will fail to germinate. Clover is frequently lost in this way.

After the seeding, use the lightest of harrows across the ribs left by the rolling. Unless conditions prevent it, do the

sowing after the dew has gone, for, if the nurse crop is wet, the smaller grass seeds will adhere to it instead of falling to the ground, and be lost.

Improving Poor and Matted Pastures. Only too often pastures are in very poor condition, and proportionately unproductive. Frequently there exists between the upper growth of grass and the soil a layer of tough fibre, formed of dead and decaying grasses. This results in the finer pasture grasses perishing while the coarse grasses flourish.

In improving poor and matted pastures it is very desirable to make sure that the drainage is good (if not, mole draining may be done) ; to test for lime and give a dressing if necessary ; and to give a hard harrowing, or light cultivating, commonly by discing. During the winter months, when the ground is soft, this is easily and cheaply accomplished. Go up and down the field one way, then across the other way. It may sometimes be useful, to ensure thorough work, to tie a couple of heavy logs of wood on the implement.

The matted fibre will gradually rot and provide humus. About March broadcast 1 cwt. of sulphate of ammonia per acre, chain harrow to level the ground, and finally give a long rolling, going up and down and then across.

Let stock graze on the field in summer. In the autumn, put on 5 cwt. of basic slag per acre.

It is often well to plough out a really poor pasture, and re-sow either at once or after taking one or two arable crops. The Government makes a grant of £2 per acre for ploughing old grass, and also pays half the cost of liming. In this connection consult the County War Agricultural Executive Committee.

H **ARROWS.** Chain, spiked, and other forms of implement used to stir and level arable land, or to scratch grass land more or less deeply and distribute animal droppings.

HAULM. Term applied to the " straw " of such crops as potatoes, peas, beans.

HAY AND HAYMAKING. Preparing Fields for Hay. Although occasionally an early bite of grass is taken from fields intended for hay, the best procedure is to shut up the meadows in March and let no stock into them until after the hay is harvested.

Before shutting up the field, give a chain-harrowing in both directions to spread the dung and stimulate the herbage, pick up large stones, and then bring on a heavy roller.

Later, it will be necessary to go round the headlands and trim back hedge growth, for it is on the outskirts of the field that damage to the mower usually occurs.

To Make Hay. Herbage for hay should be cut when in flower (whatever the weather), because in the formation or loss of seed the nutritive value of the stems is reduced.

Grass cut in wet weather may remain for a few days without taking serious harm. The swathes, however, should never be turned back upon wet ground, while the lighter they are handled with the rake or fork the better.

The secret in making hay that has been half cured, or nearly spoiled by rain, is in keeping it well opened in the swathes, which are better thrown into " wind rows," by making three rows into one, that is, drawing in one row on either side. If the crop is light five rows or swathes may constitute a " wind row ". These wind rows are commonly turned or lightened up by means of hand forks, but there are various types of machine " tedders " and gatherers. In dry weather a horse-rake is usually employed to bring the hay together into wind rows. When the moisture has evaporated and the hay is dry enough it is carted direct from the wind rows, or if need be (owing to weather conditions) made into " cocks " or the larger " pikes " or " kiles ".

A kile is in reality a miniature round hay stack, 4 ft. high and 4 ft. in diameter. The tops or roofs of the kiles must be carefully made and rounded off so as to throw off rain. Hay will often cure well in these heaps in spite of some rainy weather.

During a prospective wet season it may be desirable to make much less hay, and devote the grass at a younger stage to the making of silage (which see).

Stacking Hay. As hay is damaged if placed in close contact with the ground, the stack is built upon a base of some coarse dry material, such as logs, old peasticks, hedge-trimmings, covered with a layer of straw.

It is always well to carry up the sides of the hayrick to 10 ft. or 12 ft. in the green state, as this will then allow it to settle down to a height at the eaves of from 6 ft. to 8 ft.

Having built up the walls to the proposed height, each course of hay now put on is diminished in size, thus ensuring an even taper for the roof. Coming to the very top of the stack, two or three courses are trodden very firmly so that the ridge of the roof may be as tight as possible. The ridge is pegged or tied in place, or it may be blown off before thatching is done. Thatching should be undertaken immediately the stack has sunk to a normal level. See Thatching.

Second Crops of Hay. Some people are averse to mowing a field twice in one year, but practice has proved that no harm is done by this procedure so long as the crop warrants it. Indeed, under the modern systems of making silage and drying young grass, two or more cuts may be taken for these purposes.

HEDGES. The management of hedges is a very important item in farm-husbandry Some hedges are naturally much more difficult to deal with than others, but good work in the way of trimming, cutting and laying can be done with a hedge anything over eight years old.

Cutting and Laying. Any wood not required for laying should be cut out, and " alive " stakes should support the newly laid hedge. Live stakes should be notched fairly deeply where cut for laying or they will produce a thick unmanageable growth at the top. If new growths are few and weedy it may be desirable to use some dead stakes to give temporary strength to the layers.

The tops of dead stakes should be headed off level, but not to a point, by a clean cross-cut, and bound firmly and securely together by twisting binders or " heathers " (headers) in and out between them ; before the heathers are used, however, the small wood, or brush, should be pushed through on to the field side, where it will form some protection to the young shoots that subsequently break from the bottom of the hedge, and that would otherwise be eaten by livestock grazing in the field.

The thorns should always be laid with an upward inclination, the cut being from below upward, so that they will bend over readily without splitting and will not hold water. The growth is then bent over in the opposite direction, and woven into the general fabric of the rest.

The best time to do hedging is winter, when the wood works well. Where possible a hedge-layer should always work from the ditch side.

Making New Hedges. Hornbeam, beech, lime, maple, and hazel all grow quickly and form good hedges, but thorn, holly and myrobalan plum are the only plants that can be depended upon to resist horned stock. The first-named is the most reliable of all.

In good soils it is not essential to plant double rows ; one row of healthy welltransplanted seedlings placed 8 in. apart will suffice. These are best planted from November to the end of February.

Encourage the plants to grow freely the first year by applying a mulching of manure to the surface, and in the second year cut them back close to the ground in December or January.

The sides of the new hedge must be trimmed as growth develops, but the tops, or-leading shoots, should not be headed or cut back until they have completed each season's growth, when they may be cut to the required height of the hedge. If the sides are trimmed three times in the year while the hedge is forming, so much the better.

HEIFER. The name given to female cattle until they have their first calf. Often used up to the end of the first lactation period.

HORSES. Stabling Horses. Whatever material a stable is built of, the floor should be sound. Ordinary bricks, asphalt, and wood are not suitable for stable floors, or, at any rate, should never form permanent stable flooring.

Concrete is the best flooring material, or special hard grooved bricks.

In laying the floor, the drainage should be on the surface and slope towards a surface drain at the foot of the stall, which conveys the liquid into a well-trapped drain outside the stable.

Stalls should be not only roomy but lengthy, say 8 ft. from the head to the foot of the stall. The width of each stall should be not less than 6 ft. 6 in.

The inlet for fresh air should be by means of ventilating windows, i.e. the lower portion closed and the upper open, so that there is a constant current of pure air passing into the stable, and foul air passing out through a ventilating shaft in the roof.

A stable should be neither too cold nor overheated; a temperature of 50 to 55 degs. F. is about right.

Feeding Horses. See Feeding Stock.

The Care of In-foal Mares. The period of gestation in mares is about 48 weeks, and a mare is capable of conceiving again nine days after foaling. This fact is usually acted upon as it is a more likely period for securing conception than if mating were postponed until a later date. March and April are the usual months for mares to foal, and the latter is universally considered the more satisfactory.

In-foal mares are better worked on the land in back-band and chains than on roads between the shafts. In any event they must never be made to "back" carts or waggons or to haul timber or heavy loads. Exercise is very necessary and mares may be worked almost up to foaling-time.

When parturition is imminent, the mare should be left in a roomy horse-box and immediately after the foal is born she should be given a warm bran-mash and be allowed greenfood or roots. She should be kept in for a day or two and may then normally have the run of a pasture near the homestead.

The mare, when she is rearing her foal,

must be kept in good condition. Crushed oats, with abundance of meal in them, are excellent for their milk-making properties, and bran, too, should be given for the first few weeks.

When the mare has once more been put to work avoid overheating her; nothing upsets a foal quicker than to suck when the mare is in such a state.

Rearing Foals. See Foals.

Grooming Horses. Directly a horse comes off work, no matter whether it is summer or winter, wet or fine, sleet or snow, the feet should be washed, and this without wetting the hollows of the heels.

The feet washed, if the animal is sweating freely or if it has been out in rain give the body a vigorous shampooing with a straw wisp, first having removed excess water with a body scraper.

All parts where the skin is thin, such as around the nostrils, on the face, beneath the tail, etc., should be sponged lightly and thoroughly dried afterwards.

As soon as the body is dry, dust and mud should be removed from the limbs with the dandy brush, followed by a vigorous application of the body brush on every square inch of the body, more especially below the belly and inside the thighs.

If a horse comes in covered with mud this can either be washed off immediately or allowed to dry on.

The mane, fetlock and tail are just as important as other portions of the body.

Clipping Horses. Clipping should be done when horses have finished shedding their summer coats and the winter coat begins to get thick—in late October or early November.

For town, street or high road work clip all over, except for the saddle-patch. But for the land-working horse, the hunter, hack, etc., it is far better to half-clip only, leaving legs and belly unclipped.

Half-clipping may be done merely as a band about 18 in. wide right round the horse, to save doing the finnicky bits about the head and ears; the whole of the upper part except the saddle-patch can be clipped.

Clipped horses should be stabled and rugged up at night in winter, and a rug or some sort of covering that will not blow off with the first puff of wind should be carried to put over the horse when he is left standing in the open after a spell of work.

Diseases of Horses. Anthrax. Very deadly disease, uncommon in this country. Symptoms similar to Anthrax in Cattle (which see). No cure possible.

Catarrh. Commonest of all horse ailments. The animal is off its food and has mucous discharges from nostrils and eyes. The only treatment necessary is to take the animal off work and give a small dose of Epsom salts in the drinking water, night and morning. If applicable, clothe the body and bandage the limbs. Two or three weeks' rest is usually sufficient time for recovery to take place.

Colic. Common complaint denoted by severe pain, sweating, continual rising and lying down, rolling and wandering round the stall. In severe cases, the disease may be fatal.

Chief causes are sudden changes of food, drinking too much cold water when over-heated, prolonged feeding on dry food, worms and too much green food.

It will usually be found most satisfactory to have professional advice, as every case of colic presents an individuality. The owner can, however, try the effect of a pint of linseed oil, to which 2 oz. of turpentine and 2 oz. of sweet spirits of nitre have been added.

The draught can be repeated at intervals of three hours, until three or four doses have been given.

Feed the animal on bran and scalded oats, with a gradual return to hay.

Cracked Heels. Painful condition due to wetting the heels and failing to dry properly. Mild cases can be cured by the application of a cream composed of prepared chalk and lime-water. If the heels are badly cracked, smear them with an ointment composed of equal parts vaseline and mild mercurial ointment.

Farcy. Form of Glanders, which see.

Glanders. Dangerous and infectious disease which is notifiable (see Notifiable Diseases). Caused by a germ which invades the lungs and other parts of the body. The symptoms are a slight ulceration of the nose, discharge from the nostrils and possibly a small but painless swelling on the inner side of the lower jaw.

Any person who has upon his premises a horse suffering from, or suspected of suffering from, this disease must not remove it from the premises, but must at once report it to an inspector of the Local Authority or a police constable. Thereafter he must obey the directions of the L.A.

Grease. Disease affecting the skin between the knee and pastern or hock and pastern, characterised by a most objectionable smell.

Obtain veterinary aid in treatment. Applications of lime-sulphur ointment will sometimes give relief.

Joint-ill. See Foals.

Lymphangitis. Digestive trouble. Usually appears on the inner part of the thigh, whence it may spread to the whole of the limb.

Lack of exercise is the chief cause. Cure by administering a 5 drm. or 6 drm.

physic ball, obtainable at any veterinary chemist.

Mange. Common skin disease which is contagious and therefore notifiable (see Notifiable Diseases).

Symptoms are irritation as shown by the animal rubbing or biting the part, followed by the loss of hair in patches.

Affected animals must be isolated, clipped and singed. After this they must be washed with hot water and soda, dried and treated all over with lime-sulphur dressing.

For use as a preventive apply a more dilute form of the lime-sulphur as a spray to all parts of the horse's body. It is well to use a proprietary preparation in accordance with instructions.

Mud Rash. A rash appearing on the skin of the heels, characterised by a dry condition of the skin, on which numerous minute blisters form.

Treatment consists of giving the animal a physic ball and following this up with a liberal supply of linseed gruel along with the other forage. Plenty of good strapping is essential in order to encourage the circulation in the skin.

Pleurisy. See Pneumonia.

Pneumonia. Being infectious, the first step when this disease is suspected is to isolate the patient. The symptoms are a high temperature, dullness, lack of appetite, reddening of the eyes and nostrils, and, later, laboured breathing.

Apart from the provision of every possible comfort, warmth and plenty of fresh air, the layman should call in veterinary aid at once.

Rheumatism. Characterised by swellings of the joints.

Provide a warm and comfortable stable and keep the body clothed, the limbs bandaged, and the affected parts freely massaged once or twice a day with some stimulating embrocation, such as ordinary white oil. Epsom salts in the drinking water along with ½ oz. of bicarbonate of potash twice a day are helpful.

Ring Bone. Inflammation of the bone and bone skin of the long and short pastern bones. Causes lameness.

The effects of rest should be tried. Work on soft ground is best for affected animals. Many ring-boned horses which have been going very lame in town will continue to do good work on the land for many years.

Ringworm. This is extremely contagious and the utmost cleanliness must be observed in its treatment.

Treatment consists of clipping the hair from around the infected area and dressing with mercurial ointment or the lime-sulphur solution recommended for Mange above.

Roaring. This disease, taking its name

from the peculiar noise commonly made by the affected animals, is really an affliction of the upper part of the throat. There is no cure for the trouble and " roarers " are usually quite capable of doing a steady day's work.

Spavin. Spavin is really a chronic arthritis affecting the lower and inner aspect of the hock. One or both hocks may be the seat of the trouble.

If there is reason to believe that a spavin is forming, the animal should be taken off work and kept in a stall, as rest is of supreme importance.

Strangles. Characterised by swelling below and between the jaws, usually accompanied by a nasal discharge. The swelling increases in size and tenderness from day to day. The skin finally softens at one point and then breaks, forming an abscess. To save time a vet. should be called in to lance the abscess and keep the wound open.

Sometimes the abscess forms at the back of the jaw, just below the ear. When this occurs the animal suffers a great deal more and the course of the disease is more prolonged.

Very little treatment is necessary beyond that of clipping the hair off the seat of the swelling and rubbing in a little blistering ointment.

Wind Galls. Puffy swellings on hock or knee which should either be blistered or fired. They can also be treated by smearing with ordinary gas tar.

Very troublesome pest of horses is *bot fly.* The horse licks off and swallows the eggs, which give rise to the bots. These fasten to the walls of the stomach, and remain there till the following summer, when they are passed out with the dung in the form of chrysalides which later hatch into adult flies.

Watch may be kept for the eggs, which may be clipped off ; while the horses may be rubbed over in spring and summer with preparations that may prevent the flies from depositing eggs. Very little harm is done to a horse unless the bots are very numerous.

HOVEN. Illness of calves due to the excessive production of gas caused by over-eating wet green fodder. See page 216.

IMPLEMENTS, CARE OF. Cover is an essential in cutting costs on annual implement depreciation. The form of cover used does not matter very much. If you have a barn or open-fronted shed, so much the better. If not, you can rig up a shelter of sheets of corrugated iron and poles.

Regularly employ oil or grease on those parts of machines where fine adjustment

is required—mowing machine knives, for instance. Waste machine oil is as good as any for protecting the coarser parts of a machine. In putting it on use a brush, so that not only is the film of oil evenly distributed, but the more inaccessible places are reached.

On other parts of a machine vaseline is the best protective agent. Put on in a thick layer it will withstand any conditions of weather, and rust will never form under it.

For smaller parts, such as those that may be detached from an implement, it is a good plan to keep them lying in a shallow pan of oil when they are not in use. Mowing machine sections kept in this way will last many seasons longer.

On wooden implements and such things as harrows, cultivators and presses, paint is more important than oil. Of all the paints red lead paint is the most durable and suitable for farm implements. Before applying the paint take a small piece of coarse sandpaper and rub over the part to be painted. Small particles of dust and rust will remain on the surface after this has been done, so use a dry, clean brush to remove them—then do the painting.

See also Tractors.

KALE. Valuable farm crop much used for feeding green to livestock. Most valuable varieties are the Marrow-stemmed and Thousand-headed. See Crops.

LEGUMES. Term applied to pod-bearing plants such as peas, beans, vetches, etc.

LEY, LEA OR LAY. See Grass Land.

LIMING. Lime may be regarded as second in importance to drainage. Not only is it a plant food, but it corrects acidity, and helps to ameliorate soils and make them more suitable for growing crops.

The Amount of Lime to Use. One ton of lime is a good dressing for an acre of land. Double the quantity if you use crushed chalk or limestone.

There is not usually need to apply such a quantity of lime more than once every 3 or 4 years.

How to Apply Lime. As soon as the lime is properly slaked it should be scattered over the soil surface with a shovel and the ground at once ploughed or cultivated. A second cultivation may beneficially be given a week or ten days later, the idea being to work the lime in well.

LUCERNE. Valuable leguminous crop for feeding green or making into hay. Will yield 20 tons of green fodder, rich in protein. Sometimes called Alfalfa. See Crops.

MAIZE. A valuable green fodder crop which can be grown in favourable districts in this country. Will yield 20 to 25 tons of excellent carbohydrate crop, of great value in late summer and early autumn for stimulating the milk yield. May also be ensiled. The grain cannot, of course, be ripened here. For culture, see Crops.

MANGOLDS or MANGELS. One of the most valuable of root crops for feeding cattle in winter. It is preferable, however, not to feed them until after Christmas, as not till that time do they attain their full ripeness and suitability for feeding to stock. Under good conditions a crop of 25 tons and more may be anticipated. For method of cultivation see Crops for Farm and Holding.

MANURING. For manures for the different crops, see Crops. For manures for market vegetables, see Vegetable Section.

MARES. See Horses.

MASLIN, MESLIN, MASHLUM. Names applied to mixture of wheat and rye grown together.

MEADOWS. Term applied to grass fields intended for hay as opposed to pastures, which are the fields intended for grazing. See Grassland and Hay.

MILK. Treatment of. By far the greatest part of contamination of milk takes place in the cowshed during milking, and to reduce it to the least possible extent and obtain a sound wholesome milk, the care and treatment of milk must being at the source of production.

Milking must be done with dry, clean hands.

Cows' udders and hindquarters must be free from dirt and sediment and the udder should be washed and dried.

The hands should be washed after milking each cow.

The cowshed should be kept clean and provided with sufficient light and ventilation.

All churns and vessels should be kept free from rust, and every utensil coming in contact with milk should be thoroughly washed, and steam sterilized daily.

Straining and Cooling Milk. Immediately after milking the milk should be removed to a purer atmosphere and strained through a sieve covered with one or two layers of straining cloth, or, better still, with a small patent filter. The milk should then be passed slowly over a cooler and the temperature reduced to 60 degs. F., or under.

The importance of cooling milk immediately after it is drawn cannot be overstated, because at that time the temperature is favourable for bacterial growth, and the cooling of the milk decreases the rapidity of the multiplication of the bacteria.

Milk Recording. In view of the cost of feeding and the amount of attention that is necessary in connection with keeping cows for milk production, it is undesirable and uneconomic to retain in the herd animals that do not give a reasonable quantity of milk over the lactation period. The exact amount of milk obtained from each animal cannot be accurately noted without some reliable record being made. This merely means weighing the milk from each cow after milking. To do this, all that is necessary is to provide a spring-balance to which a pail can be hung, the milk being recorded in lbs. or pints as preferred. The spring-balance is usually of the clock-face type, on which the weight of the pail is allowed for.

Various manufacturing firms supply complete outfits for recording milk, which also include special ruled cards upon which to enrol the name of the cow, together with columns for entering the morning's and evening's milk.

MUSTARD. Valuable crop, either for ploughing in as green manure or for folding sheep. May be sown at any time between March and August at the rate of 15 lb. per acre.

NATIONAL CALF MEAL. See page 214.

NOTIFIABLE DISEASES. Certain diseases of animals, being of a very serious and infectious nature, are by order of the Ministry of Agriculture, notifiable. That is to say, upon an animal becoming infected with one of them (or suspected of being infected), immediate notice must be given to the local police.

These diseases include cattle plague (rinderpest), pleuro-pneumonia, foot-and-mouth disease, swine fever, anthrax, sheep scab, sheep pox, epizootic lymphangitis, glanders and farcy, mange in horses, rabies, tuberculosis in cattle and, in certain districts, infectious abortion in cows.

Certain plant diseases are also notifiable : Onion smut, american gooseberry mildew (on premises where gooseberries are grown for sale) and wart disease in potatoes, and others.

OATS. Valuable grain crop for all stock. For cultivation, see Crops.

ONIONS. On farms where the soil is good and easily worked, onions are a very paying crop, the chief consideration being the availability of labour for hoeing, gathering and bunching and the nearness of a reasonable market.

Onions are best suited to light soils of sandy texture but good depth. The land should be clean, well consolidated but tilthy. The crop needs heavy feeding, and dung at the rate of 15 to 20 tons per acre is desirable, unless the land is still rich from a previous dunged crop, when a mixed fertilizer may be used. Such fertilizer should contain 4 per cent nitrogen, 8 per cent phosphoric acid and 6 per cent potash, and be used at the rate of 6-8 cwt. per acre.

Seed may be drilled 1 to 1½ in. deep, in early March, in rows some 10-15 in. apart, 5 lb. of seed per acre being required. They are thinned to about 4 to 8 in. apart. The wider space the larger bulbs required.

Seed may also be sown in August and transplanted to the fields the following March; or be raised under glass early in the New Year and put out in April.

Frequent hoeing is necessary and harvesting is from August onwards, yields ranging from 6 to 10 tons per acre.

PASTURE. The name given to grassland intended for grazing as opposed to haymaking. For sowing and management, planting out and re-seeding, see Grassland.

PEAS. Two varieties of peas are commonly grown on farms—green peas for selling to market and field peas for harvesting and feeding to stock. For cultivation of former, see Vegetable Section. For cultivation of latter, see Crops.

PIGS. Breeds. There are many breeds of pigs of good type, and the choice to some extent depends upon individual preference, and perhaps on the kind mainly kept in the locality. The large Middle and Small Whites, Large Black, Essex and Wessex Saddlebacks, Berkshire and Tamworth doubtless account for the great bulk of the pigs kept in this country. Other breeds are the Gloucester Old Spots, Lincolnshire Curly-coated, Cumberland White and Dorset.

Very many herds are retained as pure breeds, but crossing is quite general, especially making use of Large White and Middle White boars with sows of other breeds. Crossing is held—and there are data to support it—to increase the stamina of the progeny, the mortality being lower and the survivors doing better after weaning. At the Pig Research Station, Wye Agricultural College, Kent, the Large White and Essex Saddleback breeds are kept, some being maintained pure and some are crossed. In grading 48 pigs of each pure-breed and crosses both ways, at about 150 lb. carcass weight, the Large White graded best, then the Large White x Saddleback boar, then Saddleback x Large White boar, and finally pure-bred Saddleback.

Feeding Pigs. For representative rations, see Feeding Stock.

One of the most important feeding rules as far as economy is concerned is never to give the pigs more at one meal than they can readily clear up. The younger the pig, the more often it needs feeding. Four times a day is not too often for feeding pigs just weaned, but three times a day is more usual.

With little pigs it is very essential that the food should be given in suitable sized troughs, increasing the size as the animals grow.

If you use house scraps, care must be taken to include in these nothing but stuff that is absolutely sweet and fresh. Soda and salt in anything but very small quantities are bad for pigs.

Pigs like variety in their meals and when possible they should be given a change.

Housing Pigs. Although pigs can never be expected to do well in a poor sty, an elaborate building is not required. If the building is of reasonable size, dry, warm and draught-proof, pigs will thrive.

As an indication of size, a sty measuring 10 ft. by 8 ft. is quite big enough for four pigs up to pork-killing age, but is hardly large enough for a brood sow and her litter when the latter are nearing weaning age.

The structure may be of any desired material, but from the point of view of drainage and sanitation, concrete floors are best.

All sorts of ills may occur, however, if the pigs are allowed to sleep on concrete. It is advisable, therefore, to cover at least a part of such a floor with a loose wooden platform.

Minerals for Pigs. Pigs of all ages benefit by being given a regular supply of minerals. A good mixture of supply consists of 4 parts finely ground chalk or limestone and 1 part of iodised salt, given at the rate of 5 lb. per cwt. in the absence of fish meal and at half this rate if 5 per cent fish meal is added.

In-Pig Sows. Whenever a sow is put

to the boar a record should be kept of the date so that the farrowing date may be known and suitable preparations made.

The farrowing sty should be thoroughly cleaned, including the removal of all old litter, scrubbing with boiling water (to which a disinfectant has been added) all boards, etc., and whitewashing walls.

Make sure that the farrowing rails in the farrowing sty (rails fixed a foot from the wall and a foot from the floor to prevent the sow crushing her piglings when she lies down near the wall) are secure against breakage.

For the littering of farrowing-pens chaffed straw is best and safest. Under it, and spread evenly over the floor, should be placed some fresh earth.

When Sows Farrow. A good plan is to have a hamper or box partially filled with hay or dry straw into which each pig is placed as soon as possible after birth and the hamper covered with an old rug or some sacks. The pigs will then dry quickly, though before putting them in the hamper it would be as well to place them to the sow's teats for a few minutes, so that they can have the benefit of a first short feed of milk.

As a rule a navel-cord breaks off at a length of 4 or 5 in. as each pig is born. Usually it is then considered to need no attention. Some breeders, however, think it worth while to tie a thread tightly round it at the body end, and touch with iodine solution, with the object of preventing the mild blood poisoning that sometimes occurs, giving rise to abdominal sores.

Occasionally, a cord nearly a foot long will be left. This should be nipped back to not more than 3 or 4 in. long, care being taken not to break it off too close to the body.

When the sow has finished farrowing and has ejected the after-birth, she should be given a little warm sloppy food. While she is drinking this, the wet portion of the " nest " can be removed and replaced with some short, dry straw, though with no more disturbance of the bed than is necessary.

The sow will return to her nest after feeding and, the piglings being then placed with her, she probably will not move again till feeding-time next morning. Then before feeding, she should be let out of the sty for a few minutes to evacuate the bowels, and will then usually settle down again with her litter for several hours.

The Care of Newly-born Pigs. When the pigs are three or four weeks old, put a low trough for them behind a partition or " creep ", with an entry too small to admit the sow. Two small daily feeds of meal, mixed preferably with skim milk, will lessen the drain on the sow, encourage stronger growth of the youngsters, and

by getting them accustomed to solid food at an early age, will diminish the check in development that often occurs at weaning.

If it is not possible to arrange things in this way, the sow can be turned out of the sty twice a day and the pigs given a feed in her absence.

In the colder months of the year there is an advantage in supplying a meal for the youngsters the last thing at night. Being induced to leave the nest in this way they will evacuate the bowels. Otherwise, in chilly, damp weather they tend to remain snuggled in the straw from early evening till feeding-time next morning—often twelve to fourteen hours. Under these conditions they do not relieve themselves as often as is desirable.

Throughout the suckling period the sow will benefit by being allowed a run out in a grass field, or even in a yard, apart from the litter. As the pigs reach six to seven weeks old, the length of time that the dam is away from them may be lengthened. This will gradually reduce the flow of milk, so that no trouble will be experienced with the udder of the sow when the pigs are weaned.

Scouring is not infrequently a big source of worry, holding up development and sometimes causing the death of weakly animals. A simple preparation that can be depended on as a great help in such an event with pigs still on the sow is sulphate of iron solution (1 oz. of the crystals to a quart of water). Give a teaspoonful for each pig every day in the food provided in the " creep " trough.

Weaning Young Pigs. The weaning process really commences with the teaching of the pigs to feed when three weeks old, as mentioned above.

The pigs can live independently of their mother when eight weeks old.

Growing Pigs on for Pork. Pigs destined to be killed as pork should be fattened up to a live-weight of about 90 to 100 lb. which should be possible in about four to five months.

As soon as the pigs are weaned, feeding must be judicious as well as generous. For rations, see Feeding Stock.

Skim milk, when available, is a great help to quick growth in porkers. We all know the merits of dairy-fed pork.

Do not make your pigs too fat or carry them beyond the desired weights, for over-fat and over-heavy pork pigs are always discounted in price.

Bacon Pigs. By far the most important branch of pig-keeping to-day is the production of bacon. This industry is now very rigidly controlled.

The prices paid for bacon pigs depend upon their weight when killed, and their " grading " as regards the measurement of back and shoulder fat and belly thickness.

Possibilities of profit for the ordinary feeder under normal conditions have commonly depended almost entirely on his getting off his pigs at under 8½ score. Above this limit they will require extra food to make a given increase of weight, and it seems clear that the amount generally needed (after 8½ or even 8 score) to produce 1 lb. gain in live weight cannot be less than 5 lb.

This means that to raise a pig from 8½ to 9½ score (dead) it will be necessary to feed about 130 lb. meal. A simple calculation shows, however, that the difference in value of an 8½ and a 9½ score pig of standard grade is normally only a few shillings.

Just as important as not marketing overweight pigs is not marketing pigs of less than 7 score dead. Such animals are fit only for pork. The factories may refuse these misfits, and even if they accept them they pay much less for them.

Every bacon pig producer should have a weighing machine, fitted with a crate for the periodical weighing of individual pigs. Then if 25 per cent of the live weight is deducted he can get a fairly accurate idea of what the dead weight will be. It should suffice to weigh them every two or three weeks.

If a feeder has contracted to sell his pigs in the best class he should send to the curer all pigs about 200 lb. live weight. Those weighing between 190 lb. and 200 lb. could have a mark made in their right ear and be sent away the next week without further weighing, while those between 180 lb. and 190 lb. could be marked in the left ear and sent off the week after the next. This arrangement simplifies matters considerably.

One point worth noting is that the belly fat is likely to be a trifle thicker if the pig is starved for twenty-four hours before slaughter—a practice always recommended by curers.

Another point urged by feeders of long experience is that it is an advantage to reduce the purely fattening part of the food in the last two weeks, on the ground that the final few pounds of weight put on by a fat pig generally goes to its shoulder, where it is least wanted.

Rations for Bacon Pigs. See Feeding Stock.

Diseases of Pigs. See chart on pages 234 and 235.

Pigs in War-time. Until normal times it is essential to bear in mind that some foods that are generally given to pigs are no longer available or must not be used in that way. Further, the allowance of meal for pigs is small and must be used in conjunction with considerable quantities of "waste" of various kinds. All waste or "swill" *must* be boiled before it is given to the pigs.

At this time, also, the sale of pigs must be in conformity with the directions given from time to time by the Ministry of Food. It is held that in order to make the most economic use of "swill" and other waste from household, town and camp, and save cereal foods, bacon pigs should be carried to heavier weig its than has been normal. To this end the maximum weight to which the "standard price" (24s. per score dead weight early in 1943) for clean pigs applied was increased from 11 to 12 score.

Full advice on feeding pigs in accordance with available foods are given in *The Smallholder,* and by the Ministry of Agriculture.

Small Pig Keepers' Council. A body formed by the Ministry of Agriculture to supervise and correlate and advise upon all matters relating to war-time pig keeping by small producers. In particular, it encourages and organizes pig clubs, which are of special value in view of facilities provided for co-operative effort.

Removal of Restrictions on Householders. During the war all restrictions on pig keeping included in householders' agreements, have been officially cancelled, and it is open to anyone to keep pigs provided that by-laws relating to health and sanitation and nuisance are duly observed.

PLOUGHING. With all heavy soils, ploughing is best done in the autumn; light soils may be ploughed either in autumn or early spring.

POISONOUS PLANTS. A considerable number of wild plants, or parts of them, are poisonous to stock, and it is well to be careful to discover if any such plants are accessible to the animals. Yew, meadow saffron, bittersweet or woody nightshade, dog's mercury, white briony, cowbane, water dropwort, hemlock, deadly nightshade, and various other species must be regarded as particularly injurious. Even horsetail and bracken must be regarded with rather more than suspicion. All should be eradicated where they occur.

POLL. Term applied to cattle that have no horns. Some breeds are naturally hornless—the Red Poll, Aberdeen Angus, Galloway, etc.—but it is quite possible to remove the horns—or rather to prevent them from developing—in any breed. This is done by rubbing the horny formation as soon as it can be felt, when the animal is very young, with caustic potash.

POTATOES. See Vegetable Section, and Crops.

PIG DISEASES AND HOW TO TREAT THEM

Diseases	Causes	Symptoms	Treatment
Catarrh ..	Improper housing conditions; exposure to cold or dampness.	Mucous discharge from nose.	Keep warm and dry; feed warm sloppy food, rub vaseline on nose daily; give teaspoonful daily of following mixed in food: ammonium chloride, 2oz.; liquorice powder, 3 oz.; sodium sulphate, 4 oz.
Diarrhœa ..	Adults: foul water: decayed food; sudden change of food; too much fruit in autumn when confined; or worms. In unweaned pigs, wrong feeding of mother.	Excrement soft and watery; sometimes streaked with blood.	*Adults:* purgative of castor oil; give a quart of thin flour gruel; feed sparingly. *Young pigs:* remove cause; give 2 teaspoonfuls thrice daily of following mixture; 2 drachms bismuth subnitrate, 80 grains of salol, and water to 2 oz.
Epilepsy ..	Teething: close confinement ; worms; too coarse, strong or innutritious food; food containing too much nitrogenous matter	Falling: rolling of eyes; champing of teeth; jerking of legs and head.	If worms suspected, treat as for these. Otherwise give teaspoonful Epsom salts followed by 10 grains bromide of potassium two or three times daily for a period.
Gastritis ..	Improper feeding ; too much salt or washing soda in wash.	Loss of appetite; vomiting; acute thirst; probably diarrhœa.	Withhold solid foods; feed on gruel; dose of castor oil (2½-3 oz.).
Lice ..	Ill-health, dirty housing and general bad management.	Unthriftiness and continual scratching and rubbing.	Pigs should be dressed with whale oil and sulphur ointment. In bad cases a bath with 5% solution of creosote is good.
Pneumonia ..	Cold, damp sties.	Cough; difficult breathing; shivering; loss of appetite.	Sloppy, flourishing diet; rub patient's sides with mustard water; house in dry, warm sty. Feed with warm food.
Rheumatism	Exposure to dampness and draught; poor and insufficient litter; innutritious feeding.	Inability to walk; joints, especially knees and hocks, swell and are tender to the touch.	See causes; purgative of castor oil; also ½ teaspoonful bicarbonate of potash in food daily for four days; wrap limbs in flannel and rub with mixture of spirits of turpentine, 2 oz.; liniment of belladonna, ½ oz.; spirit of hartshorn, 1½ oz.; olive oil, 2 oz.
Rickets ..	Worms: unsuitable food; lack of exercise.	Paralysis of hind parts; pig unable to walk.	Correct conditions in every way; add charcoal to diet of young pigs; milk also, if obtainable, with 10 grains of phosphate of calcium at each meal.
Ringworm .. N.B.—Very infectious.	Filthy conditions in house and sty; contagion.	First, red patches covered with pimples on croup, sides, flanks and sides of abdomen. Later, scabs.	Isolate patient; disinfect quarters; dress parts daily with following: Pure rape oil, 16 oz.; oil of tar, 2 oz.; paraffin oil, 1 oz.; glycerine, 1 oz.; flowers of sulphur, 2 oz.

Diseases.	Causes	Symptoms	Treatment
Scurf	Improper feeding and lack of exercise in open air.	Acute restlessness and irritation.	Give plenty of vegetable food; exercise; anoint body with mixture of pure rape oil, 6 oz.; carbonate of potash, drachm; liquid subacetate of lead, 2 drachms; water to 12 oz. Shake before applying.
Swine Erysipelas..	Febrile disease caused by a bacillus.	Variable. Nettle-rash, skin eruption, and red or violet patches, illness, high temperature, even loss of tail or ears.	Protective vaccination may be practical. If the disease is suspected isolation is essential and a veterinary surgeon should at once be called in. (May be confused with Swine Fever.)
Swine Fever	A germ encouraged by dirty sties and general bad management.	Loss of appetite; unsteady gait; purple rash on ears belly and hocks; diarrhœa; great thirst and high temperature.	Notify police immediately an outbreak is suspected and call in the vet. Isolate all affected and suspected animals and observe the closest cleanliness and antiseptic precautions.
Worms ..	Close confinement; withholding of vegetables, earth, grit, and pure water.	Restlessness; voracious appetite; poor, thin condition; scurf.	Give teaspoonful to desertspoonful per pig (according to size) of following powder: Santonin, 1 part; locust bean meal, 6 parts; areca nut (powedred), 2 parts; fine linseed meal, 6 parts.

R **APE.** Plant closely allied to the turnip and swede, but cultivated on farms for its foliage. Being hardy and therefore able to stand through the winter, is particularly suitable for feeding to lambs in February and March.

RATS, CONTROL OF. The success or failure of every rat-proofing scheme primarily depends on how completely it safeguards all possible food supplies. In the absence of food, rats will move to more favourable quarters.

It is of first importance to protect the actual buildings in which food for man or for stock is habitually stored. Stop up all holes either with cement, repair defective ventilators, seal or mend broken windows, and make every door tight fitting so that a rat cannot squeeze underneath.

Many rats are accustomed to gain entry to corrugated sheds by worming their way beneath the tin at the base. Such sheds are best rendered vermin proof by a narrow cement fender, prepared by running a trench 1 ft. deep round the building and filling it with cement to a height of about 6 in.

In rat-infested areas it will pay the farmer to protect any corn stacks whose threshing is not likely to take place for two or three months. The best safeguard is to surround each stack with galvanised wire netting of ½ in. mesh. Bury the netting in the earth and let it stand not less than 3 ft. high with the upper edge bent outward.

Hen-houses depending on the ground for a bottom invite rats. Wherever possible these dwellings should be raised off the ground and given a wire skirting.

Protective measures, however, are not enough by themselves against so formidable an enemy as the rat. There must be a constant war to the death waged against him, with ferrets, traps, guns, poisons and gases as the principal weapons against his native cunning and voracity.

Since rats have a penchant for frequenting ditches and drains, it is often possible to account for a number by setting traps beneath the water leading into pipes.

Extreme care is necessary in using poison, particularly in times when there is a shortage of water, as a rat's first instinct, after taking poison, is to seek water to refresh himself, and should he end up in a well there is no telling what horrible disease might result.

The skilful poisoner invariably " feeds " the holes with meal three or four times before mixing in the deadly dose, thus the better deceiving the rats into taking it. A trough with a little meal in it may be intentionally left outside a pigstye or duckhouse at night and if the meal disappears, one knows the bait is ready for poison. One of the best and safest poisons is squill.

The Rats and Mice Destruction Act make it compulsory on everyone to destroy rats and mice wherever found.

Note. Under war-time regulations every farmer is bound to surround stacks at threshing time with netting in such a way that rats in the stack cannot escape, and to destroy them.

ROCK SALT. All animals—particularly cattle, horses and sheep—require salt in some form or other. Some feeders supply it in the ration, but the better way is to give it in the form of a salt-lick or lump of rock salt in some place where the animals can get at it easily.

ROTATION OF CROPS. In order that it may retain its fertility arable land usually carries different crops season by season. In other words, crops must be rotated.

There are many forms of crop-rotation in use to-day varying in length from four to eight years. Which system is used depends to a great extent on the type of land and the class of farming followed.

The most-used rotation is the Norfolk 4-course system, which is as follows :

First Year: Cereal (wheat) sown in autumn.

Second Year: Fallow crop—roots, cabbages, potatoes, etc.

Third Year: Spring-sown barley or oats.

Fourth Year: Leguminous crop, such as clover, peas or beans.

In the fifth year the rotation reverts to autumn-sown wheat.

The following are some other typical crop-rotations :

5-year rotation for good loams. 1st year, swedes ; 2nd, barley ; 3rd, clover ; 4th, wheat ; 5th, oats.

5-year rotation for chalky soils. 1st year, swedes ; 2nd, barley or oats ; 3rd, seeds for mowing.; 4th, seeds for grazing ; 5th, wheat.

8-year rotation for heavy soils. 1st year, fallow crop ; 2nd, oats ; 3rd, beans ; 4th, wheat ; 5th, bare fallow ; 6th, oats ; 7th, 8th, wheat.

8-year rotation for sheep-farmers. 1st year, two catch-crops, say, vetches followed by turnips ; 2nd, swedes ; 3rd, wheat ; 4th, barley; 5th, early catch-crops such as rye followed by roots ; 6th, barley with seeds ; 7th, clover and seeds ; 8th, wheat.

RYE. Cereal not much grown in this country except for feeding green. During the war period it is being grown more extensively, for the grain is valuable for bread-making or for poultry. The straw is particularly good for thatching both stacks and houses. Rye is very suitable for light land. Seed is usually sown in autumn at the rate of 2 or 3 bushels per acre.

SAINFOIN. A leguminous crop valuable for stock. See Crops.

SALT. See Rock Salt.

" SEEDS." The name popularly given to grass and clover mixtures that are only temporarily laid down.

SEEDS. It is of considerable importance to use for sowing, seeds that are of high quality—that is, are as free from impurities as modern methods can make them, and of as good germinating capacity as possible. If a seed sample were of 100 per cent purity and 100 per cent germination it would be regarded as perfect. If the purity were 95 per cent and the germination 90 per cent its real value would be (95 × 90) ÷ 100 = 85·5 per cent. If it were really bad, say 80 per cent purity and 70 per cent germination the real value would similarly be only 56 per cent. It is easy to see that if these two samples cost the same per 100 lb. the purchaser of the one would receive 85·5 lb. of good germinable seed for his money, while of the other sample he would receive only 56 lb.

Other points relating to seed value are speed or strength of germination ; strain ; and suitability for the soil, situation and purpose for which required.

SEPARATOR. A machine for separating cream from whole milk. See Cream.

SHEEP. There are three main classes of sheep, viz., Long-woolled, Short-woolled, Mountain Sheep.

Long-woolled sheep include the Leicester, Border Leicester, Lincoln, Cotswolds, Devon Long-wool, Romney Marsh, Wensleydale and Roscommon.

Short-woolled sheep include the Southdown, Shropshire, Hampshire Down, Oxford Down, Dorset Horn, Suffolk and Ryeland.

Mountain sheep include the Blackfaced, Welsh Mountain, Gritstone, Penistone, Cheviot, Herdwick, Lank, Exmoor and Dartmoor.

Feeding. During the greater part of the year sheep obtain most of their food from grass, and provided the pasture is good, very little supplementary feeding is required. When grazing is poor a pound or two of concentrated food—cake or crushed oats or both—should be supplied (see Feeding Stock, and Feeding Stock in War-time).

Lambing Time. A clean field should be selected for a lambing fold. The soil should be dry and preferably not too low-lying. It is a mistake to have a permanent lambing fold, unless it is yearly cleansed and thoroughly disinfected.

The cause of most losses among young

lambs is storms of wind and sleet. For this reason, plenty of good, dry litter in the lambing fold is a great advantage, since it affords protection against damp rising from the ground as well as from the raging storm.

Sheep do not suffer close confinement with impunity and the less the ewes and lambs are coddled, the better. It is well, however, to provide shelter of some sort when it is raining, snowing or sleeting. The best place is an open-fronted shed, or small enclosures of thatched hurdles.

A ewe about to lamb will generally be observed to wander apart from the rest. If permitted to leave the fold she may naturally draw towards the shelter of a hedge. Keep an eye upon her, but leave her alone. Be ready to render help to the lambs if they are weak and helpless when born ; it may be necessary to steady them on their feet whilst they have their first suckle.

In a lambing fold a stove or fire is a necessary adjunct, for weakly lambs may require a warm drink, and often a warm bath is the means of reviving a half-perished lamb. It must be roughly wiped dry with straw, or a piece of sacking after its warm bath and then at once be wrapped in a woolly rug or old blanket and tucked into a box lined with straw.

When the lamb's coat is perfectly dry and it is able to run around, it may be restored to its mother. a

After Lambing. The ewes, after lambing, must not be fed too generously, but on the following day may be allowed to run out of the fold with their lambs. See that they are provided with drinking water.

At a month old the male lambs should be castrated and all of them " tailed " as well. After this operation they should be allowed to rest, preferably under cover, and on no account should they be driven any distance.

In another week or two it will be noticed that the lambs are beginning to nibble the grass, and it will now be time to provide them with separate feeding accommodation not accessible to the ewes.

For this purpose a " creep " may be used, putting behind it a low trough in which, each day, a mixture of crushed linseed cake, kibbled locust beans, and crushed oats is placed. This they will soon learn to like, and the amount should be increased until they are getting about ¾ lb. per head per day.

Weaning should be done gradually, as it is essential that the lambs receive no severe check at any time during their life.

After weaning has been accomplished, the ewes should be put on the poorer pasture, the lambs feeding on the earliest grass and receiving as much concentrate as they will clear up in two feeds. See also Feeding Stock in War-time.

At a live weight of 70 lb. to 80 lb. they should be sold.

Shearing Sheep. The only way to learn shearing is to watch an expert at work. A few hints, however, may be given.

The cutting of the wool is effected some 2 in. from the point of the shears by moving the upper blade only, the lower blade being kept on the sheep's skin and only used as a divider or runner.

Where fat sheep are shorn before selling it is customary to shear close down and well, whereas with store sheep and breeding ewes, a little more is left to afford protection.

Shearing is always done against the fall of the wool.

Custom differs as to where the start should be made. The most convenient way, however, is to set the animal on its rump and commence by opening up the head and neck, and then proceeding to clear the wool off the belly.

The fleece should be neatly rolled up by turning the sides into the middle, rolling, and then tying the neck and tail pieces to form a compact roll.

Dipping Sheep. As a precaution against Sheep Scab it is compulsory, by law, to dip sheep at least once a year. A second dipping is often advisable, but not absolutely necessary.

There are many excellent proprietary sheep dips on the market. All must be " approved " by the Ministry of Agriculture.

The dipping bath should not be less than 3 ft. in depth and it should be narrow, so that the sheep can swim through it.

The sheep must not be *re-dipped with any arsenical dip before two months have elapsed* since the last dipping, otherwise fatal results may be anticipated.

(During the war the County War Agricultural Committee should be consulted as to current dipping regulations.)

Sheep Diseases. Fluke or Liver-Rot. Affected animals lose condition, become pendulous in the belly, hollow in the flanks, yellow on the skin, and finally pass into a stage of emaciation and die.

Treatment consists in drainage of the land ; spraying or dusting herbage and by water courses with copper sulphate to destroy the small snails that act as "hosts" for the fluke; dosing the sheep, with the aid of a veterinary surgeon, with extract of male fern or carbon tetrachloride.

Foot-Rot. This disease begins beneath the horn of the hoof, either at the coronet or at the toe, and if neglected, gradually extends and the hoof may slough. There is severe lameness to an extent that the animal has to rest on its knees whilst grazing.

The affected foot or feet should be pared, and dressed with butter of antimony or a proprietary remedy.

If a number of sheep are affected, run them through a trough containing a solution made by dissolving 1 lb. of sulphate of copper and 1 lb. of washing-soda in 4 gallons of boiling water and then adding 16 gallons of water.

Sheep Scab. This disease earned its name on account of the scabby condition commonly present upon the skin, as the result of the animal rubbing itself. The first sign of the trouble is a matting of the wool.

Sheep Scab is a notifiable disease (see Notifiable Diseases).

Affected animals have to be repeatedly dipped, and a clean bill of health has to be given before the restrictions as to movement, etc., are removed. (See Dipping Sheep.)

Sturdy or Gid. Due to the presence of a Bladder Worm pressing on some portion of the brain or spinal cord.

Slaughter is the most economical method of dealing with the trouble.

Foot-and-Mouth Disease. See Cattle.

Hoven. A digestive trouble similar to that of cattle (see Cattle). A good dose for an affected sheep is ½ oz. of turpentine in ½ pint of linseed oil.

Note. Leaflets and a bulletin on various diseases of sheep (and other farm animals) are issued by the Ministry of Agriculture.

Pests of Sheep. Sheep's Nostril Fly. Eggs are laid in June on sheep's nostril and larvæ work up and feed inside cavity, being sneezed out when full grown. Treatment consists of syringing passages of nose with salt and water. Also apply coating of tar and oil to side of salt troughs.

Ticks and Keds. Live in the wool and feed on the blood. Dipping is remedy— See Dipping Sheep.

SILAGE. Food for cattle, sheep and horses, consisting of grass, clover, etc., cut and preserved in a container called a silo, whilst still in the green state. The process of ensilage is of the highest value, and even quite a small holding on which any class of stock is kept would find the produce a rich and excellent food for their animals.

Silage may be made from materials (short young grass, lucerne) that provide a fodder rich in protein, or from others (maize, sugar beet tops, potatoes) that are rich in carbohydrates. The green stuff is packed in the silo in layers from day to day and thoroughly trodden down to exclude air as fully as possible. Fermentation should suffice to encourage lactic acid. Nitrogenous materials

(lucerne, sainfoin, short young grass) may need the addition of molasses to help fermentation ; the carbohydrate fodders do not require molasses. When the material has sunk and the silo is full it should be covered with a good layer of soil on sacks or boards, and the whole covered to exclude rain.

Dairy cows may receive daily from 28 to 80 lb. (sometimes much more) of good silage, the quantity depending upon the nature of the silage and the other feeding-stuffs available. From 5-7 lb. will replace 1 lb. concentrate.

Fattening cattle may get 30-50 lb. of silage as part of their winter ration, and stores rather less; sheep 1-5 lb.; working horses up to 15 lb.; and pigs 1-12 lb., depending on their age and type.

SILO. The container used for the process of ensilage, in making Silage (which see). The Silos are of many types, being made of concrete, self-locking concrete blocks, wire-bound metal sheeting, heavy timber, wire and stave enclosures lined with stout waterproofed paper. The use of a silo is often dispensed with in favour of a stack or pit, in which silage may be made successfully, though the percentage of "waste" is usually higher.

SPAYING. This is an operation to make female animals barren by destruction of the ovaries. In general it is best carried out by a veterinary surgeon.

STABLES. See Horses.

SWEDES. Valuable crop for feeding livestock in winter. See Crops.

SWINE ERYSIPELAS. See page 235.

SWINE FEVER. See page 235.

TARES or VETCHES. See Crops.

THATCHING. This is work for an expert. Unless the smallholder has a skilled thatcher among his workers, or is himself able to do it or prepared to learn, his best plan is to employ a professional thatcher.

THRESHING WASTE. See Chaff.

TRACTORS. There must be many smallholdings on which a medium-powered tractor would be of great value, not only to do the ploughing and cultivations, but to haul harrows, roller, drill, mowing machine and the like, as well as many stationary jobs like chaff-cutting. If of the right type a tractor may also do haulage work off the farm—to and from station or market.

In addition, and because the one economic way of using a tractor is to keep it occupied every possible day and hour at its fullest capacity, the tractor may do good work on hire on neighbouring holdings that do not possess one. Many a farmer often has urgent need to get something done quickly, while good weather holds, and co-operation of this kind is of the utmost help and importance.

The type of tractor to be purchased may well depend largely upon the type of holding on which it is to be used. For example, on a market garden holding the type should be one for row-crop work—high on the wheels, with great flexibility in turning, and ability to set the wheels to fit the width between rows.

TRIFOLIUM. Another name for Crimson Clover. For cultural details, see Crops.

TURNIPS. Highly valuable root crop for feeding cattle and sheep. See Crops.

VETCHES, or TARES. A valuable crop usually grown for feeding green to cattle, sheep and horses. See Crops.

WARTIME REGULATIONS. Since the outbreak of war it has been progressively necessary to impose a wide variety of control regulations. Many of these have become more stringent owing to the heavy demands on shipping for combatant services.

It would be impossible to give here a full account of such regulations, and the more so because the position is so fluid that they necessarily change very frequently.

Among the regulations are those concerning the coupon rationing of livestock; the control of prices of all the more important farm products; control of fertilisers; control of timber, transport, fuel, marketing, crops that may be grown, and so forth.

WARBLE-FLY. Pest that plagues cattle during the summer months. See Cattle.

WEATINGS. The new name given a few years ago by the Miller's Mutual Association to the meal that was previously known as middlings, sharps, dan, pollards, thirds, etc. The new product is standardised in two qualities all over the country and carries a guaranteed analysis.

WEEDS. Most people recognise as weeds any plants that grow in a crop where they are not sown or wanted. This applies to grass land as well as arable, though with a few exceptions the species are different.

In general, weeds in grass land are reduced by a combination of draining, liming, judicious manuring—especially with nitrogen—cutting of the weeds, close grazing with different classes of livestock, and autumn mowing of any brown tufts of ungrazed material.

On arable land weeds are reduced in accordance with their kind. Annual weeds are usually readily killed by ploughing under; by cultivations such as harrowing and horse-hoeing; and in some instances by spraying. Most biennial weeds may be destroyed in the same way.

Perennial weeds vary widely, but commonly present much trouble—e.g. couch, creeping thistle, in arable land, or ragwort, bracken, creeping thistle on grass land.

In corn crops annual and biennial weeds may be destroyed by spraying with a solution of copper sulphate or sulphuric acid.

Full information about weeds is given in leaflets and illustrated bulletins issued by the Ministry of Agriculture.

WHEAT. For cultural details, see Crops.

WILD WHITE CLOVER. An important clover for the improvement of grass land, on which it may be sown at the rate of a pound or so per acre. Almost always included in "seeds" for permanent pasture.

RABBITS

ABORTION. Premature birth of litter. Due to fright, disease, over-fatness. The remedies are obvious.

ABSCESS. See Ailments.

ACORNS. See Foods and Feeding.

ADULT. In small breeds adult age is about 6 months, and in large and medium breeds about 8 months.

AGOUTI. Denotes fur of a colour resembling that of wild rabbits.

AILMENTS. Following are the chief ailments to which rabbits are subject and the best methods of treatment.

Where it is necessary to give medicine the best plan is to mix the medicine in the food or drinking water. Where this is not practicable the next best means is as follows : The measured dose is put into a medicine dropper and the rabbit is wrapped in a sack or cloth, leaving the head uncovered. The head is raised, the end of the dropper inserted at the corner of the mouth and the medicine squeezed slowly down the throat. If the rabbit chokes or coughs lower the head temporarily.

Aniseed is a flavour much liked by rabbits. A drop of oil of Aniseed added to medicine often results in the rabbit taking this naturally, so avoiding the necessity of forcible dosing. Aniseed has a carminative or soothing action.

Wounds caused by scratches, bites (or, with Angoras, accidental cuts with scissors) usually heal readily without becoming septic. No treatment is necessary beyond an application of tincture of iodine.

Abscess. Collection of pus accompanied by swelling. Due to disease or neglected injury.

Make incision in swelling after painting with tincture of iodine, squeeze out pus, and dress cavity with tincture of iodine until inflammation subsides. If cavity is deep protect with dressing of boracic lint held in place after shaving the skin.

Aspergillosis. Disease of the lungs occurring occasionally in the rabbit. There is fever, hastened or laboured breathing, usually constipation. Treat as for Pneumonia (see page 262).

Blows. Term applied to an acute and quickly fatal condition due to paralysis of intestinal muscles and consequent suspension of intestinal movements. Is particularly prevalent in damp and chilly weather. Is believed to be due mainly to mouldy food, especially clover hay.

The appetite fails, belly swells enormously ; collapse, coma and death follow.

Treatment may effect a cure *if applied in very early stages.* It consists of dosing with 20 grains powdered areca nut dissolved in 3 tablespoonfuls of boiling water and allowed to cool.

Cancer. Not a common disease in rabbits, but occasional in old age. Internal cancer, principally occurring in the uterus in old does, is frequently unsuspected until brought to light by post mortem examination. Cancer of the skin begins as a small boil or ulcer, quickly increasing in size. There is no cure and affected rabbits should be destroyed as soon as the disease is recognised.

Canker, Ear. Very common complaint caused by a parasite that invades the ear and sets up acute irritation.

The symptoms are shaking of the head, scratching at the ears and, sometimes, holding the head to one side. Examination shows redness of the inside of the ear and, later, greyish brown crusts formed by the parasites and their excretions.

Clean the ear thoroughly, removing all crusts and matter with a piece of stout wire about 3 in. long, bent at one end to form a loop, the loop being encased with swabs of cotton wool soaked in peroxide of hydrogen.

When the ear is clean, apply the following mixture on a wool swab as above, making sure that the lower parts of the ear are not neglected : Iodoform 1 part, ether 10 parts, and olive oil 25 parts. Repeat once a week until all crusts or signs of inflammation have disappeared.

Coccidiosis. This is undoubtedly one of the worst of all the diseases of rabbits, often causing a very big percentage of deaths.

The disease is caused by a minute parasite that invades the intestines, liver, or nervous tissues in large numbers. There are two distinct types of the parasite, one causing the intestinal and one the liver form of the disease. Both types may be found in the nervous form, which is possibly due to secondary infection from the liver or intestinal forms. Apparently healthy rabbits may be " carriers " of the disease. That is to say, the parasites may be present in the droppings and thus contaminate food and water taken by other stock.

Symptoms of the Intestinal Form. (*Chiefly affects youngsters from 6 weeks to 5 months*) : Dullness and loss of appetite ;

more or less profuse diarrhœa. The attack usually terminates in death within 24 hours, often preceded by convulsions or paralysis.

Symptoms of the Nervous Form. (Affects stock of all ages) : There are no definite symptoms. Death occurs suddenly, with or without convulsions.

Symptoms of the Liver Form. (Chiefly affects adults) : Listlessness, thirst, rapid wasting of the back and hindquarters, with enlargement of the abdomen. This form may run a chronic course, spread over several weeks, or it may end in death in about 10 days, preceded by coma and perhaps diarrhœa.

Treatment : Good results have been obtained in some cases by the use of iodine as mentioned below, particularly in the liver or chronic form of the disease. Some few breeders claim a percentage of definite cures by the use of crude Catechu in the proportion of 15 grains to every gallon of drinking water.

Preventive Measures. If general hygienic measures are backed up by suitable food, which implies a dry rather than a moist form of concentrated food with plenty of clean water to drink, Coccidiosis is unlikely to appear, and if, in addition, a small quantity of iodine is added daily to the drinking water, its occurrence is still more unlikely.

An iodine solution recommended for this purpose is made by dissolving 2 parts of potassium iodide and 1 part of iodine resublimate in 50 parts of water. 1 tablespoonful of the solution should be added to every pint of drinking water used.

Colds. Common complaint. Colds in the head may occur as a result of damp or draughty hutches or by chills caught on journeys.

The symptoms are sneezing and a watery discharge from eyes and nose.

To treat colds it is necessary to provide extra bedding and keep the rabbits in a dry, clean hutch away from draughts. Give from 3 to 5 drops of sweet spirits of nitre twice daily in a little warm milk and apply a little oil of eucalyptus to the nostrils, also twice daily. Isolate as far as possible from other stock as a cold may be the forerunner of Snuffles (see page 244).

Colds may be confined to the eyes, and if so bathe the eyes twice daily with boracic lotion, carefully removing any matter that may collect in the corners.

Colic. A very common complaint, due to faulty feeding. The rabbit sits hunched up, evidently in pain and often constipated. The abdomen may be swollen and tender.

Give 1 teaspoonful of the following mixture : Oil of aniseed, 1 part ; linseed oil, 4 parts ; repeating in 12 hours if necessary.

Constipation. This is not a common complaint among rabbits when they are suitably fed and housed, but is apt to occur if green foods or roots are not freely given and if drinking water is withheld. It may also occur if hutches are not sufficiently roomy to allow of free movement.

The symptoms are inclination to mope in a corner of the hutch, loss of appetite, and, possibly, swelling of the abdomen.

A good feed of greens, if the rabbit will eat, and a run on the rabbitry floor will often put matters right. In obstinate cases give opening medicine, e.g. 1 teaspoonful of linseed oil or medicinal paraffin (see Aperients). A dose of Glauber's salts given once a week in the drinking water is a good preventive measure.

Cysts. Usually caused by the cystic or intermediate form of one of the tapeworms that commonly infest dogs and occur as a result of tapeworm eggs being taken in with green or other food contaminated by dogs.

Cysts originating from the dog tapeworm, occur under the skin or in muscular tissue in almost any part of the body, the back, neck, lower jaw and behind the eye (causing the eye to bulge) being common situations.

Unless any vital function is interfered with these cysts do not cause illness, but if very large they are disfiguring. Removal is comparatively simple, but is best carried out by a veterinary surgeon under a local anæsthetic.

Another common form of cyst, also caused by the tapeworm, appears in the abdominal cavity either singly or in large numbers. Removal is impossible, but after a time these cysts tend to harden and die.

Dandruff. Non-parasitic disease of the skin, due to skin irritation from any cause, poor condition, overcrowding, unsuitable feeding.

The symptoms are the appearance of greyish-white scales on skin and coat and, occasionally, bare patches. Coat may be dry and lustreless.

Improve general health by good feeding ; provide greenfood liberally ; and give whole linseed with food on alternate days ($\frac{1}{2}$ teaspoonful for small and 1 teaspoonful for large rabbits). Clip fur or wool close to the skin and dress daily with sulphur ointment after washing away scales with 2 per cent solution of carbonate of soda.

Diarrhœa. See Scours.

Drooping Ears. One or both ears may droop as a result of paralysis of the small muscles which hold them erect. The cause is not definitely known, although some think the condition is hereditary or due to intensely hot or close weather.

Feed well and give support to the ear by means of a cardboard splint held in position by adhesive strapping.

Drooping ears may also occur as a result of tearing of the muscles or even breaking of the ear cartilege itself when rabbits are handled by the ears (see Handling). Often the damage is beyond repair, but sometimes the torn muscles may be induced to heal if the ear is supported with a splint as described above.

Eczema. A condition of the skin arising from too much heating food, lack of greenfood or drinking water.

The symptoms are greyish scales or scabs at roots of hair, generally round eyes, nose and roots of ears. Perhaps loss of hair.

Cut out heating food and give greenfood, hay and plenty of water. Cut wool or fur at affected parts and dress with sulphur ointment.

Eyes (Sore). This trouble, as distinct from Cold in the Eye (see Colds) may arise from various causes.

In Adult Rabbits. Rabbits, particularly bucks, when excited, sometimes scratch in the urine-soaked litter in the used corner of the hutch and a small particle enters the eye. Urine is highly irritating and inflammation is set up, accompanied, first by a watery and later by a thick discharge. Treatment consists of bathing with boracic lotion as directed for Cold in the Eye, applying afterwards a little Golden Eye ointment to the eyelids.

In Baby Rabbits. Baby rabbits may become affected with sore eyes (Ophthalmia) when their dam fouls her nest. Either the eyes fail to open in the normal way about the eleventh day, or they may open, but appear inflamed and swollen.

Bathe the eyes with boracic lotion, if necessary opening them very gently and carefully removing all matter. Pull down lower eyelid and squeeze lotion over eyeball. Insert a small bead of Golden ointment between eyelids on alternate days.

If the eyeball appears opaque or filmy 1 or 2 drops daily of a solution of percloride of mercury, 1 part; and water 10,000 parts, should effect a cure if applied early.

Sore eyes in baby rabbits will lead to permanent blindness if neglected.

Favus. A contagious disease of the skin and hair caused by a vegetable parasite. Most commonly affects the head, ears or legs.

The symptoms are brownish-yellow or grey scabs with slightly raised edges, often with a tuft of hair in centre. General health does not suffer if condition is treated promptly.

Isolate affected rabbits, remove all scabs with a blunt knife and dress parts daily with tincture of iodine.

Wash the hands thoroughly with disinfectant soap, after handling affected animals or, better still, wear rubber gloves when giving treatment, and wash these after use.

Gastritis. Common complaint of rabbits, due to overfeeding or food of inferior quality, particularly if fermented or mouldy.

There is loss of appetite or capricious appetite, thirst, constipation alternating with diarrhœa, dull appearance.

To treat an affected rabbit, rest the stomach by giving easily digested food in small quantities, e.g. broad bran. Give 5 crystals of Glauber's salts dissolved in milk daily. The herb known as shepherd's purse is beneficial in this condition.

Hernia (Rupture). The most common type, Umbilical Hernia, which occurs in both sexes with equal frequency, is shown by a lump of varying size in the middle line of the belly. Inguinal Hernia, occurring in bucks only, is recognisable by a swelling within the scrotum (the outer covering of the testicles), and is far less frequently met with than the first-named type.

If small, hernia is best left alone. The treatment of a large hernia is a matter for a qualified veterinary surgeon, and it is an open question whether this is worth while.

It is unwise to breed from does afflicted with hernia as death may occur suddenly at any time from strangulation of the hernia.

Hocks, Sore. Rabbits kept in dirty or damp hutches may suffer from sore hocks.

The hocks should be washed with boracic lotion, removing any matter. Dress with boracic ointment and bandage, keeping bandages in place with adhesive strapping. Give liberal greenfood, including cabbage leaves and dandelion, and, as concentrated food, give broad bran, adding a pinch of flowers of sulphur to it. Keep the hutch scrupulously clean.

Indigestion. Common complaint, especially during spells of bad weather and in periods of easterly winds, or on damp, chilly days, with little or no sunshine. The patient may refuse food altogether, but more probably will begin to eat and soon exhibit signs of discomfort.

Start treatment by giving the patient a good meal of dandelion leaves. These are an admirable pick-me-up and general tonic, acting through the liver and kidneys. If dandelion is ineffective give four or five crystals of Glauber's salts dissolved in a little milk or warm water.

For the treatment of more serious digestive troubles see Blows, Pot Belly, Colic, Gastritis and Megrims.

Influenza. This may attack rabbits but is generally of a mild type. The symptoms are fever, shivering fits, general depression, loss of appetite, and thin discharge from eyes and nose.

Keep the rabbits warm and entice

appetite with any food particularly liked. Local treatment as for Snuffles, which see.

Mange. Common complaint caused by a parasitic mite acquired through contact with infected rabbits or hutches. Parasite burrows into skin causing crusts and scabs.

The symptoms are scratching, or rubbing against hutch, sores or pimples, falling of fur and appearance of white or greyish crusts and scabs. Lips, nose, head and legs most commonly affected.

Dress affected areas with soft soap, allowing this to remain for half an hour. Wash off with warm water and rub in sulphur ointment twice daily. Repeat soap dressing every five days.

Megrims. Fits caused by digestive trouble.

Symptoms are staggering, rolling over, convulsive struggling, after which the rabbit lies exhausted with head twisted to one side and resting on the ground.

Keep in a warm airy hutch, correct errors in feeding and give 1 teaspoonful of potassium bromide 2 grains, water 2 oz., twice daily.

Paralysis. A common complaint that generally attacks the hindquarters. It may be of obscure nervous origin, when there is no known cure and speedy destruction of the rabbit is the most merciful course. The condition may also arise as a result of a damp hutch and here a cure may be effected with perseverance if measures are taken promptly.

The symptoms are powerlessness of the hindquarters which are dragged along when the rabbit moves. General health and appetite are usually unimpaired.

Keep in warm dry hutch on soft hay. Cut fur or wool and paint hind legs and lower part of back daily with tincture of iodine until flaking of skin occurs. Make small pills by mixing 2 grains camphor and 1 grain sulphate of iron with a little powdered liquorice and treacle and give one pill every alternate day. Feed well.

A type of paralysis that appears to be infectious sometimes occurs among a litter of young rabbits. Victims are best destroyed, but the infection may be prevented from spreading to the remainder of the litter if they are given for about a week a daily mash in which a 5-grain tablet of sodium salicylate has been dissolved.

Pneumonia. Fairly common trouble, due to colds, injuries, bad air, particularly if the general health is below par.

The symptoms are usually sudden in appearance and include listlessness with high temperature, increased and laboured breathing and perhaps wheezing.

The patient must be kept warm with plenty of bedding, but have *plenty of air without draught.* Tempt appetite with small quantities of nourishing food.

Bread and milk, carrot and dandelion leaves will often be eaten when other food is refused. Give 3 drops tincture of digitalis three times a day.

Pot Belly. A trouble that chiefly affects youngsters up to 6 months. It is due to overfeeding or irregular feeding, particularly of greenfood, combined with lack of exercise.

Swelling of the abdomen and sometimes diarrhœa.

Isolate the patient in a hutch as large as possible and give a run on the rabbitry floor daily. Give dry food only, cutting out greenfood until cured, with plenty of drinking water. Dose with 5 grains of ammonium benzoate in a little water twice daily.

Red Water. A fairly common complaint arising from inflammation of the kidneys. Although rarely fatal, it must not be neglected.

It is due to damp hutches, exposure to cold, errors in feeding. The urine is of a blood-red colour. There is occasionally dullness, but appetite is not lost.

Keep warm with plenty of bedding. Feed on broad bran slightly damped with warm water, hay, carrots, greenfood. Barley water to drink. Dose with 2 drops of sweet spirits of nitre in a teaspoonful of warm milk twice daily.

Rickets. A very common complaint of young rabbits, in which the bones, instead of becoming dense and hard in the normal manner, remain soft and pulpy and thus bend easily, causing deformities in various parts of the body.

The bones of the forelegs in rickety youngsters often become permanently " bowed " as a result of their being unable to support the weight of the body without giving way. Distortions of the ribs and spine are also commonly met with, and, apart from their unsightly appearance, may cause injury to or displacement of vital organs.

The chief cause of rickets is lack of fresh green leaves, and if greenfood is unobtainable during the winter, it is imperative that some substitute providing the same bone-strengthening elements be secured. It is found in both milk and cod-liver oil. From 3 to 5 drops of cod-liver oil (or 1 tablespoonful of milk in normal times) should be allowed for each youngster.

Cod-liver oil (or milk) should also be given to does in kindle if they are not receiving greenfood regularly, otherwise the youngsters may be born with a rickety tendency and may die within a few days of birth.

Ringworm. An infectious disease of the skin, in which round or oval areas appear in almost any part of the body. On these areas the coat is broken or falls out.

Give treatment as for Favus (see Favus).

Scours (Diarrhœa). Frequently a symptom of digestive trouble and may accompany various diseases affecting the digestive organs (see Coccidiosis, Pot Belly).

The symptoms are extreme looseness of the bowels.

As regards treatment, correct errors in diet, withhold all greenfood but shepherd's purse, strawberry leaves and runners or the bark and leaves of the ash tree. From ½ to 1 teaspoonful of syrup of hypophosphites, according to size of rabbit, has a very beneficial effect in this condition.

Slobbers. Common complaint resulting in a continual running of saliva from the mouth. It is due to ulceration of the mouth, generally arising from indigestion or general debility occasioned by bad housing conditions.

House comfortably with soft bedding and replace coarse dry food by bran mashes. Wash the mouth out with a 2 per cent solution of potassium chlorate once or twice daily and give twice daily 1 teaspoonful of the following mixture : To 3 oz. water add 1 teaspoonful each of potassium chlorate and powdered ginger and allow to stand for a few hours. Shake well before use.

Snuffles. One of the commonest rabbit complaints, snuffles is an inflammation of the nasal passages. If left unchecked, it spreads to the bronchial tubes and finally to the lungs, where acute and generally fatal penumonia is set up.

Predisposing causes of snuffles are the leaving of hutches unprotected from rain and cold winds and allowing the occupants to be in a perpetual state of chill and discomfort.

The trouble starts as a cold in the head, and at this stage, if the bad housing conditions are remedied and a drop of oil of eucalyptus is applied to the nostrils twice daily, a cure will be effected without further developments.

If nothing is done, however, the disease will progress, possibly ending fatally in a few days, or, more probably, running a relatively long course, during which the affected animal will be a constant source of infection to other stock.

The symptoms to watch for are frequent sneezing, thick mucous discharge from the nostrils, and sometimes more or less discharge from the eyes, noisy and laboured breathing and poor appetite.

Immediately these symptoms are noticed, isolate affected rabbits as far as possible from other stock and tempt the appetite with food that is known to be specially appreciated. Protect the patients well from damp and draughts and give plenty of bedding. Give the drinking water warm and limit the allowance, re-

moving the vessel as soon as the rabbits have had their daily drink. Keep the hutch scrupulously clean and dry.

Mix 3 drops of oil of eucalpytus with 1 fluid ounce of olive oil and apply to the nostrils twice daily with a small oilcan after wiping away all discharge from the nose and face with cotton-wool dipped in a mild antiseptic.

Spotted Liver. Yellow or white spots on the liver are almost invariably due to the chronic or liver form of Coccidiosis (see Coccidiosis).

Strangles. A disease characterised by the formation of a large " cold " abscess in the neighbourhood of the lower jaw. Such an abscess, which does not tend to burst, must be opened, with a sterilised razor-blade, cleared of its contents and the cavity dressed daily with tincture of iodine until inflammation has disappeared.

Teats, Swollen or Inflamed. Fairly common complaint. Due to a microbe that gains access through slight injury.

The affected doe appears ill, and, if nursing a litter, will neglect them. Heat, redness and swelling of teats, which are extremely tender to the touch, are often symptoms.

Apply hot fomentations constantly. If the doe is nursing remove litter, fostering them if possible, and withhold all water and greenfood in order to stop the flow of milk. A mash made by drying off tea leaves with bran often assists in this respect.

Vent Disease. A common complaint resulting in superficial inflammation of the external genital organs, accompanied by small pimples, which tend to ulcerate and, eventually, to form scabs. The disease is passed from one rabbit to another by the act of mating or by placing a healthy rabbit in a hutch formerly occupied by a victim of the disease.

Clean the affected parts thoroughly with cotton wool dipped in a mild antiseptic (weak solution of boracic crystals or permanganate of potash), removing all scabs, which come away readily without causing bleeding. Dry the parts with soft rag or lint and smear with blue (mercurial) ointment.

This treatment applied every other day will generally effect a cure in from ten days to a fortnight.

Worms. Rabbits are commonly infested with worms. Small worms appear in the droppings, and there is undue thinness.

Give 7 grains powdered areca nut and 1 grain santonin, and 1 teaspoonful olive oil an hour later.

Wryneck. A condition in which the neck is twisted and the head held permanently on one side. May occur in advanced cases of canker of the ear (see Canker).

AIR-DRYING.

AIR-DRYING. Pelts of fur rabbits sold to the furrier must always be air-dried. The method is as follows : Place fresh pelt on a board, fur side downwards. Pull into shape (mainly oblong) sufficiently to eliminate " pockets " or wrinkles, *but do not stretch tightly*, and fix with drawing pins. Hang felt in even temperature until dry, removing with a knife any flesh from fur after about 24 hours. Air-dried pelts are ready for marketing or may be stored almost indefinitely. See Pelts.

ALASKA, OR NUBIAN.

ALASKA, OR NUBIAN. A black rabbit said to have been produced originally by crosses involving Dutch and Silver Greys. Fur dense, soft and glossy. Yields a very useful pelt. Weight 5-6 lb.

ALBINO.

ALBINO. A rabbit having pure white fur or wool, and pink skin and eyes.

ANGORAS.

ANGORAS. The keeping of Angora rabbits has made very extensive strides in the last few years and the breed is now very popular.

Angoras are not kept with the main object of producing pelts or meat, although the pelt of any rabbit dying in good condition of coat has a certain market value, and unwanted Angoras may be killed off for table purposes. They are kept primarily for their wool, which has a good sale, and there are great possibilities in its production.

The wool is clipped or plucked from the living rabbit, and is used in the manufacture of yarns, materials, hats, etc.

The wool of the purely British, or exhibition, Angora, which is of great length and very fine texture, is usually less in demand than that of the newer utility type which has been obtained mainly by careful crossing of British with Continental stock. The wool of the utility Angora is comparatively short and of a texture that inclines to coarseness or " strength of fibre". There is also a fairly wide distinction between the types in other respects.

The exhibition Angora weighs from 6–9 lb. and, when sitting with the wool brushed up, gives an impression of roundness. The ears are furnished with tufts of wool, and the head, which is broad and of fine appearance, is well woolled, as are the hind legs and feet.

The utility Angora is generally larger and more massive, and its head, ears and feet often tend to be plain rather than furnished.

The Angora is naturally an albino with pure white wool and pink eyes, but coloured varieties have been " made " by crossing, in the first place, with short-coated coloured breeds. Thus we have Angoras of Blue, Brown, Black, Smoke, Golden and various other colours. It is the wool of the White Angora only that is in much demand.

There are, however, possibilities that coloured Angora wool may be required in increasing quantities for hand spinning, and the most popular colours are Brown, Black and Golden.

Harvesting the Wool. Harvesting is the term used to denote the plucking or clipping of wool from Angora rabbits.

Before harvesting commences the coat must be brushed thoroughly to free it from hay, food or other foreign matter the presence of which reduces the value of the harvested wool. See also Grooming.

There are two methods of harvesting the wool, plucking and clipping. The latter is usually preferred.

Clipping Angora Wool. Unlike plucking (see below), clipping may be done at any time when the wool has reached a suitable length, irrespective of whether the coat is loose or not. A sharp pair of scissors is absolutely essential for this operation. Tailors' cutting-out scissors answer the purpose well.

After the preliminary brushing, the coat should be parted down the middle line of the back and the wool clipped along the right side of this parting from rump to neck. When the first line or " lock " of wool has been removed the process should be repeated immediately below the clipped area and so on, until the entire right side is cleared of wool, the rabbit being placed lengthwise towards the operator with its head facing to his left.

The left side is next clipped in the same way, the rabbit's position being reversed so that its head faces the operator's right. The operator should then face the rabbit squarely and, beginning at the top of the rump, clip the wool in lines downwards in the way just described. While doing this the tail should be held away from the rump with the left hand in order that the wool behind it may be clipped and that the tail may not be cut.

After clearing the rump, trim the tail. Then stand behind the rabbit and, holding its ears in the left hand, gently raise its head so that the wool of the frontal area may be clipped. This is done on the same principle as the rest of the body— in lines, working from the chin downwards.

Finally clip the belly wool, holding the rabbit in the same way as for plucking this part. Care is needed to avoid the teats in does, and it is well to ascertain the position of these before clipping a doe. There should be eight teats, four on each side of the middle line of the belly. In an adult buck with dense belly wool be careful not to injure the testicles when clipping the lower part of the belly.

Plucking of Angora Rabbits. An Angora must be plucked only when the coat is " ripe " : that is to say, when it is falling naturally and can be taken readily from the rabbit by means of a gentle pull. If plucking is attempted when the coat is not ripe the process will cause the animal great pain and the skin will probably be torn in places. An Angora plucked at the right time feels no pain.

The wool of utility Angoras generally becomes loose when the coat has reached a length of from 2 to 3 inches, and therefore it is well to make a note of all rabbits with 2 in. of coat and to be on the watch for the time when this begins to loosen. This looseness of coat may not occur all over the body at the same time, and in this case the complete harvesting of one rabbit may be spread over a week or ten days.

To pluck, steady the rabbit with the left hand, and with the thumb and fore-finger of the right hand pluck off all wool that comes away readily.

To pluck the belly wool, sit down and hold the rabbit on its back between your knees, with its head away from you. If you hold the ears with the left hand and press the body gently against your right thigh by means of your extended left arm, the rabbit will not find it easy to spring up suddenly. Plucking of the belly wool must be done with care, as the skin here is thin and easily torn.

Grading of Angora Wool. It is essential that Angora wool be graded before it is sold. Ungraded wool—that is, wool consisting of various grades—usually commands the price of the lowest grade !

As the wool should be handled as little as possible once it has been taken from the rabbits, grading is best done at the time of harvesting.

Excluding Super Grade wool, which is very fine wool of exceptional length, Angora wool is divided into three grades.

(1) First Grade Wool must be at least 3 in. in length and perfectly clean. Any suspicion of vegetable or other foreign matter reduces the value of the wool considerably.

Further, the wool must be " free," which means that the fibres must be definitely separate one from another, and not tending to stick together. Wool that is stored for a considerable time ceases to be free, for which reason marketing as soon as possible after harvesting is advisable.

(2) Second Grade Wool consists of clean, free wool, as described above, with an average length of 2-2½ in.

(3) Third Grade Wool consists of wool that is clean, free and between 1 and 2 in. in length.

In addition to these three main grades, breeders also have to consider tangled wool and, where fine-coated Angoras are kept, webbed and matted wool.

(4) Tangled wool is that in which the fibres have stuck closely together and can only be pulled apart with more or less difficulty. The condition is seen in wool that has been stored or packed tightly and also with wool from a rabbit that has not been regularly groomed.

(5) Webbed wool is that which is harvested at the stage when the fibres of the wool cling together at the roots of the coat to such an extent that the coat, when blown apart, does not expose the skin. What actually has happened is that the fibres, whilst free at the tips and for some way down their length, cling together at the roots. If webbing of the coat is allowed to continue unchecked, the upper part of the fibres will soon be involved and the coat will become matted.

Tangled, webbed and matted wools, although saleable, realise a much lower price than those of the higher grades.

(6) Soiled wool, which includes that containing vegetable or any other foreign matter, and also that which is stained, comes into the lowest category of all, and its price is so low as to be almost negligible.

Storing Wool. Separate boxes or tins should be kept for the different grades of wool and all wool put straight into its right tin immediately it is harvested. Super and fine First Grade wool should be placed in the tins or boxes in locks or lines as it is taken from the rabbits, these locks being arranged in rows so that all the fibres lie evenly and smoothly in one direction.

If this wool has to be kept in a tin for a time before being sold it is well to line the tin with tissue or other thin paper in such a way that the wool may be lifted out intact and transferred without handling to the boxes in which it is to be sent away. Other wool of good quality need not be arranged in locks, but it should be laid carefully in the tins or boxes and in no circumstances should it be compressed.

Packing Wool for Dispatch. When packing wool for dispatch, be sure to tie all parcels securely so that there may be no risk of their coming open. Name and address, *clearly written*, should always be enclosed.

Cardboard boxes give the best protection in transit and should therefore be used for all wool of good quality.

APERIENTS. The best remedies for the common complaint of constipation are medicinal paraffin, linseed or olive oil in teaspoonful doses, or Glauber's or Epsom salts (1 tablespoonful to each gallon of drinking water).

ARGENTÉ. One of the oldest breeds of tame rabbit, kept for fur, flesh and exhibition. The three recognised varieties of Argenté are :

Argenté De Champagne. Useful breed of fur rabbit, having dense and silky bluish-white fur ticked with black with undercolour of dark slate blue. Weight 7-10 lb.

Argenté Bleu. This variety has bluish-white fur with lavender-blue undercolour, dense, silky and glossy. Weight about 7 lb.

Argenté Créme. Has creamy-white fur with orange undercolour, silky and dense. Smaller and neater than the Champagne or Bleu. Weight 4½ lb. to 6½ lb.

ASPERGILLOSIS. See Ailments.

AWNS. This term is sometimes applied to the "guard hairs" in the coat of a rabbit (see Guard Hairs).

BABY. Rabbit from birth to 3 months old.

BARLEY. See Foods and Feeding.

BASKET TRAVELLING. See Boxes, Travelling.

BEAN MEAL. See Foods and Feeding.

BEAVER. Recognised fur breed. Fur is brown resembling real beaver with bluish-grey undercolour, very lustrous, silky and dense. Weight up to 9 lb.

BEDDING. A good bed of straw or hay is advised for all breeds in cold weather, and at all times for does in kindle or nursing, and for exhibition Angoras, unless the latter are kept on wire floors. Best bedding for exhibition Angoras is wheat straw cut into 12 in. lengths. Rough hay and bracken may also be used with advantage. A base of peat moss may be used when batches of rabbits are run together in a large pen such as an old pigsty. Sawdust is often used as a base and may be quite useful, but not for Angoras; it will rot down gradually in the compost heap for garden use.

BEIGE. Exhibition and fur breed. Has silky loose-lying fur of dark chamois or light sea sand colour, faintly ticked with Vienna blue, shading upwards from light to dark. Maximum weight 5-6 lb.

BELGIAN HARE. One of the oldest and largest exhibition and table breeds. Has a fine graceful appearance, closely resembling a hare. The colour is rich deep tan or chestnut shade ticked with black.

BELLOWS resembling kitchen or smith's bellows are much used for keeping the coats of exhibition Angoras open and free from matting. See Grooming.

BEVEREN. Very popular flesh and fur breed. There are several varieties, Blue, Black, Brown, and White. An important characteristic is the "mandoline shaped" body that makes the Beveren an excellent table rabbit. The fur is soft, silky and lustrous.

Colour standards are : *Blue :* Clear shade of light lavender blue to skin. *Black* (formerly called Sitka) : Glossy jet black ; dark blue undercolour. *Brown :* Soft nut brown ; beige undercolour (dark as possible). *White :* Pure white. The White Beveren is not an albino ; the eyes should be black or blue.

BLEEDING. To ensure whiteness of flesh, table rabbits must always be bled immediately after killing. A sharp pointed knife stuck into the neck just below the ears will sever the jugular vein and if the carcase is then held head downwards, all the blood will rapidly drain away. Care is necessary not to soil fur in pelt rabbits during the operation.

BLOWS. See Ailments.

BLUE AND TAN. Small compact breed, clear dark blue in colour with rich tan markings. Fattens readily for table and yields excellent flesh. Weight about 5 lb.

BLUE BEVEREN. See Beveren.

BLUE SIBERIAN. See Siberian.

BOXES, TRAVELLING. All breeders despatching rabbits on a journey must have a supply of proper travelling boxes. These must be light, strong, well ventilated, and roomy enough to enable the occupant to turn round. They must not, however, be so large as to allow the rabbit to be thrown violently about by sudden jerks of vehicles. Standard dimensions are not less than 1 ft. long by 9 in. high and 9 in. wide for young stock ; 1 ft. 4 in. long by 11 in. wide by 11 in. high for adults.

Baskets are sometimes used in preference to boxes, as admitting more air.

BRABANCON. Very old-established Belgian table breed. Body is mandoline-shaped with abundant excellent flesh on back and hindquarters. Colours are grey, black, yellow and blue with white blaze, collar and feet. Weight varies from 5 to 11 lb. The rabbits are small eaters and fatten readily.

BRAN. See Foods and Feeding.

BREEDING. Rabbits breed readily during spring and summer (roughly February to end of August), but may be induced to breed all the year round by careful feeding and management (see Mating).

No rabbit should be bred from until fully developed, that is 6 months in small breeds (e.g. Dutch) and 8 months *at least* in large or medium sized breeds (Angora, Beveren, etc.). With very large breeds (Flemish Giant) it is an advantage to defer mating until 10 months. Rabbits bred from too early receive a check in growth and their offspring tend to lack size and stamina.

BREWERS' GRAINS. See Foods and Feeding.

BRUSHES. All rabbit-keepers, and particularly Angora keepers, need a supply of brushes for keeping the fur or wool of the stock in condition. It is particularly important to groom Angoras immediately before harvesting their wool. The best Angora brush is one having fine bristles set in a pneumatic cushion. For short-coated rabbits a small and fairly hard brush is advised (see Grooming).

BUCK.. A male rabbit. See Mating, for selection and management of bucks.

BUNCHE. A mandoline-shaped fawn breed with white blaze and chest, related to the Brabancon and noted for excellent flesh.

BUTTERFLY SMUT. Term applied to the characteristic nose marking of the exhibition English breed.

CABBAGE. See Foods and Feeding.

CANCER. See Ailments.

CANKER, EAR. See Ailments.

CARROTS. See Foods and Feeding.

CASTORREX. The earliest recognised variety of Rex rabbit. Extensively bred for fur and exhibition and used originally in the " rexing " of most fur breed (see Rex).

The colour is dark brown, toning to white towards the underpart, which is white. The fur is of a plush-like texture. The ears are large in comparison with the size of the rabbit. Weight 8-10 lb.

CHAMONIX THURINGEN. An uncommon breed of table rabbit. Colour is yellow with dark points resembling those of the Himalayan (see Himalayan). Weight about 5 lb.

CHAMPION. Term applied to a rabbit of outstanding merit after winning not less than five first prizes in open competitions under three different judges approved by the specialist club or society for the breed, and of which the owner of the rabbit is a member.

Championship certificates are granted by the club in payment of a registration fee, and the owner of the rabbit thus becomes entitled to use the prefix " Champion " (generally abbreviated to Ch.) before the rabbit's name in advertisements, pedigrees, etc.

CHICORY. See Foods and Feeding.

CHIFOX. Uncommon breed kept mainly for exhibition. It is produced in all colours usual to fur breeds and its chief feature is fur of 2½ in. in length, which must be soft and dense and free from any suggestion of woolliness. Size should be as large as possible without being too heavy in bone.

CHINCHILLA. One of the earliest, and at one time the most popular, of the fur breeds. The fur resembles real Chinchilla fur and its characteristic marking is due to four distinct colours in each hair, which is slate blue at the base, the intermediate portion being pearl grey merging into white tipped with black. The body fur from nape of neck to flank is interspersed with longer hairs of jet black. The fur is exquisitely soft in texture and about 1 in. in length. Although prices for Chinchilla pelts have fallen considerably in recent years, the breed is worth consideration, as a combined flesh and fur rabbit, as the rabbits are neat and " cobby " in shape with small bones. Weight 5½ to 6½ lb.

CHINCHILLA GIGANTA. Fur closely resembles that of the Chinchilla, but the rabbit is much larger and may weigh 11 or 12 lb.

CLEANLINESS. Scrupulous cleanliness of hutches, food vessels and, most particularly, all kinds of food, will do a great deal to prevent the occurrence and spread of disease. (See also Hutch Cleaning.)

Food and water vessels should be scalded out once or twice a week with hot water in which soda has been dissolved.

All foods should be stored in receptacles which will prevent contamination by dirt and to which mice and rats cannot get access. Care should be taken to ensure that greens, roots and hay are not given to the stock in a dirty condition.

CLOVER. See Foods and Feeding.

COBBY. Term used to convey a compact roundness in shape, particularly of the back and haunches.

COCCIDIOSIS. See Ailments.

COLDS. See Ailments.

COLIC. See Ailments.

COLONY SYSTEM. In the colony system of rabbit-keeping the stock are run together in large or small batches instead of each rabbit being housed in a separate hutch. It is practised with wool rabbits as well as with table and pelting rabbits.

Colonies may be accommodated in the open air on waste land or in meadows, or under cover in sheds, barns, outhouses or other suitable buildings.

Outdoor Colonies. In open-air colonies the area to be used must be enclosed with a 5-ft. wire-netting fence continued 2 ft. below ground. Wire netting of not more than 1 in. mesh is advised, as a larger mesh will enable youngsters to squeeze through and escape.

Shelter of some kind is provided into which the rabbits can, if necessary, be shut at night and where they can take cover from wind or heavy rain.

As regards space in relation to numbers, from 6 to 7 sq. ft. per rabbit should be the minimum in outdoor colonies; in the night shelters 2 sq. ft. of floor space for every adult and half that area for every youngster should be the rule.

One of the great advantages of outdoor colonies is that the rabbits require comparatively little attention, and, where grass and herbage are good, are practically self-supporting.

Indoor Colonies. In all indoor colonies the floor space per rabbit should be 6 sq. ft. for adults and at least 4 sq. ft. for youngsters.

Barns and large sheds may be used to house indoor colonies, but as the rabbits will not be grazing, food must be provided, and therefore an indoor colony costs more to run than an outdoor one unless the stock can be fed entirely upon greenfood. Here also, however, a great deal of labour is saved, as food and water can be put down in troughs for a large number of rabbits in far less time than where each hutch door has to be opened and closed and each rabbit has its individual food and water vessels.

For the " Garden " Rabbit-keeper. It is possible to moderate the colony system to suit the small man with little space. Garden sheds can be adapted, or outdoor pens and runs with shelters attached may be used by those who feel disposed to devote part of their gardens to miniature colonies.

Colony Breeding. In colony breeding some rabbit-keepers run rabbits in the proportion of 1 buck to every 10 or 12 does and allow breeding to proceed unchecked during the natural breeding season (from early spring to late summer). Boxes or, in large outdoor colonies or warrens, mounds of earth wherein the does might burrow, are provided for nesting purposes.

Under these conditions the rabbits breed freely, does often mating as soon as they have kindled, and this in time tends to decrease the size of the stock.

For this and other reasons the most up-to-date colony breeders exercise some supervision over breeding activities. Colonies of does, therefore, are made up and a buck or bucks, according to the number of does, allowed to run with them for 28 days only. At the end of this time the bucks are removed and nesting boxes provided for the does.

An alternative course is to transfer the does to hutches to kindle and rear their litters.

CONSTIPATION. See Ailments.

CROSS BREEDING. Term applied to the mating of two individuals of different breed. The progeny of such matings are termed *cross-bred* rabbits. The object of crossing one breed with another is to blend in the offspring certain desirable characteristics. Crossing is widely practised by table rabbit breeders with the object of combining such qualities as size with those of fineness of bone, minimum weight of offal and delicacy of flesh.

CURING OF SKINS. See Pelts.

CYSTS. See Ailments.

DANDRUFF. See Ailments.

DEWLAP. Term applied to the large fold of flesh beneath the chin which is sometimes developed in advancing age, particularly in does inclined to be fat. In most exhibition breeds a dewlap constitutes a fault. In exhibition Flemish Giant does, however, a dewlap is definitely required.

DIARRHŒA. See Scours under Ailments.

DISEASES. See Ailments.

DOE. A female rabbit.

DOMESTIC RABBIT CLUBS. As part of the Government plan to encourage domestic rabbit keeping during the war a Domestic Poultry Keepers' Council was set up and clubs formed in all parts of the country. These clubs aided members by means of the co-operative purchase of requirements, the provision of assistance and advice on doctoring, feeding, management, etc., and other such facilities. In addition, domestic club members were enabled to draw, through their club, a bran ration for does up to seven (those with eight does and over were regarded as commercial rabbit producers, and drew the bran ration through the Ministry of Agriculture Rationing Division). A condition of the issue of a bran ration was that fifty per cent. of the progeny of the does must be sold to a hospital, canteen, or other approved buyer.

DRESSING SKINS. See Pelts.

DROOPING EARS. See Ailments.

DRESSING OF CARCASES IN "OSTEND" STYLE. After skinning (see Skinning) and paunching (see Paunching), the back feet should be cut off to within one inch of the joint and the remaining portions threaded through the sinews. The front feet should be removed and the ends of the legs twisted back and inserted in slits made for the purpose between the ribs on each side. The carcase, fully extended, should be hung up until thoroughly cold and set.

DUTCH. The pure-bred Dutch is kept in this country almost exclusively for exhibition, as its small size (about 5 lb.) and bi-coloured pelt prevent it from being a good proposition for table and pelt production. The flesh of the Dutch, however, is very white and delicate in flavour and its percentage of offal and bone small in proportion to size. When crossed with other breeds, these characteristics are passed on to the offspring with remarkable consistency, and the Dutch is therefore extensively used for crossing with large breeds for table purposes.

In a good exhibition specimen the forepart of the body or "collar" and front legs and feet are white and the hinder part or "saddle," back legs and tail coloured, the colour stopping about 1½ in. behind the toes of the hind feet, which are white. The cheeks and ears are coloured with a white blaze which is continued under the colour on the cheeks to join the collar. All lines of demarcation between white and colour must be clean and unbroken. Recognised colours for Dutch are : Black, Blue, Tortoiseshell, Steel Grey, Brown Grey, Light Grey, Yellow and Chocolate.

EAR CARRIAGE is important in exhibition rabbits. In a normal healthy rabbit at rest, the ears lie parallel to one another, resting on the back, but at the slightest sound or movement they are brought up to an erect or semi-erect position, giving an appearance of alertness.

EARS, DROOPING. See Drooping Ears, under Ailments.

ENGLISH. Primarily an exhibition breed, but being large (6 to 8 lb.) is a good table rabbit either pure or in crosses with other breeds. Being a marked variety, the pelts are of no commercial value.

The exhibition English is judged principally on its head and body markings. The head markings include the Butterfly Smut, which is a coloured mark shaped like a butterfly with wings extended, the body running vertically up the nose, the eye circles (coloured circles round the eyes) and a spot of colour in each cheek. The ears are coloured.

Body markings include the saddle, an unbroken coloured mark running from base of ears to root of tail ; body or loin markings, which are spots below and distinct from the saddle ; and chain spots which start near the saddle behind the ears, running down to the body markings. There should be one spot on each leg and six belly, or teat, spots, on each side of the underpart.

Recognised colours are Black, Blue, Grey and Tortoiseshell.

ERMINEREX. The white variety of the Rex (see Rex). Pure white in colour, the fur is extremely dense, velvety in texture and glossy. Eyes may be inky red or bright blue. The Erminerex is graceful, inclined to "raciness" in shape, and weighs from 6 to 8 lb. The fur of this breed is one in which furriers are particularly interested, and it is a good proposition as a fur and flesh rabbit.

EXHIBITIONS. Shows with classes for rabbits are organised by the various specialist clubs and by local agricultural or fancier's societies. Breeders may enter their stock in appropriate classes on payment of a small entry fee for every rabbit exhibited. Cash prizes are awarded to those adjudged of highest merit.

Rabbits that are to be exhibited should be good typical specimens of their breed, corresponding as far as possible to the standard of points laid down. Winning exhibition rabbits and their progeny may be of high value.

Preparing Rabbits for Exhibition. During the last week before the show, animals should be groomed more frequently than usual in order to remove loose hairs and get the desired gloss to the coat.

If any pieces of straw or other material have become tangled up in the coat, they can easily be removed by placing a spot of glycerine in the palm of the hand and gently rubbing the coat in the direction in which it lies. See also Table Shows.

EYES, SORE. See Ailments.

FALSE CONCEPTION. It is only after the stimulus provided by mating that the female element of reproduction is discharged from the ovary, and it does not meet with the male element until about 10 hours after mating. It occasionally happens that these elements fail to meet and a condition known as False Conception is set up in the doe. The same changes occur in the uterus (womb) as take place in true pregnancy and the doe shows all signs of being in kindle. The condition lasts from 15 to 21 days, at the end of which the doe plucks herself and makes a nest and very frequently is found to be producing milk. No litter is produced, but if the doe be given young to foster she will rear them. A doe introduced to the buck at this time will invariably accept service and eventually produce a litter.

FANCY RABBITS. Rearing exhibition rather than utility rabbits.

FAT RABBITS. Over-fatness is definitely a fault ; fat rabbits seldom remain healthy for long. Sudden death from heart failure is by no means uncommon in fat rabbits. Reproductive activity is considerably hindered by excess of fat, which may cause disinclination to mate, or sterility in spite of mating.

If rabbits become over-fat, immediate reduction in quantity and quality of food is desirable. Iodine as advised for Coccidiosis (see Coccidiosis) is beneficial in obstinate cases.

FAVUS. See Ailments.

FITS. See Megrims, under Ailments.

FLEAS. These pests are seldom found on rabbits kept in hygienic conditions, but may be present in newly-bought stock. Such should be freely sprinkled with powdered naphthalene (which should be rubbed into the coat thoroughly) and placed for an hour or so in a closed box, after which they should be placed in a clean hutch.

FLEMISH GIANTS. Excellent breed of table rabbits and also popular for exhibition. Often used for crossing with other breeds to produce large rabbits for table. The pelts are usually of small commercial value (but war conditions have altered this, and every skin should be carefully dried for sale).

The Flemish is a very large breed, running from 12 lb. to as much as 20 lb. in weight, and therefore the adults are too large for present-day carcase requirements. Young Flemish, killed at 10-14 weeks old, however, make good table rabbits, although the proportion of bone and offal is fairly high.

Exhibition specimens must be steel-grey in colour, the under parts and under side of tail being white. Does must have a dewlap.

FOODS AND FEEDING. The main points in deciding upon a method of feeding rabbits are that the food be such as will (1) induce and maintain perfect health and condition ; (2) promote satisfactory wool, fur or flesh production ; (3) ensure free growth in young stock and fertility in breeding stock ; and (4) prove economic without being inferior in quality. (This refers to normal times. Of necessity the position in war time, when household scraps and the like must take the place of wheat, oats, etc., is different. See War-time Feeding, page 254).

Such feeding is best attained by making greenfood the main item of the diet, for this, plus a little hay (see Hay) supplies all that is necessary to stock animals if in sufficient quantity and good variety (see Greenfood).

If green or other succulent food is unobtainable in sufficient quantitity the deficiency must be made good by a certain amount of concentrated food. Clipped oats or, in cold weather, clipped oats and wheat in equal proportions, varied by the best broad bran, make the ideal concentrated diet for all rabbits. (Such concentrates are almost unobtainable under war conditions.)

For Breeders and Young Stock. Breeding stock and young stock up to 3 months, even if they are receiving a liberal green diet, should always have some concentrated food in addition. This may be suitably supplied by the materials mentioned above, with the exception of wheat, which is not good for youngsters, or for does when in kindle or nursing litters. Wheat tends to bring does " in season " while, with youngsters under 6 months its use encourages over-early maturity.

For Wool Rabbits. Greenfood and hay, with a regular supply of water, if the former is fed in sufficient quantity, contain all that is necessary to good health and a satisfactory wool yield. Oats and broad bran added to these will introduce the factors essential to the brood doe both while she is in kindle and during the period of lactation.

Wheat is of value to adult stock as a "conditioner" in cold weather, and is of particular value in promoting fertility in breeding stock. A course of wheat-feeding before mating of brood does before mating will help to ensure a high percentage of fertile matings, and a certain amount of this grain keeps stud bucks up to the mark as regards fertility, vitality and general condition.

Appropriate quantities of concentrated food fed in addition to a reasonably good green ration for various types of stock, in normal times, are as follows :

for the trouble of gathering and all green waste from the kitchen and much from the garden can be made use of.

Frosted greenfood, or frosted roots, must not be given to rabbits, for it causes severe internal derangement, the result of which is almost invariably fatal, particularly in young stock.

During the winter it is a good plan to have the greenfood in hand the day before it is required and place it in the food-store.

When greenfood must be gathered whilst it is frosted it should be placed in a bucket of cold water for half-an-hour to thaw before being given to the stock.

Mashes. A mash is a blend of soft

	Oats.	Wheat.	Bran.
Brood Does	1 to 2 oz. (according to size of litter)	—	1 or 2 handfuls
Stud Bucks	1 oz.*	1 oz.*	1 handful
Stock	½ to ¾ oz.*	½ to ¾ oz.*	1 handful
Young Stock			
(2 to 3 months)	½ oz.	—	1 handful
(3 to 6 months)	¾ oz	—	1 large handful

* Or same amount oats and wheat in equal proportion.

This system of concentrated feeding is advised as being preferable to a "mash" system in that it is more suited to the physiological needs of the rabbit. Mashes are best reserved for the final fattening of table rabbits and, under certain conditions, for occasional use with breeding stock.

Greenfood. This is the best of all foods for rabbits, and is sufficient can be obtained for each rabbit to receive from ¾ lb. to 1 lb. daily all stock, save babies and breeding stock, will thrive on this alone, plus a little hay.

It is important, however, that a good variety be arranged for, and that the food be fed as fresh as possible. The leaves of cabbage, cauliflower, and other cultivated greens, and such wild greens as grass, clover, dandelion, plantains, sow thistles and coltsfoot, hedge clippings, to give only a few examples, are all useful and valuable greenfoods.

Acorns, beechmast and horse chestnuts may also be used in small quantities (say 5 per cent of the food) either whole, or crushed, or dried and ground to meal. See under their respective heads.

Greenfood as Medicine. Many plants have tonic and medicinal properties and can be used instead of medicines in various diseases and minor ailments. In addition to being a good and wholesome food greenstuff also has the advantage of being cheap. It can be grown at home or obtained wild

ingredients, and may be fed dry or wet (i.e. moistened with water). Mashes are not recommended for general use, as, being soft, they give insufficient work to teeth, jaws and digestive organs, but may be used advantageously for fattening rabbits required for table use and occasionally for breeding stock when below par or reluctant to mate.

The principal ingredient of a good mash should be the best broad bran. This may be used in a proportion of up to 50 per cent of the whole mash. The balance may be composed of middlings and such meals as barley meal, maize meal, Sussex ground oats, and, for breeding stock, small quantities of soya bean meal.

Storage of Food. All rabbit foods must be kept clean, dry and secure against the inroads of vermin. Corn, bran and meals should be stored in receptacles with close-fitting lids. Old milk churns, if obtainable, make excellent "bins," as do tea-chests if fitted with hinged lids. Trusses of hay should never be stood on the ground on an earth floor, but should be rested on wooden-slatted racks.

Feeding Details. *Acorns.* Useful food if given in small quantities—5 to 10 per cent of the ration. May be given whole or ground, but must not be used unripe.

Apples. These are appreciated by rabbits and can be fed as a change from roots in winter. Apple peelings and cores may be used. Never give unripe apples.

Artichokes, Jerusalem. Useful crop to grow for rabbits. Tuber, stem and leaves are all readily eaten. The tubers are the equal of potatoes.

Barley. Not a good food for rabbits except as final preparation for killing. Best used in form of barley meal mixed with other meals in mashes.

Bean Meal. A valuable protein food, usually the meal of the horse bean. See also Soya Bean Meal.

Bran (Broad). A very valuable food for all rabbits. Can be given either alone or as chief ingredient in mashes. Possesses gently laxative properties and is also valuable for does during lactation. (Limitation of feeding stuffs during the war resulted in bran being allowed only for breeding does).

Brewers' Grains. Good as occasional food if used fresh. Best used as " dried grains " and given after soaking for a few hours.

Cabbage. Valuable as greenfood. Has a blood-cooling effect. The stalks are also readily eaten but should be split into strips before use.

Carrots. Good winter food—rich in vitamins and possessing tonic properties. Young carrot tops and thinnings may be used freely as greenfood but matured tops should not be fed in excess.

Chicory. A valuable and safe greenfood much liked by rabbits, and useful as a " conditioner ".

Clover. One of the best greenfoods for rabbits. Flowers, leaves and stems may be used freely and are of particular value when mixed with grass.

Clover Hay. A very valuable food, as in addition to providing the bulk and " roughage " necessary to rabbits it is of definite feeding value. Clover hay should find a place in the diet of all breeding stock, as it contains the " fertility " vitamin (Vitamin E) in good proportion.

Cod-liver Oil. A valuable food for rabbits, especially those having an insufficiency of greenfood, or even of sunlight. Correct proportions are 3-5 drops according to size of rabbit or 1 pint of oil to 1 cwt. of broad bran. The best method of mixing is to place the bran in a pail and pour the oil into a hollow in the middle, then letting it stand for an hour or two. By that time the oil will be absorbed into the immediately surrounding bran and can be worked into the remainder without undue trouble.

Dandelion. When used judiciously, dandelions are one of the most valuable green foods for rabbits, possessing special tonic properties and being useful in staving off debility and loss of appetite in very hot weather. They should not, however, be used constantly unmixed with other greenfood as in this way they may have

an unduly laxative action and an over-stimulating effect upon the kidneys.

Hay. A very important item in the diet of rabbits. Clover hay and clover mixture contain much protein and are materials of high feeding value (see Clover Hay), while good meadow hay is also an excellent food, containing much nourishment and many herbs of a health-giving nature, rich in minerals.

All hay should be sound and sweet, free from dust, mould or musty smell. Hay is best supplied in wire racks attached to the outside of the wire door.

Hedge Cuttings. These make useful food provided hedges do not include shrubs or trees which are poisonous. See Poisonous Plants in Smallholding Section.

Ivy, unlike most evergreens is not poisonous to rabbits, but is useful in winter and early spring as occasional greenfood. In summer and autumn it must be avoided as it has a harmful effect at the flowering and fruiting stages.

Kale. Where crops are grown specially for rabbits kales are a good choice, offering a good means of providing an all-the-year-round supply of greenfood. They are an excellent food, yield a heavy crop, and may be fed safely to stock of all ages. Best varieties are Thousand Headed, Marrow-stem, Giant and Perpetual Kales.

Kohl Rabi. A plant allied to the cabbage, having a bulbous leafy stem with a taste resembling that of the swede. Useful and valuable winter food.

Linseed. A useful occasional food when given with discretion. Linseed imparts a gloss to the coat and on this account is used by fanciers, both whole and as linseed meal in mashes, when preparing stock for exhibition.

Care is necessary in feeding linseed, as it has a tendency to loosen the coat. This property may be turned to good account in the condition known as " sticking in moult " (see Moulting). Linseed is also a laxative and therefore extra hay should be given with it. Roughly ½ to 1 teaspoonful of whole linseed twice a week will serve for coat conditioning, but individual capacity for taking this food is best judged by observation.

Lucerne. A valuable protein crop to grow for rabbits, where land is available. It is cousin to the clovers, and will stand for years, giving several cuttings yearly. A safe and well-appreciated greenfood.

Maize. An imported food, not recommended for rabbits, being too heating and fattening. May be used under certain conditions for breeding stock (see Mating) or as a fattening agent (see Mashes).

Mangolds. Useful root food when the mangolds are " ripe " about January to February. Before this time mangolds are unripe and too acid to make safe food.

I*

Must be used sparingly as they are definitely laxative and inclined to fatten.

Milk. Cow's or goat's milk, although not *necessary* foods for rabbits of any age, may be used with advantage in normal times (*not* in war-time) for brood does and young stock if available at a low cost. Milk is valuable, where greenfood is scarce, in the periods of gestation and of growth, but its use has no great influence upon the actual production of milk in nursing does.

Nettles (Stinging). These, when dried, make useful winter food for rabbits. They should be cut and dried in the sun like hay and stored for use in the same way.

Oats. A valuable grain food, best used in the form of " clipped " oats. Crushed or rolled oats are not recommended as a staple concentrated food as they lack certain desirable properties and do not provide work for the teeth and jaws. If obtainable at a reasonable price, however, rolled oats are beneficial as an occasional variation in the diet of brood does and young stock. All oats used should be clean and free from dust and the clipped form should appear bright and plump.

A regular supply of sound clipped oats is one of the best means of encouraging and maintaining good milk production in brood does.

Potatoes. Although potatoes are inclined to be fattening if fed in quantity, war-time feeding has made their use very necessary. They must always be cooked; raw potatoes should never be used. Potato parings may be boiled and form part of the general ration at any time.

Salt. It is a decided advantage to provide rabbits with salt. The best way of doing this is to place a lump of rock salt in the hayrack for the animals to lick at will.

Shepherd's Purse. A small wild herb with heart-shaped seed vessels much used as a cure for Scours (Diarrhœa). Is a good greenfood for all rabbits.

Soya Bean Meal. Very useful for feeding in small quantities in mash. Believed to promote fertility (see Mashes).

Sow Thistle. Useful weed the stems of which contain a milk-like juice. Makes excellent greenfood, particularly for nursing does.

Sugar Beet Pulp. A good food for occasional use. Can (normally) be purchased in dry condition from sugar factories. The pulp must be soaked before use and swells to several times its original bulk in the process. After soaking in an equal measure of water for about 12 hours, it may be dried off with broad bran: each rabbit may receive a good handful.

Swedes. A very useful root food, related to the turnip.

Wheat. A good " conditioner " but,

as it possesses heating and stimulating properties, should be used with care.

Weatings. Name introduced some years ago in place of the terms middlings, sharps, etc. Breeders should always ask for weatings in preference to using the other names, for weatings have a guaranteed food value.

War-time Feeding of Rabbits. As concentrate rations were not allowed for rabbits, save for a minimum of 4 breeding does owned by one person or society, it became necessary to feed the great bulk of rabbits on greenfoods, hay, and various forms of waste—household scraps, boiled potato parings and chats, roots and root parings, odd pieces of dry bread, and such unrationed materials as were both suitable and obtainable from time to time. Where care was taken in the preparation of the food, such feeding was entirely successful, and this was demonstrated by the great war-time increase in rabbit keeping.

FUR BREEDS. Rabbits kept primarily for the production of pelts—which usually also make good table rabbits. Recognised fur breeds include Argentés, Beaver Beige, Beveren, Chinchilla, Havana, Lilac, Rex (many varieties).

FURNISHINGS. The luxuriant wool about the head, face and feet in Angora rabbits. Also the ear-tufts in this breed.

GASTRITIS. See Ailments.

GESTATION PERIOD. The period of pregnancy. This period in the rabbit is usually 31 days, but there may be some variation and the litter may be born as early as 28 days, or as late as 36 days, after mating.

GREENFOOD. See Foods and Feeding.

GROOMING. Fur rabbits require only occasional grooming except during the moult, or when being prepared for show (see Moulting). Utility Angoras must be regularly groomed with a brush (see Brushes), and especially before harvesting to free the wool from foreign matter. Fine coats of exhibition Angoras require daily grooming from the time the animals are 5 weeks old, this in order to keep them from matting.

How to Groom Angoras. After freeing the coat from pieces of hay, etc., successive partings should be made, beginning along the middle line of the back, and the coat brushed on each side of the parting. Next all the wool on one side should be brushed upwards and held against the rabbit with the free hand. The wool should be released a little at a time and brushed downward. Repeat on opposite side and rump.

Hold the head up by placing the hand under the chin and brush the chest upwards and downwards. Face furnishings, ear-tufts and crest on head must be brushed.

To brush the belly coat and the furnishings of hind legs and feet sit down and hold the rabbit on its back on your knees.

Always after grooming blow upon the coat in all directions to make sure that it is " clean brushed," that is, that the skin can be seen clearly wherever the coat is blown upon. If bellows are used in preference to a brush (see Bellows) the cost should be freely blown in all directions in every part of the body.

Grooming Table. A small table of a convenient height and of a size to accommodate a rabbit is a great convenience in grooming rabbits, enabling the animal to be " got at " from all sides without difficulty.

GUARD HAIRS. Term used to describe the relatively long hairs interspersed among shorter ones in the coats of the majority of rabbits. Proportion of guard hairs to shorter hairs varies from approx. 1-20 to 1-70. The fur of rabbits with few guard hairs is more silky, and therefore superior to that in which they exist in greater numbers. In the fur of the Rex varieties guard hairs are almost or entirely absent. Guard hairs are sometimes referred to as " Awns ".

HANDLING. It is an advantage to handle young rabbits as much as possible. Youngsters seldom handled are inclined to become nervous, and perhaps develop a spiteful or vicious disposition in adult life.

The correct way of handling a rabbit is to lift it by the scruff of the neck, supporting the rump with the free hand if the animal is heavy. Does, from the time of mating onward, must always be supported in this way when lifted, whatever their weight. Handling by the ears is incorrect and cruel ; it is always painful to the rabbit and there is risk of breaking the ears themselves or tearing the small muscles controlling their movements (see Drooping Ears).

HAVANA. A self-coloured breed of a rich chocolate or liver brown. Popular in this country as an exhibition breed and, in addition, one of the best combined pelt and table rabbits. The fur is thick and very glossy, showing a purplish sheen over the brown and having a pearl-grey undercolour.

The Havana is compact and " cobby " in shape and these qualities, coupled with its markedly broad and rounded loins and well-developed hindquarters, ensure a well-shaped fleshy carcase. Weight 5-7 lb.

HAY. See Foods and Feeding.

HEAT. See Season.

HIMALAYAN. A small breed (weight about 5 lb.) with white body and legs, pink eyes, and black nose, ears, feet and tail. Chiefly bred for exhibition purposes, although the fur is very silky and occasionally used in imitation of ermine. The flesh is of a very delicate flavour, somewhat resembling that of chicken.

HOCK JOINT. The juncture of the foot with the leg, forming the heel.

HOCKS, SORE. See Ailments.

HOUSING RABBITS. Rabbits are housed in hutches that may stand in the open or be placed under shelter in a shed. Any existing shed can be used as a rabbitry if it is light, airy and weatherproof, but if nothing of the kind is available, it is a simple matter to erect some type of shelter that will keep wind and rain from the hutches and enable necessary work to be carried on irrespective of weather conditions.

Hutch Dimensions. Wood is undoubtedly the most economical and practical material of which hutches can be made. Hutches properly made of asbestos sheeting are a good alternative, and the war-time shortage of timber brought them into considerable use.

All hutches should be as simple in pattern as possible with a view to saving time and labour. Roominess is a point of importance and minimum hutch dimensions are as follows :

For large breeds (Flemish, Belgians, etc.):
Single hutches—Length, 4 ft. ; width, 2 ft. ; height, 1 ft. 6 in.
Breeding hutches—Length, 5 ft. ; width, 2 ft. ; height, 2 ft.
Stud Bucks' hutches—Length 4 ft. 6 in.; width, 2 ft. ; height, 2 ft.
For medium breeds (Angoras, Chinchillas, etc.) :
Single hutches—Length, 2 ft. ; width, 2 ft. ; height, 1 ft. 6 in.
Breeding hutches—Length, 3 ft. 6 in. ; width, 2 ft. ; height, 2 ft.
Stud Bucks' hutches—Length, 3 ft. ; width, 2 ft. ; height, 2 ft.
For small breeds (Dutch, etc.) :
Single hutches—1 ft. 6 in. square.
Breeding hutches—Length, 3 ft. ; width, 2 ft. ; height, 1 ft. 6 in.
Stud Bucks' hutches—Length, 2 ft. 6 in. ; width, 1 ft. 6 in. ; height, 1 ft. 6 in.

Building Hutches. Hutches may be made conveniently in " stacks " of two or more tiers. This combines economy of material with neat appearance.

The simplest and most economical method of building hutches is to use grocery boxes. A good stack of three breeding hutches can be made from three boxes of a uniform size, say, 4 ft. by 2 ft. by 2 ft. Stack the boxes lengthwise, one above the other, and firmly screw them together by means of four lengths of 2 in. by 1 in. battens, two at each end, vertically at the corners.

The four battens should be extended for at least 1 ft. beyond the lowest box and will form " legs " upon which the stack will stand, raising it well above the ground.

Make a door to cover the stack of three boxes, the frame being of 2 in. by 1 in. battens, and nail boards from the lids of the boxes over one-third of this, covering the other two-thirds with ½ in. mesh wire netting. This door should be hung on three hinges, and fastened with two metal buttons. Give the finished stack one or two coats of creosote inside and out, and allow to dry thoroughly before use.

At a little extra trouble the stack can be fitted with a sloping roof and reinforced with felt for outdoor use.

Built hutches are necessarily more expensive, but even in this connection money can be saved by the use of boxes. Wooden glass-cases are well worth the trouble of taking to pieces, and yield a quantity of stout, useful wood, suitable for outdoor hutches. Egg-cases and other light boxes will provide wood well adapted to the construction of indoor hutches.

Provided the hutches have a good foundation of stout framework, those constructed from this light wood both look and wear well.

Stacks of hutches may be built against existing walls or fences, thus economising in wood.

Breeding Hutches. For breeding does it is well for the end of the hutch covered by the boarded part of the door to be half-boarded off from the rest of the hutch, to form a nesting compartment.

HUTCH CLEANING. This important routine job must be carried out regularly. All hutches should be thoroughly cleaned at least once a week, and those containing does with young litters not less than twice.

Rabbits usually select one corner of the hutch in which to urinate, and, as it is mainly the urine that is objectionable, it is an advantage to remove the litter from this corner daily and replace it with fresh.

When cleaning out a hutch thoroughly remove the rabbit first of all. Rabbits are very often upset and made nervous and bad-tempered when shovels or brushes, which have possibly been used for other hutches and which, therefore, smell of other rabbits, are inserted rudely into their domain. Secondly, it is not easy to do the job thoroughly while the rabbit is in the hutch.

All floor litter should be removed with a shovel, or, if preferred, raked with a proper hutch scraper.

When all soiled litter has been removed the hutch should be swept out—floor, sides and top—with a small, stiff hand brush. Next, a little disinfectant should be sprinkled on the floor or, better still, the whole of the inside of the hutch sprayed with disinfectant, and after scattering a little peat moss over the floor the fresh litter may be given.

HUTCHES. See Housing Rabbits, Morant System.

HUTCH LITTER. See Bedding.

IMPERIAL. A medium-sized self-blue rabbit of English origin, not widely bred.

IN-BREEDING. This term means the pairing of nearly-related rabbits. It is practised extensively by exhibition breeders in order to fix and intensify desirable characteristics ; but inexperienced rabbit keepers are advised to leave it severely alone, as undesirable characteristics are also fixed and intensified by this means.

INDIGESTION. See Ailments.

INFLUENZA. See Ailments.

JAPANESE. Breed of rabbits kept in this country for exhibition, but having a good reputation on the Continent as a table rabbit. It is a yellow rabbit more or less heavily marked with black bands or patches, compact and short in shape. Weight about 5 lb.

KALE. See Foods and Feeding.

KILLING. Rabbits should be killed without being frightened beforehand, as undue excitement tends to depreciate the quality of the flesh.

The best method of killing is to strike a sharp blow at the back of the neck. If the rabbit is hit in the right spot death is instantaneous and that with only one blow. If several attempts are made not only is unnecessary suffering caused to the rabbit but excessive bruising will result, spoiling the appearance of the skinned carcase.

The right spot in 'which to hit a rabbit in order to kill it is just below the point where the ears join the head. The rabbit should be allowed to sit naturally upon a box or table, the ears being held forward with the left hand, while the blow is struck with a short blunt instrument, such as a stick, held in the right hand, or even with the edge of the hand itself.

Although this blow kills almost instantaneously, there is always, in a healthy rabbit, a certain amount of kicking and struggling immediately afterwards. This is due to muscular contraction after death and not to any pain felt by the animal.

Immediately the rabbit is killed, if it is for table, it should be bled (see Bleeding).

KINDLE. Term used to denote pregnancy in a doe. A pregnant doe is described as being " in kindle," and when young are born the doe is said to kindle.

KINDLING. The act of giving birth to a litter. Whenever a rabbit is due to kindle preparations should be made for the arrival of the litter.

On the twenty-seventh day after mating clean the hutch thoroughly and put in plenty of sawdust or peat moss as floor litter, and also a layer of straw. At all times also provide soft hay as nesting material. If the doe has begun to make her nest before the cleaning-out is given, leave this alone, provided it is clean.

Never disturb a doe while she is kindling. See that she has plenty of nesting material and drinking water and leave her to herself.

After the doe has kindled give her a meal of greenfood, hay, oats and fresh water. Does drink a great deal before and after kindling, because they are probably slightly feverish and also because the plucking of wool from sides and breast with which to cover the nest may make the mouth dry.

When feeding the doe, a glance should be given to the nest, which will appear as a fluffy mass in a corner of the hutch, probably moving slightly. Do not touch the nest; it is best to wait for a day or two before doing this.

For the care of the young after birth, see Litters.

LILAC. A rabbit with fur of a pinkish-dove colour, bred for exhibition and to some extent for fur, although the pelts are not of great interest to furriers. Being compact and cobby in shape, the Lilac makes a good table rabbit.

LIME. When cleaning sheds, etc., used as indoor colonies, or night shelters in outdoor colonies, it is useful to sprinkle the floors freely with lime, which is a good cleanser and preventive of disease.

LIMEWASH. Useful for coating the walls of hutches, etc. A good wash is made by mixing lime with enough water to make a moderately thick " paint ". Limewashed interiors are an advantage in rabbitries inclined to be dark. A handful of salt added to the wash at the time of mixing will prevent any tendency to " flaking " when dry, and a little paraffin will add to the disinfectant properties of the wash.

LINSEED. See Foods and Feeding.

LITTER (HUTCH). Material used on hutch floors to provide warmth and to absorb urine (see Bedding, Hutch Cleaning).

LITTERS. The number of young produced at one birth by a doe rabbit is referred to as a litter. The average number in a litter may be taken as from 5-7, but there is wide variation and a doe may produce any number from one to twelve or more at a birth.

When young rabbits are born they should remain undisturbed for a day or two. Then they should be examined to see that all is well.

Before a litter is examined, remove the doe and shut her in a box with a little greenfood to occupy her attention. Then carefully and quickly remove the layer of wool covering the nest and take a look at the babies within.

Remove any dead ones, and make sure that no strands of wool are wound round the necks or limbs of living ones, for such, if allowed to remain, may cause strangulation or stoppage of circulation and consequent withering of the limbs. For this reason it is advisable to examine the nests of Angoras daily.

After a day or two a doe will probably not object to the examination of her family while she is present. In order that she may perceive no unfamiliar smell about her nest, before you touch it or the babies stroke the doe herself a few times.

When returning the doe to her hutch after your inspection, handle her gently and give her food to interest her. Also see Mutilation of Litters and Weaning Young Rabbits.

LONG CLAWS. The claws of rabbits grow as their age advances, and if allowed to remain long may result in casualties. For example, they may be caught in the wire netting of a hutch door, and the rabbit's efforts to free itself may lead to the claw being pulled out. Rabbits also sometimes injure eyes or ears with long claws while washing their faces, as they frequently do.

Thus, whenever claws are long they should be cut by means of a sharp pair of wire clippers. In performing the operation care must be taken to cut the claws well above the " quick " or growing point. If the claw is cut cleanly about half an inch above the fleshy sheath from which it grows, all risk of this will be avoided.

If the rabbit is disposed to struggle much, wrap it up in a cloth or piece of sacking, getting somebody to hold it during the operation on the claw.

LOP-EARED RABBIT. A purely fancy rabbit, not suitable for fur or flesh production. Its principal feature is the great length of ears, these sometimes measuring as much as 27 in. or more from tip to tip across the head.

LUCERNE. See Foods and Feeding.

LYNXREX. One of the Rex varieties (see Rex), having fur that gives an effect of orange shot with silver.

M **AIZE.** See Foods and Feeding.

MANGE. See Ailments.

MASHES. See Foods and Feeding.

MATING RABBITS. In mating rabbits it is important that the right procedure be followed and buck and doe watched *all the time they are together.*

The doe should always be put into the buck's hutch for mating, not *vice versa*, as many does resent the presence of the buck in their own hutches and may turn on the intruder and injure him.

As soon as the doe is introduced into the buck's hutch he will advance towards her, when, in all probability, she will stretch herself out and raise her hindquarters. The buck will then seize her by the back of the neck and serve her. He will always fall over backwards or sideways after service. The service itself only occupies a few seconds, and is not complete unless the buck falls over in the way described. Some breeders allow a second service, but this is unnecessary and undesirable. If the buck has fallen over it may be assumed that an effectual mating has taken place, and that, provided both buck and doe are fertile, a litter will duly arrive.

Some does—particularly maiden and excitable does—will not immediately lie down to receive the attentions of the buck when introduced into his hutch. They will stamp with their hind feet and race round the hutch a few times with the buck chasing them before finally lying down and accepting service. One need not, therefore, conclude that a doe is not ready to mate if she seems disposed to run away from the buck at first ; if she *really* is not ready, she will soon crouch down, probably in a corner, with her hindquarters well tucked in. This means removal to her own hutch and a fresh trial another day.

It may be necessary to try a maiden or shy doe several times before service is accepted, but do not be tempted to shut the buck and doe up in one hutch and leave them for a time. It cannot be too forcibly emphasised that this practice is bad for both buck and doe. (Also see Winter Breeding).

Choice of Buck. The buck should be in perfect condition, well fed and in good health. He should be a typical specimen of his breed and strong in all breed points. For management see under Stud Buck.

Reluctance to Mate. It is a common experience when attempting to breed rabbits in winter to meet with difficulty in getting does to mate. This is invariably due to incorrect feeding. Where the food is of such a nature as will maintain good breeding condition, and at the same time prove sufficiently stimulating to keep up a good tone of the reproductive organs, " shy breeders " are an exception even during November and December, the " worst " months.

Wheat and oats in equal proportion, make the best concentrated food for general use with brood does up to the time of mating. A certain amount of variety is decidedly advantageous, as this makes for greater enjoyment of the food, leading in its turn to greater benefit. Thus, a soft feed or mash may occasionally take the place of the hard corn.

A Mash to Encourage Mating. A good mash (when the ingredients are available) for the purpose of overcoming reluctance of does to mate is : Broad bran, 7 parts ; middlings, 4 parts ; alfalfa meal, 2 parts ; soya bean meal, 1 part—all by weight. This may be used not more than twice weekly as a change from the wheat and oats mixture for does previous to mating, and should be discontinued after mating has taken place. The mash may be fed dry, or, if preferred, in a crumbly moist state.

Another variation may be made by giving an occasional feed of broad bran (2 parts) and rolled oats (1 part). This feed, which is invariably enjoyed and should be given dry, may be continued throughout pregnancy.

Naturally " Shy " Does. In addition to does that are reluctant to mate merely because of the weather, there are also does that seem to be naturally shy breeders and with which, no matter how they are fed, there is likely to be trouble. If does of this type are to be used for winter breeding, it is a good plan to give, for a week before mating, a small quantity of kibbled maize on alternate days.

Various kinds of greenfood have heating and stimulating properties, and the use of these is a distinct help in bringing shy or backward does into season. Parsley and groundsel have both been successfully used for this purpose, and so also have celery tops.

A Last Resort. Should a doe still prove unwilling to mate in spite of stimulation by means of food, she may frequently be prevailed upon to do so by arranging that she occupies a hutch in which a buck has been kept previously, particularly if the hutch is placed in such a position that she is able to see a buck. The hutch used for this purpose should not be cleaned out, the doe being put straight on to the litter the buck has used.

An alternative scheme is to use a large hutch having two compartments divided one from the other by means of small-mesh wire netting. The doe is placed in one of these compartments, while the other is occupied by the buck with which it is proposed to mate her. The close proximity of the buck usually has the effect of bringing a doe in season in a very short time.

Reluctant Bucks. Bucks are seldom reluctant to mate unless below par ; a healthy, vigorous buck will almost invariably give service when called upon. Sometimes, however, after a busy period at stud a buck's powers may show signs of flagging. He should then be allowed to rest for a week or two, while fed well and given the stimulating mash advised for brood does twice weekly.

MEAT RABBITS. From September to March, in normal times, there is a very wide demand for rabbits for meat. Enormous numbers of carcases have been imported in the past from the Continent, and, in addition, London wholesale dealers alone have taken hundreds of thousands of carcases from British breeders. In almost every town there are butchers and poulterers who are open to take regular supplies of " tame " carcases throughout the season. Since the outbreak of war the great development of domestic rabbit keeping (see Domestic Rabbit Clubs) has gone far to replace former imports, and provide large quantities of carcases for the ordinary consumer.

What Buyers Want. Table rabbit production will always be profitable provided efficient management and economy are duly practised and care taken that the products meet trade requirements. Buyers chiefly demand young carcases

weighing $2\frac{1}{2}$-3 lb. when dressed, and this preference applies to private as well as trade buyers.

Best Varieties for Table. The majority of fur breeds make good table rabbits, but where the aim is purely carcase production, most breeders agree that cross-breeds give the most satisfactory results. Cross-breeds provide just what can be most profitably produced—rabbits that will yield a dressed carcase weighing from $2\frac{1}{2}$ to 3 lb. at approximately 3 months old. As the dressed carcase weight is generally about two-thirds of the live-weight, this means youngsters turning the scale alive at from $3\frac{3}{4}$ to $4\frac{1}{2}$ lb.

Good Table Crosses. The most popular table cross is that of a Flemish Giant buck with Belgian does, by means of which one may obtain youngsters that will be ready for killing at 10 or 12 weeks. Youngsters from a Belgian buck and Blue Beveren does are also good, turning the scale at $4\frac{1}{2}$ lb. at 14 weeks old.

A Flemish buck crossed with an English doe again gives large youngsters which come forward quickly and show plenty of firm flesh at 3 months old.

A rabbit that deserves more consideration in crossing for table purposes is the Dutch. The pure-bred Dutch rabbit is too small to be satisfactory for market purposes, but judiciously used with other breeds it greatly improves the carcase. A buck of the Belgian-Flemish cross mated with Dutch does produces rabbits that are ideal for table purposes, combining fleshy back and thighs and the delicately flavoured white flesh of the Dutch with the large size of the Flemish and the Belgian.

Fattening Meat Rabbits. Rabbits required for table purposes may be " finished off " by intensive fattening for 10 days before killing. A longer fattening period is not advisable, as rabbits tend to lose condition if over-fed for long.

During the fattening process the rabbits should be kept either singly or in batches in small hutches in which exercise is much reduced, and fed on hay, greenfood and mash. One of the best mashes for this purpose in normal times has been composed as follows: Boiled potatoes or house scraps, 50 per cent, well mixed with an equal quantity by measure of the following mixture : Broad bran, 3 parts ; 1 part each of barley meal, Sussex ground oats and fine weatings (parts by weight). Mix with sufficient warm water to produce a thoroughly moist, although not " sloppy," consistency and give as much as is cleared up quickly twice a day.

MEGRIMS. See Ailments.

MILK. See Foods and Feeding.

MORANT SYSTEM. This system of keeping rabbits is distinctive, and has a great number of adherents. It was originated many years ago by a Captain Morant, and consists in housing the rabbits in wooden hutches the floors of which, except in the sleeping compartments, are made entirely of wire netting. The hutches are placed on the ground so that the rabbits can crop the grass, etc., through the meshes of the wire, and are moved to fresh " pitches " twice or three times daily as the grass is eaten.

The system has a great deal to recommend it, and makes a strong appeal to the man with a fair amount of land who, for some reason, does not want to run his rabbits in colonies, but wishes to reduce his feeding costs to the minimum.

The fact that rabbits in Morant hutches are always nibbling at greenfood means that they will need far less bought food than is necessary for rabbits housed in ordinary hutches.

Morant Hutches. Morant hutches are quite simple to make at home. Suitable dimensions for the accommodation of a doe and litter, four weaned youngsters, up to 3 months old, or two from 3 months up to breeding or pelting age, are 4 ft. long by 2 ft. wide by 1 ft. 6 in. high in front sloping to 12 in. at the back, to give the roof a fall of 6 in. A hutch of this type can be made by nailing weather boards or box boards to a frame of 2 in. by 1½ in. timber. One quarter of the entire hutch, including the bottom, should be boarded to form the sleeping compartment, and the remainder, with the exception of the roof, covered only with wire netting to form the run or feeding compartment. Half-inch wire netting should be used for the sides of the run, but for the floor 2-in. mesh netting should be used.

The entrance from the sleeping compartment to the run may be fitted with a sliding door so that the rabbits may be shut in if necessary. Ventilation should be provided by half a dozen holes, one inch in diameter, bored about 4 in. from the top of the hutch on each side.

It is best to dispense with a door in a Morant hutch and instead to have the roof hinged, thus forming a lid that may be raised when necessary.

MOULTING. Rabbits, like most other stock, pass through a period during which the old coat falls out and is replaced by a new one. During the first year of life, in the rabbit, there are two moults, one about the fifth month and one about the eighth month. Thereafter a complete change of coat occurs once a year, the climax of the process occurring in the autumn.

It is *all-important* that pelt rabbits be not killed whilst there is any trace of moult in the pelt. Pelts from moulting rabbits are useless to the furrier.

Rabbits moult in different sequence according to their age. The change of a coat that has been carried for about a year is a more trying affair than the acquisition of the adult coat at 8 months. Adult bucks make a long and gradual process of the moult, beginning with the head quite early in the summer. Then the fur breaks behind the ears and the coat is gradually shed in the direction of the tail, the last place to finish being on the rump low down just above the tail.

A normal body moult takes, on an average, about six weeks to complete superficially, and longer to complete to the pelting stage. Heavy-coated rabbits of the fur breeds can be helped by the removal with thumb and fingers of much of the loose, dead fur, and by stroking with a damp hand and brushing. By these means much hair can be collected.

It is important to keep the hutches of moulting rabbits reasonably clear of dead fur by a daily brushing-up of the floors, etc., during the heaviest part of the moult, otherwise the animal, unable to escape from the débris, will be apt to get intestinal troubles from swallowing some of it.

" Stuck in Moult." When the moult is unduly protracted a rabbit is said to be " stuck in moult ". This condition may be the result of a low state of health, and is frequently assisted by attention to the bowels and the feeding of greens with known tonic properties, such as dandelion and cabbage. Half a teaspoonful of whole linseed, given with the concentrated food on alternate days, works wonders in bringing away the old coat and imparting gloss and " life " to the new one. Daily grooming with brush or bare hands is also of assistance.

Breeding from Moulters. There is considerable controversy as to the wisdom of breeding from rabbits in moult. There does not appear conclusive evidence to support the theory that breeding when in moult is productive of bad results to parents or offspring, but it is well to be on the safe side and avoid using breeding stock of either sex when the coat is falling.

MUTILATION OF LITTERS. The most common cause of this trouble is neglect to supply the doe with water. The slightly feverish condition of a doe who has just kindled makes her excited and creates in her a thirst which is augmented by dryness of the mouth caused by plucking fur or wool for her nest. If the doe cannot quench her thirst she may become wild, turning on her babies and mutilating or perhaps eating them. Fright from unusual noises or disturbance at kindling time may also cause a doe to maltreat her litter. Does that are kept near bucks will sometimes desire to mate soon after their litters are born. At such times they may mutilate or eat their babies, or they may stamp on them and crush them to death.

NEST. A doe makes her nest by forming a hollow in a mass of hay and lining this with fur or wool plucked from her body. When the young are born she covers them with more fur or wool plucked for the purpose. Some does make their nests early and others leave all preparation to the last moment, but usually the nest is made after the twenty-first day of pregnancy.

NEST BOX. Term applied to a specially made box given to a doe in which to make her nest. Nest boxes are not essential and are not now so popular as formerly, but they may be used with advantage for winter breeding in cold or exposed situations, as they ensure that the young cannot crawl out into the hutch where they may die from exposure. The box should be placed in the hutch during the second week of pregnancy. It should be fitted with a hinged lid for examination and a hole about 6 in. in diameter should be cut out in one end for the entrance and exit of the doe. An appropriate size for a nest box is 16 in. by 12 in. by 10 in. high.

NETTLES. See Foods and Feeding.

NEW ZEALAND RED. A rabbit of a reddish-buff or golden sable colour originating in California where it is extensively bred for flesh and fur. Not popular as a fur rabbit in this country, possibly because the fur is somewhat harsh in texture. Adults weigh 7-9 lb. Youngsters average 4½ lb. at 3 months, but are rather too heavy in bone to be ideal table rabbits.

NORMANDY GIANT. A rabbit resembling the Flemish Giant but smaller and finer in bone and with a broad well-developed back. Crosses of this breed with the Chinchilla and the Havana give good table rabbits.

NUBIAN. See Alaska.

OATS. See Foods and Feeding.

OFFAL. Term used to denote organs removed in paunching (see Paunching).

OSTEND RABBITS. Term applied to the carcases of any breed when paunch-

ed and skinned for market and prepared in the Ostend style. Term arose from the fact that carcases were first marketed in this way from Ostend in Belgium (see Dressing, for method of preparing table rabbits in the Ostend style).

OVARIES. Organs which produce the female element of reproduction. Two in number, situated on either side of the upper part of abdomen, near the kidneys.

OVUM. Female reproductive element. Ova are discharged from the ovaries in response to the stimulus of mating and travel thence to the uterus (womb) where they are fertilised by the male element about 10 hours after mating.

PACKING (OF CARCASES). Carcases must not be packed for market until cold and set. It is well to use light non-returnable boxes or crates into which the carcases fit snugly. In normal times boxes should be lined with greaseproof paper, paper also being placed between each layer of carcases. Carcases must not be wrapped separately in paper. Nail down all packages securely and enclose name and address *clearly written*.

PARALYSIS. See Ailments.

PAUNCHING. Term applied to the removal of stomach, intestines, gall bag and bladder. Paunching may be done before skinning if the pelt is not required; otherwise it should be done after to avoid risk of soiling the fur.

When paunching a skinned carcase, slit the belly in the middle line from the vent to the junction of the ribs with the breast bone. Cut through the food pipe above the stomach and remove stomach and intestines and bladder intact, taking care to remove the last piece of intestine under the pelvic bone. Some buyers require that this bone be split and requirements in this respect should be ascertained.

The gall bag, which is a small blackish-green bag beneath the upper lobe of the liver should be removed carefully: if broken the flesh will be bitter.

Heart, lungs, liver and kidneys should be left in the carcase. When paunching a rabbit in its skin the offal is removed through a 4 in. slit made downwards from the breast bone.

PEDIGREE. A record of ancestry. Valuable for reference in planning breeding operations, and kept by all exhibition breeders. It is customary to furnish with exhibition stock pedigrees for three generations at least. These pedigrees are statements, signed by the breeder selling the

I**

rabbits, setting forth the names of the sire and dam of the rabbit in question together with those of the grandparents and great-grandparents on each side.

PELTS. After a boom period in which rabbit pelts were in good demand at a good price, pelt-production has not been very profitable during the last few years. The war brought a great change, however, vast quantities of pelts of all kinds being urgently required for home use and export, at prices attractive to breeders.

Securing the Pelts. In order to obtain a first grade pelt it is necessary to kill a rabbit when in full coat. The least trace of moult in the coat will render the pelt valueless commercially.

To ascertain whether a rabbit is ready to kill for its pelt blow upon the coat all over the body so that the skin may be seen beneath the fur. This should be clear and of a uniform pinkish or greyish colour, according to the breed. Black or dark areas on the skin mean that the coat has not completely broken through the skin in these particular spots. One such area only, even though it be no bigger than a wheat grain, means that the rabbit is not ready to kill for its pelt.

Wherever one of these " moult spots " appears there will be a corresponding dark spot on the " flesh " side of the pelt.

Dressing Pelts. Pelts intended for sale should be dressed professionally. Home dressing must necessarily fall short of the trade standard, which is very high and buyers also demand uniformity of dressing. Those who wish to try dressing pelts at home for their own use should get good results by carefully following these directions :

Fix fresh pelt to a board as for air-drying (see Air-Drying) and remove with a sharp knife all fat and flesh and the thin layer of wafer-like tissue adhering to flesh side.

Put 1 oz. saleratus with ½ lb. alum in a jug and pour in slowly 1½ pints of boiling water, stirring until both ingredients are dissolved. Put aside until cold.

Sponge pelt, still attached to board, with this solution daily for a week, drying in a warm room for about an hour afterwards and storing between applications in a fairly cool place.

When the pelt is removed from the board at the end of a week it will be hard and may be softened by rubbing, beating and pulling vigorously in all directions. Rubbing gently with pumice stone will give the surface a good finish.

Storage of Pelts. Pelts to be stored must be laid flat in pairs, fur to fur or skin to skin. They should be wrapped in grease-proof paper and placed in a receptacle that will admit no damp and as little air as possible, and will exclude mice

or moths. Bags of powdered naphthalene should be placed among the pelts, which should be taken out and shaken from time to time.

Packing of Pelts. Air-dried pelts must be packed flat and not rolled up. Pack in pairs, either fur to fur or skin to skin, and give protection with a sheet of cardboard at each end of the package. Dressed pelts should be rolled up and enclosed in tissue paper and finally in stout brown paper securely tied. Pelts must on no account be dispatched while " fresh ".

PNEUMONIA. See Ailments.

POISONOUS PLANTS. See Smallholding Section.

POLISH. Essentially a fancy breed as its small size (2½ to 3 lb.) rules it out as either fur or table rabbit. The Polish is pure white with blood red eyes and of neat appearance.

POTATOES. See Foods and Feeding.

POT BELLY. See Ailments.

RABBIT COURT. A closed-in yard where rabbits are allowed comparative freedom, hutches being provided as sleeping quarters. This type of accommodation resembles the modern Colony System (which see).

RABBIT WARREN. A large enclosed area where rabbits are run and allowed to breed naturally, much on Colony lines but without control at mating and breeding time (see Colony System).

REARING. See Young Stock.

RED WATER. See Ailments.

REX RABBITS. The term " Rex " does not apply to any one breed of rabbit but to numerous varieties, of comparatively recent introduction, possessing certain definite coat characteristics. " Rex " is part of the name given to the first rabbits of this type, these being called Castorrex, which means, literally, King Beaver. The Rex coat is much shorter than that of the normal fur rabbit and almost, or entirely, devoid of guard hairs, which gives a pelt an appearance of plush rather than of fur.

The original Rex rabbit from which the many recognised varieties have sprung was discovered by a Frenchman among a litter of doubtful origin. Experimental breeding in which this rabbit played a leading part led eventually to the production of the Castorrex.

After the introduction of the Castorrex further experiments resulted in the " rexing " of almost every known variety of rabbit, so that we now have, to name only a few varieties, Chinchilla-rex, Havana-rex, Lilac-rex, Angora-rex, etc.

In some of the Rexed varieties we get a wide range of beautiful colours and shadings. The Lynxrex, for example, gives the effect of an orange-coloured rabbit shot with silver, while the fur of the Opalrex has a surface colour of soft blue with a layer of golden tan beneath, which, in its turn, gives place to a slate blue undercolour.

Rex fur has distinct commercial value, but care must be taken to select only the best stock for the purpose and to work with those varieties that furriers most desire to handle.

There is a demand for pelts of the Erminerex, which is pure white, at satisfactory prices. This is probably the best variety of Rex to adopt from a financial point of view, as any demand for white fur is more likely to remain stable than for fur of any definite colour. With all coloured furs fashion changes must be taken into account.

There should also be definite sales for pelts of the Havanarex, the Nutriarex and the Black Alaskarex, as these yield good utility shades.

Beginners must be sure to start with high-grade stock. Inferior or harshcoated Rex is useless from the furrier's point of view.

RICKETS. See Ailments.

RINGWORM. See Ailments.

RUSSIAN. Continental name for the Himalayan rabbit. See Himalayan.

SABLE. The Sable is a breed of rabbit having a sepia-coloured pelt. It occurs in two distinct varieties, the Marten Sable and the Siamese Sable, which may be described broadly as " dark " and " light " respectively. The colour standard required for the Marten Sable insists upon " rich dark sepia brown " and for the Siamese Sable " soft dark sepia ". The colour in each must shade off gradually to a paler sepia on the flanks and, in the Siamese Sable, the belly. The belly in the Marten Sable is white and the chest, flanks and rump are ticked with white hairs. The fur is soft and dense. Weight of both varieties 5-7 lb.

SCOURS (DIARRHŒA). See Ailments.

SEASON. Term applied to the period when a doe will accept service by the buck. This period, or " heat," lasts 3 or 4 days and recurs regularly in healthy does approximately every 3 weeks from early spring until late summer. As autumn approaches the periods become less frequent and during the winter are often absent altogether.

Most does give definite signs of being in season, such as stamping with the hind feet, carrying about mouthfuls of bedding, rolling, plucking the fur or wool. Others give no sign whatever, and their condition can only be ascertained by trying them with the buck. See Mating.

Generally, but not invariably, the external organs of does in season are swollen and when examined the mucous membrane covering these organs appears pinkish-purple in colour.

SELECTIVE BREEDING. Term applied to the careful choice of stock for breeding purposes, with the object of producing or accentuating desired characteristics in the offspring (see In-Breeding).

SEX (DETERMINATION OF). The sex of adult rabbits is easily recognised by a casual inspection of the under part of the body. The teats of a doe, usually 8 in number, can be seen or felt beneath the fur or wool, 4 on each side of the middle line of the body. The testicles of an adult buck are apparent at the extreme lower end of the abdomen.

With youngsters sex is not easily obvious, but with a little practice one may soon learn to distinguish a buck from a doe. If the skin covering the external sexual organ is pulled back and gentle pressure applied, the organ itself can easily be seen. The opening of the male organ is circular, while that of the female organ is V-shaped, the point of the V being directed downwards and backwards.

Examinations for sex should be conducted as quickly and gently as possible, as, if it is unduly prolonged, there is risk not only of hurting the rabbit, but of setting up irritation that will react unfavourably upon its disposition.

SIBERIAN. A breed that yields a saleable pelt at 5 months—2 or 3 months earlier than with most fur breeds. This, added to the fact that the Siberian is a good table rabbit, brings the breed well to the fore as a dual-purpose one.

There are three varieties, Blue, Black and Brown.

The Siberian possesses a coat the fur of which, when turned in a reverse direction, gives the appearance of being without guard hairs and thus somewhat resembles Rex fur. Weight 5-7 lb.

SILVER. There are three recognised varieties of this breed, Silver Fawn, Silver Grey and Silver Brown. Silvers were formerly bred for both table and pelts, but are now chiefly of interest as exhibition rabbits. The self-coloured fur of Fawn, Grey or Brown must be evenly ticked with bright silver hairs, and this "silvering" must extend to feet, tail, nose, chest and ears.

SITKA. See Beveren.

SKINNING. Skinning a table or pelt rabbit may seem a difficult operation to the novice, but experience soon makes it easy. The procedure is as follows:

After paunching, lay the rabbit flat on a table and start on it by separating from the skin the thin layer of flesh—the "flank"—on each side of the incision made for the purpose of paunching. This will come away quite easily.

(If it be desired to skin before paunching, an incision should be made along the belly as for paunching, but less deep—only sufficient to cut through the skin itself. Then proceed as described below.)

Now insert the thumb between the flesh and the skin and work it gently towards the back of the rabbit and up to the hind leg. Reaching this, grasp the leg in the free hand and push it towards the rabbit, working the skin loose right round it with the thumb, right up to the knee joint.

Next, take the leg between the thumb and first finger (both inserted between it and the skin), then, with the thumb and finger of the other hand, take hold of the skin and give a sharp pull. This will draw the leg out of the skin and it will break off at the ankle joint of the leg. Free the other hind leg in the same way.

This done, separate the tail from the body with a knife, grip the two hind legs firmly in the left hand and, with the right, pull the skin away from the carcase as far as it will go. A long, slow pull will bring it off easily up to the forelegs; and these legs can be freed like the hindlegs. Afterwards, pull the skin again, when it will come off right up to the ears.

Both hands will be required to skin the head, so the rabbit should be hung up. To this end, make a slit, with a sharp knife, between the tendon and bone of one hind leg, and pass the other leg through it. You can now hang the rabbit up on a nail, by means of the loop formed by the legs.

To skin the head, take the skin in the left hand and, pulling it gently all the time, cut off the ears close to the head; as the skin is stopped by the eyes, cut round them. Continue to use the knife on the forehead, etc., as it becomes necessary, until the skin is right off.

SLOBBERS. See Ailments.

SMOKE PEARL. Flesh and fur breed with soft dense fur, smoke-coloured, shading to pearl grey on the back and pale fawn on the flanks and chest. Head, ears, legs and the upper side of the tail match the neck. Chest, rump, flanks and feet are well ticked with long white hairs. Neat in shape and weighing from 5-7 lb.

SNUFFLES. See Ailments.

SPORT. Term applied to any unexpected variation of type appearing in a litter.

SPOTTED LIVER. See Coccidiosis, under Ailments.

STAMPING. Alarm will cause rabbits to stamp loudly with their hind feet. Stamping is also a sign of sexual excitement in both does (see Season) and bucks. Bucks are much given to stamping in the spring, particularly if they can see, or even smell, does.

STORAGE OF FOOD. See Foods and Feeding.

STRANGLES. See Ailments.

STRAW. This makes good bedding for all stock if obtainable cheaply. Rabbits eat a good deal of their bedding and therefore oat straw, which has some feeding value, is best for short coated breeds. Wheat straw should be used for Angoras, as it is less likely to stick in the coats.

STUD. Term applied to any collection of rabbits where breeding takes place.

STUD BUCK. Term used to describe a buck used for breeding.

The buck in a breeding stud is at least as important as the does. A buck in his lifetime sires, and therefore sets his stamp upon, many times the progeny for which any one doe is responsible, and it is essential, therefore, that his welfare be duly considered and no pains be spared to keep him in condition.

A stud buck's hutch must be large enough to allow him free movement, and if he can sometimes be given a run it will be greatly to his advantage.

On no account must the stud buck be overworked. As a general rule he should not be used more than twice a week. If however, his services are in great demand for a time, special attention must be paid to his feed in order to prevent his vitality from flagging. At such times a daily ration of corn (⅔ wheat and ⅓ oats), green-food and hay *ad lib.*, and an occasional feed of broad bran will keep him up to the mark. (During war-time, of course, such generous treatment is impossible).

When he is " resting," cut down his corn and, in warm weather, leave out the wheat altogether, making up the deficiency with green-food.

Bucks lacking in sexual vigour, provided they are otherwise healthy, may be " toned up " by the occasional use of a stimulating mash (see Mating Rabbits, Reluctance to Mate).

STUD FEES. Fees charged for the services of a stud buck. The amount varies according to the excellence of the buck. As much as £1 1s. may be expected for a winning buck who is a proved sire of good youngsters, while a moderate specimen might command 5s. Does are always sent to the buck, never vice versa.

SUGAR BEET PULP. See Foods and Feeding.

SWEDES. See Foods and Feeding.

TABLE RABBITS. See Meat Rabbits.

TABLE SHOWS. These are informal shows held at the periodical meetings of local Domestic or other Clubs, where a small entry fee is paid and rabbits are awarded prizes according to merit. With the development of domestic rabbit keeping table shows have become a regular feature of club meetings.

TANS. (Black and Tan, Blue and Tan, Chocolate and Tan, Lilac and Tan). Exhibition rabbits resembling the Dutch in shape and size. Tans are too small to be popular as table rabbits, although they fatten readily and carry flesh of good quality. Their colour is black, clear dark blue, deep chocolate or a rich dove-colour with bright tan markings on neck,, insides of ears, edge of lower jaws, round eyes and nostrils ; chest, belly, flanks and underpart of tail all to be bright tan. Weight about 5 lb.

TEATS, SWOLLEN. See Ailments.

TEETH. It is highly important that a stock rabbit's teeth be normal. The chief point is that the upper and lower rows should meet evenly. If they do not the rabbit will be unable to bite the hard substances that keep the teeth " filed " and prevent their growing to an abnormal length.

If the teeth do not meet evenly there is deformity of the jaw, and this is usually caused by Rickets. There is no cure for the condition once the bones have hardened, and a rabbit with this disability should be allowed to live only just long enough to get pelt and carcase ready for killing.

Cutting Over-long Teeth. Rabbits tending to grow exceptionally long teeth must be fed on soft mashes, and it will be necessary to cut the teeth down to the normal length every 3 or 4 weeks. If left uncut these long teeth usually show a tendency to curve inwards, the upper ones over the tongue and the lower ones over the top lip, preventing the rabbit from opening its mouth.

To cut the teeth the rabbit should be placed on a table, the head held firmly by one person, and the lips pulled well back from the teeth, while a second person nips off the ends of the teeth with sharp pliers or strong wire cutters.

UTILITY RABBITS. Term applied to rabbits kept with some definite profit-making purpose in view (that is, wool, flesh or fur production) as opposed to those kept only for fancy or exhibition.

VENT. Term used to describe the region including the anus and external genital organs.

VENT DISEASE. See Ailments.

VISITING DOES. Does sent to the owner of a stud buck for mating. Immediately on arrival a visiting doe should be placed in a clean hutch and given food and water. She should be allowed to rest for some hours before being put with the buck and should not be allowed to travel by rail unattended for at least 12 hours after mating.

WATER, DRINKING. Drinking water is essential to the well-being of rabbits of all ages and should be provided fresh daily. It assists in digestion and milk production, increases resistance to disease and is necessary to the satisfactory growth and development of young stock. Water is of the greatest importance to in-kindle does, particularly during the latter half of pregnancy and at the time of kindling. From the fourteenth day after mating water should be in the hutch night and day. Lack of drinking water is the commonest cause of does eating their young.

Drinking water is best supplied in earthenware vessels, preferably of the type that can be hung on a hook on the wall of the hutch. Tins are not suitable. Whatever is used should be so situated that it cannot be upset by the rabbit. If the water contained is upset the rabbit must-go thirsty until filling time comes round.

Water vessels should be kept scrupulously clean, being washed out regularly and occasionally disinfected.

WEANING YOUNG RABBITS.

Young rabbits should be left with their dam until they are 8 weeks old. Then they should be weaned, or separated from their parent. Before weaning the youngsters should be taught to fend for themselves.

In a normal doe, the milk supply declines gradually after about the fourth week following the arrival of her litter. Thus the babies day by day get just a little less natural food and so must start looking round for something to make up the deficiency. Day by day they eat more and more supplied food, until by the end of the eighth week, the withdrawal of the small amount of milk they are then receiving is of no consequence. They are, in effect, entirely self-supporting.

Premature Weaning. It is occasionally imperative that babies should be weaned early. For instance, a doe will sometimes become unkindly disposed towards her litter when they are about a month old, because she comes in season again and desires to mate. In such an event the babies must be taken from the doe and given special attention as regards food.

Normally, a daily drink of cow's or goat's milk, or a dish of bread and milk morning and evening, has been given. In war-time cod-liver oil should be given, mixed with whatever dry food is available. Good hay and some broad bran should always be in the hutch, and it is a wise plan to mix syrup of hypophosphites with the drinking water in the proportion of half a teaspoonful to half a pint of water.

Caution should be exercised as to the amount and nature of the greenfood at first, care being taken to avoid too great quantities of any plant possessing laxative properties, but the babies should be accustomed gradually to a large daily supply. (Also see Young Stock).

WHEAT. See Foods and Feeding.

WILD RABBITS do not do well in captivity and even when caught and confined in colonies rarely give satisfactory results.

WINTER BREEDING. Rabbits do not breed naturally during the winter and trouble may be experienced in getting the stock to mate. This may be overcome (see Mating, Reluctance to Mate). Special attention also is necessary for does at kindling and afterwards to ensure that the young do not die of cold.

In cold weather it is a good plan to bed a breeding hutch with straw, in addition to a good layer of sawdust or peat-moss litter. A doe nearly always scratches a hole in the floor litter when beginning to make her nest, and if plenty of material is used as floor litter added depth will be given to the nest. A good deep nest considerably minimises the risk of draughts. This does not mean that the doe will not need some additional softer nesting material, which is essential and should be supplied in considerable quantity. In frosty weather many does pile hay on top of their nests until the mound nearly reaches the roof of the hutch.

Hutches used for winter breeding, while being well protected from rain, snow and draughts, should be so placed that they admit plenty of light and air. Direct sunlight, whenever possible, is beneficial to both doe and young. Every precaution must be taken to prevent drinking water from freezing. The container is best embedded deeply in hay or wrapped in flannel.

WOOL PRODUCTION. See Angoras.

WOOL-SWALLOWING. Owing to the peculiar nature of its coat, the Angora rabbit is open to a form of illness which affects no other breed. Wool may be swallowed, either with the food or as a result of the habit some Angoras have of licking their coats.

If the digestive system is kept in a state of healthy activity, the quantity of wool taken may pass through the system and do no harm. It is, therefore, important that the diet of Angoras shall include, in addition to plenty of greenfood, such things as broad bran, which is gently laxative, and sufficient hay to give the stomach and intestines bulk upon which to work.

On the other hand, incorrect feeding, encouraging a sluggish action of the digestive system, will allow wool that happens to be swallowed to collect in the stomach, where it will in time either form a solid mass, blocking the passage into the intestines, or collect upon the walls of the stomach, first hindering and finally stopping the action of the digestive glands. The result in either event will be fatal.

Such a state of affairs is, happily, preventable. In addition to providing a suitable diet, it is a wise precaution in every Angora rabbitry to add Glauber's salts once or twice weekly to the drinking water—1 tablespoonful per gallon.

Any rabbit that habitually licks its coat, exhibits symptoms of dullness and loss of appetite, accompanied by constipation, should be given a fairly drastic aperient. One dessertspoonful of medicinal paraffin, if given at the first appearance of the symptoms, will probably result in the appearance of a considerable quantity of wool in the droppings and the rabbit will speedily recover health.

WORMS. See Ailments.

WRYNECK. See Ailments.

YOUNG STOCK (REARING AND CARE OF). Rabbits at birth are hairless and have closed eyes. The coat breaks through the skin a few days after birth and the babies are well coated by the end of a week. The eyes open about the eleventh day, and the babies leave the nest usually between the fourteenth and twenty-first days.

While they are in the nest the babies receive all necessary food from their dam, but as soon as they emerge they will feed for themselves, sharing their dam's food. The quantities of food supplied must therefore be increased as the babies' appetites develop.

It is not essential to modify the nature of the food; but it is an advantage, with a large litter when the doe's milk is deficient, to incorporate cod-liver oil with all bran used (see Cod-Liver Oil).

The way in which greenfood is fed to young stock is important. If the doe's ration is increased slightly every day after the babies begin to feed they will become accustomed gradually to liberal supplies, greatly to their benefit. It is when large quantities of greens are given irregularly that harm results.

Under no conditions must greenfood be fed to young stock on an empty stomach. Hay or other dry food should always be given beforehand. When greenfood is scarce or unobtainable babies should be given cod liver oil; otherwise they will become rickety (see Rickets).

Young rabbits are particularly susceptible to Coccidiosis, and as a risk always exists that a doe may be a " carrier " of this disease, scrupulous cleanliness of hutches is a very important point indeed (see Coccidiosis).

When Does Kill Their Young. See Mutilation of Litters.

GOATS

GOAT-KEEPING had made great headway in the years immediately preceding the war. It has received a further big fillip since 1939, and the coming of restrictions on the distribution and sale of cows' milk still further emphasised the advantage of keeping goats for milk production. One or two goats might well be regarded as part of the normal stock of each and every holding to provide milk, and where milk supplies are sufficient, butter and cheese. These can be made from goat's milk without difficulty and both are tasty in the extreme.

BREED, CHOICE OF. Choice of breed or type must be influenced to some extent by the conditions under which the goats will be kept. If they will have to " rough it " more or less a hardy type such as the shaggy-coated Old English goat, should be chosen. The short-coated types will not endure the same exposure to extremes of weather conditions.

The heaviest milkers are usually to be found in the British breed. This covers all goats which through cross-breeding or uncorrectness of type are not considered officially as belonging to such breeds as the Toggenburg, Saanen and their British types, British Alpine and Anglo-Nubian.

Numbers of pedigree, heavy-milking goats are long lactationers—which means that it is possible to miss mating them one season, but they continue milking well for two years. Non-pedigree goats will not often milk continuously over a long period. Frequently they will not milk right through to the next kidding, so that it becomes necessary to keep more than one, mating one to kid in January and another in May.

BREEDING GOATS, SELECTING. The time to start studying goats with the view to selecting animals for autumn mating is in July.

Discard for breeding purposes any crossbred animal which has the characteristics of two or more pure breeds totally unlike each other—say the drooping ears of an Anglo-Nubian and the short legs of a pure Toggenburg.

Poor feeders, and animals lacking in vitality, should not be bred, nor a nanny that milks at the expense of her condition. The chosen nanny in each case should be a long-lactation animal and a consistent milker, giving not less than 2,000 lb. of milk a year.

The ideal goat for breeding has a massive, lengthy frame, deep, well-sprung ribs, plenty of stamina, a silky coat, full, bright eyes, telling of sound health and vitality, and, of course, high-class milking ability.

DOCTORING GOATS. When it is necessary to give a goat medicine back the animal into a corner of her house and hold her steady with the knee, pressing it gently but firmly into her side. Raise the muzzle with the left hand and, by inserting the thumb and finger and pressing them between the teeth or gums, open her mouth and administer the required dose.

A good way to give the medicine is by means of a small teapot, placing the spout in the goat's mouth. Of course, the liquid must be given slowly or the goat will not be able to swallow quickly enough to prevent choking.

The following are the main troubles to which goats are subject, and the recommended methods of treatment :

Broken Horns. First wash the broken parts clean with lukewarm water and dry well with a clean cloth. Then anoint the fractured parts with stockholm tar and finally fix the horn in position by means of a suitably shaped stick bound securely to the sound horn. Leave the horn like this until the fracture sets satisfactorily.

Bronchitis. Denoted by dry, husky cough, soon becoming moist ; discharge from nose, disturbed breathing and cold extremities. Treatment : Steam head with a few drops of friar's balsam in hot water. Give tablespoonful doses 3 times daily of following : Ammon. carb., 80 grains ; tinct. camph. co., 6 drachms ; tinct. senegal, 4 drachms ; water to 8 ounces. Bandage extremities. Give warm sloppy food. Keep dry and warm.

Colds. A rough and ready, yet very sure treatment, is to use carbolated vaseline. Place about half a teaspoonful on the back of the tongue three times a day. Also smear a little in each nostril twice a day.

Constipation. Give two tablespoonfuls of medicinal paraffin. Correct the diet. If liver is inactive, give 2 tablespoonfuls of mag. sulph. dissolved in warm water.

Diarrhœa. Give dry, easily digested food. Limit water and fluid supply. Give flour gruel. Also give tablespoonful doses of following : Chlorodyne, $1\frac{1}{2}$ drachms ; carbonate of bismuth, 80 grains ; syrup of ginger, I ounce ; water to 8 oz. The dose must be given 3 or 4 times daily.

Foot and Mouth Disease. Although not so susceptible to this disease as cattle, goats are liable to contract it during outbreaks. Exactly the same steps must be taken as when cattle are affected. See Cattle in Smallholder Section.

Foot-Rot. It is when they are out at pasture that goats are most likely to contract foot-rot, particularly if the ground is wet. In case this trouble occurs it is as well to keep a good foot-rot ointment on hand. One can be made by mixing together 3 oz. vaseline, 2 oz. verdigris powder, 2½ oz. resin ointment, ½ oz. oil of turpentine, 1 oz. linseed oil, ¼ oz. dried alum, 1 drachm carbolic acid and 1 oz. of prepared chalk. A chemist will make this up for you. The above ointment should be kept in a tightly closed tin when not in use. Before the ointment is applied the hoof should be first washed and dried and pared, if necessary. Repeat the dressing twice a week until the rot has gone.

Garget, or Caked Udder. This is an extremely dangerous trouble, and may mean the loss of the udder if not promptly attended to.

At the first sign of hardening give a cooling dose of medicine, say milk of magnesia, and put the goat in a warm, dry, well-bedded stall. Foment the udder with hot salted water, drying very carefully with old linen.

Well massage the udder and extract all the milk or curd that can be got out, then well rub with an ointment made of camphor mixed into lard. Continue to milk.

Repeat the treatment every two hours until the udder returns to normal. Special care must be taken thoroughly to dry the udder after fomenting so that there is no risk of it becoming chilled.

Gastritis. A goat may contract gastritis if she eats anything green that is the least bit wet. A proved dose which can be quickly prepared to give relief in such a case is this : Half-ounce Epsom salts dissolved in 2 tablespoonfuls of hot water. To this is added 1 tablespoonful of essence of ginger and 1 tablespoonful of sugar. Add this mixture to a ¼ pint of pure raw linseed oil.

Hoven. Denoted by swelling up in the left flank ; grunting ; persistent standing ; harassed breathing. Treatment : give 3 tablespoonfuls of mag. sulph. and 1 drachm of powdered ginger in ½ pint of warm linseed tea. Also give a tablespoonful of whisky. Apply hot packs round the abdomen. Sparse feeding on warm mashes and gruel is necessary.

Lice. A good cure for lice is a solution of quassia chips and half an ounce of cheap shag tobacco. Heat to boiling-point a quart of water and pour this over a muslin bag containing the chips and shag in a pail. When this liquid has become lukewarm, it is ready for application. Use a sponge and rub against the lay of the fur—not in the same direction as it grows.

Mammitis, or *Inflammation of the Udder.* Denoted by hot and tender udder ; discoloured milk ; rise of temperature ; constipation. Treatment : Rub the udder with camphorated oil. Dress any sores with carbolised oil. Give 3 tablespoonfuls of mag. sulph. and follow up with powders of 40 grains of pot. nit. and 1 drachm of soda bicarb., 3 times daily in water, as a drench with ½ pint water, or in mashes. Support udder with a clean cloth with holes in it for the teats to pass through.

Poisoning. This is a trouble which, if not attended to at once, may end fatally. For most plant poisons a satisfactory antidote is a drench consisting of ½ lb. of melted margarine. Linseed oil and hot coffee may be used afterwards.

Simple Cough. Give 3 times daily a teaspoonful of an electuary of tannin, 1 drachm ; pot. chlor., 1 oz. ; treacle, 4 oz. Rub throat with camphorated oil.

Sore Eyes. Apply boracic ointment both in the morning and at night. If the soreness has gone very far and the use of the ointment has no effect, the eyes must be bathed night and morning with milk and warm water mixed in equal parts. After bathing with this solution, dry well with the aid of some clean, soft rags or cotton wool. Then apply boracic ointment. If the goat is inclined to make things worse by rubbing the sore parts, lightly bandage the eyes up for a few days.

Stiff Joints. For ordinary stiffness, rub in a good embrocation. If rheumatism is suspected, rub in a liniment made to this old-fashioned and good recipe : ½ pint vinegar, ½ pint turpentine, whites of 2 eggs, thoroughly shaken till well mixed. Also give a 5-grain aspirin tablet once or twice daily.

Worms. Give a tablespoonful of castor or olive oil late at night upon an empty stomach ; also, in the morning, whilst the animal is still fasting, give her three grs. of santonin in a little warm milk. Fast for a further hour, then repeat the oil dose and immediately follow with a warm bran mash. Repeat this treatment a week later, if required.

FOODS AND FEEDING. Considerable study was given in pre-war days to the question of which method of goat-feeding produces most milk having regard to its cost. The following is typical of the rations given to heavy and long milking animals—those producing say 1½ gallons daily :—

Bean meal and soya bean meal, 15 parts ; decorticated cotton seed meal and decorticated ground nut meal, 12½ parts ; wheatfeed, 30 parts ; barley meal, ground oats and maize meal, 32½ parts; locust bean meal and molasses, 10 parts; mineral mixture 2½ parts.

War-time Feeding. As with all other stock, considerable modifications in feeding methods became necessary with war-time restrictions allowing goat-keepers only limited amounts of protein foods and cereals.

Among the foods that are valuable for goats are roots—especially carrots, swedes, beetroot, mangolds and sugar beet; green leaves from such crops as kale, and pea haulm; and leafy twigs from hedgerows. When available, hay has remained a staple food. It should be added that dried young grass and silage; see pp. 223 and 238) are both very valuable food for goats. People who keep only one or two goats may well dry for winter use all the lawn mowings, which will be rich in protein.

Goats have also shown readiness to eat acorns, horse-chestnuts and beech mast in small quantities after these foods have been properly dried, ground or crushed and mixed with other foods.

Following is a typical war-time feeding method : morning meal, half the daily ration of meals, followed by an allowance of hay.

During the day two feeds of greenstuff or other alternative foods, but where circumstances permit the goats should be allowed to browse instead.

Evening meal : the other half of the meals, followed by more greenfood.

GESTATION. The gestatory period with goats ranges between 150 and 163 days, with 156 as the average.

GOAT CLUBS. As with pigs, poultry and rabbits, goat clubs exist for the encouragement of goat keeping, the provision of information and advice, and the improvement of the stock. Guidance in organizing new clubs may be obtained from the Secretary of the British Goat Society, The Cottage, Roydon Road, Diss, Norfolk.

GROOMING ROUTINE. Goats require grooming every day at a regular hour, the best time being directly after the morning milking.

Start with the head, work down on to the neck and shoulders, and then do the foreleg of one side, the one side of the back and body, the hind leg and the tail. Then, starting again at the head, treat the other side of the animal in the same fashion.

If there is much dirt on the coat, start with the curry-comb and follow on with the dandy-brush. Otherwise start with the dandy-brush, going completely over the body of the animal and brush away,

working in the direction of the lie of the hair, until both skin and coat are clean and all of the dust has been removed. Then go over again with a straw whisp to remove any dust that may have settled on the coat.

After this comb out longer hair on the head, tail, flanks or legs. Should eyes, ears or nostrils be at all dirty or encrusted, use the sponge, slightly moistened, on them. Should there be dirt on the feet, especially between the claws of the hoofs, wash clean and dry them.

Goats' hoofs must be attended to regularly. If neglected, they may cause considerable trouble.

Every few days take a look at each hoof and see whether the outer rim of horn is growing beyond the level of the centre part or frog. If overgrowing has taken place even to the extent of, say, ¼ in., take a knife with a strong, sharp blade and shave off thin strips of horn.

HOOFS, TRIMMING. See Grooming.

HOUSES, TYPES OF. Where one goat is kept it should have a stall measuring at least 8 ft. long and 4 ft. wide. On a larger scale, a house measuring 20 ft. long by 16 ft. wide would thus accommodate eight goats, with a 4 ft.-wide passage between two parallel rows of four stalls.

Kids, goatlings and milkers could all be housed under the same roof in a building measuring 32 ft. by 20 ft. A space of 12 ft. by 16 ft., partitioned off, would form the goatlings' house; another, 8 ft. by 16 ft., the kid's house; and the remainder making the space for the eight stalls and passage-way already mentioned.

In the kids' portion of the house there should be a sleeping bench 2 ft. 6 in. wide, and about 12 ft. long, raised a foot or so off the floor; while the goatlings should have a similar bench, but 3 ft. wide.

Large windows, made to open, in the sides of the building, particularly on the south side, will ensure plenty of health-giving light and ventilation.

An excellent addition to the building would be a roofed-over but open-sided veranda 6 ft. wide, running the length of the west side. Here the goats could shelter in the event of rain.

KIDDING TIME MANAGEMENT. Some little time before kidding is expected allow the goat to run loose in a suitable building at night, though she may run out by day as usual if the grazing season is on.

Feed her on the usual lines as regards hay and greenstuff. As regards concentrated food, feed only bran mash if available for a week before kidding is expected and until kidding is over. The average gestation period is 150 days.

The goat will exhibit signs of restlessness two or three hours before the kids are actually born, will show signs of discomfort, will lie down and get up again a time or two, the parts will become enlarged, and the animal may bleat a little. Then kidding will soon begin.

If the kidding occurs before the pasturing season, or at night at any time, then the animal should kid indoors.

After Kidding Care. All that need be done is to see the kids reach the mother's teats, then clear up, spread out the bedding and make the mother clean and comfortable, giving her a drink of oatmeal water, slightly warmed.

After kidding all that need be done is to see the kids reach the mother's teats, then clear up, spread out the bedding and make the mother clean and comfortable, giving her a drink of oatmeal water, slightly warmed.

If the kids are to be hand-reared, take them away at once and as soon as possible after kidding milk the mother, relieving the udder of about half its contents. Afterwards milk regularly morning and night in the usual fashion. For the first three days do not strip when milking ; afterwards strip as clean as possible at each milking.

KIDS, DISHORNING. Clip the hair off the top of the animal's head and you will see little smooth spots where the horns will come. Rub grease or lard into the skin around these hornbuds, leaving the buds untouched as well as a space around them about the size of a sixpence. Also put a ring of grease round the top of the clipped head.

Now take a stick of caustic potash wrapped in a piece of paper to protect the fingers, one end being left uncovered. Moisten the stick on a damp rag and then rub it on each horn-bud until the skin is broken. Rub a little more and gradually widen the circle until it is about the size of the sixpenny piece already mentioned. Simply putting a dab of the caustic on the horn-buds will not prevent them from growing. The buds must be well blistered, to do the job properly.

KIDS, REARING. Before kids are born, the goat-keeper must decide whether to hand-rear them or rear them on the dam. Hand-rearing is naturally somewhat of a tie, especially for the first month, but its advantages make it worth while.

One important point in hand-rearing is that kids can be put on to meal-feeds much more easily. Other advantages are that the mother milks much better, for goats suckling kids are apt to hold up some of the supply ; there is no upset at weaning

time ; and the kids are much more docile and easily managed than those reared on the dam.

Also when a goat is suckling a kid she is likely to develop uneven teats which make her very unsightly.

Where the kids are to be hand-reared, the owner must have in readiness a bottle and teat. To start with, a ½-pint olive-oil bottle is suitable, and as the kid grows older and takes more, a wine bottle may be substituted. The teat may be of the soft pull-on variety, but a type of black rubber plug teat made for lambs is very popular.

One method of hand-rearing is to remove the kids at once, before the mother has had time to lick them, taking them out of sight and hearing.

Another plan very suitable for the small goat-keeper who owns only one or two goats is to have ready a fairly large box or packing case about 2 ft. high lined with straw. As soon as the kid or kids are born and fairly dry, place them in the box, which should stand in a corner of the mother's pen. She can then lick and clean the kid, but it cannot suck her.

The First Feed. The time to give a kid the first feed is when it is standing and looking for food.

Bring a jug and the bottle filled with hot water to keep them warm, pour away the water, and milk into the jug, taking just enough to ease the udder. Fill the bottle and feed the kid.

For the first feed it will be easier to take the kid on one's knees and gently open its mouth. Kids usually suck at once.

If a kid refuses to suck, wait an hour or two more and try again ; when hungry, it will drink readily. Kids should be fed four times daily to start with, the last feed being as late in the evening as possible.

Let the kid take as much as it will, but never force it to take milk, if unwilling.

The milk must be given at blood heat; wrong temperature is a common cause of scour, also due to dirty bottles or teats. Bottles and teats *must* be kept scrupulously clean. Whatever teat is used the hole must be neither too large nor too small : the former would cause the kid to " bolt " the milk, with risk of colic or scours.

Take the kids out of the box to exercise after each feed, and whenever working in the stable. Do not leave them alone with the mother until at least a week or they may learn to suck, and if they do it will be very hard to break them of it.

After a week they may be loose with her by day, but kept in the box at night. In this way the dam will not fret, and the kid, if it is a single one, will also have company. If it does not worry her unduly by jumping on her back when she is resting, it may remain in her pen a long time, but two or more kids are best kept by themselves after about 10 days with the mother.

The Daily Ration. By the end of the month the kids will be taking up to a pint at a feed. Then they are cut down to three feeds. A wine-bottle holding 1¼ pints is about right, making 3¾ pints in the day.

Plain goats' milk will rear the best kids, but where the milk is urgently required for other purposes, meal may be gradually introduced at a month old. A kid should never be put straight on to meal, however, or it may be upset, and perhaps refuse to take it. If added in gradually increasing quantities to milk the change over may be effected without any upsets.

From a fortnight old onwards the kids will begin to pick at hay, grass and other foods.

MATING GOATS. Female goats usually come into season from August to March at intervals of twenty-one days. The usual signs are restlessness, almost continuous switching of the tail, noisy bleating, together with a perceptible swelling of the external organ. Often the goat may show little desire for food.

Secure the service of the male without delay, as although in some cases the condition will persist for perhaps three days, this is not always so.

Milking before Mating. It not infrequently happens that goats may yield milk without having been mated and produced kids. If so, it may be possible to take advantage of it, but not omitting to mate the goat normally.

MILKING METHODS. To start milking, both teats should be grasped gently but firmly, one in each hand, and the fingers should squeeze the milk out with an opening and closing movement, but without pulling the teats down at all. This is important, as pulling down will in time cause an ugly pendulous udder and long teats.

When there is only a little milk left, "stripping" begins. Rub and massage the udder all over, drawing the hands over it with a downward movement. Then it is necessary to draw the finger and thumb down the teats to get the last drops. Thorough stripping is most important, as any milk left behind will be absorbed back and if persisted in will encourage the goat to dry off.

Always milk with both hands, steadying a small pail getween the knees or feet. Milking with one hand is a slow process and may in time cause a lop-sided udder. Speed, thoroughness, cleanliness and gentleness are the main points of a good milker, to whom a goat will let down her milk more freely than to a slow, laborious milker.

When a goat has much long hair under the body and on the thighs, it is an aid to clean milking to trim some of this back. Apart from being unhygienic, the long hairs are apt to catch in the milker's hands, resulting perhaps in spilt milk.

It is important to fix regular hours for milking, and there should be as nearly equal intervals between them as possible.

STUD GOAT SCHEME. A scheme organized by the Ministry of Agriculture and the British Goat Society, under which stud goats became available for cottagers and smallholders. The object is to improve the milking capacity of goats generally. (The scheme is in abeyance during the war).

TETHERING GOATS. Goats should only be tied up when conditions make it imperative. In such cases see that the tethers cannot catch upon anything or get wound round tree stumps and the like or the animals will not be able to get enough free play for grazing.

The tethers, and the goats, must be moved on sufficiently often to ensure that plenty of herbage is always within reach.

Have water and rock salt within reach and, in warm weather, some sort of shelter must be provided.

If tethering is started early in the year, care must always be taken to see that the goats are never left out when cold winds or rain are about.

BEES

ACARINE DISEASE. See Diseases.

ANTS. As enemies of bees. See Pests.

APIARY, SITE FOR AND AR-RANGEMENT OF. In choosing a site for an apiary, the nature and resources of the surrounding country must be carefully considered.

The Ideal Site. Among fruit-plantations for an early nectar-yield ; close to sheep-farms where large areas of flowering crops are grown for sheep-feed, supplying the material for the main honey-harvest of the year ; and having a range of heather-clad hills within practicable distance, whence late honey would be obtained.

Aspect. Hives of an apiary should, if possible, face south with enough east in it to allow a certain amount of early sunshine. The hives should be sheltered from cold or rough winds. A good plan is to arrange them in a row about a yard apart as near to a dense hedge or close fence as may be, while leaving room for a path wide enough to take a wheel-barrow behind them without difficulty.

All hives, of course, should face forward. No traffic should ever pass across the bees' line of flight.

APPLIANCES. Apart from the hives themselves (see Hives) the tools and appliances required are simple and few. The hives will need to be opened for various reasons from time to time, and when this need arises the bees must first be subdued or quietened. For this object smoke is generally used and a contrivance called a bellows-smoker will be required. The same object can be effected by fumes of carbolic-acid, a cloth damped with dilute acid being employed.

Tools. A good strong screwdriver, a couple of painter's scrapers, a fairly soft, small, fibre hand-broom, and an article called a " hive-tool " comprise about all that will be needed for ordinary manipulations.

Protection from Stings. For the protection of his face from stings and to prevent the irritating effect of bees crawling about on the skin, the bee-keeper should accustom himself to wear a veil of fine black silk net which may be permanently fixed upon a straw hat with wide brim.

Gloves are not desirable as they interfere with the proper manipulation of hives. A little vinegar rubbed on the hands just before going among the bees, will usually obviate the risk of stings.

BEE ESCAPE. A contrivance fitted to Escape Boards (see under Extracting), etc., allowing bees to pass downwards through it but barring their passage upwards.

BEE-PARASITES. There are several parasites of bees, the chief being the following :

The Blind-louse. This is a very minute reddish-brown creature that may live on all bees of a hive, but chiefly upon the queen. The Blind-louse does not do much harm unless it becomes very prevalent in a hive ; then it may be attacked by gently puffing sulphur fumes over the crowd of bees on the combs by means of the ordinary smoker, a little flowers of sulphur having been first incorporated with the fuel used.

If only a few of these parasites are observed in a hive from time to time, this can generally be ignored, as the creature commonly disappears on the advent of cold weather. If the queen is seen to be markedly affected, she should be captured and her assailants driven off by a little tobacco-smoke.

The Wax-Moth. There are two species of this pest, one large and one fairly small. Both enter beehives and prey on the combs. It is only in weak stocks, however, that the Wax-moth constitutes a danger. Strongly populated hives will generally keep their hives clear of this pest.

The Wax-moths lay their eggs almost anywhere in the hive, but chiefly on the combs. On hatching, the larvæ proceed to devour the comb and everything that comes in their way, driving tunnels through the combs and lining them with tough silk forming a very serviceable protection against any assailant.

The hiding larvæ should be captured and killed whenever seen. A convenient tool for pulling their silken tubes out of the comb, is an ordinary fine steel crochet-hook. When going through a hive, watch should always be kept for Wax-moth larvæ.

The moth is also apt to get at stored combs. If such combs be found to be infested, they should be placed in supers stacked one over the other. On the top of the pile put a quilt of old clean flannel and sprinkle this generously with fluid carbon-disulphide. Cover over with a piece of waterproof sheeting, or an old mackintosh coat.

The fumes from the drug (which are highly inflammable) will travel down through the supers and kill the moth in all stages of its life. Allow twenty-four hours for the process.

BEE STOCKS. The apiary may be stocked either by purchasing established hives of bees or by procuring swarms and running them into empty hives.

Buying Stocks and Swarms. The best time to buy stocks of bees is the early spring, when the risks of wintering are over. Swarms are obtainable from May to the beginning of July, but the best and strongest swarms are always those first thrown by the hive.

Choice of Strain. Honey-bees vary greatly in working qualities, and in appearance, most strains being dark brown. Bees of Italian blood have yellow bands, either so broad and bright that the bee appears all yellow, or narrowed to thin lines as with hybrids.

The best policy for the prospective buyer is to disregard the question of species altogether, and obtain his stocks or swarms, of whatever colour, from a reputable bee-breeder under a guarantee that they are hardy, healthy and good honey-getters.

Life in the Hive. A normal hive of bees contains one queen or mother-bee who is the parent of the whole colony. The *worker-bees*, of whom there may be between ten thousand and fifty thousand or more, according to the time of year, are females undeveloped sexually, and so incapable of reproducing their species.

Drones. The complement of the hive is made up by five or six hundred drones, or male bees, whose sole function is the fertilization of the young virgin queens when these fly abroad on their mating flight in summer.

A queen-bee has only one coition with a drone, this sufficing for her fertilization for the whole of her active life.

The drone dies after having thus fulfilled his function. The queen returns to the hive where she soon commences egg-laying, thereafter, as far as is known, never coming out of the hive until she accompanies a swarm in the spring of the following year.

Queen Bees. A queen-bee has, apparently, the power of producing at will eggs that will hatch out into either male or female bees. The eggs which she deposits in the small or worker comb-cells produce worker-bees, and those laid in the somewhat larger drone-cells result in the male of the kind.

A queen-bee herself is not raised in the ordinary cells of the comb, but a special structure of very much larger size is made for her nurture, generally on the edge of one of the central combs.

Length of Life. A queen-bee's life is much longer than that of the worker or drone. A queen has been known to live for five years, whereas the average life of a worker-bee is scarcely more than six weeks. Drones would doubtless live to a fair age were it not for the fact that they are ruthlessly slaughtered by the worker-bees at the end of each summer.

BEESWAX. Supposed by the ancients to be a substance which the bees gathered from flowers, but actually a product of the bees' own bodies.

Beneath the horny segments of the abdomen of the worker-bee on the under side are found certain pentagonal shallow depressions where scales of raw wax are formed, these being removed by the bee and masticated, when what we know as beeswax is formed.

BLIND LOUSE. See Bee Parasites.

BROOD FRAMES, to wire. See Wiring.

———

CANDY. See Feeding of Bees.

CASTS. Term applied to any swarm that leaves the hive after the first. Casts should always be returned to the mother-hive, or otherwise utilised.

CLOTHING FOR THE BEE-KEEPER. Bee-keepers are advised to wear a washable tunic of white linen, with a collar buttoning to the neck, an elastic in the sleeve-hems at the wrists, and several roomy pockets.

It is a good plan to wear trouser-clips, to prevent bees crawling up the legs.

———

DEMAREEING. Term applied to an operation carried out for the prevention of swarming. The time to choose for the operation is when the weather is settled fair and the hive to be treated is " boiling over " with bees.

All the combs in the brood-chamber, except one with some unsealed brood and eggs, are removed from the brood-box and put into an empty brood-chamber stationed at one side of the hive. The queen should be on this single comb left in the old brood-box, and the box should be filled up with nine empty combs, or frames furnished with full sheets of foundation.

A queen-excluder is now placed over the old box and the new brood-chamber placed upon the excluder, with an extra comb added to make up its complement of ten frames.

In three weeks' time, all the brood in

the upper box will be hatched out and probably the combs nearly filled with honey.

It is important to prevent the building of queen-cells in the upper box, or swarming will take place in spite of the operation. This upper box must be opened in from seven to ten days after manipulation, and all queen-cells carefully searched for and removed. If any sign of a queen-cell remains failure of the device is certain.

Should the hive have honey-supers on it when the demareeing is attempted, these may be placed between the excluder and the upper box, or they may be emptied of bees and given to another hive to complete. In any event the supering of the demareed hive should be attended to as soon as it appears necessary.

DISEASES OF BEES. The only really important diseases to which the honey-bee is subject are Acarine disease, Foul-brood and Dysentery.

Acarine. This is caused by the invasion of the breathing-tubes of the bee by a minute parasitic mite, which multiplies therein. It is highly contagious.

Symptoms of Acarine Disease. When a lot of bees are seen to be crawling about on the ground obviously in a weak condition, that is the time to suspect acarine disease. Examine such bees, particularly noticing their wings. Bees affected by the disease carry one wing, or both, at an odd angle, as if the wings had been dislocated; also the abdomen is extended.

Such bees may be noticed falling from the alighting board on to the ground, or climbing to the tops of grass stems and huddling together in small groups.

Cure is only possible when the disease is detected in an early stage, when only a few affected bees have been seen.

Obtain from the chemist a special preparation called Frow Cure. It is poisonous to humans, also it is inflammable, so take due precautions. The mixture consists of 2 parts petrol, 2 parts nitrobenzene and 1 part safrol.

Frow Cure is used by measuring out one-half drachm of the fluid and sprinkling it over a piece of absorbent material, like felt, 4 or 5 in. square, which is then pushed into the hive entrance as far back under the brood frames as possible. Do this one evening about dusk, after the bees are in for the night.

On the following evening, about the same time, withdraw the pad of felt, sprinkle a second dose of the mixture upon it and replace the pad.

Repeat this process every subsequent evening for the next four evenings. Then leave the pad in the hive, without

further sprinkling, for four days longer, when it should be withdrawn altogether.

When undergoing treatment, the bees are in a very drowsy state, so entrances of hives must be guarded from attacks by robber bees.

The normal death-rate among the bees is sure to be increased somewhat during treatment. Do not be alarmed at this. But be sure to rake the floor-boards clear of dead bees, with a hooked wire, every day while the treatment is in progress and for some time after. Otherwise suffocation of the stock may occur.

There is another treatment, the "Modified Frow" treatment, the mixture consisting of 5 parts ligroin, 6 parts nitrobenzene and 2 parts methyl salicylate (all parts by volume).

For a normal colony wintered in a standard hive, a flannel or cotton wool pad should be placed over the feedhole in the crown board or quilt, and on it should be well distributed—drop by drop with a marked dropper—half a drachm (30 minims) of the mixture.

Every other day repeat the dose on the same pad until in all seven 30-minim doses have been given. Leave the pad in the hive for three days after the last dose has been given, then remove it, the treatment being complete.

Where Acarine has reached a serious stage, the only thing to do is to kill and burn all inmates of the hive, apparently healthy specimens as well as crawlers.

It is quite safe to use a hive which has housed a diseased colony, after the lapse of three weeks. All risk of infection will have died out then.

Dysentery. This complaint occurs mainly in winter or early spring. It is commonly caused by uncapped food absorbing moisture from the air, and fermenting. Dampness within the hive is another common cause of it.

Dysentery is indicated by the bees voiding their excrement over everything, inside and outside of the hive. For this reason, it is often taken for Acarine. Combs not covered with bees should be removed and those combs which remain should be closed up with the division board. The hive must be kept dry, more ventilation being allowed. A new cake of warmed candy (or combs of sealed honey if available) should be put into the hive. The disease will generally disappear on the return of favourable flying weather.

Foul Brood. This disease, as its name implies, affects the brood—the immature bees in their larva state in the cells. There are two forms of the disease, commonly known as American, and European Foul Brood. The more common of the two in this country is the American type.

The disease is most infectious. Furthermore it is regarded by the authorities as incurable. Thus complete destruction of affected stocks of bees is the only safe measure.

The bee-keeper should always keep a watchful eye on the brood frames, in order to detect this disease as early as possible.

Where foul brood is present there is a very patchy arrangement about the sealing of the cells, and some of those which are sealed have a shrunken or concave capping. Healthy grubs lie firmly curled up in their cells ; those which are diseased lie in unusual positions.

Where the disease is advanced, the grubs are dead in the cells, are partly dried up, are of a dark colour, and give off an offensive smell. Dried remains may even be seen in some of the open cells.

Bee-keepers who see evidence of the disease in any stock should at once get into touch with The Director, Rothamsted Experimental Station, Harpenden, Herts. Directions will then be given as to the best means of destruction of the affected stock, and the disinfection of apparatus to prevent infection of remaining stocks.

Under the Foul Brood Disease of Bees Order of 1942 County Committees may investigate suspected outbreaks of the disease and, if confirmed, may require the destruction of bees in affected colonies, and disinfection of hives and apparatus.

DISINFECTION OF HIVES. See Disease, under Foul-brood.

DRIVING BEES. See Bee-Driving.

DRIVING IRONS. See Bee-Driving.

DRONES. See Bees.

DYSENTERY. See Diseases.

ENEMIES OF BEES. See Pests.

EXCLUDER, QUEEN. Device for preventing the queen from leaving the breeding-quarters of the hives and passing up into the honey-storing departments and there laying her eggs. Adventitious patches of brood in honeycomb are always a nuisance and when they occur in sections of comb-honey the latter are spoiled entirely.

The best of the queen-excluders is a square made up of fine steel parallel wires in a wooden frame, the wires being such a distance from one another that the worker-bees can easily pass, whilst the larger-bodied queen is prevented from doing so.

Another type consists of a square of zinc-plate perforated all over with slots of the right width. This is cheaper but needs care in handling as it is apt to buckle, when its utility may be destroyed.

With the framed wire-excluder it is necessary that the surface of the frame should not stand up above the plane surface of the wires more than $\frac{1}{8}$ in. on either side of the excluder ; otherwise the space taken up by the device will be too much between the top of the brood-frames and the bottom of the frames in the super-rack above ; and the bees are likely to block this up with comb.

EXTRACTING HONEY. The ancient method of separating honey from the comb was to crush the latter and place it in sieves whence it could be drained and then bottled. This method, however, involved the destruction of the comb. This, in modern bee-keeping, is saved intact and returned to the hives, to be re-filled with honey by the bees, saving both time and money. (While the bees are making a pound of comb, they consume anything from 4 to 12 lb. of honey, and also require many days in which to effect the work.)

The modern practice is to extract honey by centrifugal force with an appliance known as an extracting-machine. This consists of a circular vat in which revolves a cage holding either two or four comb-frames. Geared mechanism causes the cage to spin rapidly thereby throwing the honey out of the cells against the sides of the vat. The honey is drawn off from time to time by means of a tap at the bottom of the vat.

The Extracting Outfit. In addition to the machine above mentioned, the honey-extracting outfit should contain the following appliances and tools :

Two uncapping knives for uncapping the combs before they are emptied of their contents. Two knives are required because they must be heated for use. Whilst one is employed, the other stands ready in a vessel of hot water. A good pattern of knife-heater, consisting of boiler and stove combined, is listed by all bee-appliance makers.

An uncapping tray, a vessel usually square, made to stand on an ordinary table, with a grid above and over this a cross-bar. The frame is rested on its corner upon this bar and the sheet of cappings is sliced off the whole of one side of the comb with one upward sweep of the knife, the sheet falling upon the grid below whence the honey sliced off with it drops into the vat. The frame is then turned round and its other side is treated in the same fashion.

The Extracting Operation. The un-capped combs are placed in the cage of the extracting machine, care being taken that, when revolving, the comb goes round with its bottom-edge foremost. The comb-cells have a slight upward in-clination which would retard the outflow of the honey if the combs were put in the reverse way.

' *The Honey-Ripener.* When the vat of the extracting machine is nearing full-ness, the honey is drawn off and trans-ferred to a storage-vessel, which is called a honey-ripener. This is merely a metal drum with lid and tap below, its capacity usually a hundredweight or more. In this ripener, the honey is left in a warm place for about twenty-four hours, by which time any scum and particles of wax will have risen to the surface. The honey may then be bottled right away, or be first submitted to a straining process.

It will be found that the thinner part of the honey will have risen to the top of the ripener. This should not be sold as prime honey but be set aside for bee-food or kitchen use.

Emptied Combs, How to Deal With Them. The emptied combs should be dealt with in the following way :

Return them to the supers and after the bees have ceased flying for the day, dis-tribute these supers among the hives. Wet honey-frames, however, must not be given back to the hives without taking the following precaution, as otherwise the stocks may be upset :

Place upon each hive a Porter escape-board, a framed board of the right size to cover in the whole of the top of the brood-chamber, with, in its centre, a bee-escape (which see) and also fitted with a by-pass so contained that it can be closed at will by a slide of tin. Before placing a super of wet combs on a hive, close the by-pass so that the bees of the stock cannot get up into the super.

Leave the hive thus until the late even-ing ; then go out and open the by-pass as quietly as possible. By this procedure no excitement will be occasioned to the bees, but they will gradually take possess-ion of the super of wet combs, clean them up, and eventually fill them again with honey.

If it be desired to take away the super of cleaned combs for storage until the following year as, for instance, at the end of the nectar-flow, all that is necessary is to close the by-passes in the Porter-boards and, by the morning, the supers will be entirely emptied of bees and can be taken away.

FEEDING OF BEES. When a stock of bees is short of food and natural supplies are insufficient or non-existent, the bee-keeper must make good the deficiency, by giving the bees either syrup or candy.

Preparing Syrup. The density of the syrup given varies according to the season and the purpose in giving it.

If the object is to build up as quickly as possible the amount of stores in the hive, thick syrup should be supplied and the bees allowed to take down as much as they will store. The syrup should consist of best white crystallised sugar and water in the proportion, by weight, of 2 parts of sugar to 1 part of water. This should be put in a deep wide enamelled-iron pan and stirred con-stantly until it boils. It should then be allowed to boil very gently for four or five minutes.

As this syrup, or much of it, will prob-ably remain in the combs for a lengthy period and possibly will not all be sealed, it is advisable to incorporate with it an antiseptic. The agent recommended is Thymol, which is obtainable in the form of colourless crystals. About three grains of Thymol should be dissolved in a little water, and added to every 10 lb. of syrup in the batch.

It is important to stir the syrup thor-oughly after the solution of Thymol has been added, and the syrup pan should not thereafter be allowed to remain on the fire.

If the feeding is being done to stimulate the activity of the bees in spring, the syrup should be relatively weak ; the sugar and water being weighed out in equal quanti-ties, or even with a preponderance of water (1 part by measure to double that amount of water.) Preparation is as for thicker syrup, but boil only for a minute or two, and omit the drug.

The object in syruping is not to supply a quantity of food for storage but to ad-minister it in attenuated but regular doses, never allowing the supply to exceed the amount which the stock can consume for the time being. If enough is given to permit storing it in the combs, the object of the feeding will be defeated.

The whole art of this stimulating spring feeding is to lead the bees to accept it as a regular supply of natural nectar coming into the hive, and that will be necessarily thin and scanty.

For methods of supplying the syrup to the bees, see Feeding Devices.

Candy, Disadvantages of. Candy is chiefly used as a preventive from starva-tion in the off-season. A cake of candy is placed over the cluster of bees, so that when they require food they may resort to it.

Food in this form has the drawback that the bees will need water to liquify it ; also candy is too stimulating a food at a time when the bees should be kept as inert as possible. Where practicable, candy should generally be avoided and good thick syrup medicated with Thymol supplied instead.

Preparing Candy. Weigh out the required quantity of white crystallised sugar, and to every 12 lb. used, add one quart of cold water and a teaspoonful of cream of tartar. Place the ingredients together in an enamelled iron vessel, with 4 or 5 in. of its depth to spare.

Set the vessel over very slow heat and let the mixture gradually warm up, stirring occasionally with a wooden spoon until the sugar is completely dissolved. Then turn on the heat and bring the stuff briskly to the boil, stirring steadily the whole time. Now the heat can be somewhat reduced again, but let the brew boil for a minute or so, still keeping up the stirring operation.

To Test Candy. The mixture must always be tested to decide whether it has been boiled enough. Drop a little upon a cold plate and try it with the finger. It should not stick to the skin. If it *does* stick, boil a little longer.

After the mixture has been judged to be ready, cool it as quickly as possible.

A good way is to have a large tub of very cold water standing close at hand and to plunge the vessel containing the candy straight into it up to the brim. Watch it carefully, stirring constantly until you see the stuff beginning to cloud and thicken.

You must not overdo this period or you will not be able to pour the candy easily into the moulds awaiting it. The right moment to choose can only be learnt by practice. But if accurately judged, the candy ought to set in the moulds snow-white and firm, yet without a trace of hardness and glassiness.

When cold, it should be easy to score the face of the candy-cake by a gentle sweep of the back of the finger-nail.

Bee-candy is usually run into shallow boxes with glass-bottoms, holding 1 or 2 lb. A cheaper plan is to use for the purpose the round glasses in which ready-cooked tongues are sold.

War-time Sugar Allowances. The importance of maintaining the honey supply was recognised in the early days of the war, and special sugar allowances were made available to bee-keepers on application to their local food office.

FEEDING DEVICES. Home-made Feeder. When it is necessary to give any large quantity of syrup to bees, a very simple and effective home-made feeder may be made in a few minutes out of a " lever-lid " tin, such as treacle is sold in.

The lid of the tin is punctured with a number of very fine holes, more or less according to the rapidity with which it is

desired the bees should take down the syrup. The tin should then be filled and the lid jammed down, the tin being inverted over the brood-nest. The syrup will not run out of the holes in the lid, but the bees can suck it out.

Pierce the holes from the upper side of the lid, so that the burred edges will be turned towards the interior of the tin.

This easily prepared and inexpensive home-made feeder should be put over a hole in the lowermost quilt a little less in diameter than the tin itself.

All feeders should be packed round and covered up with plenty of old blanket and the like.

Bottle-and-Stage Feeder is obtainable from any appliance-dealer. It is useful mainly when small, regulated quantities of syrup have to be given, and consists of a bottle with a ring of holes perforated in the lid and an indicating arm attached thereto.

The neck of this bottle is inserted into a wooden stage, the stage-gap being fitted with a sunken-tin lining having a slot cut in it for half its circumference.

As the bottle is revolved, the indicator shows how many holes are uncovered through the feeder-slot, and exposed for the bees' use. Thus either only one hole may be made available, or the entire half-ring of holes be exposed.

In use, the wooden stage is first settled over a hole in the lowest quilt. The bottle is then filled with syrup as warm as the finger can easily bear, the perforated cap put on securely and the bottle then inverted over the stage and firmly pressed down into its seating.

———

HIVES. The most suitable size for a bee-hive is one having a capacity of from 1,800 to 2,000 cubic inches. The hive most favoured is the standard hive of the British Bee-Keepers' Association. This contains ten comb-frames hanging side by side in a brood chamber, the frames measuring 14 in. long by 8½ in. deep. This is for the domestic quarters of the stock, honey-chambers (or supers) being added above according to the needs of the colony.

In certain low-priced hives the walls are made of a single thickness of timber. The usual plan, however, is to have hive-walls double throughout, those of the brood-chamber being packed between with some good heat-retentive material such as chaff or wood-wool. The upper storeys or " lifts " of the hive are commonly single cases, the honey-racks which they contain adding, in themselves, second walls.

A good roof, preferably of the " hip " pattern, with wide, overhanging eaves, is essential to a bee-hive, and the whole should be constructed of well-seasoned yellow-deal, the outside being given three coats of paint compounded of English white-lead, boiled linseed-oil, and genuine turpentine. Cheap paint soon turns to dust and rubs off when exposed to British weather.

Modern hives stand well away from the ground on oak legs, or oak stumps driven into the soil and sawn off true to a spirit-level. The latter plan is the best, as there is no chance of the hive getting out of the perpendicular afterwards, either through subsidence of the ground or the burrowing of moles.

Home Hive-making. Many bee-keepers prefer to buy their hives ready-made. It is quite feasible, however, even for the novice, to manufacture his own hives at home out of used boxes obtainable from grocers for a very small price.

Only those boxes made from stout timber should be chosen and each box should be taken apart before starting work.

The only dimensions which must be strictly adhered to in making a hive, are the interior measurements of the brood-nest ; and this only because it is best to use the B.B.K.A. standard-frames. Ten of these frames side by side and fitted with 1½ in. spacers, or metal-ends, require a brood-box measuring inside exactly 14½ in. by 14½ in., by 9 in. deep.

All other measurements necessarily arise from the needful size of the brood-box, interior, but when constructing the outer case make it of such a size that plenty of room exists between the two cases.

The honey-racks and their lifts or outer cases will only be on the hives during the summer season. In winter, the roofs close down well over the edge of the outer wall of the brood-box. Where the various storeys of a hive meet, the junctions should be covered by plinths, to exclude rain and draught.

A hive-roof should be made of good height so that a bottle-feeder may be stood within it without interfering with its seating.

The capacity of a hive-entrance is important. The entrance should be so contrived that it may open to the full width of the brood-chamber, or be contracted to a couple of bee-ways. This is usually effected by slides.

Alighting-boards should have a good slope away from the entrance, and the whole of the alighting-board should be sheltered from rain by a fairly deep porch.

In home hive-making, the greatest care should be taken to cut the parts accurately to ensure rectangular fitting. Otherwise the necessary ¼ in. bee-ways round and between all frames may suffer either extension or contraction. If the former,

the bees are sure to build comb in the increased space ; if the latter, it is equally sure that the diminished bee-passages will be closed altogether.

The side-ledges upon which the ends of the comb-frames rest must be perfectly straight or the frames will not hang plumb, when the combs may be welded together and their chief virtue of movableness entirely lost.

HONEY, PRODUCTION OF. The most important thing, as soon as the nectar-flow begins, is to keep super-space well in advance of the bees' requirements. If a stock is allowed to get anywhere near to being cramped for want of honey-room, the incoming foragers are sure to deposit their loads in the brood-chamber, and the queen will thus be robbed of some of her laying-space, with the inevitable result that the progress of the colony will be impeded. The supply of young bees will be arrested and after-swarming will be induced just when the colonies are wanted to keep their full working-strength unimpaired.

Increasing Supers. Every couple of days or so all supers should be inspected and another super placed underneath the first directly the original super is approaching the half-full mark. Third supers should be ready to go on the hives and be put under the second racks as soon as needed.

By this means the activity of the stock in nectar-storage is perceptibly increased. Towards the end of the flow, additional supers should be placed on top of the pile, not next the brood-nest as above advised.

Managing the Racks. As soon as a rack (whether of shallow-frames for extracting or sections for comb-honey) is filled and completely capped over, it should be removed from the hive.

The Porter clearing-board (see Extracting) should always be employed when carrying out this operation. It should be thoroughly cleansed and scraped, and the bee-escape device in its centre in good working-order.

The escape is best removed from the board and boiled for a few minutes in water containing a small piece of washing-soda, rinsed under the tap and set to dry. It should then be tested to see that its converging springs have their points the right distance apart to allow bees to pass down without possibility of returning to the super.

When bringing the clearing-board into use, have a box near the hive and lay the device on this, making sure that it is the right way up. Then remove the full rack and centre it carefully upon the board. Now lift rack and clearing-board together and return the whole to the hive above the other supers, if any, and cover up warmly. If this be done on an evening, the rack should be emptied of bees by the early morning, when it may be carried off to the honeystore, or extracting-house.

The clearing-board should be removed from the hive without delay, as, if left on for any time, the escape-device may be clogged with propolis and thus thrown out of action.

Extracting Honey. See Extracting.

Extracted Honey, Preparing for Sale. All run-honey should be put up in Ministry of Agriculture jars, holding 1 lb. or ½ lb. These are sold by all appliance-dealers at the same price charged for the old types of jar, but are in many ways superior. The M. of A. jar does away with the old unsatisfactory screw-caps and substitutes a cap closed down on the quarter-section-turn principle ; and the jar is practically shoulderless, thus allowing the honey to be much more easily removed, especially when candied.

Every jar should be neatly labelled, and be rolled in a sheet of wax-paper or cellophane.

When filling the jars, allow the honey to reach a little into the neck so that, if the honey scums after bottling, the scum may be spooned out without materially reducing its contents. Do not forget to close the jars tightly.

Granulated Honey. The best form in which to offer bottled honey for the market is that of a crystal-clear liquid of good density. Most users of run-honey prefer it in this condition.

On the other hand, there are many people who like their honey candied to an almost solid, opaque state. Such honey should be of fine grain, fondant-like, and entirely free from any tendency to " greasiness " of texture.

Fineness of grain may be secured by stirring the jars after the candying process has manifestly begun ; a second stirring a few days after the first will promote still further the formation of very fine crystals in the honey. Fine-grain candying is also assisted by storing in a cool, dark place. A high temperature and exposure to much light tends to favour coarse granulation.

At the same time, the rapidity and type of graining appear to depend in no small degree on the particular crops from which the honey is derived. In a district where dandelion, charlock, or sainfoin abound, all the honey will be subject to rapid and complete granulation. Mustard-honey also candies very quickly, and admixtures of any of these with other sources will cause the whole batch to granulate. Honey made from nectar absolutely from a single source, such as pure white clover or ling-heather, however, does not seem to granulate at all.

Honey that is too pronounced in colour or somewhat dingy of hue will, if allowed to candy, assume a lighter and more agreeable shade. This, of course, increases its saleability.

The Cause of Streaks. An annoying blemish that sometimes shows itself in jars of granulated honey is the formation of whitish streaks and blotches against the inner surface of the glass, considerably detracting from its appearance.

This defect, which often ruins the appearance of so many bottles of otherwise beautiful candied honey, is caused mainly by imperfect cleansing of the bottles themselves.

To Clean Honey Jars. Honey jars should always be thoroughly cleaned with hot soap-suds inside as well as out, followed by a good rinsing out with a couple of changes of pure cold water before being inverted on a grid-table to drain and dry. When dry, the bottles should be stood upside-down on a newspaper-covered shelf, to protect them from dust until needed for filling. Thus treated, the objectionable " frosting " ought never to occur.

How to Re-liquify Candied Honey. Unless the batch to be cleared is a very considerable one, the best and simplest way to deal with it is to heat it after bottling.

The bottles should have their caps put on quite loosely and then be stood, close together but not touching, in a large preserving-pan with a flat bottom, the jars not resting on the bottom but on thin laths. The pan should then be filled with cold water up to the shoulders of the jars and placed over gentle heat. The heat may be gradually increased, if necessary, until it reaches 130 deg. F. Honey should never be heated beyond this point or it will lose much of its aroma and flavour.

Time should be allowed for the honey to clear. When this has been effected, the pan should be stood on a wooden table or floor, and covered over with old blankets or the like so that the cooling process may be as gradual as possible.

When quite cold, the caps may be lifted and any scum which may have formed on the surface removed, after which the caps can be screwed down firmly.

Honey thus treated will be beautifully clear without having suffered any perceptible deterioration in the process, and will not granulate again.

Candied Sections. Sections wherein the honey was either partially or wholly granulated owing to exposure to cold or inherent tendency of that particular kind of honey to candy, are less attractive for sale. The old practice with such sections was usually to render them down into run-honey. This is wasteful and expensive, and it is quite possible with care to restore the honey in sealed comb to something like its original fluidity without greatly affecting the structure of the comb by subjecting it to a temperature of not more than 100 degs. F.

The process should be very gradual and effected in a water-jacketed container, so that there may be no risk of overheating, or the comb may break down.

Seeing that the melting point of pure beeswax is about 147 deg. F., even higher if rendered from quite new comb, and most honeys that have granulated may be re-liquified at about 130 deg. F., there is, theoretically, a sufficient margin for safety. The length of time necessary for the man-ipulation to complete itself, however, may sometimes be weeks, and renders it im-practicable on anything like a large scale.

Section Honey. The secret of suc-cessful comb-honey production is to work only with the biggest and best stocks. The preparation of such stocks for section-work should be commenced at the very beginning of the rousing of the cluster soon after Christmas. These stocks, picked out for their size and vigour, as well as the proved ability of the bees for even capping of cells, should be stimu-lated to early strength by every means in the bee-keeper's power, so that the brood-nests may be packed with worker-bees by the time the nectar-flow commences.

The great art is to prevent the stocks breaking themselves up by swarming. The Demaree swarm-prevention system (see Demareeing) is invaluable here.

Foundation of a specially thin gauge, light in colour, is used in sections. Small triangular pieces held in the split top of the section and hanging vertically therein are often used for economy's sake, but it is always cheaper in the long run to employ full-sized sheets for sections. These save time and the sections are finished sooner. Also the combs are built more regularly and their appearance thus enhanced.

The best section to use is what is known as the " two-way, split-top."

Preparing Sections. First give all the joints a wipe with a damp rag at the back. This prevents them from breaking. Fold carefully, leaving half the top un-folded. Make sure that the folding is done square, and the dove-tailing at the corner well and truly made.

Prepare the whole required batch of sections thus far ; then proceed with the foundation-fixing. The little squares should be cut so that they do not quite touch the sides or bottom of the section. This is to allow for expansion of the foundation under the heat of the hive.

Sections are sometimes made with in-ternal grooves at bottom and sides, to receive the edges of the foundation. This looks very neat, but the foundation is almost sure to buckle, and make a lop-sided comb.

Hold the waxen square in position while you engage the other piece of the split-top, the upper edge of the sheet being nipped between the splits. Then bend the foundation with a small piece of straight wood so that it hangs quite centrally in the section. Overdo this bend slightly as the sheet always recovers a little.

Fixing in the Sections. The sections go into the racks in rows of three, with dividers separating them. Be careful that these dividers—which should be of thin tin—are accurately placed and that, when you are adjusting the wedging-up boards, the sections are really tightly pressed up together.

If you have to keep filled racks any time before putting them on the hives, store them in a warm place safe from dust and vermin.

Never omit using a queen-excluder under all honey-racks, either in comb or run-honey production. There is always the risk, when a big crowded stock is in question, that the queen may go up and lay in the super.

When the first section-racks are about half filled with honey, second racks should be put on—under the first racks, next the excluder.

Be very chary of giving whole new section-racks to stocks as the nectar-flow draws to a close, or you may be left at the end of the season with a lot of unfin-ished, and so unsaleable, sections on hand. Third and subsequent racks should be made up with the uncompleted sections contained in the racks already taken off. Probably most racks will show one or two of these.

HONEY RIPENER. See Extracting.

ISLE OF WIGHT DISEASE. Former name for Acarine disease, which see under Diseases.

LAW IN REGARD TO BEE-KEEPING. Bees may not be kept in such a situation that they con-stitute a nuisance to adjoining occupiers of land.

Swarms may not be followed by their owner over land in private occupation without permission to do so being first obtained, though should that permission be unreasonably withheld, and the swarm lost to its owner, or suffer detriment by his enforced inability to secure it, the owner of the land so refusing his permission to the owner, is probably liable for dam-ages ; that is to say, the value of the swarm.

In a case of disputed ownership of a swarm, if the owner has followed it and kept it in sight or hearing, his rights in it will be probably be upheld. Where a swarm is seen to enter, or can be proved to have entered, a hive belonging to another man, a Court of Law will most likely confirm the claim of the owner of the swarm to a right in the bees.

PARASITES. See Bee Parasites.

PESTS AND ENEMIES OF BEES. Next to wax-moths (see Bee Parasites), mice, ants, wasps, and various insect-eating birds are often troublesome to the hives.

Mice will enter hives containing weak stocks and do havoc among the combs. They will also get up among the quilts, and there excavate a snug nest either in the spring for breeding or in winter for shelter. They should always be dealt with promptly.

Hives of the W.B.C. type, consisting of two cases with air-spaces between, are specially apt to harbour mice where the bees cannot attack them.

Ants. In certain districts ants are likely to invade beehives. The remedy is to stand each hive-leg in a tin containing a little water, or preferably paraffin.

Wasps may be a great nuisance in summer, especially with weak stocks. They go boldly in at the flight-hole, trusting to their superior fighting powers and tough skins to win them immunity.

As a remedy, the hive-entrances should be contracted to about an inch, when a strong colony of bees will usually cope with the nuisance. Occasionally, when wasps are very numerous, they will completely gain the upper hand, and the stock will be wiped out. The best course is to search for the wasp-nests and destroy them.

Birds. There are many small birds that habitually prey upon bees, perching upon the alighting-boards and picking up the bees as they emerge from the hives, or catching the bees when flying. In the latter event, it is hardly possible to do more than scare the birds from the vicinity of the hives by hanging jagged pieces of broken window-glass in wire slings from the trees. This dodge is very efficacious at all times of year. The glass twirls round in the wind and sends flashes of light in all directions, effectually frightening off the smaller birds.

When birds have formed the habit of perching on the alighting-boards, the most useful thing to do is to cage-in the door-steps of all hives with bent wire-netting. These cages are quick to fashion and easy to apply with a few thumb-tacks, and they do not seem to interfere with freedom of flight as far as the bees are concerned.

POLLEN. The principal source of food for bees. It is, of course, gathered mainly from flowers.

Artificial Pollen. The scarcity of natural pollen during the early spring months, when breeding is going forward in all healthy hives, and ever-increasing quantities of food are urgently needed, has led many bee-keepers to look for some substitute for natural pollen to give the bees, and which they would accept for the purpose. Though bees will readily enough carry into the hives such farinas as pea-flour, however, it is extremely doubtful whether they in any way serve as alternatives for pollen brought in fresh from flowers.

Expert opinion is that this so-called artificial pollen is of such dubious value that it would be advisable to avoid it altogether, and concentrate effort to help the bees by surrounding the hives with as many early pollen-bearing plants as can possibly thrive on the space available.

Should it be thought urgent, in default of natural pollen, to supply bees with some substance that might augment the body-building and subsistence content of their food, a much more likely expedient would be to incorporate white-of-egg with the sugar-syrup administered for stimulative purposes in early spring (though not in war-time).

PORTER ESCAPE BOARD. See Extracting, also Honey.

PROPOLIS. This is a resinous material used by bees for a variety of purposes—as a waterproof varnish, a stopping for crevices in the hive, a strengthening or adhesive substance, a coating for comb, etc. It is supposed to be gathered from certain trees that exude resinous matter, and it is a fact that bees will collect such matter on occasion. Properly, however, it is a product of the bees' own bodies.

QUEEN BEES, LIFE OF. See Bees.

QUEEN BREEDING. This has now become a specialist art in this country, our foremost bee-masters taking up the enterprise in increasing numbers.

No doubt the grand secret of successful honey-production is to re-queen all stocks of bees at frequent intervals, perhaps every year, from the best obtainable strains of these intensively bred mother-bees. Such super-queens, under proper management, will double the population of an ordinary stock within a few weeks of their installation ; and, given a favourable nectar-flow, will equally double the average honey-yield from the stock.

These enormously prolific queens are, however, apt to wear themselves out, and be nearing the end of their resources before the bee-keeper well realises the fact. It is safer therefore, to observe the " one season rule " if you go in for Star Queen bee-keeping.

QUEEN EXCLUDER. See Excluders.

QUILTS. Squares of flexible fabric used for covering frames in a hive. They serve to confine the bees and, at the same time, to keep the chambers warm and easily get-atable.

Quilts may be roughly divided into two kinds—the porous and the non-porous. The latter are now practically obsolete, having been generally discarded in favour of coverings that allow a certain upward percolation of air, this tending to prevent dampness of the brood-nest.

Glass Quilts. Almost the only form of non-porous quilt at present in use is that consisting of a sheet of glass in a wooden frame, which permits of ¼ in. bee-way over the tops of the frames and has a central feed-hole pierced through the glass.

The convenience of a glass quilt is undeniable, as it allows the bee-keeper to obtain a view of the interior of his brood-nests in cold weather, or during the winter lying-up season, without disturbing the bees. If always kept warmly covered up, except when being momentarily used for observation purposes, the admitted defect of the glass quilt—that it is apt to get bedewed with moisture and so impede visibility—may be ignored in practice.

On the whole, however, devices of this kind had best be avoided on the score of expense, as there is no actual necessity for them.

Porous Quilts. The simplest and cheapest of these consists of an 18-in. square of carpet-felt, or old blanket, or any sort of soft woollen carpet. Old Wilton carpet which has been well freed from dust fulfils the purpose excellently. As many thicknesses of these materials as desired may be used together for overhead winter-packing and if any contract the slightest dampness they can be easily replaced, as their cost is negligible.

The quilt next to the frame-tops should, however, always consist of a square of new unbleached calico that has had the dressing washed out of it in the piece before cutting. This calico quilt should have a feed-hole in its centre, covered, when feeding is not in progress, by a small square of the same material. The bees do not attempt to chew holes in this cloth and its strength is necessary to withstand the rough usage it is bound to receive when being removed.

Cushion Quilts. Another form of upper quilt in very general use is the cushion type. Bags of at least 20 in. square when in the flat, should be made and strongly machined on all four sides, leaving, of course, an unsewn space for the insertion of the filling material. The best cloth of which to make the cushion covers is butter-cloth (not what is called butter-muslin, which is quite another thing).

The common practice of making cushion quilts of what is called " art-muslin," should be avoided.

The bags are best filled with granulated cork, such as can be very cheaply obtained from grape-importers. Chaff, grain-husks, old bed-flock, autumn leaves, and many other materials of the like nature will also serve for the purpose. Whatever is used, must be bone-dry and as little as possible able to attract moisture.

Cushions are excellent for creating a warm hive in winter. Their only defect is that, if they once contrive to get damp, it is not easy to dry them again thoroughly.

For this reason, in exceptionally wet seasons, or where bees are kept in very humid situations, cushions are best avoided and a tray quilt used.

Tray Quilts. The tray quilt consists of a framework of light laths covered with butter-cloth. The top is left entirely uncovered, it being, as its name suggests, merely an open tray full of any sort of insulating material that can be emptied out and replaced by new dry filling in a few moments. The tray quilt should be made to fit accurately into the lift of the hive. Its cloth sides will bulge a little in all directions, thus making a close fit.

All hive-quilts, of whatever nature, should have the square of new calico next the frames, as described above.

RE-QUEENING. With stocks worked under the modern high-pressure system, a queen should never be kept after her second full season of activity. Some very successful bee-masters make a practice of re-queening after every complete season of the queen's work. This, of course, only applies when the hive has not re-queened itself by swarming or by supercession.

Re-queening is also necessary when a queen has proved herself deficient in vigour or fecundity, or when undesirable traits, such as viciousness of temper, have shown themselves in the stock.

The best time to replace the queen of a hive is immediately after cessation of the honey-flow, and this may be as early as the middle or end of July. Where there is a clover-flow with a heather-flow following and a break in between, the best time for re-queening with a young fertile queen will be just after the first flow has failed, so that the new queen may get to work early enough to provide a good head of young worker-bees for the heather.

When the New Queen Arrives. Assuming that the queen is obtained from a professional breeder, she will arrive by post in what is called a combined travelling and introducing cage. She will be accompanied by an escort of about 100 worker-bees. The cage will be panelled with wire gauze for ventilation and will have a hole at one end plugged with sugar-candy.

On the queen being received, the bee-keeper should go to the hive which is her destined home and remove from it the old queen, plus any queen-cells, or signs of them, replacing the quilts as before, leave the stock for twenty-four hours to enable it to realise its loss, then take the new queen, just as received with her attendants in her cage, place the cage over the central frames of the brood-nest under the quilts and at once cover up all snug and warm.

The bees of the hive will be able to lick the queen through the wire-mesh panel of the cage, but will not be able to join company with her until they have eaten away the plug of candy which fills the hole at the cage-end by which she must emerge. This they will at once begin to do, but they will not succeed in opening a passage for the queen to get down into the brood-chamber for some time, probably a day or more. During the interval the imprisoned queen will be acquiring the characteristic odour of the hive. By the time she is free to mingle with her prospective subjects, there will be nothing about her to excite suspicion of strangeness, and she is nearly certain to be accepted as ruling sovereign of the whole stock.

One precaution, however, must never be omitted. After the cage has been got into place upon the frames and the hive closed, the stock must not be interfered with in any way for at least a week, and it would be safer to wait still longer. Any attempt, during this critical time, to look into the hive to see if the queen has been favourably received by the bees, is nearly certain to result in their destroying her.

ROBBING. In the spring, when stocks are getting strong but nectar is still scanty, and again in late summer when the honey-flow is done, the robbing of the stores of one stock by another stock is no uncommon incident. It is usually the weak hives that are attacked, strong ones being generally able to protect themselves.

Directly signs of robbing are observed, measures should be take to stop the trouble, or the stock attacked may be wiped out.

The first thing should be to get a water-ing-can with a large fine rose, fill the can with cold water to which a strong dash of carbolic acid has been added, then give the hive and surrounding soil as good an imitation of a storm-shower as can be contrived. Hold the can as high as possible and let there be no mistake about the thoroughness of the job. The robber-bees are pretty sure to be temporarily driven off by this manœuvre and the home-bees will take refuge within the hive.

Now get two slips of perforated zinc and fasten them with drawing-pins, one on each side of the hive entrance, so as to reduce the bee-way to ½ in. or even less.

A further plan is to pile up wet grass or hay loosely upon the alighting-board.

A cloth hung over the face of the hive and sprinkled plentifully with 1-in-10 solution of carbolic is also very useful in repelling invaders.

Good results often follow if a large piece of window-glass is stood on the flight-board, leaning against the face of the hive so that it allows a passage-way round its sides but prevents direct entry. The glass has a confusing effect on the robbers.

The attack will sometimes be renewed at intervals during the whole day, so that the protective gear must not be removed until after sundown, when the entrance may be opened to its fullest extent to give the stock an airing. It must, however, be contracted again in the early morning of next day, robber-bees being extremely early risers.

Even when the onslaught is not renewed, frequent visits to the apiary should be made during the next few days to see that all is right.

SKEP. A dome-shaped hive made of basket-work, not often used nowadays —except for taking swarms.

SMOKING BEES. See Subduing.

STARTER. A narrow strip of foundation fixed to the top bar of each frame from which the bees will begin their comb-building.

STINGS, PREVENTION OF. See Appliances.

STOCKING THE APIARY. See Bees.

SUBDUING BEES. By injecting smoke into their dwelling or submitting the stock to the fumes of carbolic acid bees may be rendered quiet and docile, and unlikely to interfere with the bee-keeper in any way when he opens their hive.

Bee Smokers. Of the many patterns of bee-smokers at present on the market, the most suitable for the man with only a moderate number of hives is that known as the British pattern. For a big apiary, however, the American pattern smoker is to be recommended, as it can be kept alight for a longer time and can be re-charged with fuel without having to be emptied, re-filled and lighted again.

Smoker Fuel. Practically any material that will smoulder without flame, can be used in the smoker. Most bee-keepers employ ordinary corrugated brown-paper, rolling it up fairly tightly in the form of cartridges of the right size to fit the barrel of the smoker, these being secured by tacks pushed in at each end.

Another very suitable fuel is old sacking that has been steeped in a strong solution of saltpetre and then thoroughly dried before being cut up into strips for rolling. In the American smoker, close-packed wood-chips, old hay, cotton-waste, or dry leaves may be used to good effect.

Care of Smokers. No matter what fuel be used, there is always a tendency for the interior of the smoker to become fouled with a black, greasy deposit. This should be cleared out at frequent intervals, particularly if the smoker be much used.

When the smoker is put away at the end of the season, it should be thoroughly cleaned, its bellows oiled and its metal parts well vaselined to prevent rust.

The Carbolic Method. Carbolic acid as a subduing agency for bees has much to recommend it, though occasionally one meets with a stock that, far from being quietened, is only reduced to furious excitement by it. In such instances, luckily rare, only smoke should be employed.

The best way to use carbolic is to have a square of thick calico moistened with a solution consisting of two parts of the acid in twenty parts of water with one part of glycerine added to keep the cloth from drying up. The calico should be large enough to cover the top of the brood-chamber.

Many bee-keepers make these squares with a ½-in. iron rod sewn into the hem of the cloth on two opposite sides. These rods hang down over the hive-sides and serve to keep the cloth in place.

Whether using smoke or carbolic, it is essential, in subduing bees, to wait a full couple of minutes after giving the first few puffs of smoke in at the hive-entrance, or putting the cloth in place, before commencing to open up the hive.

SWARMS AND SWARMING.

Swarming is the method of the honey-bee for multiplying colonies of its species. As the population of a hive grows and approaches inconvenient crowding, the stock begins to prepare for the sending off of an emigration-party, which will choose a separate home elsewhere. It is always the old queen who leads off this party, a new queen being raised by the stock to take her place.

Ordinarily several queen-cells are begun, maybe a dozen, so that the future of the hive may not depend on a single life. When the swarm has gone off, and the new queen has been hatched and mated, the surplus young queens are destroyed, generally by the new head of the stock.

When after-swarms occur, it is because one or more of these young queens have been permitted to survive, the second swarm usually issuing about nine days after the first. Sometimes a third or even more emigration-parties may be thrown off. These after-swarms should always be deprived of their queen and returned to the mother-hive on the evening of their day of issue ; otherwise the stock will have so reduced its working-strength as to render itself useless for surplus honey-getting.

When to Expect Swarms. First swarms may be expected after the first week in May, earlier or later according to the weather and forwardness of the season. Generally they may be looked for between noon and three o'clock in the afternoon. After-swarms may go off almost at any time in daylight on warm bright days.

Taking a Swarm. To "take a swarm" the bee-keeper will need first, a good hiving-skep of the dome-shaped pattern, with a stout ring on top by which it can be securely held. Next he must have a square board large enough to cover the mouth of the skep and a bit over all round. Further, in his pocket he must have handy the outer part of an ordinary match-box, a few yards of good string and a couple of clean sacks or other material with which to cover the skep after the swarm has been hived. The smoker should be at hand ready charged so that it can be lighted at a moment's notice.

The bees may hang together in a cluster on the alighting-board of the hive for some time before the swarm goes off or they may show no previous signs of swarming at all, just changing in a moment from an orderly community to a wild mob of roysterers. Whichever way it is, there is seldom any need for hurry or concern on the part of the bee-keeper.

On hearing the commotion and seeing the cloud of bees pouring up into the air, he should take a leisurely way to the honey-shed and secure the prepared kit, then return to the swarming hive and watch the proceedings.

After the last of the emigrating-party has rushed from the hive, the whole concourse may remain in the air for a good ten minutes, veering about hither and thither in apparently aimless fashion before there is any tendency of the throng to set in one direction. At last will be detected the beginnings of general agreement as to a direction of flight.

At times, swarms pitch in the most unlikely and most awkward of places, but it is usually upon the bough of a tree, at no great height from the ground, that the bees will come to rest at last, forming themselves with incredible rapidity into a dense mass. When this happens proceed as follows :

Wait ten minutes or so to allow the last flying bees time to join the settled throng. Then, with skep grasped by its handle in one hand and the board in the other, approach the swarm, bring the inverted skep up beneath it so as to enclose as great a mass of the bees as possible (standing the board on the ground temporarily). Now with the disengaged hand, give the branch a series of sudden strong shakes. If this is properly done, the third shake at most should have discharged almost all the bees into the skep. Before many of them have time to get on the wing, clap the board over the skepmouth, turn the skep right-side-up, and place it upon the ground beneath the tree.

Now comes in the use of the match-box case. The skep has to be raised in front, to give the hived bees their liberty. Lift the fore-edge of the skep with the fingertips of one hand, while, with the other hand, thrust the little case just beneath the skep-rim, end-on to the emerging stream of bees.

Of course, about half the swarm immediately crowds out of the skep, either rising on the wing or climbing upon the outside of the skep. If you have secured the queen in your shaking, a good proportion of the bees will not come out when you thus open the door for them. Many are sure to remain ; and in fifteen or twenty minutes those that have flown will return to the queen and be safely back again either in or round the skep.

Your last care should be to cover the skep up, all except its front, with the sacking, to shelter it from the sun.

The hived swarm should be left thus undisturbed until about sundown of the same day or until the bees are in for the night. Then it must be transferred to its permanent quarters in the apiary.

Preparing the New Hive. Much time and work for the bees is saved if one can at once give the swarm a home in a hive whose brood-nest contains ten frames of fully-drawn-out worker-sized comb. There will then be no need, as far as the queen is concerned, for the bees to build comb for her, as she will have a whole broodnest of empty cells all ready waiting for her use. The bees of a swarm are always frantic to commence comb-building at the earliest possible moment, and one can take advantage of this impulse by giving them at once a second story, or super, of frame

filled with sheets of foundation, or a rack of sections, not forgetting to put a queenexcluder under this.

If a honey-flow be on, you will find the bees immediately take possession of the super and there give vent to their combbuilding furore, the super being soon filled with honey.

If, however, you have no drawn-out combs in store, new frames with full sheets of foundation should be used.

The new hive should be stationed in the place the swarm is intended to occupy as a stock. All the quilts should be put on, with the roof in place, but the entrance set open to its widest extent. The alighting-board should have a second board so placed that it forms an extension to the first, gently sloping to the ground. The two boards should be covered with a piece of old sheeting uniting the two and held in position with drawing-pins.

Hiving the Swarm. Go to the skep, where the bees should be now all safely in for the night, lift it carefully and carry it over to the prepared hive. Now, after blowing a sharp puff or two of smoke into the skep, pour the bees out upon the cloth in front of the hive-entrance, when they will at once begin to run in and in ten minutes or so they will all be inside the hive. Should any appear reluctant to enter, a little smoke will keep them on the move.

The new stock normally needs no attention thereafter until it requires supering. If, however, the weather turns out cold and wet immediately after the hiving it is as well to give the bees a bottle of syrup to help them along.

SWARM BOARD. See Swarms.

SWARM-CATCHERS. Several devices exist for automatically trapping a swarm on its issue from the hive, the object being to prevent its loss during the absence of the owner.

All such appliances, however, are open to the serious objection that, as they have to be got into position well before a swarm is expected, they greatly interfere with the traffic and ventilation of the hive.

All swarm-catchers, with a few exceptions, follow one general principle in design.

They consist of a box that clips on to the face of the hive. The box is open below, back and front, but the lower half of the front is covered with a sheet of queen-excluder attached and bent in a special manner.

The upper part of the box is really a comb-frame chamber containing two or more full-size standard frames with drawn-out comb or full sheets of foundation.

When this type of catcher is in action

and a swarm is issuing, the bees rush out at the hive-entrance. The worker-bees pass through the queen-excluder and out into the open air. When the queen reaches the excluder, however, she is prevented from following them. She then climbs up the excluder into the comb-chamber above and there settles. She is immediately followed by a great number of her bees and eventually those that are flying outside track her down and join her.

In a few minutes the whole swarm is clustered together partly within the comb-chamber and the rest outside.

At this point or soon after, the swarm-catcher should be unhooked from the face of the hive and removed so that the mother-hive may be restored to its normal conditions of free entrance and exit.

SYRUP, TO MAKE. See Feeding.

THYMOL. Antiseptic agent included in Bee Syrup. See Feeding.

TOOLS FOR BEE-KEEPERS. See Appliances.

UNCAPPING KNIVES. See Extracting.

UNCAPPING TRAY. See Extracting.

UNITING STOCKS. As the success of bee-keeping depends in the main on having all hives strong in population, any existing weak stocks should either be united, two together, or joined up with a stock of normal strength, whether in spring or autumn.

The best and safest way to unite two stocks is as follows :

The two hives, unless side by side, must be brought into that position gradually, moving them towards each other no more than a foot at a time per day. When they are close up side by side, one must decide which of the two queens is to be kept, and then find and remove the other.

About 3 or 4 p.m. on a fine day, smoke and open both hives, lay a single sheet of ordinary newspaper over the brood-frames of the hive containing the selected queen, make a few holes in the newspaper with a fine nail, then remove the brood-box of the other hive and stand it bodily upon the newspaper. The two lots of bees thus cannot get at each other except through the paper when they have succeeded in enlarging the holes in it sufficiently to allow them to pass through.

The doubled hive should now be closed up warmly and left entirely alone for a couple of days, by which time the two colonies will have peaceably united. The sheet of newspaper can be disregarded, as the bees will remove it piecemeal on their own account.

VEIL. See Appliances, protection from stings.

WAX MOTHS. See Bee Parasites.

WAX-RENDERING. Even in the smallest apiary, it is worth the trouble to save every scrap of comb, foundation, etc., that turns up in the course of each day's work, and put it aside for melting down as soon as enough of these odds-and-ends has accumulated. The best method of melting is to use what is called a " Solar Wax Extractor," the heat of the sun only being employed to do the melting.

There are many different forms of " Solar," but they are all designed on the same principle. The appliance consists of a box with glass lid, much like a gardener's frame. Inside the box is a tray-like, curved receptacle for holding the combs, odd scraps of wax, etc., and there is also a vessel to receive the wax as it melts. Useful dimensions are 4 ft. long, 2½ ft. wide, and 1 ft. deep.

The lid of the Solar extractor should be double-glazed, if possible, as this retains heat longer.

The box should be made of stout, well-seasoned material, not nailed but screwed together, the woodwork of the lid also being screwed ; otherwise the heat of the sun may draw the joints. The lid should fit closely over the box with a deep flange. All outside surfaces should be painted dead-black for heat-absorption.

The curved comb-tray should be made of bright sheet-tin. The strainer may be either riveted or soldered on to the curved lower lip of the comb receptacle. Both the tray and the vessel which receives the melted wax should be removable.

Wax rendered by the " Solar " system will need no further treatment ; it will be quite pure and ready to be caked up for use or sale. Moreover, its colour is clearer and brighter than wax rendered in any other way.

Winter Rendering. The Solar extractor above described is applicable only to wax-rendering during summer heat. When fairly small quantities of waste-comb have to be dealt with at other seasons, it may be done by using a " double saucepan," the lower one about one third filled with water, while the upper one receives the comb to be melted. The whole is then stood over a moderate fire and allowed to boil until the melting process is complete.

Beeswax should never be rendered in an ordinary pan directly touching the source of heat; treated thus it is sure to burn and the colour and quality be inevitably ruined.

To Render Wax in Quantities. When the quantity of old combs, etc., to be rendered down, is considerable, the following method will be best :

Cut the combs from their frames and remove any wires that may remain in them. Then make a bag of butter-cloth and pack the combs, etc., in it. The ordinary domestic copper is now used. The bag is placed upon some cross-bars of wood at the bottom in such a way that it does not touch the metal of the copper at any point. A couple of bricks may serve as weights upon the bag to prevent it floating.

The copper should now be filled up to such a mark as will well cover the bag with water to which has been added commercial sulphuric acid in the proportion of 1 part in 100 parts of water.

The water may then be boiled until the contents of the bag have yielded up the wax in them.

The wax will rise to the surface, whence it may be removed in a cake when cold. It is then re-melted and strained before it can be run into moulds. Very serviceable metal moulds with divisions each holding 1 oz. of wax may be obtained from bee-appliance makers. The 1-oz. cake is the most convenient form in which to offer beeswax for sale.

When the old combs to be treated are much clogged with pollen and pupa-skins they should be broken up and soaked in rain-water for a day or so. This prevents the debris absorbing wax during the melting process, and more cake-wax is secured. The soaking water is best acidulated like the boiling water above mentioned, the colour of the wax recovered being greatly improved.

Storing Wax-Scraps. Where the bee-keeper adopts the plan of saving wax-scraps through the summer season, for rendering down in a single large batch at the end of the year, these should be stored in a bee and vermin-proof box, and care should be taken that no wax-moth grubs or eggs are enclosed with the scraps. All must be looked over before being put into the box, and some naphthalene balls in a muslin bag may be kept in the box.

Failing such care, when rendering-time comes round, the contents of the box may be merely dust and rubbish.

WIRING FRAMES. When comb-foundation is fixed into frames it has to be wired in for safety. It is risky to handle unwired frames when heavy with honey.

For wiring, the orthodox fine tinned wire, obtainable from all appliance-dealers, is quite satisfactory.

Take a frame and proceed as follows : Drive in a ¾-in. brass gimp-pin at each of the points marked A.B.C.D. in the sketch, the pin going through the side-bars and projecting inside the frame. Then take a pair of small round-nosed pliers and turn the point of each pin into the form of a hook, as shown in the sketch.

Now make an eyelet on the end of the wire. Engage the eyelet with hook A. Pass the wire tightly across to hook B., down over hook D., across to hook C., then upwards and over centre of wire A.—B., and pull the wire firmly down until the top cross-wire assumes the position A.—E.—B. Finally bring the wire down and over hook D. again, where the end should be fastened off securely by twisting.

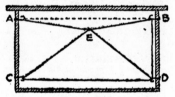

Method of wiring comb-foundation in a frame.

Having done this, take the wire off this trial frame, straighten it, and cut as many pieces of the same length as you have frames to treat. The entire batch of frames can now be wired as above described.

It is important to stretch each piece of wire before using. This is most conveniently done by passing its eyelet over a nail driven into the wall and hauling on the wire, when it will be found that its length can be increased by 2 or 3 in.

When a frame has been properly rigged, the wire, when twanged, ought to give out a high note like a treble harp-string.

WORKER BEES. See Bees.